Compartive Vertebrate Endocrinology

The long-awaited third edition of this popular textbook retains the successful format of previous editions, dealing with the nature, actions and roles of hormones among vertebrate animals. Special emphasis is placed on the evolution and origins of hormones and their receptors, on the role of hormones in the physiological coordination of vetebrates, and on dealing with each endocrine process in the context of the organism's physiology, ecology and evolution.

Comparative Vertebrate Endocrinology discusses the intimate physiology of the endocrine system and the pivotal role of hormones in coordinating basic body processes such as nutrition, reproduction, calcium metabolism and osmoregulation, as well as their contributions to animal coloration, molting and development. The species included range from lower chordates through to mammals, including marsupials.

Peter Bentley graduated from the University of Western Australia where his first degree was in Zoology and his PhD in Physiology.

In a distinguished career which has seen the publication of over 200 original research articles and five books, Professor Bentley has held academic positions at University College London, the University of Western Australia, University of Bristol, Duke University, and the Mt Sinai School of Medicine of the City University of New York. Prior to returning to Australia, he was Burroughs Wellcome Distinguished Professor of Pharmacology in the College of Veterinary Medicine, North Carolina State University.

Comparative Vertebrate Endocrinology

Comparative Vertebrate Endocrinology

THIRD EDITION

P.J. BENTLEY

CAMBRIDGE
UNIVERSITY PRESS

PUBLISHED BY THE PRESS SYNDICATE OF THE UNIVERSITY OF CAMBRIDGE
The Pitt Building, Trumpington Street, Cambridge CB2 1RP, United Kingdom

CAMBRIDGE UNIVERSITY PRESS
The Edinburgh Building, Cambridge CB2 1RU, United Kingdom
40 West 20th Street, New York, NY 10011-4211, USA
10 Stamford Road, Oakleigh, Melbourne, 3166, Australia

First published 1976
Reprinted 1980
Second edition 1982
Third edition 1998

Printed in the United Kingdom at the University Press, Cambridge

Typeset in Garamond 11/13pt [VN]

A catalogue record for this book is available from the British Library

Library of Congress Cataloguing in Publication data

Bentley, P. J.
Comparative vertebrate endocrinology / P. J. Bentley. – 3rd ed.
    p.   cm.
Includes bibliographical references and index.
ISBN 0 521 62002 3 (hb). – ISBN 0 521 62998 5 (pb)
1. Encorcrinology, Comparative.   2. Vetebrates – Physiology.
I. Title.
QP187.B46   1998
573.4'16–dc21   97–27048   CIP

ISBN 0 521 62002 3 hardback
ISBN 0 521 62998 5 paperback

Dedicated to the memory of Hans Heller and Harry Waring,
who introduced me to comparative endocrinology

Contents

Preface to the third edition

This edition has been extensively revised and contains nearly 1000 new references and over 60 new or modified figures. This proliferation reflects advances in our knowledge of hormones and their actions, and a persistent interest in the application of this knowledge to the domain of comparative endocrinology. Both of the dedicatees of this book are now deceased. When they first became interested in this subject over 60 years ago they utilized contemporary pharmacology to help to lay the foundations of the discipline. Its recent propagation largely reflects the use of the techniques of molecular biology and the unravelling of the genome to decipher the interrelationships of hormones in both vertebrates and their invertebrate progenitors.

It would not have been possible for me to prepare this edition without the collaboration of The University of Western Australia, particularly the Physiology Department and Professor Trevor Redgrave. All members of the department helped by making me feel welcome there. The Biological Sciences Library is the principal repository of the new information that I have used. Its comprehensive collection and helpful staff provided a pleasant venue for many hours of searching.

Peter Bentley
The University of Western Australia
December 1997

Preface to the second edition

It is over 6 years since the first edition of this book went to press. Interest in comparative endocrinology has not wanted in that time, as shown by the steady stream of papers and the organization of meetings and symposia on this subject. Several new hormones have been identified and described in the interim. Information about the synthesis of proteins that act as prohormones has provided enlightenment about the existence of more "hormone families" with consequent speculation about their evolution. An increased utilization of radioimmunoassays and immunohistochemistry has promoted many of these advances. There has also been an increased appreciation of commonalities of the endocrine and nervous systems, as described in the discipline of neuro-endocrinology. However, because the basic information about the endocrine system has not really changed, it has been unnecessary to alter significantly the conclusions at the end of each chapter.

In view of the great expansion of the literature, the preparation of this edition has been especially challenging. I have generally refrained from substituting new references for old ones, a practice that would ignore the seniority of discoveries and distort the historical perspective of the subject. There are thus many more references in the text. I hope that this does not distract the students for whom this book is primarily intended. They should "read around" the references and use them as a source if necessary. More senior readers may find the expanded bibliography more useful. Finally, I would like to apologize to the many endocrinologists whom it has not been possible to quote but without whose discoveries our knowledge of this subject would be much poorer.

P. J. Bentley
New York
February 1982

Preface to the first edition

This book has been written primarily for use as a textbook by undergraduate, as well as graduate, students. It is hoped that it may serve as a basis for course work in comparative endocrinology and also as an auxiliary text to aid in the teaching of comparative animal physiology. In order to gain the most from this book, the reader should have a basic knowledge of zoology and animal physiology. I have nevertheless attempted to put the endocrinology that is described into a broader biological framework by relating it to the animal's physiology, ecology, and evolutionary background. This is one of the reasons why I have departed from the more usual format of previous textbooks in this area, which generally deal with each endocrine gland in succession, chapter by chapter. Instead, I have attempted to describe certain broad and basic biological processes, the functioning of which is often coordinated by the secretion from several endocrine glands.

No attempt has been made to describe invertebrate endocrinology, as the rapid growth of this area really justifies a separate textbook. The book by K. G. Highnam and L. Hill (*Comparative Endocrinology of the Invertebrates*, Elsevier: Amsterdam, 1970) deals admirably with this subject.

It has not been possible in a book of this nature to give a complete list of original references. There are far too many of these, and many of the earlier observations are already a part of the "classical literature". Instead, I have attempted to refer the reader to more recent papers and reviews that contain references to the material described and can act as useful "starting points" for the students who wish to study the subject further. In order to keep abreast of developments in the various subject areas described, the current literature should be consulted. The principal journals where papers on these subjects are published are *General and Comparative Endocrinology, Journal of Endocrinology, Endocrinology*, and *Comparative Biochemistry and Physiology*. In addition, many papers appear in the standard physiological journals, especially *Journal of Physiology* and *American Journal of Physiology*.

P.J.B.
Mount Sinai School of Medicine of The City University of New York
September 1974

Some abbreviations used in endocrinology

ACTH	corticotropin (adrenocorticotropic hormone)
ADH	antidiuretic hormone
AMH	antimullerian hormone
ANP	atrial natriuretic peptide (s)
AVP	arginine-vasopressin
cyclic AMP	cyclic adenosine-3'5'-monophosphate (cAMP)
CaBP	calcium-binding protein
CBG	corticosteroid (cortisol)-binding globulin
CCK	cholecystokinin
cDNA	complementary DNA
CG	chorionic gonadotropin
CGRP	calcitonin gene-related peptide
CRE	cyclic AMP response (or regulatory) element
CREB	cyclic AMP response element binding protein
CRF	corticotropin-releasing factor
CRH	corticotropin-releasing hormone
CT	calcitonin
DAG	diacylglycerol
DHEA	dehydroepiandrosterone
DHT	dihydrotestosterone
EDLF	endogenous digitalis-like factor
ER	estrogen receptor
FSH	follicle-stimulating hormone
GABA	γ-aminobutyric acid
GH	growth hormone (somatotropin)
GH-RF (or -RH)	growth hormone-releasing factor (or hormone)
GIP	gastric inhibitory polypeptide (glucose-dependent insulinotropic poly-peptide)
GLP-1, GLP-II	glucagon-like peptide I, II
GMP	guanosine 5'-monophosphate
GnRH	gonadotropin-releasing hormone (LHRH)
GR	glucocorticoid receptor
GSD	genotype sex determination
GTH-I, GTH-II	gonadotropic hormone (in fish) I, II
hCG	human chorionic gonadotropin

hCS	human chorionic somatomammotropin (hPL)
hPL	human placental lactogen
HRE	hormone-response element
5-HT	5-hydroxytryptamine (serotonin)
-IF	-inhibiting factor
IGF-I, IGF-II	insulin-like growth factor (s) I, II
-IH	-inhibiting hormone
IP$_3$	inositol-1,4,5-trisphosphate
IT	isotocin
JAK	janus kinase
KT	ketotestosterone
LH	luteinizing hormone
LVP	lysine-vasopressin
MAO	monoamine oxidase
MCH	melanin-concentrating hormone
MIP	molluscan insulin-related peptide
mRNA	messenger RNA
MR	mineralocorticoid receptor
MSH	melanocyte-stimulating hormone (melanotropin)
MT	mesotocin
NHP	nasohypophysial protein
NPY	neuropeptide Y (or neuropeptide tryosine)
OT	oxytocin
PACAP	pituitary adenylate cyclase activating protein
PG	prostaglandin
PIP$_2$	phosphatidylinositol-4,5-bisphosphate
PKA	protein kinase A
PKC	protein kinase C
PL	placental lactogen
PLC	phospholipase C
PMSG	pregnant mare's serum gonadotropin
PMY	peptide MY (methionine–tyrosine)
PNMT	phenolethanolamine-N-methyltransferase
POMC	proopiomelanocortin
PP	pancreatic polypeptide
PR	progesterone receptor
PRL	prolactin
PTH	parathyroid hormone
PTHrP	parathyroid hormone-related protein
PYY	peptide YY (or peptide tyrosine–tyrosine)
-R	receptor (suffix)
RAS	renin–angiotensin system
-RF	-releasing factor
-RH	-releasing hormone
-R-IH	-release-inhibiting hormone
rT$_3$	reverse T$_3$

SHBG	sex hormone-binding globulin
SRIF	somatotropin release-inhibiting factor (somatostatin)
STC	stanniocalcin
T_3	triiodothyronine
T_4	thyroxine
TBG	thyroid hormone-binding globulin
TDF	testis-determining factor
TGF	transforming growth factor
TNF	tumor necrosis factor
TR	thyroid hormone receptor
TRH	thyrotropin-releasing hormone
TSD	temperature-dependent sex determination
TSH	thyroid-stimulating hormone (thyrotropin)
VIP	vasoactive intestinal peptide
VNP	ventricular natriuretic peptide
VP	vasopressin

1 Introduction

The book describes a method of transferring information within vertebrates. Such communication is necessary in order to coordinate physiological processes with each other and to the happenings in the external environment. Even unicellular organisms synchronize their various internal life processes. In such small creatures, however, local accumulations of metabolites may exert a direct control on biochemical reactions, whereas external stimuli have relatively widespread effects so that specialized pathways for communication may not be as necessary. Therefore, when the distances involved are short, physical processes such as conduction, convection, and diffusion may be adequate for the integration of the physiological processes. Nevertheless, even unicellular organisms possess specific coordinating systems such as that seen in the protozoan *Tetrahymena* (Blum, 1967), which possesses epinephrine. This hormone has similar metabolic actions in this protozoan to those that it has in vertebrates.

The problems of communication and coordination are greater in multicellular than in unicellular organisms. There are several reasons for this, especially their larger size. As the linear distances between the different parts of an animal increase, simple physical communications become relatively slower and less precise, and so not as effective. In multicellular organisms, the cells are usually specialized and perform different functions that, in combination, are essential for the animal's life. Thus, some tissues may be concerned with the formation of reproductive germ cells, several others with the preparation of suitable nutritive materials, and yet others with building morphological structures. The ultimate successful completion of these processes will be determined by the effectiveness of the communication between the tissues themselves and the external environment.

The transfer of information in animals

There are three principal ways by which cells in multicellular organisms can communicate with each other. First, when they are in close juxtaposition and are only separated by narrow fluid-filled spaces, direct electrical and chemical interactions can occur. Cells also maintain some structural connections with each other and secrete special excitants by which they may also communicate.

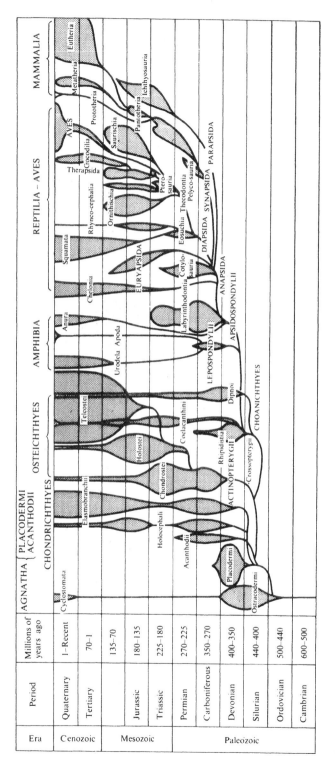

Era	Period	Millions of years ago
Cenozoic	Quaternary	1–Recent
	Tertiary	70–1
Mesozoic		135–70
	Jurassic	180–135
	Triassic	225–180
Paleozoic	Permian	270–225
	Carboniferous	350–270
	Devonian	400–350
	Silurian	440–400
	Ordovician	500–440
	Cambrian	600–500

Fig. 1.1. A classification of vertebrates in relation to their phylogenetic origins and a time scale in terms of paleontological periods. (From Torrey, 1971.)

Such excitants are often called local hormones; if they affect a different type of neighboring cell their action is described as a paracrine one or, if it is a cell of the same type, it is described as an autocrine one. Second, contact between more remote cells can be maintained along tracts of nerve cells that are merely tissues specialized for such exchanges of information. Third, chemicals may be released, for example from the endocrine glands, into the blood, which carries them to special sites that are physicochemically programed to react and respond to them. The actions of hormones are slower than those of nerves but as they may persist in the plasma and at target sites their effects generally last longer. Nerves follow defined anatomical pathways but hormones are diffusely distributed. However, hormones attain precision in their actions by utilizing specific receptor molecules at their intended target sites.

The endocrine glands are tissues that, unlike exocrine glands, have no ducts but release their secretions, called "hormones," directly into the blood passing through them. It is with the diversity of such hormonally controlled processes that we will be principally concerned in this book. It should always be recalled, however, that the endocrine glands represent only a single facet of the animal's communication network and that nerves are also important. Nerves and hormones are often mutually interdependent and may even act together to control a single process. Nerve cells can respond to hormones in a manner that influences behavior, and endocrine glands often receive information and directions from the brain. Both hormones and nerves can act together to control the melanophores in certain fishes. Many hormones, including epinephrine, vasopressin, oxytocin, and the hypophysiotropic hormones, are even made by nerve cells.

What is comparative endocrinology?

Comparative endocrinology concerns the study of the endocrine glands in different species of animals, both vertebrates and invertebrates. Its aims are analogous to the older and more classical disciplines of comparative anatomy and comparative physiology. The prime academic objective is to reconstruct evolutionary pathways by the study of extant species. Figure 1.1 shows the phylogenetic relationships of the vertebrates and this emphasizes the extant groups that may be particularly interesting in such studies. The mere examination of the endocrine system of some bizarre and exotic vertebrate does not alone constitute "Comparative Endocrinology" (it may be "Animal Endocrinology") unless the data can be considered in relation to that in other, phyletically related species. Such information can be used to help to confirm, complete, and even extend our knowledge of the phylogenetic relationships between vertebrates and to follow the evolution of endocrine mechanisms. The lungfishes (Dipnoi) may afford us an example. These fishes have long been considered, on the basis of morphological information, to be close to the original line of evolution connecting the bony fishes (Osteichthyes) and the Amphibia. As we shall see later, homologous vertebrate hormones often exhibit considerable differences in their chemical structure. Many such differences are apparent between the hormones in fishes and tetrapods. The structure of several hormones present in lungfishes, however, shows a greater similarity to those in tetrapods than those in other fishes. For instance, a neurohypophysial hormone called mesotocin is present in amphibians, reptiles, and birds, but in bony fishes the homologous hormone is isotocin (which differs from mesotocin by a single amino acid substitution), with the exception of the lungfishes, which have mesotocin. It has also been found that the growth hormone and prolactin present in lungfishes are more like those in tetrapods than in other fishes.

Apart from contributing to the overall phyletic study of vertebrates, the comparative endocrinologist aspires to reconstruct the lines of evolution within the endocrine system itself. This can be done by examining and comparing in different species, the morphology of the endocrine tissues, the structures and biological activities of their secreted hormones, receptors and their different physiological roles.

Occurrence of vertebrate hormones in other organisms

It is now generally accepted that vertebrate hormones and their receptors may have evolved from identifiable counterparts among the invertebrates. The increasing awareness of this concept has been promoted by the use of immunocytochemical techniques to identify minute amounts of related molecules at specific sites in various invertebrates, especially among the molluscs and insects. Such an identification of related hormones has been aided by microbioassays and recombinant DNA techniques. The latter have been used to produce probes to identify coding nucleotide sequences that can be deciphered to describe the polypeptide molecules. Often, the genes themselves have been identified and chemically dissected so that their structural organization can be compared with those of their putative vertebrate relatives. Even the most sceptical may be impressed by their possible relationships. Molecules that are identical or are close chemical relatives to vertebrate hormones have been identified among the invertebrates (see for instance, LeRoith *et al.*, 1986; de Loof and Schoops, 1990; Pertseva, 1991; Renaud *et al.*, 1991).

Vertebrate hormones have even been identified in plants (Le Roith *et al.*, 1986). Thyrotropin-releasing hormone has, for instance, been found in alfalfa (Jackson, 1981a). Such occurrences obviously do not indicate a phylogenetic relationship though they could reflect the reutilization of a pool of available genetic material. When they have been identified in plants, the role of such vertebrate hormonal substances is usually obscure. However, if consumed by animals they may influence their physiology. Examples include a toxic deposition of calcium in the tissues of cattle caused by vitamin D in their forage (Wasserman *et al.*, 1976) and an inhibition of seasonal reproduction in desert quail by plant estrogens (Leopold *et al.*, 1976).

The presence of molecules related to vertebrate hormones among the invertebrates is not surprising when one recalls that the principal neurotransmitters acetylcholine, catecholamines, 5-hydroxytryptamine, and γ-aminobutyric acid (GABA) are all shared on many occasions by both of these major groups of animals. Insulin-like molecules have been identified in the gut tissues of some molluscs and a honeybee (Plisetskaya *et al.*, 1978; Moreau, Raoelison, and Sutter, 1981), calcitonin and somatostatin in nerve tissue of a sea squirt (a tunicate, Protochordata) (Falkmer and Ostberg, 1977; Girgis *et al.*, 1980), and cholecystokinin (CCK) in the brain of a blowfly (Duve and Thorpe, 1981). The occurrence of iodinated tyrosine molecules with activity like that of thyroid hormones in protochordates is also well known. A molecule that cross-reacts with antibodies to mammalian vasopressin (antidiuretic hormone, ADH) has been found in the brain and other tissues of a locust (Proux and Rougon-Rapuzzi, 1980). The physiological roles of such substances in these invertebrates are uncertain, but it has been suggested that they may act as neurotransmitters or hormones and, in the instance of insulin and epinephrine, they may even have comparable metabolic effects to the ones they have in vertebrates.

It is tempting to suggest that the presence of vertebrate types of hormone in invertebrates may reflect some evolutionary relationship. In the instance of the protochordates and the molluscs this may be so. Their presence in insects, however, makes such common origins appear rather tenuous. It seems likely that nature has been somewhat frugal or has a limited inventiveness in its provision of molecules and their encoding genes, that are suitable for roles in chemical coordination mechanisms. They may then have been reutilized by nature on several separate occasions. However, the possibility that such similar molecules may arise owing to convergent evolution should not be forgotten. Common molecular requirements may have limited structural solutions.

Molecular evolution and the endocrine system

The phyletic distribution of hormones and their precursors, their synthesizing enzyme systems, and receptors among the vertebrates reflects the processes of evolution. While such components of the endocrine system have retained characteristics in common, they frequently display changes in their chemical structures. The structural homologies of the hormones, their precursor prohormones, preprohormones, and their receptors often permit the construction of molecular phylogenetic trees (cladograms) and trace their evolution among vertebrates, from cyclostome fish to mammals, and sometimes even include invertebrates. Estimates of the time when such changes may have occurred can sometimes even be proffered.

The evolution of protein families (see for instance Wilson, Carlson and White, 1977; Doolittle, 1981, 1989; Niall, 1982) began in prokaryotic organisms which, probably, possessed a 'library' of less than 1000 proteins. These archetypes are thought to be the progenitors of the 'superfamilies' to which many contemporary proteins belong. Their evolution and proliferation have involved various types of genetic event but were probably initiated by gene duplication followed by mutations on their DNA sequences to code for novel proteins. Such modifications include point mutations and rearrangements such as tandem duplications, gene conversion, and exon shuffling. Viruses may be involved in such transposition of genetic material. There appears to be a tendency to increase the amount of DNA in the genome as a result of such events. Such a process may provide the organism with more opportunities to produce proteins, including hormones, that may aid their adaptation to the environment.

Homologies in the amino acid sequences of the proteins of the endocrine system reflect the nucleotide sequences of their genes. When describing such relationships in the endocrine system it is more usual to compare the sequences of their amino acids, but those of their nucleotides are also used. Similarities are usually expressed as a percentage (%) identity or homology. Thus, two identical polypeptides will have a homology of 100%. Lesser values suggest a more distant relationship but may still be consistent with a common ancestor. Two proteins of similar length and amino acid composition would be expected, on average, to display a 5% identity by chance alone. (For comparisons using nucleotide sequences, this value would be 25% as only four bases are involved, compared with 20 amino acids.) Greater values than these may reflect an ancestral relationship, though it may sometimes be quite remote. The reliability of low homology values depends on various factors and can be influenced by the particular amino acid composition of each molecule (the proportions may vary), the chain length, and the alignment used for such comparisons. 'Tails' of unmatched amino acids may be present and there may be interruptions, or 'gaps', in the sequences. Such factors can, however, be compensated for. Computer programs are available for making such comparisons and can provide statistical limits of the reliability of a particular value (Doolittle, 1990). Criteria for assessing the relationship of molecules vary but an identity of between 15 and 25% in amino acid sequence is usually considered as being 'suggestive' and over 25% is highly likely. Such pronouncements can be aided by statistical information. The possibility that convergent evolution may be contributing to similarities in molecular structure should be considered (Doolittle, 1989). Hormones and receptors may on the basis of such information be classified into families and superfamilies. The latter involves clear evidence of a homology, though with percent identities often less than 50%. Several families may make up a superfamily and an arbitrary percent identity of greater than 50% has been suggested as one such criterion for a family, but there are others.

The usefulness of phylogenetic information derived from differences in the amino acid sequences of peptide and polypeptide hormones varies depending on the size of the molecule and the relative number of conserved amino acid sites that it contains (see Hedges, Moberg, and Maxson, 1990; Dores, Rubin and Quinn, 1996). The latter may be integral, and indeed essential, to the molecule's biological activity and, therefore, they may be invariant over a broad phyletic range of species. As a result, such sites can be relatively 'uninformative' with respect to the detailed tracing of phylogenetic trees. (Though they are very useful for identifying functionally important regions of peptides and polypeptides). Small molecules such as the neurohypophysial hormones, gonadotropin-releasing hormones, and even insulins have a high proportion of such conserved amino acids. The phylogenetic information such molecules provide may then even be inconsistent with the fossil record and comparative morphology. In such instances, the structures of their preprohormones and prohormones may provide more useful cladistic information. Larger hormones such as growth hormone and prolactin usually have a greater number of variable or 'informative' sites for such analyses.

The concept of a 'molecular clock' has provided a method of utilizing observed differences in amino acid or nucleotide sequences among homologous proteins or polypeptides to estimate the period of time that may have elapsed since they diverged. Such a clock has been utilized to describe evolution in the endocrine system. It is calibrated from information available in the fossil record in units such as the rate of exchange of amino acids (or nucleotides) per year. By counting amino acid substitutions one can then estimate the time since the divergence of the two molecules occurred from a common ancestor. These estimates have sometimes been quite contentious (owing to possible reversions or back mutations) but they are, nevertheless, often very productive (see Lewin, 1988).

The methods used in comparative endocrinology

Many of the endocrine glands were described morphologically long before their functions were recognized. Establishment of a physiological role is usually initiated by observations of the effects of its surgical removal. Administered extracts of such tissues ("replacement") should be able to correct any observed deficiency. The particular active component in the extract then must be identified. The final criterion for a hormonal status is its identification in the blood and its observed release in response to appropriate stimuli. The mere presence of a response to administration of a tissue extract or a hormone from another species need not reflect an endocrine role.

The scientific techniques utilized in the study of comparative endocrinology have usually been derived from the contemporary technology. At first it was anatomical, histological, and surgical extirpation. Pharmacology provided a variety of techniques, including ways to prepare glandular extracts and measurement of their activity using biological preparations ("bioassays"). It also furnished the concepts of "receptors" and the relationships between chemical structure and biological activity. Chemical methods were used to separate out and purify the hormones and often to determine their structures. Immunology provided antibodies to hormones and so fostered their immunocytochemical detection. It also offered methods for comparing the serological relationships of large protein hormones. The availability of radio-isotopes has influenced virtually all aspects of comparative endocrinology by making possible the "labeling" of hormones; this facilitated their measurement by radioimmunoassay and their identification at tissue sites. The most recent technological contribution has been molecular genetic methods (recombinant DNA and molecular cloning techniques), which have provided knowledge of the nucleotide sequences that encode hormones, their precursors, and their receptors. Nucleotide sequences for hormone-like peptides and proteins that were not known to occur naturally have also been uncovered in this way. Such codes for peptides may be for as yet undetected hormones, "unused" ones, or even long extinct ones. This gene technology has even resulted in the identification of the genes that express the hormones and their receptors and has provided a view of the characteristic organization of their DNA and the patterns of the exons and introns present.

The uses of comparative endocrinology

The classic, or academic, aims of comparative endocrinology have been described. The provision of such intellectual satisfaction is not, however, sufficient justification for all! There are, indeed, a number of other contributions that such studies can make to biology, and some examples of these are given in this section.

The process of reproduction in vertebrates is dependent on the endocrine secretions, and an understanding of this relationship can provide information that may be usefully applied when, for esthetic or economic reasons, we may wish to increase, or decrease, the fecundity of a species. This type of study, therefore, constitutes a contribution to the field of "biological control" (Bern, 1972).

Knowledge of the endocrine system in humans has largely been made possible by experiments on other animals. This has principally involved mammals like rats, rabbits, and monkeys but also some more exotic and bizarre creatures. Quantitative measurements of gonadotropins and melanocyte-stimulating hormone (MSH) were originally made using the responses of the clawed toad (*Xenopus laevis*), and prolactin levels can be measured by its effects on the pigeon's crop-sac or on the behavior of a newt. Oxytocin is assayed by utilizing its ability to decrease the blood pressure of chickens, and the rate of water movement across the toad's urinary bladder can be used to distinguish between two, chemically different, mammalian ADHs.

The responsiveness of a toad's urinary bladder to ADH and aldosterone is used to study the 'mechanism of action' of these hormones on membrane permeability. Such preparations provide useful "models" of hormonal effects on the mammalian kidney.

The relationship of the structure of a molecule to its biological activity is a field of considerable interest to biologists. The diversity, or polymorphism, in the structure of vertebrate hormones, together with their disparate effects on different tissues and in various species, offers a natural "laboratory" for such studies. Nature has had a long time and wide opportunities to experiment with the effects of changes in molecular structures on the activities of such excitants. At present, this is most clearly seen among the neurohypophysial hormones, of which there are 12 known chemical variants among the vertebrates. These hormones are peptides containing nine amino acid residues and often only differ from one another by a substitution at a single chemical locus. They are very reactive molecules and can exert actions at many different sites ranging from the uterus and mammary gland to blood vessels, the kidney, and the amphibian skin and urinary bladder. Analogous effector tissues in different phyletic groups exhibit different abilities to respond to each such hormone, be it a natural one or a variant made in the chemist's laboratory. There are available, and in use, more than 20 different effector preparations that can be used to study the effects of changes in chemical structure among these hormones on their biological effectiveness. Natural variants of hormones, in which the biological activity has been altered in some way, may be of potential use to humans or may provide information about structural modifications that may be medically advantageous. Calcitonin (a hormone concerned with the regulation of calcium in the body) from the ultimobranchial bodies of salmon differs from the hormone present in humans by 16 of its 32 amino acid residues and yet it is much more potent in humans than the human hormone itself. Salmon calcitonin is used therapeutically to treat Paget's disease of the bones. ADH from pigs is effective in humans. It differs from the human hormone by a single amino acid substitution. Early preparations of the synthetic pig ADH were found to be more chemically stable than the homologous human hormone and provided the blue print for a manufactured drug used to treat human diabetes insipidus.

Table 1.1. *The secretions of the endocrine glands*

Gland	Hormones	Principal target tissues/effects
Pituitary		
Adenohypophysis		
Pars distalis	Follicle-stimulating hormone (FSH)	Ovary and testis
	Luteinizing hormone	Ovary and testis
	Thyrotropic (TSH)	Thyroid
	Corticotropic (ACTH)	Adrenocortical tissue
	Growth hormone (GH) (somatotropic)	Liver (forms IGF-I), muscle, bone, adipose tissue, gills (teleosts). Proliferation, metabolism
	Prolactin	Mammary glands, fish gills, tadpole metamorphosis, corpus luteum, skin, etc.
Pars intermedia	Somatolactin (teleosts)	?
	Endorphins	Nerve and endocrine cells
	Melanocyte-stimulating hormone (MSH)	Melanocytes, pigmentation and color change
Neurohypophysis		
Pars nervosa	Vasopressin (ADH), vasotocin	Kidney, amphibian skin, urinary bladder
	Melanin-concentrating hormone (MCH)	Teleost melanocytes (other?)
	Oxytocin	Mammary gland, uterus
Hypothalamus	Pituitropins; GnRH, CRH, TRH,[a] somatostatin, etc.	Release of hormones by the adenohypophysis
Thyroid gland	Thyroxine (T_4), triiodothyronine (T_3)	Tissue metabolism and differentiation: thermogenic (homeotherms); morphogenetic (larval amphibians, teleosts)
Parathyroid glands	Parathyroid hormone (PTH)	Bone and kidney
Ultimobranchial bodies (C-cells in mammalian thyroid)	Calcitonin (CT)	Bone and kidney (gills?)
Adrenal glands		
Cortex (interrenals in sharks and rays)	Cortisol, corticosterone, cortisone, 1α-hydroxycorticosterone	Tissue metabolism (liver, muscle), proteins to amino acids, gluconeogenesis; gills, intestine (various)
	Aldosterone	Na^+ and K^+ in kidney, sweat and salivary glands, gut, amphibian skin and bladder
Medulla (chromaffin tissue)	Norepinephrine (noradrenaline), epinephrine (adrenaline)	Tissue metabolism (liver, muscle, adipose tissue), glycogenolysis, mobilization fatty acids, calorigenic, constriction and relaxation of smooth muscle
Islets of Langerhans		
A-cells	Glucagon	Liver (glycogenolysis); adipose tissue (fatty acid release); gluconeogenesis
B-cells	Insulin	Liver, muscle and adipose tissue (amino acids to protein, glucose to fat and glycogen)

Source	Hormone	Function
D-cells	Somatostatin	A- and B-cells
PP-cells	Pancreatic polypeptide	Increases exocrine pancreatic secretion
Gonads		
Ovary		
Ovarian follicle	Estrogens (estradiol)	Female sex organs and characters, mammary glands, brain
	Inhibins	Inhibit FSH
Corpus luteum and interstitial tissue	Progestins (progesterone)	Uterus and mammary glands
Testis		
Interstitial tissue (Leydig cells)	Androgens (testosterone)	Male sex organs and characters, sperm maturation, brain
Sertoli cells	Antimullerian hormone (AMH)	Inhibits mullerian duct development (embryos)
	Inhibins	Inhibits FSH
Placenta (pregnant eutherian mammals)	Estrogen (estriol), progesterone	Uterus, mammary glands, fetus
	Chorionic gonadotropin (hCG)	Corpus luteum
	Placental lactogens (PL)	Tissue metabolism
Gut		
Stomach (pyloric mucosa)	Gastrin	Stimulates secretion of gastric juice
Intestine (mucosa)	Gastric inhibitory peptide (GIP)	Inhibits secretion of gastric juices, promotes insulin release
	Secretin	Stimulates secretion of pancreatic juices from exocrine pancreas
	Cholecystokinin (CCK)	Enzyme secretion from pancreas (and hormones from endocrine pancreas)
	Glucagon-like peptide (GLP)	B-cells (promotes insulin release)
Kidney		
Tubular cells	1α,25-dihydroxycholecalciferol ($1,25(OH)_2D_3$)	Intestine, bone, etc.
Juxtaglomerular cells	Renin	Plasma α^2-globulin→angiotensinogen→angiotensin (targets: adrenal cortex, vascular smooth muscle, thirst center)
Pineal gland	Melatonin	Hypothalamus (hormone-release rhythms), melanocytes (larval anurans, cyclostomes?)
Corpuscles of Stannius (some bony fishes)	Stanniocalcin, teleocalcin	Calcium metabolism in fish (gills, etc.?)
Urophysis (some fishes)	Urotensins I and II	Osmoregulation, corticosteroidogenesis?
Heart	Natriuretic peptides (ANPs)	Kidney (natriuresis), blood vessels (dilatation)

[a]For explanation, see List of abbreviations

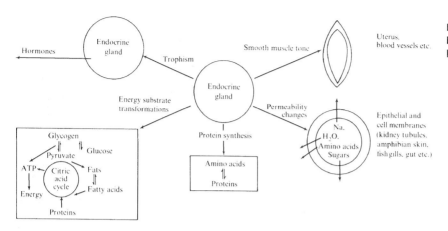

Fig. 1.2. Summary of the basic types of action of hormones in vertebrates.

The diversity of vertebrates as a background for endocrine variation

There are some 42 000 extant species of vertebrate animals. The vertebrates originated some 500 million years ago as creatures who apparently lived in the sea or, possibly, in fresh water. They subsequently evolved and occupied almost every conceivable habitat in the oceans, in freshwater rivers and lakes, and on the land. Their abodes range from the cold polar regions to hot equatorial ones, from deserts to swamps, from high mountains to the ocean deeps. The considerable morphological and physiological diversity of verte-brates mirrors their success in this multitude of environmental conditions. It is therefore not surprising to find that the endocrine system exhibits inter-specific differences that reflect adaptations to such different environments. Nevertheless, it is also somewhat unexpected to find that considerable similar-ities are still apparent in the endocrine systems of species as distantly related as the hagfish (Cyclostomata) and humans.

The endocrine glands of vertebrates have special roles to play in the regulation of many types of physiological process that include reproduction, osmoregulation, intermediary and mineral metabolism, and growth and development (Table 1.1). The nature of the responses to hormones differs considerably but can be classified into several major groups including their actions on membrane permeability, muscular contraction, the transformation of substrates involved in intermediary metabolism and growth, and a con-trolling (or tropic) action on other endocrine glands (Fig. 1.2). Such re-sponses are not direct but involve the activation of chemical mechanisms that mediate the final effect. They initially include interactions with specific receptor molecules that may be present in the cell plasma membrane or associated with the genetic machinery of the nucleus. A cascade of biochemi-cal reactions usually follows resulting in the hormone's final response.

Many, though not all, of the endocrine glands are essential for life and the reproduction and survival of the species. In other instances, however, their immediate importance for survival is not clear. Animals cannot reproduce if the endocrine function of their gonads is compromised, and death soon follows complete destruction of the adrenal cortex. Life may be shortened if the islets of Langerhans fail to produce sufficient insulin, and normal growth, development, and maturation of the young will not occur if the secretion of pituitary growth hormone or thyroid hormone is inadequate. By comparison, ADH from the neurohypophysis is not essential for life though in its absence very large volumes of urine are secreted by the kidney. In humans this is an annoying condition as prolonged sleep is not possible and even during the waking hours it can lead to social difficulties; but it is not fatal. If drinking water were in limited supply, however, dehydration could be a potential problem and absence of this hormone might then affect survival. It should also be remembered that whereas too little of a hormone can constitute a problem, too much may also result in physiological difficulties. Hormone imbalances can result from genetic abnormalities, the presence of tumors, and accidental disruption of the events controlling secretion of the hormone.

Endocrine glands, or tissues, have been identified among all of the vertebrates. Those common to the major groups (from the Cyclostomata to the Mammalia) are the pituitary, pineal, thyroid, endocrine pancreas, adrenal chromaffin and cortical tissues, and gonads. The parathyroid glands have been found in the tetrapods but not in the fishes. The ultimobranchial bodies of their homolog, the thyroid "C"-cells, have been identified in all groups except the cyclostomes. The corpuscles of Stannius are more restricted and are only present in some of the bony fishes. The urophysis, still a putative endocrine gland, is confined to the caudal regions of teleost fish and although some similar tissues are also present in elasmobranch fish it is not present in other vertebrates.

Such tissues secrete more than 40 different hormones in a mammal. If we include all the naturally occurring analogs of these hormones that occur among the vertebrates, we can account for at least twice this number of hormones and there are undoubtedly many more.

It is conceivable, indeed likely, that other endocrine glands exist among the vertebrates. It was only in the early 1980s that the role of the mammalian heart as an endocrine organ was unexpectedly recognized. The muscle of the atria had been previously known to contain secretory-like granules. These organelles were found to contain an excitant that was called atrial natriuretic peptide because of its ability to promote the excretion of sodium in the urine. The long-suspected presence of a natriuretic hormone was thus confirmed. This peptide belongs to a related group that exists throughout the vertebrates from cyclostome fish to mammals.

Despite their anatomical and embryological homologies, the endocrine glands of vertebrates display considerable diversity in their morphological arrangements, the chemical nature of their secretions, and even in their physiological role in the body. It is principally about these differences that we will be concerned in the succeeding chapters.

Conclusions

Physiological processes are coordinated with the aid of both nerves and hormones. Each of these mechanisms has special characteristics that may be suited to the needs of the particular process involved and they often operate together. During the course of geological time vertebrates have evolved and acquired morphological features and physiological processes that have permitted them to adapt to changing environments and to occupy a variety of ecological habitats. Such biologically important changes are accompanied by the neural and humoral processes necessary for their coordination. Contemporary species of vertebrates exhibit considerable structural and functional diversity that can be related to the nature of the life they lead and to their ancestry. They are classified into systematic groups that are also thought to reflect their evolution. Thus, a comparison of the endocrine function of contemporary species of vertebrates is of importance not only in fully understanding how they live today, in a particular environmental situation, but also because it may tell us how such hormonally mediated processes evolved.

2 Comparative morphology of the endocrine tissues

Endocrine glands and tissues display a diversity in their gross morphological and histological patterns. This is particularly apparent when comparing species from phyletically distant groups. In some instances the physiological significance of these differences has been recognized but in most this is not so and may be related to the initial pattern of embryonic growth. If, however, one intuitively suspects a close relationship between structure and function, then the lack of a known correlation may merely reflect our ignorance.

The endocrines may display several different types of morphological variation (see, for instance, Pang and Schreibman, 1986; Matsumoto and Ishii, 1992). The positions of the endocrine glands in the body may not be the same. The variation can be of a minor nature, such as that seen with the ultimobranchial bodies, which can be situated near the heart or the thyroid gland. In some fish, however, thyroid tissue may vary in position from the branchial region to the kidney. Endocrine cells may show varying degrees of association and be scattered as individual cells, in small segments, or "islets." or be closely associated as a compact gland enclosed in a capsule. Such aggregation of an endocrine tissue is commonly seen as one ascends the evolutionary (or the phyletic) scale. In addition, different endocrine tissues may display diverse associations with each other, as for instance the conglomeration of chromaffin and interrenal (or adrenocortical) tissue in the adrenal gland. Their relationship to the neural and vascular tissues can be very important. Pituitary tissues usually cannot function properly if they are transplanted to other parts of the body (ectopic transplant) or if the small blood vessels between the gland and the brain are cut. The major blood vessels not only carry hormones away from endocrine tissues but also supply them with nutrients and controlling stimuli. The pattern of the vasculature within the gland can also be important for its correct functioning.

The types of cell that make up an endocrine gland are, not surprisingly, similar in homologous glands among the vertebrates. Such similarities, as reflected by their microscopic anatomy (size, shape, the presence of inclusions, granules, etc.) and their reactions with dyes (tinctorial relationships), serve to aid in their identification. Antibodies to specific hormones are used to identify the cells where they are formed. These antibodies may be labeled with radioactive materials or fluorescent dyes so that the precise locus where they react can be seen. The histological appearance of endocrine cells may change somewhat at different times depending on their secretory state. This characteristic can be used to predict their activity and physiological role. Inactive thyroid cells, for instance, have a flattened form, rather than columnar appearance, which is typical of their active state, whereas neurohypophysial tissue that is depleted of its hormone has little stainable (with Gomori chrome–alum hematoxylin) neurosecretory material.

The pituitary gland

The pituitary is a conglomerate of tissues and cells that reflect the 10 or more major hormones it secretes. These hormones help regulate the activities of the thyroid, adrenal cortex, and gonads and contribute to the control of various other physiological activities, including water and salt metabolism, growth, lactation, parturition, and the pigmentation of the skin. A comparative account of the anatomy of this gland has been provided by Holmes and Ball (1974) and a beautifully illustrated summary by Mikami (1992).

Embryologically, the pituitary arises as a result of a downgrowth of tissue (the infundibulum) from the brain and an upgrowth (the hypophysis) from the roof of the mouth. Enclosed within these tissues is a piece of mesoderm that forms a net of blood vessels, sometimes called the "mantle plexus." The pituitary lies in close apposition to the hypothalamus at the base of the brain. In mammals it is usually enclosed in a small, bony chamber, the sella turcica, and it is connected by a stalk of nervous tissue to the brain, just behind the optic chiasma (Fig. 2.1). The hypophysis partly differentiates into the adenohypophysis, which secretes seven or eight hormones that are formed by a number of distinctive types of cell. These are most descriptively labeled by the name of the hormone they secrete followed by the suffix *trope*. We thus have thyrotropes, gonadotropes, somatotropes, and so on. The adenohypophysis can be divided on a gross morphological basis into three or four sections: the pars tuberalis, the pars distalis (sometimes with a rostral and caudal section), and the pars intermedia. The last gives rise to MSH and endorphins; in teleost fish it also produces somatolactin. The rest of the hormones come from the pars distalis. The pars tuberalis lies between the pars distalis and the brain (in the region of the median eminence) and is associated with the blood vessels that connect the two.

The neural, or infundibular, tissue forms the neurohypophysis, which basically lies caudally to the adenohypophysis, hence the terms anterior and posterior lobes of the pituitary. The neurohypophysis is connected to the brain by the *infundibular stalk*. The two hormones (ADH and oxytocin in mammals) it secretes are formed in nerve cells (by a process called neurosecre-

Fig. 2.1. The pituitary gland in humans. It lies in a bony chamber at the base of the brain to which it is connected by a stalk. The pars intermedia is quite small in humans but may be much larger in other species. (From R. Guillemin and R. Burgus, *The Hormones of the Hypothalamus.* Copyright © 1972 by Scientific American, Inc. All rights reserved.)

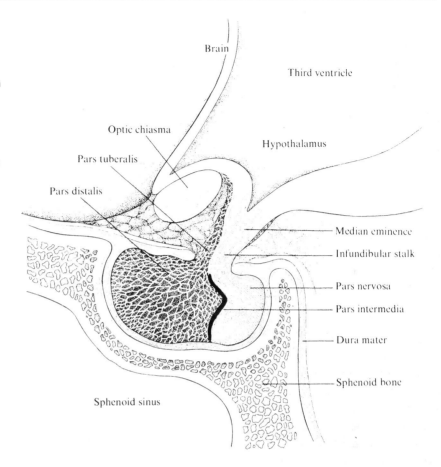

tion) that originate in the supraoptic and paraventricular nuclei in the brain of amniotes or the preoptic nucleus of amphibians and fishes. The axonal tract running from the bodies of these nerve cells in the nuclei to the periphery, where the hormones are stored and released, is called the supra-opticohypophysial tract. The neurohypophysis, therefore, consists of the distal parts of nerve cells interspersed with glial cells and pituicytes.

Three regions of the neurohypophysis can be distinguished. The rostral *median eminence* is part of the wall of the hypothalamus and lies in close conjunction with the adenohypophysis to which it is usually connected by a system of portal blood vessels that originate from the mantle plexus. The median eminence is contiguous with the infundibular stalk, which connects it to the most prominent part of the neurohypophysis, the *pars nervosa* (or *neural lobe*). The latter is much more highly developed in terrestrial tetrapods than in the fishes. The phyletic development of the amniote neurohypophysis is shown in Fig. 2.2.

Primitive type as in some reptiles

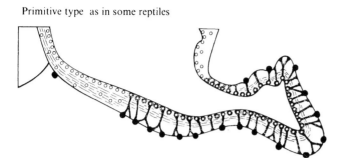

Birds and reptiles

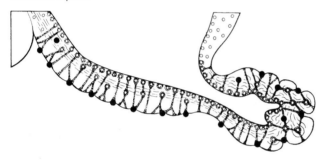

Mammals

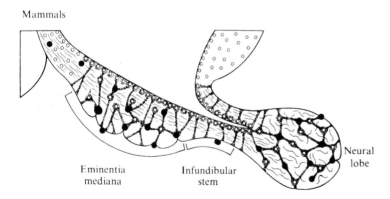

Eminentia
mediana

Infundibular
stem

Neural
lobe

Fig. 2.2. Histological differentiation of the amniote neurohypophysis. The primitive form is seen in reptiles such as the Rhynchocephalia. Chelonia, and some Lacertilia. Solid black lines are the blood vessels; nerve fibers are thinner lines. (From Wingstrand, 1951.)

Comparative morphology of the pituitary

The diverse morphology of the vertebrate pituitary provides us with some information (albeit equivocal) about the nature of the evolutionary changes that may have taken place in this gland. Attempts have been made to choose or construct the pituitary that is considered most typical of each major

Fig. 2.3. The pituitary glands of fishes. Diagrammatic representation from midsagittal section. Small dots, nervous tissue; black, neurohypophysial tissue; large open dots, pars distalis; horizontal lines, pars intermedia: thick black lines, blood vessels. III indicates the third ventricle. (From Ball and Baker, 1969, modified slightly according to Holmes and Ball, 1974.)

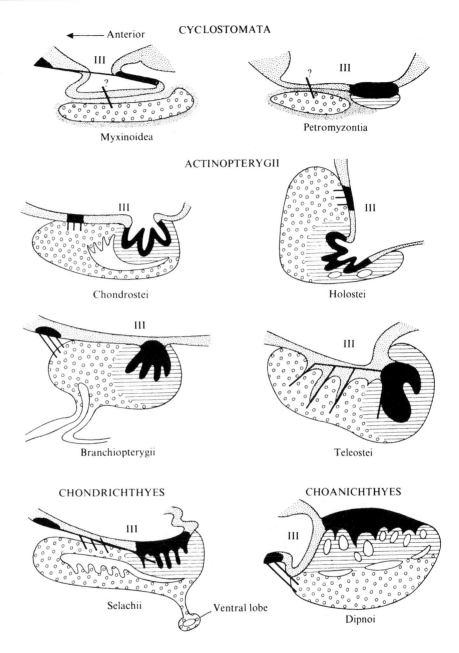

phyletic group. Considerable differences from a "median gland" may nevertheless exist among various species within each systematic group.

The structures of the pituitaries of fishes, from the Cyclostomata to the Dipnoi, are shown in Fig. 2.3. The cyclostomes have a simple type of pituitary in which the different regions are only loosely associated with each other. The parts of the adenohypophysis in these phyletic prototypes are often termed the pro-, meso-, and metaadenohypophysis. They are thought to correspond, respectively, to the cephalic part of the pars distalis (or possibly the pars tuberalis), the caudal pars distalis, and the pars intermedia of other vertebrates. The adenohypophysis in cyclostomes is a relatively thin layer of tissue. Unlike all other vertebrates, except teleost fish, no vascular connections between the hypothalamus and adenohypophysis have been identified. It appears that a diffusion of hypothalamic neurohormones may be adequate for the needs of any local control processes that may exist (Nozaki, Gorbman, and Sower, 1994). Indeed, the close proximity of the adenohypophysis to the brain may not be functionally essential in these lowly fishes. Considerable intraspecific variation occurs, and ectopic transplants of the adenohypophysis to other parts of the body do not appear to compromise its function, at least in hagfishes (Myxinoidea) (Fernholm, 1972).

The actinopterygian fishes possess a pituitary in which there is a close association between the various component tissues. The homologies of these tissues to those in tetrapods have on occasion been difficult to recognize but they undoubtedly exist. The neurohypophysis is not a very discrete tissue in fishes (there is no distinct neural lobe) and shows considerable admixture with the pars intermedia into which it sends finger-like projections and with which it shares a common blood supply (Fig. 2.3). Portal blood vessels connecting the median eminence and adenohypophysis have been described in all groups of actinopterygians except the teleosts. Considerable variation has been observed among the latter in which the blood supply to the adenohypophysis passes initially through the neurohypophysis. No clear portal system, as seen in other actinopterygians, is apparent in teleosts. In what appears to be a secondary adaptation, hypothalamic neurons enter the pars distalis and either directly innervate the hormone-producing cells or release the neurohormones in their close proximity (Peter *et al.*, 1990).

At least five distinct types of cell have been identified in the fish pituitary, as shown in Fig. 2.4, which is that of a teleost, the eel *Anguilla anguilla*. These cells are present in distinct zones. A greater intermingling of the different types of cell tends to occur as one ascends the phyletic scale but even in mammals, where distribution is relatively homogeneous, some zonation occurs.

The chondrichthyean fish, which include the Elasmobranchii (sharks and rays) and the Holocephali (chimaeras), have a pituitary that on superficial examination looks rather different from that of other fishes. It displays a similar basic structure, including a hypophysial portal system, but some gross differences, such as a rather large pars intermedia, are often apparent. In the Elasmobranchii there is a characteristic and distinct lobe of the adeno-hypophysis that lies below the pars distalis to which it remains connected by a strand of tissue (but no blood vessels). It is called the ventral lobe. In the cloudy dogfish, *Scyliorhinus torazame*, gonadotrope and thyrotrope cells have

Fig. 2.4. The midsagittal section of the pituitary of a teleost fish (the eel *Anguilla anguilla*). The cells (tropes) that produce the hormones in the pars distalis can be seen to lie generally in distinct zones. (From Olivereau, 1967.)

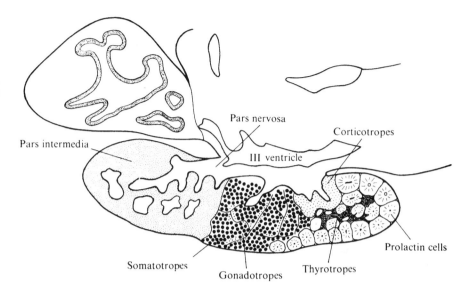

Pars nervosa

Corticotropes

Pars intermedia

III ventricle

Prolactin cells

Somatotropes

Gonadotropes

Thyrotropes

been identified immunocytochemically in this tissue (Honma and Chiba, 1996). The Holocephali also possess an aberrant lobe of the pars distalis. In the rabbitfish, *Hydrolagus colliei*, this lobe is encased in tissue in the buccal cavity and it has no direct connection with the pituitary. It is called the buccal lobe or rachendachhypophysis. Both the ventral lobe and the buccal lobe have been shown to contain gonadotropin-like activity and so may be involved in regulating reproduction (Dodd *et al.*, 1982). The coelacanth *Latimeria chalumnae* is a relict bony fish that bears a number of similarities to the chondrichthyeans. The coelacanth (Lagios, 1975) has a separate ventral lobe-like segment connected to the pars distalis by a tubular cavity. It has a separate blood supply and has been called a pars buccalis. Such aberrant self-sufficient lobes of the pars distalis occur sporadically among the fishes, where they apparently have an endocrine role that is not directly under the control of the hypothalamus. Possibly they receive hypothalamic secretions in the general systemic circulation or they may have an autonomous relationship with their target organs.

The pituitary of lungfishes shows more similarities to that of tetrapods than to those of other fishes. This is especially interesting in view of the special phyletic relationship that is usually considered to exist between lungfishes and tetrapods. The gross similarities between the amphibian and lungfish pituitaries can be seen in Fig. 2.5. The different types of cell in the adenohypophysis are more intermingled in lungfishes (not as separated as in other fishes), just as in the tetrapods. The neurohypophysis of the lungfishes also displays the beginnings of the differentiation of a neural lobe.

The neural lobe, which is a characteristic of the tetrapods, is formed by the enlargement, in a posterior direction, of the neurohypophysis. Wingstrand

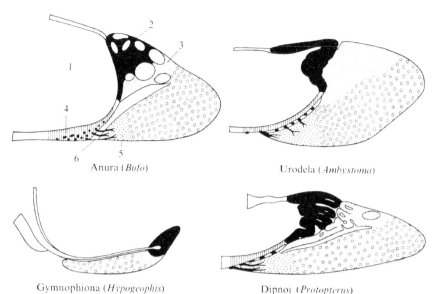

Anura (*Bufo*)

Urodela (*Ambystoma*)

Gymnophiona (*Hypogeophis*)

Dipnoi (*Protopterus*)

Fig. 2.5. The pituitaries of the three main groups of the Amphibia compared with that of a dipnoan (lungfish). 1, saccus infundibuli; 2, neural lobe; 3, pars intermedia; 4, median eminence; 5, zona tuberalis; 6, portal blood vessels. (From Wingstrand, 1966. Originally published by the University of California Press; reprinted by permission of the Regents of the University of California.)

(1966) has suggested that this change may be related to a terrestrial manner of life in which the secreted hormones had a special significance. This theory is consistent with measurements showing a much greater amount of stored hormonal material in the neurohypophysis of tetrapods than in that of fishes (Follett, 1963).

The basic morphological pattern of the tetrapod pituitary is well exemplified in the reptiles (Fig. 2.6). It can be seen, however, that even here differences between the major systematic groups occur. Such variations usually reflect the relative degree of development of the neurohypophysis and the presence, reduction, or absence of the pars tuberalis. The reptilian adenohypophysis has distinct cephalic and caudal zones.

The pituitaries from more than 100 species of birds have been examined and, as in the reptiles, the pars distalis has two distinct regions. It is interesting that a pars intermedia has not been identified among the birds. The hormone typically secreted by this tissue, MSH, has nevertheless been identified in the pituitary of the domestic chicken (Shapiro *et al.*, 1972). Whether this MSH is a normally secreted hormone in birds is unknown, however. This absence of a pars intermedia is not unique to birds; it is not present in elephants or whales, either.

Among the mammals, considerable morphological differences exist in the intimate arrangement of the tissues within the pituitary (Hanstrom, 1966). The detailed embryonic development of the pars distalis differs from that of other amniotes. In addition to being absent in whales and elephants, the pars intermedia is very much reduced in adult primates, including humans. The simplest type of pituitary is seen in the echidna, *Tachyglossus aculeatus* (an

Fig. 2.6. The pituitaries of the five main groups of Reptilia. 1, median eminence; 2, infundibular stem; 3, neural lobe (pars nervosa); 4, pars intermedia; 5, pars tuberalis; 6, portal blood vessels; 7, pars tuberalis interna; 8, cephalic lobe of pars distalis; 9, caudal lobe of pars distalis. (From Wingstrand, 1966. Originally published by the University of California Press; reprinted by permission of the Regents of the University of California.)

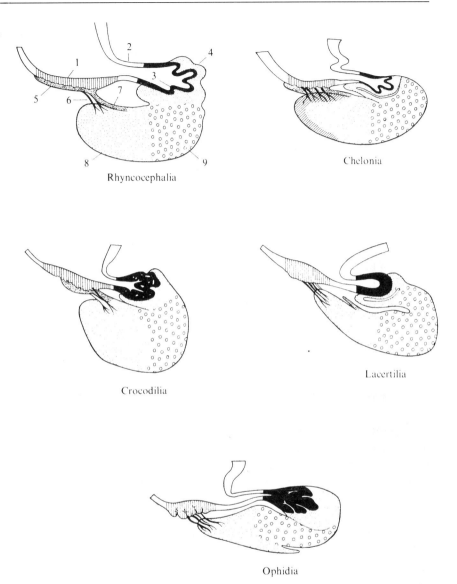

Rhyncocephalia

Chelonia

Crocodilia

Lacertilia

Ophidia

egg-laying monotreme), and in some rodents and insectivores. It is interesting that the echidna shows a pattern that is considered to be like that of a "primitive" mammal. The echidna's pituitary, however, is typically mammalian in its embryonic origins. It nevertheless has some features, including a prominent portotuberal tract between the median eminence and pars distalis, that are seen more often in birds and some reptiles.

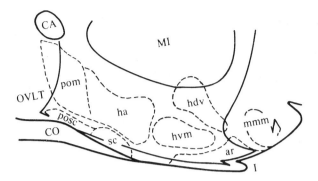

Fig. 2.7. Parasagittal section through the hypothalamic region of the rat, showing the distribution of nuclei of nerve cells. OVLT, organus vasculosum lamina terminalis; 1, median eminence; ar, arcuate nucleus; ha, nucleus anterior hypothalamus; hdv, nucleus dorsomedialis hypothalamus; hvm, nucleus ventromedialis hypothalamus; pom, nucleus preopticus medialis; posc, nucleus preopticus, pars suprachiasmaticus; mmm, nucleus mammilaris medialis, pars medialis; sc, nucleus suprachiasmaticus; CA, commissura anterior, CO, chiasma opticum; MI, massa intermedia.

The hypothalamus

The activities of the pituitary gland are mainly controlled from the hypothalamus, both as a result of nerve impulses and the secretion of hormones. The hypothalamus is that region of the brain that lies on its ventral surface just below the third ventricle. It is situated just behind the optic chiasma and in humans it is about 2.5 cm long and weighs approximately 4 g. It consists mainly of nerve cells, some of which originate in other parts of the brain and the autonomic nervous system, but it also contains its own complete neural network. These nerve cells originate in "nuclei" of tissue: aggregations of cells which have been carefully mapped (Fig. 2.7). Each nucleus appears to have special functions. The axons of the nerves, whether they originate inside or even outside the hypothalamus, tend to converge in the ventral peripheral region of the tissue and make up the median eminence.

The hypothalamus contains two main types of nerve cell: those that transmit nerve impulses in the usual way and others that have a neurosecretory function. Various neurotransmitters have been identified in the hypothalamus, including acetylcholine, dopamine, norepinephrine, 5-hydroxytryptamine (5-HT), histamine, and GABA. The presence of these substances is thought mainly to reflect the normal neural activities of the tissue. The hypothalamus has also been shown to contain a variety of peptides formed by the process of neurosecretion. Some of the neurons that make them originate in the supraoptic and paraventricular nuclei outside the hypothalamus. These are called the *magnocellular neurons* and, in mammals, form ADH and oxytocin. Within the hypothalamus are smaller, *parvicellular neurons* that form such peptides as somatostatin, thyrotropin-releasing hormone, and gonadotropin-releasing hormone. These products of neurosecretion can be released from their nerve terminals in the median eminence and pass into the blood vessels of the hypophysial–portal system, in which they flow to the pars distalis. However, as described above, such a portal system is absent in cyclostome and teleost fish.

The release of such peptides is regulated within the hypothalamus as a result of the activities of its nerve network and also in response to the presence of hormones and metabolites in its blood supply. In the latter instance, these may arrive via the internal carotid artery or (see Bergland and Page, 1979) they may pass in a retrograde direction from the pituitary gland up certain of the hypophysial–portal vessels. The neurons in the hypothalamus may synapse with each other or with those nerves that arise from areas outside of the hypothalamus, and either of these nerves may also synapse with the neurosecretory cells. The hypothalamus is clearly a primary site where the activities of the nervous and endocrine systems intermix.

Endocrinology and the evolutionary origins of the pituitary gland

The pituitary gland consists of an association of two types of tissue, epithelial cells derived from the anterior part of the gut, and neural tissue from the floor of the forebrain. As described earlier, the process of development can be seen in the embryo to involve an upgrowth of the pharyngeal (stomodeal) epithelium from a small indentation in the roof of the mouth called Rathke's pouch. This tissue forms the adenohypophysis, which usually consists of the pars distalis, pars tuberalis, and pars intermedia. A downgrowth from the diencephalon forms the neurohypophysis, which consists of the infundibular (hypophysial) stalk, the pars nervosa and the median eminence. The pars nervosa abuts onto the pars intermedia and the median eminence onto the pars distalis. Embryological variations on this basic pattern of development can result in morphological differences in pituitary architecture. The pars nervosa in tetrapods enlarges posteriorly to form the infundibular process or neural lobe. The delineation of the median eminence is also enhanced in tetrapods and it is not as readily apparent in the fish, especially cyclostomes. The anterior (oral) lobe of Rathke's pouch forms the rostral (cephalic zone) pars distalis and the posterior (aboral) lobe forms the caudal pars distalis and pars intermedia. Development of the last appears to depend on its anatomical apposition to the pars nervosa. In birds and some mammals such a favorable juxtaposition does not appear to occur and the pars intermedia is absent. In mammals, a zonation of the pars distalis into a rostral and caudal lobe is also not clear and this may reflect a failure on the development of the anterior lobe of Rathke's pouch (Wingstrand, 1966). The pars tuberalis develops from the lateral lobes of Rathke's pouch and envelopes the median eminence. Such development fails to occur in fish and some lizards, which lack a pars tuberalis. However, in elasmobranchs, the ventral lobe of the adenohypophysis (see above) appears to result from the development of the lateral lobes of Rathke's pouch.

A remnant of Rathke's pouch appears to persist in some bony fish, where there is a canal connecting the rostral pars distalis and the buccal cavity. This channel is seen in two genera of the subclass Branchiopterygii: the bichir,

Polypterus (from the Nile), and the reedfish, *Calamoichthys* (from the Niger delta). This structure is also seen in some teleosts that are considered to be "primitive": the milkfish, *Chanos chanos*, and the tarpon, *Megalops atlanticus* (see Olsson, 1990). It is called the buccohypophysial or orohypophysial canal. The hypophysial sac, which lies beneath the brain of cyclostomes, appears to be another such vestige (de Beer, 1928). In the teleosts, immunocytochemical interactions suggest that epithelial cells lining the canal in the region of the pars distalis are prolactin-secreting cells. They are open-type cells that send projections into the lumen of the canal and may thus also be chemosensory cells that can respond to changes in the composition of the communicating external medium. Prolactin cells in the pituitary of teleost fish exhibit a sensitivity to the osmotic concentration of their bathing extracellular fluids. There has been speculation that this ability may reflect one present in ancestral fish where a more direct exposure of such cells to external solutions may have occurred (Grau, Richman, and Borski, 1994).

The evolutionary origins of the pituitary gland have been the subject for speculation for over 100 years but an answer has been elusive. The availability of immunocytochemical methods for detecting pituitary hormones has recently resulted in a resurgence of efforts to solve this riddle (Thorndyke and Georges, 1988; Olsson, 1990; Gorbman, 1995). Earlier and current attention has focussed on the morphology, embryology, and histology of epidermal and neuroectodermal structures emanating from the brain and buccal cavities of protochordates, such as ascidians (Urochordata) and amphioxus (Cephalochordata), and extant vertebrates such as cyclostomes. The ammocoete larvae of the last have been of particular interest. Structures with features that are reminiscent of the vertebrate pituitary gland and Rathke's pouch have been observed in protochordates, but whether they are actually ancestral homologs has usually been in doubt. The neural complex of an ascidian, *Ciona intestinalis*, consists of a gland-like structure (the neural gland) that is connected by a canal to the buccal cavity. In this species, it lies ventrally to the cerebral ganglion so that it bears a superficial resemblance to the developing pituitary gland. This neural complex may have an endocrine role and influence ovarian function in these organisms. Immunocytochemical corticotropin-like and gonadotropin-releasing hormone-like (GnRH-like) activities have been identified in this gland-like tissue (Thorndyke and Georges, 1988). The possible morphological analogy of the ascidian neural complex to the pituitary gland was first described by C. Justin in 1881 (cited by Thorndyke and Georges, 1988). However, its embryonic origins appear to be solely from neural tissues and, therefore, it cannot be homologous to the composite neuroectodermal and buccal epithelial tissues of the pituitary. A small neotenic planktonic larvacean, *Oikopleura dioica* (Urochordata), possesses a ciliated tube that runs from the pharyngeal cavity to the brain and then empties into the body cavity (Olsson, 1990). Histological observations suggest that its cells, which originate in the brain, secrete a product into the body cavity, but attempts to identify this secretion have not, it seems, been successful. Amphioxus (*Branchiostoma*; Cephalochordata) possesses a pit-like structure in the dorsal region of its pharynx that is called after its discoverer: Hatschek's pit or groove (Sahlin, 1988). The similarity of this structure to Rathke's pouch may be more than a superficial one. In 1917, E. S. Goodrich suggested that it may be homologous to the vertebrate hypophysis. Hatschek's pit is an evagination of the pharynx that extends dorsally, but it does not reach the brain. It is considered likely that it has a chemosensory function. The columnar epithelium that lines it has secretory granules, suggesting that it also has an endocrine function (Olsson, 1990). (This channel is reminiscent of the buccohypophysial canal of the branchiopterygean fish.) Several attempts have been made, using immunocytochemical techniques, to identify putative hormonal products in the cells lining Hatschek's pit. A gonadotropin-like substance has been identified in both Chinese (*B. belcheri*) and European (*B. lanceolatum*) species of amphioxus (Chang *et al.*, 1984; Gorbman, 1995). A GnRH-like material was also identified in the Chinese species. These observations further suggest that there may be a homologous relationship between the adenohypophysis in vertebrates and Hatschek's pit in protochordates. However, a Swedish immunocytochemical study on *B. lanceolatum*, using an antiserum to a gonadotropin, as well as to growth hormone, prolactin, thyrotropin and corticotropin, failed to find evidence of the presence of any of these adenohypophysial hormones in the epithelial cells of Hatschek's pit (Sahlin, 1988). A cholecystokinin–gastrin-like activity was, however, identified, which would be consistent with the affinities of Hatschek's pit with the anterior part of the gut, where such hormones exist in vertebrates. Immunocytochemical procedures can be fickle, especially when using heterologous hormones to produce antisera. The definitive outcome of these fascinating observations will be keenly awaited. The future application of molecular genetic procedures may be needed to resolve the possible relationship of the putative secretions of Hatschek's pit and the pituitary gland.

Endocrine glands may respond directly or indirectly to internal physiological changes, but external environmental stimuli usually follow a neural route. The pituitary has a special role in transducing such information. Neuroendocrine cells, such as are found in the brain, may have an ability to respond to such stimuli and also to secrete hormones. Such cells are usually "internalized" in extant vertebrates, but in ancestral forms they could have had a more direct communication with the external environment. As we have seen, vestiges of such chemosensory hormone-secreting cells possibly exist in the orohypophysial ducts of branchiopterygean fish and some so-called primitive teleosts, where they appear to be remnants of Rathke's pouch. Such cells may also be present among protochordates, where they may be associated with

Hatschek's pit. It has, therefore, been of special interest to observe that some hypothalamic and pituitary hormones have been identified in the olfactory system of embryonic and larval vertebrates. GnRH has been identified in nerve cells in the nose of fetal mice (Schwanzel-Fukuda and Pfaff, 1989). During development, these cells migrate back to the brain and the hypothalamus. Such GnRH-containing neurons have also been identified in the nasal epithelium of embryonic chickens, from where they also migrate back to the forebrain as development proceeds (Norgren and Lehman, 1991; Akutsu et al., 1992). In the sea lamprey *Petromyzon marinus*, a putative hormone that is present in the pars distalis and plasma of the adult lampreys is also found in the olfactory organ and nasohypophysial canal of their ammocoete larvae (Sower et al., 1995). This substance is a homodimeric glycoprotein, which has been called nasohypophysial protein (NHP). It is encoded by a gene that also produces the precursor for corticotropin and several other pituitary hormones (POC, see Chapter 3).

Gorbman (1995), after reviewing such information, has suggested that the pituitary gland may have evolved 'from a chemoreceptive olfactory structure'. The primogenitor could have been Hatschek's pit, which still exists in some contemporary protochordates.

The endocrine glands of the pharynx: thyroid, parathyroids, and ultimobranchial bodies

Apart from the adenohypophysis, which has its origins in the roof of the mouth, three (or four, if one includes the thymus) other endocrine glands arise from the pharyngeal tissues: the thyroid gland from the floor of the pharynx, the parathyroids from the II, III, and IV gill pouches, and the ultimobranchial bodies from the last (VI) pair of these.

The thyroid

Thyroid tissue is present in all vertebrates, though its gross morphological arrangement varies somewhat. Its hormones have been identified in all vertebrates and have a ubiquitous presence in such common events and processes in the life cycle as growth, differentiation, metamorphosis, reproduction, hibernation, and thermogenesis. Their precise effects and roles, however, are not always clear, especially in the cold-blooded vertebrates. A propensity to react with cell nuclear receptors, and thus influence genetically controlled processes, may be basic to an understanding of their roles. The "thyroid unit' is a follicle in which a group of epithelial cells surrounds a central cavity that is filled with a glycoprotein secretion called thyroglobulin (Fig. 2.8). The encompassing cells have a columnar appearance when they are most active and a flattened one when they are least active. Thyroid follicles

Fig. 2.8. The thyroid gland of the laboratory rat showing the follicles surrounded by epithelial cells. (*a*) The inactive condition where the cells are flattened and the follicles are distended with 'colloid', which contains the thyroglobulin. (*b*) The active condition where the epithelial cells are columnar and little colloid is present.

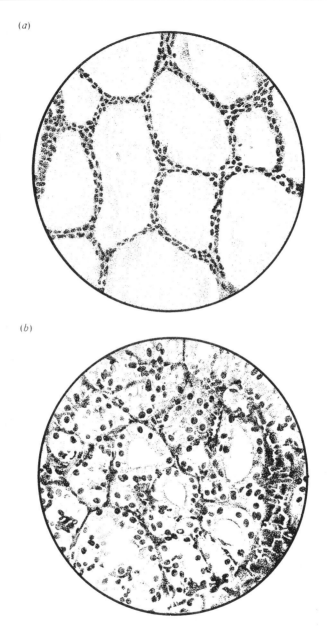

(*a*)

(*b*)

have a remarkable ability to trap inorganic iodide, which can be stored and incorporated into hormones that are, in turn, stored in the follicle cavity. It is probably the only endocrine gland that stores its products outside the cells.

In humans, the thyroid gland is situated in the region of the neck and it has a generally comparable position in other vertebrates. In cyclostomes and most teleost fishes, the thyroid follicles lie scattered along the blood vessels under the pharynx. Occasionally they may be found further afield (heterotopic), even in the kidneys. In chondrichthyean fish (sharks and rays), some teleosts, like the Bermuda parrot fish and tuna, the lungfishes, and the coelacanth (Sarcopterygii) (Chavin, 1976), the follicles are aggregated into a distinct glandular mass. This pattern persists in higher vertebrates; there are two such aggregates in amphibians, birds, and many reptiles. In lizards, these two lobes are joined, which is also usually characteristic of mammals. The thyroid gland, especially in mammals, has close morphological associations with the parathyroid glands and the C cells of the ultimobranchial bodies. The latter are present within the structure of the thyroid gland itself (see next section). Whether such a juxtaposition has functional consequences or is merely the result of a fortuitous ontogenetic event is unknown.

The thyroid appears to have the longest phylogenetic history of any endocrine gland (Table 2.1). It is present not only in vertebrates; tissues that may be homologous, though not having the characteristic follicular units, have also been identified in protochordates, including amphioxus (the lancelet, *Branchiostoma lanceolatum*; Cephalochordata) and various ascidians (sea squirts, Urochordata). The development of the thyroid in lampreys can be followed during the metamorphosis of its ammocoete larva. This beast collects small particles of food by filtering water that passes, with the help of ciliary action, through its pharynx. This process is aided by a ventral outgrowth from the floor of the mouth called the endostyle or subpharyngeal gland. An analogous tissue also exists in amphioxus and ascidians. It secretes a sticky mucus that traps the food particles before they can pass out across the gills. This action has been likened to that of "moving flypaper." Embryologically, the endostyle of the lamprey ammocoete larva differentiates to form the adult thyroid. This has given rise to speculation as to whether the endostyle in the ammocoete and in protochordates has some thyroid function.

The endostyle does not contain thyroid-like follicles. It has, however, been shown (Barrington, 1962), like the thyroid, to be able to accumulate selectively and concentrate radioactive iodide. This has been demonstrated not only in the lamprey ammocoete but also in amphioxus and several ascidians. The iodine formed is bound in organic form with tyrosine and organoiodine compounds (*iodothyronines*), including, possibly, small amounts of thyroxine (for a summary see Table 2.1). The endostyle of amphioxus can also form a thyroglobulin-like molecule in which thyroid hormones can be synthesized (Monaco *et al.*, 1981). Immunological evidence also suggests that ascidians (Urochordata) can form thyroglobulin and thyroid-like hormones (Thorndyke, 1978; Dunn, 1980). It would appear that the Protochordata have an ability to synthesize thyroid hormones, though thyroid follicular structures have not been observed in nonvertebrates.

Iodine readily reacts with proteins containing the amino acid tyrosine. Indeed, in one extensive investigation in which cows were being fed experimental diets containing thyroid compounds to improve their milk yields,

Table 2.1. *The thyroid in the phylum Chordata*

Subphylum	Class	Species[a]	Thyroid gland	Thyroid-like[b] activity
Hemichordata[c]		*Glossobalanus minutus*	—	—
Protochordata				
Urochordata (tunicate)	Ascideacea (sea squirt)	*Ciona intestinalis*	—	+
		Clavelina lepadiformis	—	+
	Larvacea		—	+
	Thaliacea	*Salpa maxima*	—	+
Cephalochordata	Amphioxi	*Branchiostoma lanceolatum* (amphioxus)	—	+
Vertebrata	Agnatha (cyclostomata)	(Lamprey)		
		ammocoete larva	—	+
		adult	—	+
		(Hagfish)	+	+
	Chondrichthyes (elasmobranch)	(Shark)	+	+
		(Skate)	+	+
	Osteichthyes (teleost)		+	+
	Amphibia		+	+
	Reptilia		+	+
	Aves		+	+
	Mammalia		+	+

[a] Not a complete list. Common names are given in parentheses.
[b] That is, synthesis of iodothyronines.
[c] Not usually classified as Chordata at present time but included for reference in the light of earlier discussions.
Source: Rall, Robbins, and Lewallen, 1964.

these were made by incubating proteins, such as casein, with iodine at an appropriate pH and temperature. Therefore, the spontaneous formation of organoiodine compounds in nature would not be surprising. Indeed, among the ascidians, the outer tunic or coat contains scleroproteins that combine with iodine. Iodinated tryosines also have been isolated in many other nonvertebrates, including coelenterates. Barrington (1962) has conjectured about the possibility that the spontaneous occurrence and availability of such compounds in nature may have led to their use as hormones. Subsequently, their formation may have become more localized in tissues such as the endostyle.

Parathyroid glands and ultimobranchial bodies

The parathyroid glands make their phyletic debut in the amphibians and persist in the reptiles, birds, and mammals. One to three pairs of these glands are present in the upper thoracic and neck regions. In mammals, they are often near the thyroid (hence their name). Two pairs are usually present in amphibians, but larval and neotenic forms lack these glands. Among the Reptilia, the Crododilia have one pair, the Chelonia and Ophidia two pairs and the Lacertilia one to three pairs. Birds and mammals have one or two pairs. Initially during development (Roth and Schiller, 1976), there are three pairs of parathyroid glands originating from the II, III, and IV gill pouches. (Those derived from the II pouch are referred to as parathyroids II, from the III, parathyroids III, and so on.) The glands derived from the second gill pouch, however, usually disappear and if only one pair persists it is generally parathyroids III. The absence of parathyroid glands from aquatic larval and neotenic amphibians may reflect a prior morphological commitment of the gill pouches. Alternative sites for such genetic expression of a parathyroid hormone in such species and the fishes do not appear to have become available. The parathyroid glands consist of a parenchyma containing two main types of cell. The chief cells contain granules and secrete parathyroid hormone, but the role of the oxyphil is unknown. The histochemical appearance of the chief cells has resulted in their classification as ''light cells'' with few granules and low activity, and ''dark cells'' with many granules signifying that their activity is high.

As their name indicates, the ultimobranchial bodies (Copp, 1976; Pearse, 1976) are embryologically derived from the last (VI) pair of branchial pouches. The glandular cells are granulated and form follicles. Ultimobranchial gland tissue is present in all vertebrates with the apparent exception of the cyclostomes. The glands are situated in the region between the esophagus and the heart, sometimes in the proximity of the thymus or thyroid gland, and in mammals the glandular tissue is usually incorporated into the thyroid gland itself. A pair of ultimobranchial bodies is retained in many species but in elasmobranchs and some urodeles and lacertilians the left gland is much larger, while in bony fishes they fuse to form a single gland.

The association of the thyroid, parathyroids, and ultimobranchial tissues may be somewhat complex. The last two tissues are histologically different from the thyroid and they lack the typical follicular structure of that gland. Their positions in the domestic fowl are shown in Fig. 2.9. The morphological distribution of these three glands has often made it difficult to dissociate the effects of the latter two, which both elaborate secretions having opposite effects on blood calcium concentrations. The parathyroids, especially in mammals, are usually closely associated with the thyroid, though not to the same extent as the ultimobranchial tissues. Removal of the mammalian

Fig. 2.9. The position of the thyroid gland, the parathyroids, and the ultimobranchial bodies in the domestic fowl. *Gallus domesticus.* (From Copp, Cockcroft, and Keuk, 1967b. Reproduced by permission of the National Research Council of Canada.

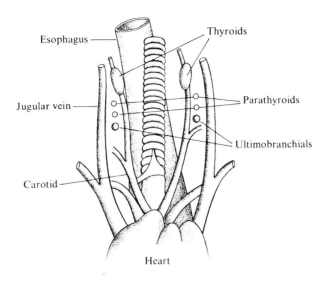

thyroid, including that in humans, is often associated with low blood calcium levels and an associated muscle tetany. This is the result of removal of or damage to the parathyroid gland, an observation that furnished an important clue as to its possible endocrine significance. The concomitant absence of the C-cells was not initially apparent and only became so after examination of the effects of thyroid extracts on plasma calcium levels.

In order to elucidate the respective roles of the C-cells and the parathyroids in mammals, morphological variations between species have been usefully exploited. In dogs there are two pairs of parathyroids, one embedded deeply in the thyroid. Rats, however, only have a single pair of parathyroids, which are at the surface of the thyroid and so can be destroyed with a cautery. In neither species is it possible to isolate the blood supply of the parathyroids from that of the thyroid, which contains the C-cells, so that many crucial endocrine experiments cannot be performed on them. Sheep and goats have two pairs of parathyroids and one of these is situated near the thymus, where it has a separate blood supply from that of the thyroid. These animals have played an important role in elucidating the respective roles of the endocrines in calcium metabolism (Hirsch and Munson, 1969). In addition, the pig has also proven to be useful and in this species the thyroid has no attached parathyroid tissue so that one can deal with the C-cells in relative isolation from the former.

In nonmammals the ultimobranchial bodies are usually separated from the thyroid. Nevertheless, in birds, for example the domestic fowl, they may contain parathyroid tissue (Copp, 1972). In pigeons, calcitonin is found not only in the ultimobranchials but also in the parathyroids and thyroid.

The admixture of these three distinct endocrines may have some fundamental significance but this is unknown. It has been suggested that the differentiation of the C-cells is aided by their association with the thyroid tissue. It is also possible that, as has been observed in the adrenals, some functional symbiosis may occur. Much of the variation, however, would appear to be the result of embryological complications. Although this has certainly helped to hide their effective roles from endocrinologists, it has provided some fascinating intellectual exercises during which much of the interspecific variation that initially served to deceive has been productively utilized.

The adrenals

The adrenal glands are so named from their position adjacent to the kidneys. In mammals, they are a composite gland made up of two distinct tissues arranged in two zones, an outer cortex surrounding an inner medulla. The cortex is mesodermal in its origins and secretes several steroid hormones involved in the regulation of intermediary and mineral metabolism. It is called the interrenal or adrenocortical tissue. The medulla is neural tissue homologous to that of the sympathetic ganglia. Because it stains dark brown with chromic acid it is called chromaffin tissue. It secretes the catecholamine hormones epinephrine (from A-cells) and norepinephrine (from NA-cells). These hormones are also called, respectively, adrenaline and noradrenaline. They have several roles, including the mobilization of fats and carbohydrates, as well as exerting an influence on the tone of the muscle surrounding many blood vessels and the heart. Although the chromaffin and adrenocortical tissues are closely associated in mammals and other tetrapods, they are often quite separate in fishes. In cyclostome fishes, the chromaffin tissues are widely dispersed in small islets along certain blood vessels. Putative adrenocortical tissue ("yellow bodies") has been identified embedded in the posterior cardinal veins of the pro- and mesonephroi (Idler and Burton, 1976). This tissue, however, has not been conclusively shown to secrete steroid hormones, but it does appear to be a target organ for corticotropin, which usually stimulates the activity of such tissue. In bony fishes, including teleosts, holosteans, and the coelacanth, adrenocortical tissues lie along the posterior cardinal veins in the anterior part of the kidney (the head kidney) (Lagios and Stasko-Concannon, 1979). In the lungfishes, it is dispersed more widely. In sharks and rays (Chondrichthyes), the adrenocortical tissue forms a more compact glandular mass lying between the kidneys; hence the name interrenals. In the dogfish (Fig. 2.10a) islets of chromaffin tissue lie along the inner borders of the kidneys, whereas the interrenal forms a fairly complete mass between them. The adrenocortical tissue of the skate, by comparison, forms several lobules (Fig. 2.10b).

Fig. 2.10. Adrenal tissues in the Chondrichthyes. (a) The smooth dogfish (*Mustelus canis*). Double row of black dots indicates chromaffin tissue lying between the two kidneys. Interrenal (adrenocortical) tissue is stippled. This lies in several pieces between the kidneys. (b) The skate (*Raja laevis*). The broken U-shaped interrenal is shown lying between the kidneys. (From Hartman and Brownell, 1949.)

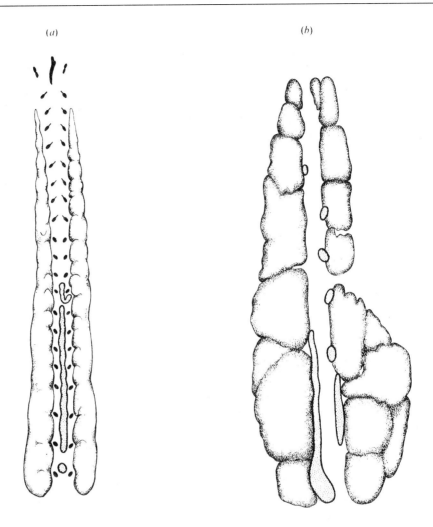

(a)   (b)

In the Amphibia, chromaffin and adrenocortical tissues are usually associated with each other, lying in islets on the ventral surface of the kidney (Fig. 2.11). Considerable differences can be seen among various species. In urodeles, they are in scattered groups; in *Siren* (Fig. 2.11a) they lie in rows between the kidneys, and in *Necturus* and *Amphiuma* (Fig. 2.11b,c) they are on its surface. Anurans, like the leopard frog (*Rana pipiens*), have contiguous strips of adrenal tissue (Fig. 2.11d). It is interesting that in the African lungfish (*Protopterus*), adrenocortical tissues lie in islets along the postcardinal veins and on the ventral surface of the kidney, a pattern similar to that seen in urodeles (Janssens *et al.*, 1965).

In anamniotes, the adrenocortical tissues and the mesonephric kidney have a common embryological origin so that their close association is not unexpected. Amniotes, however, have a metanephric kidney (the mesonephros is

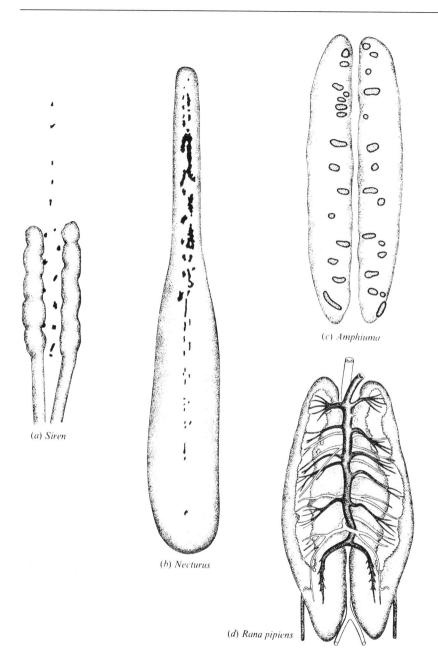

(a) Siren

(b) Necturus

(c) Amphiuma

(d) Rana pipiens

Fig. 2.11. Adrenal tissues in the Amphibia. Urodela: (a) *Siren*; (b) *Necturus*; (c) *Amphiuma*. The adrenal tissue is shown as the dark area lying on the ventral surface or between the kidneys. Anura: (d) *Rana pipiens*. The adrenal tissues lie in two strips (light color) along the outer ventral border of each kidney. (From Hartman and Brownell, 1949.)

not seen in adults) so that the kidneys and adrenals are less intimately connected. The adrenals, more predictably, form separate compact masses of tissue lying near the kidneys. Considerable variations nevertheless still exist, as seen among the different major groups of the reptiles (Fig. 2.12). The chromaffin tissues of reptiles are more closely intermingled with the adrenocortical tissues than they are in amphibians. This admixture of the two

Fig. 2.12. The adrenal tissues in the Reptilia. The adrenals are shown in black in relationship to the kidney(s) (shaded). (From Hartman and Brownell, 1949.)

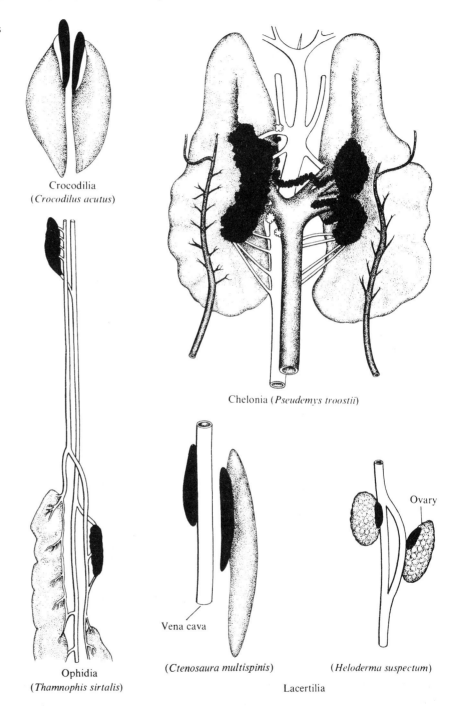

Crocodilia
(*Crocodilus acutus*)

Chelonia (*Pseudemys troostii*)

Ophidia
(*Thamnophis sirtalis*)

Vena cava

(*Ctenosaura multispinis*)

Lacertilia

Ovary

(*Heloderma suspectum*)

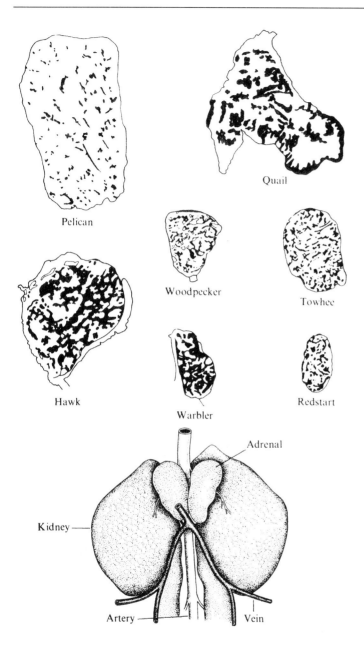

Fig. 2.13. The adrenals in birds. *Top.* Cross-section of the adrenal glands from various species showing the distribution of the chromaffin tissue (black) and the adrenocortical tissue (white). *Bottom.* The adrenals of the herring gull (*Larus argentatus*). (From Hartman and Brownell, 1949.)

tissues is even more apparent in birds (Fig. 2.13) where the adrenals may be fused to form a single gland. Mammals have paired adrenal glands (Fig. 2.14) with a distinct cortex and medulla. It is interesting that this is not as well defined in the echidna *Tachyglossus aculeatus* (an egg-laying monotreme), whose adrenals are considered to be similar to those of reptiles (Wright, Chester Jones, and Phillips, 1957).

The relative amounts of adrenocortical and chromaffin tissues vary in

Fig. 2.14. The adrenals in mammals. *Top*. The position of the adrenals in relation to the kidneys in a variety of species. *Bottom*. Cross-section of the adrenals from various species showing the chromaffin tissue (black) and adrenocortical tissue (white). (From Hartman and Brownell, 1949.)

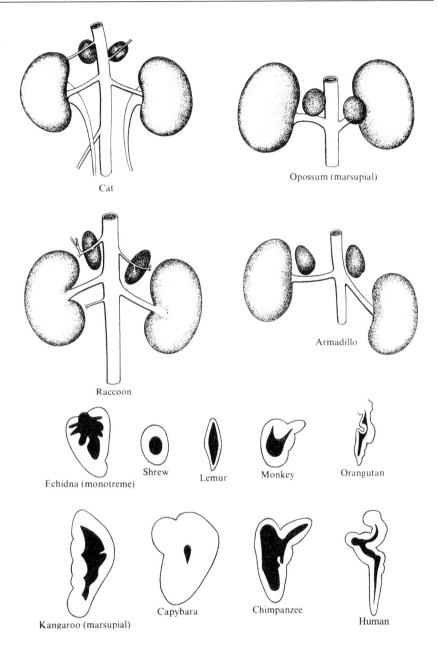

Cat

Opossum (marsupial)

Raccoon

Armadillo

Echidna (monotreme) Shrew Lemur Monkey Orangutan

Kangaroo (marsupial) Capybara Chimpanzee Human

different species. As shown in Table 2.2, there are similar amounts of both in the domestic fowl; in the dog, adrenocortical tissue is five times more predominant, and in the guinea pig there is more than 60 times as much adrenocortical as chromaffin tissue. Adrenocortical tissues may also show considerable variability depending on the season, diet, and physiological condition of the animal. It is well known that in many reptiles and amphibians the adrenocortical tissue regresses in winter and proliferates in summer

Table 2.2. *Weights of the adrenal medulla and cortex in various species*

Animal	Weight (g)	Medullae (g)	Cortexes (g)	Proportion
Fowl	2 000	0.1	0.1	1:1
Dog	15 000	0.25	1.25	1:5
Cat	3 000	0.02	0.35	1:17.5
Rat	200	0.002	0.04	1:20
Rabbit	3 000	0.01	0.4	1:40
Guinea pig	500	0.008	0.5	1:62.5

Source: Hartman and Brownell, 1949.

(Chan and Phillips, 1971). This change can also be related to breeding and has been observed in teleost fish (see, for instance, Robertson and Wexler, 1959; Chan and Phillips, 1969; Lofts, Phillips, and Tam, 1971). Birds from marine habitats, where a lot of salt is present in the diet, have larger adrenals than those species where fresh water is freely available. Glaucous-winged gulls (*Larus glaucescens*) reared with only salt solutions to drink have much larger adrenals than those given fresh water (Holmes, Butler, and Phillips, 1961).

The mammalian adrenal cortex is histologically composed of three types of cell situated in three layers or zones. These are the outer *zona glomerulosa* (round cells, rich in mitochondria and poor in lipids), an intermediate *zona fasciculata* (columnar cells, rich in lipids), and a smaller inner *zona reticularis* (flattened cells poor in lipids) (Fig. 2.15). The zona glomerulosa is not apparent in all mammals, such as some mice, lemurs, and monkeys. It appears that the three zones are each principally (though possibly not exclusively) involved in the formation of distinct hormones: the zona glomerulosa forms aldosterone and corticosterone and the zona fasciculata cortisol and corticosterone. The zona reticularis may secrete androgenic steroids; this zone hypertrophies in certain conditions associated with an excess production of these hormones in humans. Aldosterone assists regulation of sodium metabolism in mammals. It is, therefore, not unexpected to find that the zona glomerulosa can undergo considerable hypertrophy in mammals, such as rabbits and kangaroos, that live in areas where the salt content of the diet is low (Fig. 2.16) (Blair-West *et al.*, 1968).

It should be emphasized that there is a crossover in the abilities of each zone of the adrenal cortex to produce particular steroids. It seems likely (see Tait and Tait, 1979) that the basic difference in the cells is in their ability to respond to the different types of stimulus rather than an innate biochemical distinction. Thus, the zona fasciculata strongly responds to corticotropin while the zona glomerulosa increases its steroid secretion only slightly in the presence of this pituitary hormone. In contrast, the zona glomerulosa secretes lots of aldosterone in response to angiotensin II. Possibly, the cells in each

Fig. 2.15. (*a*) Histological section of the adrenal of a mammal, the racoon, showing the zonation of the adrenal cortex. (*b*) Enlargement of the capsular glomerular zone. (From Hartman and Brownell, 1949.)

(*a*)

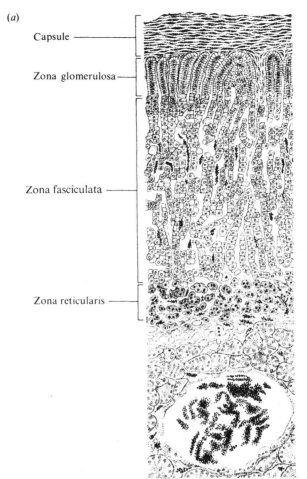

Capsule

Zona glomerulosa

Zona fasciculata

Zona reticularis

(*b*)

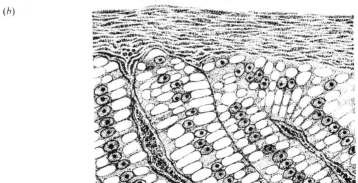

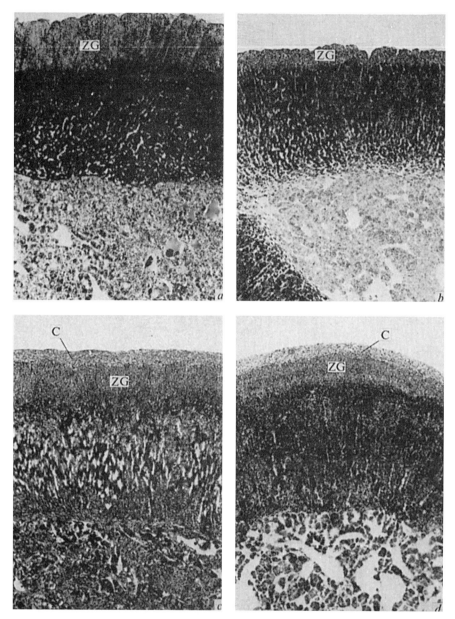

Fig. 2.16. The adrenal glands, in section, from two marsupials from sodium-deficient and sodium-replete areas. *Macropus giganteus* (kangaroo): (*a*) sodium deficient; (*b*) sodium replete. *Vombatus hirsutus* (wombat): (*c*) sodium deficient; (*d*) sodium replete. Note that the zona glomerulosa is wider in sodium-deficient animals. In the wombat, a thicker capsule lies at the outer border of this cell layer. C, capsule; ZG, zona glomerulosa. (From Blair-West *et al.*, 1968, and J. R. Blair-West, personal communication.)

zone possess different types of hormone receptor. Despite a lack of discrete tissue zonation, it seems likely that the production of aldosterone by non-mammalian tetrapods also occurs in distinct types of cell, as shown in the bullfrog *Rana catesbeiana* (Varma, 1977).

Some mammals possess an additional type of adrenocortical tissue that usually lies between the medulla and definitive cortex. The fetuses of some primates have an enlarged region of tissue in this position, which regresses after birth. In humans, this "fetal zone" is at least partly under the control of corticotropin and can secrete dehydroepiandrosterone sulfate (DHEAS). This steroid can be utilized by the placenta as a substrate to form an estrogen. The fetal zone may contribute to the process of pregnancy. Several rodents, especially mice but also some voles, shrews, and squirrels, have been shown to possess a special layer of adrenocortical tissue at this site that has been called the "X-zone" or "transient" zone (Idelman, 1979). This tissue regresses at puberty in the males, owing to the effects of androgens, and in the females during pregnancy (Chester Jones and Phillips, 1986). It persists following castration and its activity can be promoted by the administration of a gonadotropin (luteinizing hormone (LH)) but not corticotropin. Evidence for steroidogenesis by this tissue is lacking. In some marsupials, especially the brush-tail possum (*Trichosurus vulpecula*) there is, following sexual maturity, a gross unilateral enlargement of the adrenal in the females. This hypertrophy reflects the presence of a "special zone" of tissue lying between the zona fasciculata and medulla (Chester Jones and Phillips, 1986). Development of this zone can be promoted by the injection of a gonadotropin (follicle-stimulating hormone (FSH)), even in castrated males (Weiss and Carson, 1987). It can apparently secrete androgen-like steroids (Weiss *et al.*, 1989). Their function is uncertain but it has been suggested that they may have a pheromonal role (Chester Jones *et al.*, 1994).

Among the vertebrates, there appears to be an evolutionary trend toward a more intimate association of the adrenocortical and the chromaffin tissue. This tendency may partially reflect their embryogenesis, such as the tissue aggregation that follows the loss of the mesonephros. The relationship of the two endocrine tissues has functional significance. The adrenocortical tissue can certainly function in the absence of the chromaffin tissue, as seen *in vitro* in the laboratory. However, the chromaffin cell's ability to convert norepinephrine to epinephrine depends largely on the presence of corticosteroids. The NA-cells (which secrete norepinephrine) can even be converted to A-cells during neonatal life if they are exposed to corticosteroids (Coupland, 1968). The adrenal medulla contains one enzyme, phenylethanolamine-*N*-methyltransferase (PNMT), that by methylating norepinephrine converts it to epinephrine. The formation of this enzyme is induced by steroid hormones from the adrenal cortex (Pohorecky and Wurtman, 1971). The concentrations of the steroids must be high, far higher than normally present in the systemic circulation. This is achieved by the direct transfer of the steroids to the medulla through a local portal blood system. PNMT activity has also been identified in nonmammals, and the association of chromaffin and adrenocortical tissue in other vertebrates may have an important, even determining, role in their abilities to form epinephrine.

There has also been speculation that secretions from the adrenal medulla may influence adrenocortical functions (Hinson, 1990). As the blood flow through the adrenal is in a centripetal direction, a vascular pathway for such substances is difficult to envisage. However, nerve fibers with cell bodies in the medulla are present in the cortex and the products of such nerve cells, catecholamines and neuropeptides, have been found in the cortex. Sporadic observations suggest the possibility that such catecholamines and neuropeptides, including neuropeptide Y (Chapter 3), may influence cortical steroidogenesis, blood supply, and tissue hyperplasia. The peptide adrenomedullin, which has been identified in the adrenal medulla (and several other tissues, Chapter 3), has a potent vasodilator action and can also inhibit the steroidogenic effect of angiotensin II on the adrenal cortex (Mazzocchi *et al.*, 1996). A possible paracrine function for it within the adrenal gland has, however, not apparently been investigated.

The endocrine hormones of the gut

The gut is the longest and most diffuse of the endocrine organs and it has sometimes even been called the "diffuse neuroendocrine system" (DNES). This description (see, for instance, Vigna, 1986) has resulted from the identification of numerous types of hormone-containing cell dotted along the epithelial lining of the gastrointestinal tract. In 1938, F. Feyrter described such cells, which he called "clear cells" as they did not stain using the methods that were then available. However, more recent immunocytochemical procedures (Larsson, 1980) using antibodies to various known peptides have facilitated the endocrine assignment of these gut cells. At a recent count, 14 such cell types have been identified, though not all as established sources of known hormones. These cells contain peptides, which include secretin, gastrin, somatostatin, glucagon-like peptide I, cholecystokinin–pancreozymin, gastric inhibitory peptide (GIP), and vasoactive intestinal peptide (VIP). In addition, similar cells in the gut of some fishes, as well as endocrine cells in the pancreas, secrete insulin, glucagon, and pancreatic polypeptide. These cells are, therefore, also referred to as members of the gastroentero–pancreatic or GEP system. The presence, distribution, and morphological arrangement of such cells varies somewhat among different vertebrates but the general picture is similar throughout and can often be traced back to invertebrates. It is of special interest that many such peptide-secreting cells have also been identified in the nervous system, especially the brain. These peptides include cholecystokinin, gastrin, vasoactive intestinal peptide and somatostatin, but apparently not insulin, secretin, or gastric inhibitory peptide. The common occurrence of such neuropeptides in both neural and gut tissues led to the suggestion that they may share a common neuroectodermal origin. This hypothesis was called the APUD concept (the acronym stands for *a*mino content and/or *p*recursor *u*ptake and *d*ecarboxylation), referring also to the presence of amines, such as dopamine and 5-hydroxytryptamine in such cells (Pearse, 1968). The broad application of this interesting concept is, however, in doubt as the endocrine cells associated with the gut appear to have an endodermal origin. The presence of such peptides at sites in both the gut and brain may reflect the acquisition of an ability to express the same gene in both types of tissue.

The different peptide-secreting cells have been named in various ways. Attempts at uniformity have been the subject of several meetings, starting in 1969, at such venues as Wiesbaden, Bologna, Lausanne, and Santa Monica. The simplest notation has been to call the cells secreting secretin, S-cells, gastrin, G-cells, and pancreatic polypeptide, PP-cells. However, historic and literary constraints have added other codes such as A-cells (glucagon), B-cells (insulin), D-cells (somatostatin), M-, I-, or CCK-cells (cholecystokinin) and L-cells (glucagon-like peptides). Differences from this notation unfortunately still exist and no doubt will evolve in the future.

The primeval cells of this type, which are associated with the GEP system, appear to have originated in the lining of the gut. They are seen at this site in contemporary molluscs, protochordates (such as amphioxus), sea squirts (ascidians), and the ammocoete larvae of cyclostome fishes (see, for instance, Fritsch and Sprang, 1977; Falkmer, Ostberg, and van Noorden, 1978; Plisetskaya *et al.*, 1978). Aggregations of such cells that lie adjacent to the gut, often in the pancreas and liver, appear to be subsequent endocrine specializations. The cells lying in the lining of the gut, however, usually exist along its various segments (Fig. 2.17). Many of these cells may send tufts of microvilli into the lumen of the gut so that they can "taste" its acidity and the products of digestion. These open-type cells may then, if appropriate, secrete their hormones into an adjacent blood supply. Alternatively, secretion may occur into the extracellular space, where the peptide may have a paracrine role and influence the activities of neighboring cells. Other such glandular cells are of the closed type and do not directly communicate with the gut contents.

In mammals and most vertebrates, the secretion of insulin and glucagon arise, respectively, from the B-cells and A-cells that are associated with the acinar tissue of the enzyme-secreting exocrine pancreas. These endocrine cells usually form small aggregates (Falkmer and Patent, 1972; Falkmer and Ostberg, 1977) of tissue, originally called the islets of Langerhans in mammals but also referred to as the endocrine pancreas, principal islets, pancreatic islets, and, in some fish, Brockmann bodies. In most species, these islets are also associated with D-cells, which secrete somatostatin, and PP-cells, which secrete pancreatic polypeptide. The particular admixture and arrangement of such cells in the islets varies between species (Table 2.3). In cyclostome fishes, the tissue aggregates appear to contain B-cells and D-cells (or D-like cells) (Hilliard, Epple, and Potter, 1985; Youson *et al.*, 1988) and they are not associated with the pancreas. In northern hemisphere lampreys, three such

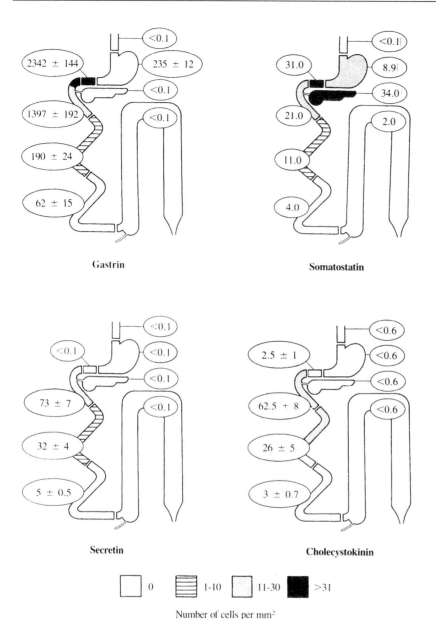

Fig. 2.17. The distribution of some gut hormones in the different regions of the gastrointestinal tract of humans. The number refers to the mean concentrations (in pmolg) in the whole thickness of gut sections. The shading indicates the density of each cell type, as observed by quantitative immunocytochemistry. (From Bloom and Polak, 1978.)

aggregates have been observed along the gut, one in the liver, while in the antipodean species *Geotria australis*, there is a single discrete organ lying near the esophagus. In hagfish, there is a single gland lying around the bile duct near its entrance to the intestine (Fig. 2.18*a*). In lampreys, the B-cells form into the follicles of Langerhans (Barrington, 1942). The apparent phyletic relationship between the exocrine pancreas and the endocrine islets does not appear, therefore, to have been an early ancestral arrangement.

Table 2.3. *Evolution of the endocrine pancreas in vertebrates*

Pisces	Agnatha	Myxinoidea (hagfish)	Endocrine pancreas is a separate organ derived from the biliary duct; exocrine pancreas partially incorporated into the liver
		Petromyzontia (lamprey)	Endocrine pancreas is a separate organ derived from the duodenum
	Chondrichthyes	Elasmobranchii (shark, torpedo)	Endocrine pancreas is formed of cells located around the medium- and small-sized pancreatic ducts; exocrine pancreas completely separated from the liver
		Holocephali (chimaera)	Endocrine cells are integrated in the exocrine tissue; they are accumulated in clusters, which are not vascularized
	Osteichthyes	Crossopterygii Coelocanthini (*Latimeria*)	Endocrine cells located around the ducts as in Elasmobranchii. First evidence of islet formation: some B-cells are organized around the capillaries between the acini; exocrine pancreas separated from the liver
		Teleostei (95% of the living fishes, the most evolved fishes)	Endocrine pancreas either forms a separate organ (principal islets or Brockmann bodies) (goosefish, toadfish, etc.) or is integrated to exocrine pancreas.
		Dipnoi	Forms a separate organ, 'principal islets', as in teleost
Amphibia	Urodela		Endocrine cells scattered in the exocrine tissue or gathered in islets; some species have no A-cells
	Anura		Endocrine cells originate from the pancreatic ductule during larval stage and accumulate in islets, which persist during the after metamorphosis
Reptilia			Endocrine cells for islets; sometimes they accumulate around duct as an external coat (in some snakes)
Aves			Endocrine pancreas mostly found as vascularized islets free of exocrine tissue
Mammalia			

Source: Pictet and Rutter, 1972.

The Chondrichthyes also have different distribution patterns of A-, B-, and D-cells. In the Elasmobranchii (sharks and rays), the cells are usually situated near or around the ducts of the exocrine pancreatic tissue. In the Holocephali (chimaeroid fishes), this tissue extends more among the exocrine cells though it retains its association with the ducts. (Fig. 2.18*b*). There is also considerable diversity in the arrangement of the A-, B-, and D-cells among bony fishes (Osteichthyes). Some teleosts (toadfish and goosefish) show an aggregation of these cells into two pea-sized glands called principal islets or Brockmann bodies (Fig. 2.18*c*). Others, such as the eel (Fig. 2.18*c*), have islets of tissue

(a) CYCLOSTOME (*Myxine*)

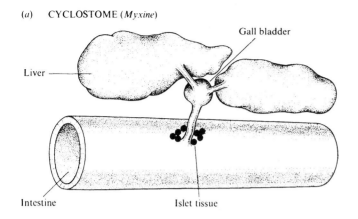

Gall bladder

Liver

Intestine

Islet tissue

Fig. 2.18. The various types of pancreas in fishes. (a) Cyclostome type (*Myxine*): ring-like arrangement around the bile duct. (b) Chondrichthyean types. Left, many elasmobranchs; right, Holocephali.

(b) CHONDRICHTHYES

Islet tissue

Exocrine pancreas

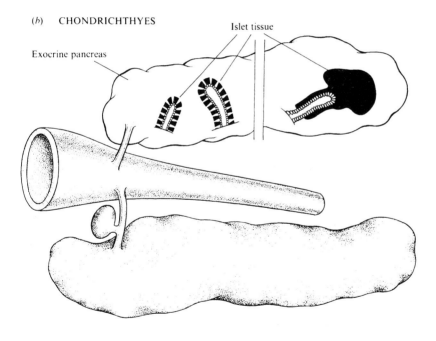

(c) Actinopterygian types. (i) *Anguilla*; this is present in a few teleosts and is similar to that in tetrapods. (ii) The more general teleost with Brockmann bodies or 'principal islets.' (From Epple, 1969.)

(*c*)   ACTINOPTERYGII (type i)

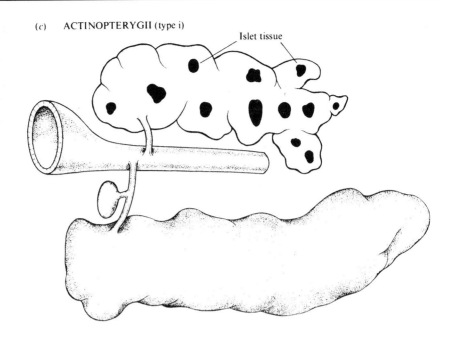

Islet tissue

ACTINOPTERYGII (type ii)

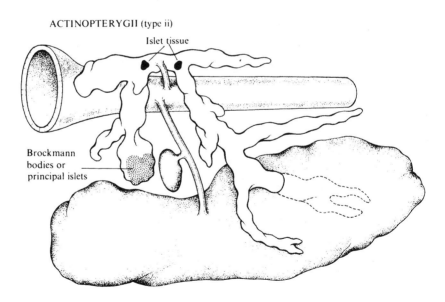

Islet tissue

Brockmann bodies or principal islets

scattered among the exocrine tissue as in mammals. In lungfishes (Dipnoi), the tissue is congregated into teleostean-like principal islets; the coelacanth (Crossopterygii) has an elasmobranchin-type arrangement where the tissue is associated with the ducts of the acinar cells. The islets of Langerhans in the Amphibia are usually quite small and contain mainly B-cells; some urodeles

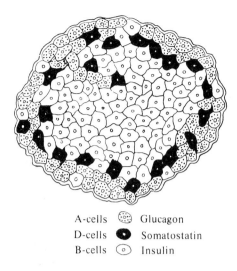

Fig. 2.19. Schematic representation of an islet of Langerhans in humans showing the distribution of glucagon, somatostatin, and insulin-containing cells. Islet cell types for which a function has not yet been positively established are omitted. (From Orci and Unger, 1975.)

A-cells   Glucagon
D-cells   Somatostatin
B-cells   Insulin

may even lack A-cells completely. The reptiles and birds have a much higher proportion of A-cells, which probably accounts for the large amount of glucagon that can be extracted from their pancreases. In birds three types of islet are present: "dark" islets containing mainly A-cells, "light" islets with B-cells, and mixed islets with both types. As in mammals, the islets in birds contain D-cells at their periphery, but the relative distribution in the three types of islet does not appear to have been described. While the proportion of B-cells in the mammalian pancreatic islets is generally high, 60 to 90% of the total, there are some exceptions (White and Harrop, 1975). In sheep, the content of B-cells is only about 13% of the total and similar values have been observed in three species of kangaroo. However, the B-cell content in the latter does not appear to be a marsupial characteristic as it was 53% in the Australian brush-tail possum and high values have also been observed in the American opossum. Possible metabolic reasons for such differences are not apparent.

The association of the hormone-secreting islets and the acinar tissue in the pancreas of most vertebrates has evoked the question 'Why are the islets of Langerhans?' (Henderson, 1969). Has the relationship of the endocrine and exocrine tissue any functional significance or is it merely a fortuitous ontogenetic event? It has been suggested that the secreted hormones may contribute to the integrity of the acinar tissue (Henderson, 1969). On the one hand, somatostatin and pancreatic polypeptide have been found to inhibit secretion of pancreatic juices, but whether they can exert such actions locally is unknown. On the other hand, the blood vessels of the exocrine pancreas may merely provide a convenient lattice among which the islets can lodge and secrete their products into the liver (Epple and Brinn, 1986). Functional vascular relationships between the exocrine and endocrine pancreas are quite tenuous so that such interactions are uncertain.

An idealized reconstruction of a human islet of Langerhans is shown in Fig. 2.19. There is a core of B-cells surrounded on the peripheral surface by A-cells. Underlying the latter and also in contact with the B-cells are D-cells. It has been suggested that the inner B-cells supply the basic insulin needs and that finer control of blood glucose concentration is exerted by the outermost B-cells and glucagon released from the A-cells. This latter process is inhibited by the release of somatostatin from the D-cells, whereas both hormones can increase or decrease, respectively, the release of insulin from their adjoining B-cells. A similar distribution of peripheral A-cells and a core of B-cells is seen in rats, rabbits, and cattle (White and Harrop, 1975). However, B-cells predominate in the periphery of the islets in the horse and the Australian brush-tail possum (White and Harrop, 1975; Epple and Brinn, 1986). Information about the arrangement of the blood vessels supplying these tissues may provide an explanation of such differences.

The gonads

Although the ovaries and testes are not essential for the survival of the individual, they are for the propagation of the species. They have a dual but related function: the production of ova and sperm as well as several hormones that are concerned with the development of the germ cells and the fertilized egg.

The gonads are formed from the dorsal celomic epithelium. The adrenocortical (or interrenal) tissue has a similar origin, and both secrete related steroid hormones. The primordial gonadal tissue consists of a cortex and a medulla. The latter differentiates into the testis, the former the ovary. The steroidogenic cells that are present in the gonads (as well as the adrenals) have a distinctive structure. They contain a very well-developed endoplasmic reticulum, mitochondria that have tubular cristae, and usually lipids that histochemically behave like cholesterol. A notable characteristic that helps histological identification of steroidogenic cells in the gonads is the presence of the enzyme 3β-HSDH (Δ^5-3β-hydroxysteroid dehydrogenase). It is responsible for the conversion of certain precursors of the steroid hormones into progesterone and androstenedione, which may subsequently be converted to other hormones.

The gonads are usually paired structures lying in the body cavity near the kidneys. The testes of most mammals (except a few such as whales, the elephant, and guinea pig), however, are suspended outside the abdominal cavity in the scrotal sac. Some species have only a single gonad, which usually reflects the degeneration of the other or possibly their fusion. The cyclostome fishes only have a single testis and ovary in the median line, whereas in nearly all birds only the left ovary reaches full development. A single ovary is also sometimes present in teleost and chondrichthyean fishes.

Considerable diversity exists in the relationship of the gonads to their excretory ducts. These are lacking altogether in cyclostomes, where the eggs and sperm are released into the body cavity and thence through pores to the exterior. Some fishes, especially teleosts, have gonoducts that are merely extensions of the ovaries. In most vertebrates, including many bony fishes, sharks and rays, and tetrapods, the germ cells pass through a homologous series of ducts derived from the Wolffian duct in the male and the Müllerian duct in the female.

Testis

The testis is formed from the medulla of the primordial gonad. Its morphology displays considerable uniformity among different vertebrates but there are some variations of this basic theme (Dodd, 1960; Lofts, 1968; Nagahama, 1986). The testis is usually surrounded by a tunica albuginea enclosing a parenchyma mainly composed of seminiferous tubules, which in some non-mammals forms lobules. Hormone-secreting Leydig cells (interstitial cells) usually lie in the interstitium between the seminiferous tubules (Fig. 2.20). The formation of spermatozoa may be a continuous process or, in many nonmammals, it may occur in bursts. The latter results from their synchronous formation and release from cysts associated with lobular structures. Spermatozoa are formed in a progressive process starting with the mitotic divisions of spermatogonia to form primary spermatocytes. The spermatogonia lie adjacent to the basal membrane of the seminiferous tubules. Also present at this site are the columnar Sertoli cells. During their development, the germ cells become enmeshed amongst the Sertoli cells and move centripetally along their outer walls. The Sertoli cells are also called 'nurse cells' as they foster the development of the spermatozoa. They produce several secretions including steroid hormones, inhibin, and antimullerian hormone (Chapter 3).

Leydig cells have not been identified in the interstitium of the testis of all vertebrates. Some urodele amphibians appear to lack such cells at this site. A possibly analogous type of cell has been identified (Fig. 2.20) in the walls of the seminiferous tubules and cysts. They are called wall cells or boundary cells. Such cells have also been found in species that possess ordinary Leydig cells. A morphological homology between the two types of cell is contentious, but they may still perform similar hormone-secreting functions.

The testes undergo considerable structural changes associated with the periodic, or cyclical, breeding behavior. These changes are reflected in the size and lipid content of the interstitial and Sertoli cells.

Fig. 2.20. The testis. (*a*) Sections from a reptile, the viper, showing the lipid-filled interstitial (Leydig) cells. (*b*) A teleost fish, the pike, showing the boundary cells in the wall of the testicular lobule. These are the homologs of the interstitial cells that are absent in most of the fishes. (From Gorbman and Bern, 1962, based on data from B. Lofts and A. J. Marshall.)

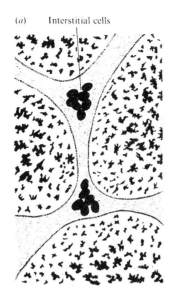

(*a*) Interstitial cells

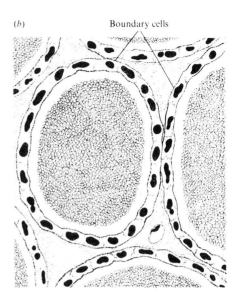

(*b*) Boundary cells

The ovaries

The ovary is formed from the cortex of the primordial gonadal tissue. Its morphology displays many variations among the vertebrates but it exhibits a number of common structural features (Fig. 2.21). The ovary (see Dodd, 1986) is usually encased in a tunica albuginea, which is underlain by a surface (or "germinal") epithelium. In the genetic female, the primordial germ cells (gonocytes), which are not derived from the so-called "germinal" epithelium, give rise to nests of primary and secondary oogonia through a series of mitotic divisions. The latter enter meiotic division, usually early in gonadogenesis, to form the primary oocytes. These cells are lodged in the outer cortical zone of the ovary and provide future sources of ova. When reproduction is imminent, folliculogenesis commences and selected oocytes are enveloped by granulosa cells and are encased in a well-vascularized two-layered theca. The theca interna can synthesize hormones (steroids and inhibin) while the theca externa forms a firm fibrous coat containing collagen. The theca externa of the mature Graafian follicle ruptures at the time of ovulation and the ovum is extruded. The collapsed follicle may regress and become a corpus albicans or, as occurs in mammals, the granulosa cells proliferate and a hormone-secreting corpus luteum is formed. Nonmammals can form similar structures, which are called postovulatory follicles; they are also commonly referred to as corpora lutea. This process occurs in ovoviviparous and viviparous species of reptiles, amphibians, and fish, especially elasmobranchs. Unovulated follicles may degenerate and are called corpora atretica. It has been suggested, but not confirmed, that the corpus atreticum may secrete hormones. The ovary,

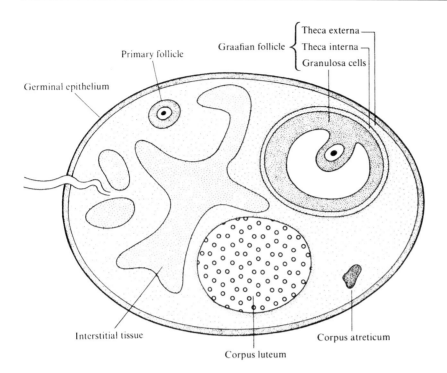

Fig. 2.21. The ovary. Diagrammatic representation of the various tissues present. (From Lofts and Bern, 1972.)

especially among mammals, sometimes contains interstitial tissue (the interstitial gland, which secretes hormones). The various ovarian components including nerves and blood vessels are contained in connective tissue called the ovarian stroma.

The juxtaglomerular apparatus

The kidney is the site of formation of a protein called *renin*, considered by some to be an enzyme and by others a hormone, which can be released into the circulation. The secretion initiates the formation of a peptide called *angiotensin*, in the plasma, which contributes to the regulation of sodium retention in the body and also increases blood pressure. Renin is formed by cells situated near the renal glomerulus at a site called the juxtaglomerular apparatus. In mammals (Barajas, 1979), it consists of *juxtaglomerular cells* (on the afferent glomerular arteriole) and the *macula densa*, which is a thickening of the distal renal tubule in the region where it abuts onto a glomerular area between the glomerular arterioles called the *polkissen* (or extraglomerular mesangium) (Fig. 2.22*a*). In nonmammals the situation is less complex (Fig. 2.22*b*) as the macula densa and polkissen are apparently absent (the former may be present in birds). However, structures that appear to be homologous to the mammalian macula densa and polkissen have also been observed in elasmobranch fish (Lacy and Reale, 1990). Many species have juxtaglomeru-

Fig. 2.22. The juxtaglomerular apparatus. (*a*) The laboratory rat. (*b*) The bullfrog (*Rana catesbeiana*), showing the absence of a macula densa and pilkissen. (From Sokabe *et al.*, 1969.)

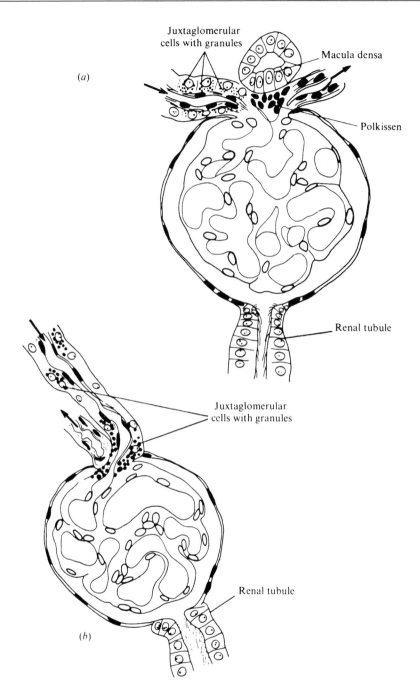

lar cells (Fig. 2.23). They contain "granules" that can be stained histologically in a distinctive way (Bowie's method). Such juxtaglomerular granules occur in arterioles often distant from glomeruli in teleost, dipnoan, and coelacanth fishes, and even in aglomerular fishes, but in most vertebrates they occur in

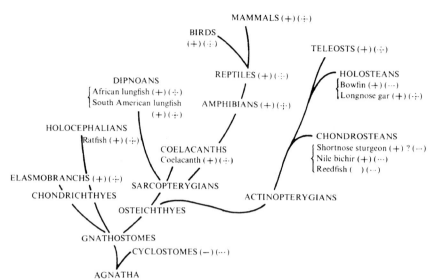

Fig. 2.23. Phylogenetic distribution of juxtaglomerular granules (stained by Bowie's method) and renin activity in the kidneys of vertebrates: (+) or (–), presence or absence of renin; (-⦂-) or (---), presence or absence of granules. It can be seen that while the two are always associated with each other in tetrapods and most fishes, this is not so in all holosteans or chondrosteans. (From Nishimura, Ogawa, and Sawyer, 1973, modified by H. Nishimura. For data on elasmobranchs, see Hazon *et al.*, 1989, Lacy and Reale, 1990.)

the afferent glomerular arteriole. Juxtaglomerular cells are present in birds, reptiles, amphibians, teleost and elasmobranch fishes. These observations parallel the identification of renin in the kidneys of these groups. An inconsistency exists in that juxtaglomerular cells with granules stainable by the Bowie method have not been found in the Chondrostei or Holostei, even though there is evidence to suggest the presence of renin in such fish. Renin has also been tentatively identified in the corpuscles of Stannius in teleost fishes (Chester Jones *et al.*, 1966; Sokabe *et al.*, 1970), but it is not clear if Bowie's granules are also present in these tissues (compare for instance Krishnamurthy and Bern, 1969; Sokabe *et al.*, 1970).

An additional putative component of the juxtaglomerular complex has been identified (Ryan, Coghlan, and Scoggins, 1979). Granulated epithelial-type cells encircling the polar region of the glomerular tuft have been observed in several mammals, including sheep and humans, though they are much smaller and are difficult to identify in rats and mice. They have been called granulated peripolar epithelial cells. It has been suggested that these cells could be secreting a product into the glomerular filtrate that may influence the absorption of fluid from the proximal tubule.

The macula densa may be concerned with the regulation of the release of renin in mammals (and birds?). Renin and angiotensin contribute to the regulation of sodium levels in the bodies of mammals. It has been suggested that local changes in the sodium and/or chloride permeability of the macula densa may regulate the release of renin from the juxtaglomerular cells (Peach, 1977). Sodium depletion increases the release of renin and this response has been shown to be dependent on a renal "vascular receptor" (Gotshall *et al.*, 1973). In this respect it is noteworthy that a reduction in renal blood flow, such as results from hemorrhage or stimulation of the renal nerves, also promotes renin release in mammals. This response can be blocked by a β-adrenergic-blocking drug (propranolol) that acts distally to the renal vascular muscle (Coote *et al.*, 1972). Therefore, several steps appear to be involved in the release of renin but more precise evidence as to the role (if any) of the macula densa is lacking. It is interesting that nonmammals may also utilize the renin–angiotensin system to aid sodium regulation in the body and as these animals may lack a macula densa it presumably does not have an essential role in this process in all vertebrates. It remains possible, however, that it could mediate some effect on the release of renin that is special to the mammals.

The pineal gland

The pineal (see Ariëns Kappers, 1970; Wurtman, Axelrod, and Kelly, 1968; Korf and Oksche, 1986; Sato and Wake, 1992) originates as a sac-like evagination from the dorsal part of the brain (the diencephalon) (Fig. 2.24). It lies beneath the cranium in the midline position and is also called the *epiphysis cerebri*. Embryologically, its hollow stem maintains contact with the III ventricle and this pattern persists in many nonmammals. The primeval role of the pineal appears to have been that of a photoreceptor. This function is still apparent in contemporary vertebrates, especially fish, amphibians, reptiles, and, possibly, even birds. The pineal gland forms and secretes a hormone called melatonin (Chapters 3 and 4) in a diurnal pattern, so that the hormone's levels are elevated in darkness and are suppressed by light. These events occur in cells called pinealocytes, which are described as photo-neuroendocrine cells. Melatonin is an indoleamine that can act as a transducer for signaling the length of the day and night and, hence, also the time of the year. It may thus play an important role in influencing biological rhythms, including seasonal reproduction. The pineal may be the site of synthesis of other hormones, apart from melatonin, and peptides such as vasoactive intestinal peptide and vasotocin have been tentatively identified there.

The pineal gland is present in most vertebrates. It is absent, however, in hagfish (myxinoid cyclostomes), crocodiles, at least two species of chondrichthyeans (*Torpedo ocellata* and *T. marmorata*), and several mammals, including whales. It is possible, however, that pineal tissue may persist in the brain of such vertebrates. The pineal is very small in the elephant and rhinoceros. In humans, the pineal is a relatively simple knob of tissue, but in other vertebrates it is more complex. It appears to have undergone considerable changes during its evolution (Fig. 2.24). In lampreys and teleost fish, there is an adjacent lobe of tissue called the parapineal with which it shares a common site of origin and connections in the brain. In some fish there is a pineal

LAMPREY

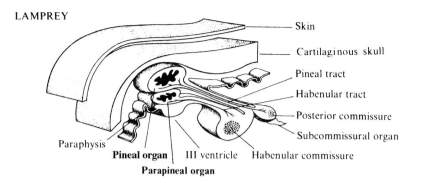

Skin

Cartilaginous skull

Pineal tract

Habenular tract

Posterior commissure

Subcommissural organ

Paraphysis

Pineal organ | III ventricle | Habenular commissure

Parapineal organ

Fig. 2.24. The pineal of various vertebrates in relation to the dorsal diencephalic roof region. (From Wurtman, Axelrod, and Kelly, 1968.)

TELEOST FISH

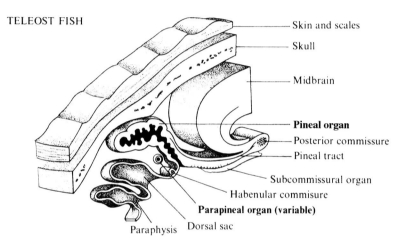

Skin and scales

Skull

Midbrain

Pineal organ

Posterior commissure

Pineal tract

Subcommissural organ

Habenular commisure

Parapineal organ (variable)

Paraphysis   Dorsal sac

ALBINO RAT

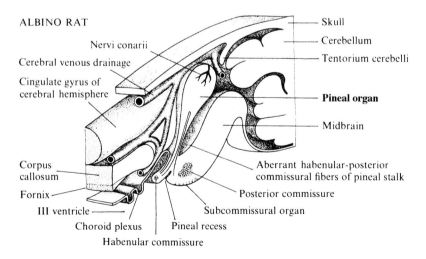

Nervi conarii

Cerebral venous drainage

Cingulate gyrus of cerebral hemisphere

Skull

Cerebellum

Tentorium cerebelli

Pineal organ

Midbrain

Corpus callosum

Fornix

III ventricle

Choroid plexus   Pineal recess

Habenular commissure

Aberrant habenular-posterior commissural fibers of pineal stalk

Posterior commissure

Subcommissural organ

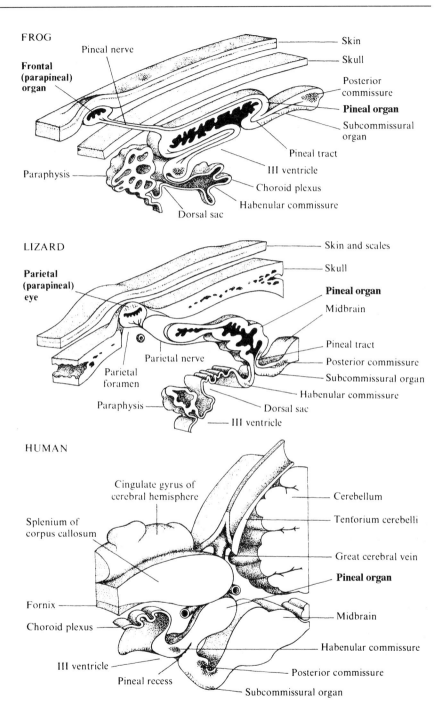

FROG

Frontal (parapineal) organ

Pineal nerve

Skin

Skull

Posterior commissure

Pineal organ

Subcommissural organ

Pineal tract

III ventricle

Choroid plexus

Habenular commissure

Paraphysis

Dorsal sac

LIZARD

Parietal (parapineal) eye

Parietal foramen

Parietal nerve

Paraphysis

Dorsal sac

III ventricle

Skin and scales

Skull

Pineal organ

Midbrain

Pineal tract

Posterior commissure

Subcommissural organ

Habenular commissure

HUMAN

Cingulate gyrus of cerebral hemisphere

Splenium of corpus callosum

Cerebellum

Tentorium cerebelli

Great cerebral vein

Pineal organ

Fornix

Choroid plexus

III ventricle

Pineal recess

Midbrain

Habenular commissure

Posterior commissure

Subcommissural organ

window or fontanelle, overlying the pineal, and analogous structures have been observed in the skulls of fossil fish. In anuran amphibians the pineal is connected by a nerve tract to an extracranial frontal organ on the dorsal surface of the skull, lying under the skin. This structure contains photosensory cells. Among the reptiles, some lizards and the tuatara *Sphenodon* (Rhyncocephalia) have a well-developed parietal eye (the "third eye"), which contains a cornea, lens, and retina-like structure connected to the pineal. The main central part of the pineal is called the epiphysis or pineal body and the whole structure makes up the pineal complex.

The principal type of cell in the pineal body is the pinealocyte, which is supported by glial cells. The pinealocyte displays an evolution in its structure from the fishes to the mammals. In cold-blooded vertebrates, it retains a photosensory-like structure that is similar to retinal cone cells. In birds, this morphology has changed somewhat and pinealocytes are referred to as "modified photoreceptor" cells. They still, however, retain a rhodopsin-like photoreceptive pigment and even *in vitro* can still change their activity in response to light (Deguchi, 1981). A typical photoreceptive structure has been lost in mammalian pinealocytes, but their fine structure and immunocytochemical properties suggest that remnants of a photoreceptive system (the S-antigen photoreceptive marker) persist.

The pineal retains commissural connections to other parts of the brain and in mammals and birds receives a sympathetic innervation. It is the site of many nerve cells and ganglia that make connections to other parts of the brain, including the hypothalamus. In mammals the sympathetic innervation arises from the superior cervical ganglion retina. The sympathetic nerves pass through the suprachiasmatic nucleus, which receives an imput from the retina and which is thought to be the site of a "biological clock", to the pineal. Cutting this sympathetic nerve to the pineal in mammals abolishes the diurnal rhythm of its activity. This effect is not, however, observed in birds. The pineal has a copious blood supply and an intrapineal capillary network. Attempts to identify blood portal systems directly connecting the pineal to other parts of the brain have not been successful.

The pineal gland has formerly been described as a "vestigial" organ. Despite the implied slight about its current usefulness, it has shown considerable phylogenetic persistence and may have evolved from a primarily sensory organ to a photoneuroendocrine one.

The corpuscles of Stannius

The corpuscles of Stannius are named after their discoverer H. Stannius (Stannius, 1839). They are associated with the surface of the kidneys in two groups of the bony fishes: the Holostei (bowfin and garpike) and the Teleostei. Corpuscles of Stannius are absent from the Chondrostei (sturgeons) and all other vertebrates (Krishnamurthy and Bern, 1969; Wendelaar Bonga and Pang, 1986). Large numbers of these small oval or irregularly shaped bodies may be present, 40 to 50, or even more in the bowfin, *Amia calva*, but smaller numbers are usually seen in the teleosts where only one or two such glands may be present. Some teleosts appear to lack these tissues altogether. The morphological position of the corpuscles of Stannius may reflect their embryonic origins from the pronephric or, sometimes, mesonephric ducts. They are well vascularized and have an autonomic nerve supply that appears to be confined to their blood vessels. Two principal types of secretory cell have been identified: Type I (PAS positive) and type II (PAS negative). Type I cells which contain numerous secretory granules, predominate. They are probably the site of synthesis of a glycoprotein hormone called stanniocalcin (formerly called hypocalcin and, sometimes, teleocalcin) (Chapter 3). Stanniocalcin lowers plasma calcium concentrations and so may be especially useful to fish in sea water. The type I cells are activated when euryhaline teleosts are transferred from fresh water to sea water. Another glycoprotein has been identified in extracts of corpuscles of Stannius and it may also be a hormone. It has been called teleocalcin. A renin-like material has been found in extracts of these tissues but it is suspected by some that it may be a contaminant from associated kidney tissue. Stanniocalcin has recently been identified in the plasma of sharks, salamanders, and humans (Wagner *et al.*, 1995). Immunocytochemical observations indicate that it is present in renal tubular cells in humans and rats (Wagner *et al.*, 1995; Haddad *et al.*, 1996). Its mRNA has also been found in human ovaries, prostate, and thyroid (Chang *et al.*, 1995). The presence of stanniocalcin in the mammalian kidney may reflect the embryological origins of the corpuscles of Stannius in bony fishes.

The urophysis

Tucked away beneath the vertebrae in the tail of teleost fishes is a lump of tissue that has been called the urophysis and which may influence their osmoregulation and assist smooth muscle contraction in the urinogenital tract. This "gland" was first described in 1813 by A. Arsaki and has since been identified in some 400 different species of teleost fishes. Despite its widespread distribution and systematic persistence, its physiological role is still not understood but it has captured considerable attention from several endocrinologists (Fridberg and Bern, 1968; Kobayashi *et al.*, 1986).

The urophysis, like the neurohypophysis, is composed of neural tissue, where the cell bodies are situated in the posterior part of the spinal cord. Axons of these cells pass outside the spinal column ventrally where they make contact with blood vessels (that lead through the kidneys) to form a neurohemal junction. Such an arrangement is admirably suited to the discharge of endocrine secretions into the circulation. These nerve cells contain granules and appear to be typical neurosecretory cells such as those seen in the neurohypophysis, though their tinctorial characteristics differ. Extracts of this tissue show several biological activities: they can alter the permeability of some membranes to water and sodium, contract certain smooth muscle preparations, increase the blood pressure of eels, and lower the blood pressure of rats. The biological activity results from at least two peptides, called urotensin I and urotensin II (Chapter 3). The former has an homologous structure to hypothalamic corticotropin-releasing hormone while the latter bears some similarities to somatostatin. Immunocytochemical observations indicate that both urotensins may be present in the same cell.

A distinct neurohemal urophysis has only been identified in teleost fish. The elasmobranchs, however, also possess neurosecretory-type cells in the caudal part of the vertebral column. These cells are giant neurons, about 20 times the size of an ordinary motor neuron and, in *Raja batis*, extend along the last 55 vertebrae. They are called *Dahlgren cells* after their discoverer and send their axons out of the ventral part of the spinal cord to make contact with blood vessels there. The tissue is, therefore, more widespread along the spinal cord. Small Dahlgren cells have been found in some bony fishes: the Holostei and the Chondrostei. Immunocytochemical observations using antisera to urotensin I and II have identified diffuse homologous neurons in the caudal spinal cords of members of the Holocephali and Dipnoi. However, they are poorly developed or absent in cyclostomes (Owada, Yamada, and Kobayashi, 1985; Onstott and Elde, 1986). Immunocytochemical studies suggest the presence of urophysial-like peptides at extraurophysial sites including parts of the brain and anterior spinal cord (Kobayashi *et al.*, 1986; Yulis and Lederis, 1988).

The types of structure present in elasmobranchs and teleosts are shown in Fig. 2.25. The diffuse distribution of tissue seen in the sharks and rays may represent a primitive pattern that has subsequently evolved among teleosts to form a discrete aggregation. The widespread distribution and systematic persistence make one suspect that the urophysis serves a physiological role. Its histological and cytological appearance, similarity to the neurohypophysis, and its neural and vascular connections suggest an endocrine gland. Histochemical changes are also apparent when the fish are transferred between fresh water and sea water. Neither of the urotensins have been identified in the plasma, so their hormonal status remains a putative one at present.

Unconventional endocrine tissues

Many hormones can be synthesized in tissues with functions that are not usually considered endocrine ones. Such an expression of genes that code hormones may reflect their localized paracrine actions, but the hormones may also enter the systemic circulation and thus have true endocrine roles. Such

Fig. 2.25. The urophysis. *Top.* Proposed evolution of the teleost urophysis from elasmobranchs, which have neurosecretory Dahlgren cells. Longitudinal section through the tail. *Bottom.* (a) to (d). Different configurations of the Dahlgren cells among the elasmobranchs. Transverse sections through the spinal cord: the vascular beds are shaded and the menix is represented by a heavy line. (From Fridberg and Bern, 1968.)

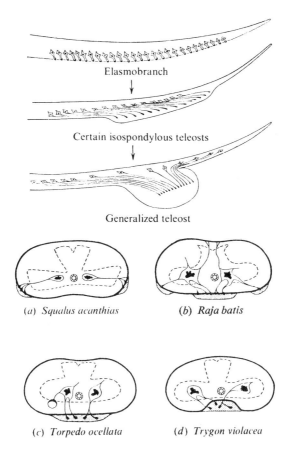

Elasmobranch

Certain isospondylous teleosts

Generalized teleost

(a) *Squalus acanthias*

(b) *Raja batis*

(c) *Torpedo ocellata*

(d) *Trygon violacea*

tissues include the heart, blood vessels, adipose tissue, mammary glands, and the placenta. The natriuretic peptides (Chapter 3) are synthesized by the atrial muscle of the heart, and sometimes also by the ventricles. The latter can also form adrenomedullin, which is also formed in the adrenal medulla and endothelial cells lining blood vessels. Adipose tissue can secrete a hormone called leptin (Zhang *et al.*, 1994), which contributes to the regulation of appetite and the body fat mass (Chapter 5). The syncytiotrophoblast of the mammalian placenta can synthesize steroid hormones, such as estrogens and progestins, gonadotropins, and placental lactogens. The particular types of hormone secreted by the placenta depends on the species and the stage of gestation. Such hormones contribute to the maintenance of pregnancy and the development of the embryo (Chapter 9). The mammary glands have also been shown to produce a variety of hormones, especially during lactation, including growth hormone, parathyroid hormone-related protein, estrogens, progestins, and prolactin (Peaker, 1995; Gabou *et al.*, 1996).

Conclusions

It can be seen that although the endocrine glands display considerable interspecific differences in their morphology many of these variations can be placed into categories that correspond to major systematic groups of vertebrates. It would thus appear that the endocrines have evolved in a relatively orderly manner which may be influenced by broad structural considerations, such as the animal's shape, size, and pattern of embryonic differentiation, as well as its particular hormonal requirements. Such evolutionary changes are not, however, confined to the glands' morphology for, as we shall see in the next chapter, considerable variation also occurs in the chemical structures of the hormones themselves.

3 The chemical structure, polymorphism, and evolution of hormones

In a mammal the endocrine glands secrete more than 40 distinct hormones. In addition, different species may form many hormones that although structurally analogous nevertheless display chemical differences. Such natural variants are usually characteristic of a single species and represent a polymorphism of the excitant's molecular structure. This change has a genetic basis. It may only be the substitution of a single amino acid residue in the molecule of a peptide hormone or it may be much more extensive. The biological effects of such differences can be considerable or negligible.

Vertebrate hormones belong to two principal classes of chemical compound. Some are made from cholesterol. These are the steroid hormones from the adrenal cortex and the gonads. The others are made up of amino acids and range in complexity from those, like epinephrine, that are derived from a single tyrosine molecule, to others like the pituitary growth hormone that contain about 190 such units. The molecular weights can vary from about 200 to 30 000.

What properties do these molecules have that make them suitable to be hormones? What characteristics may be important for their utilization as such? Armed with considerable hindsight about endocrine physiology some answers can be offered. The basic requirements will not be the same for all hormones but will depend on what they do. The steroid hormones are poorly soluble in water but readily soluble in lipids. This will facilitate their penetration into the cell and fixation at intracellular sites. Such lipid solubility will also be important if a hormone is to penetrate the blood–brain barrier. Transport in the blood is essential for a hormone to fulfill its physiological role so that, if they are hydrophobic molecules, they must either be effective at very low concentrations or be attachable to protein components that carry them to their sites of action. This binding is especially prominent among the steroid and thyroid hormones but also may involve proteinaceous hormones. An ability to interact with other biological molecules is also important for "triggering" the excitant effects of hormones. They must be capable of interacting with a receptor molecule in, or on, the effector tissue. Such an interaction must not be of a strong covalent nature but must involve chemical forces whereby an equilibrium of a reversible nature occurs. Above all, a hormone must have a high degree of specificity towards its target receptor site. Necessarily this is a property of both structures. The manner by which it is accomplished is still largely conjectural. Hormones have complex three-dimensional structures that contain various components which may be electrically charged, hydrophilic or hydrophobic, acidic or basic, and so on. Such properties together may constitute a "key" to which the receptor acts as a complementary "lock."

In order to function optimally, a hormone molecule needs to possess some other properties consonant with its physiological role. For adequate control, hormonal responses often need to be rapidly terminated. The excitant can either be readily excretable in the urine or bile, or, by virtue of the presence of chemical groups that can be changed by metabolic processes, be converted to an inactive form. The synthesis of hormones is not always rapid enough to meet the immediate demands for their release, so that their accumulation and storage in glandular tissues may be necessary. In this instance the molecule should possess a considerable measure of innate stability and be able to interact with cellular (or even extracellular, as for thyroxine) binding proteins that facilitate this storage. Related to such a process is an ability to undergo rapid mobilization from such storage sites so that the hormone can be released into the blood.

Structural differences between hormones are tentatively assumed on the basis of differences in their biological actions. They are confirmed by the demonstration of variations in their chemical and immunological behavior and ultimately by the determination of their molecular structure. It is usually a comparatively simple procedure to show that two hormones differ from each other. Tests for biological activity, for instance changes in blood glucose levels, an ability to alter blood pressure, decrease urine flow, and so on, are reasonably straightforward laboratory procedures. Broad chemical differences in even very impure preparations can often be seen when, for instance, one compares their stabilities at different temperatures and pH values, solubilities in different solvents, relative rates of destruction when incubated with various enzymes, chromatographic mobilities, and so on. Such biological and chemical characterization can be used to identify and measure the relative quantities of the hormonal material present in an extract. Determination of chemical composition and structure is a more complex procedure and, before this can be done, highly purified preparations of hormones and their receptors must be made. These can be used directly to determine the structure, such as a sequence of amino acids, or to identify genes that may encode the molecule. Gene cloning procedures can provide DNA that is complementary (cDNA) to messenger RNA (mRNA) encoding the prohormone or its receptor. The amino acid sequence can be deduced from its coding nucleotide sequence. This contribution of molecular biology to endocrinology has been very important as it facilitates the determination of amino acid sequences and is especially useful when large molecules are involved. Related molecules may also be identified in this way; some of these have not previously been known to exist. Indeed they may be encoded by nucleotides but may not be normally expressed in significant quantities in the animal.

Although the chemical structure of many hormones is known, this knowledge is mainly confined to the mammals, especially with respect to the larger protein hormones. In addition, although the disposition of chemical groups and the sequence of amino acids may be known, less information is available as to their three-dimensional (tertiary) arrangement. Such data are ultimately required if we are to understand properly how the hormones work.

Steroid hormones

Steroids are chemical compounds derived from cholesterol. They consist of a series of carbon rings, the basic unit being the cycloperhydrophenanthrene nucleus. Such compounds occur widely in nature and are not confined to the animal kingdom. Plants contain many steroids and some of these may even exhibit activities reminiscent of those of the mammalian hormones.

Several different types of steroid function as hormones in vertebrates. These and the parent cholesterol molecule are shown in Fig. 3.1a,b. They are often classified in the following manner. (1) Those based on pregnane and containing 21 carbon atoms (C_{21}). These include the adrenocortical steroids and progesterone, which, apart from being a metabolic intermediate in the formation of most steroid hormones, also acts as a sex hormone, especially during pregnancy. (2) Androstane compounds with 19 carbons (C_{19}); this includes the androgens, which have the actions of male sex hormones. (3) Estrane (C_{18}) compounds, which have actions of female sex hormones (estrogenic). (4) Vitamin D (Fig. 3.1c), a group of sterols the precursors of which are commonly obtained in the diet and which can be converted (Chapter 6) into hormones that influence calcium metabolism.

The hormones from the gonads and adrenal cortex are all derived from cholesterol compounds. In the instance of C_{18}, C_{19}, C_{21} steroids, various metabolic pathways, usually involving several hydrolase enzymes, lead to the formation of the ultimate hormone product (Fig. 3.2). These include the female sex hormones, called *estrogens* (C_{18}), estradiol-17β, estrone, and estriol; the *androgens* (C_{19}), mainly testosterone but also its metabolic precursor androstenedione and more active metabolite 5α-dihydrotestosterone; *progestins* (C_{21}), progesterone and the *adrenocorticosteroids* (C_{21}) cortisol, corticosterone, aldosterone, and 1α-hydroxycorticosterone. Some other adrenocorticosteroids, such as cortisone, are also sometimes found in the blood.

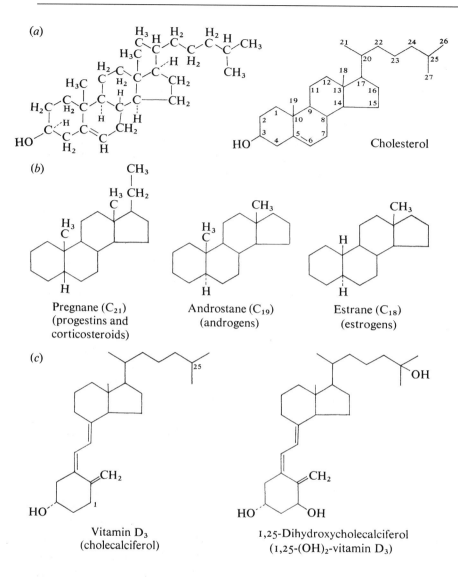

Fig. 3.1. (a) The chemical structure of cholesterol and the conventional manner of numbering the carbon atoms. (b) The parent steroid compounds for the progestins and corticosteroids (C_{21}), androgens (C_{19}), and estrogens (C_{18}). (c) Vitamin D_3 and its active metabolite $1\alpha,25$-dihydroxycholecalciferol.

Cholesterol

Pregnane (C_{21})
(progestins and
corticosteroids)

Androstane (C_{19})
(androgens)

Estrane (C_{18})
(estrogens)

Vitamin D_3
(cholecalciferol)

1,25-Dihydroxycholecalciferol
(1,25-$(OH)_2$-vitamin D_3)

Many other steroids are found in the steroidogenic tissues, where they constitute intermediates of the hormones; others may represent products of steroid catabolism. The chemical structures of these hormones are shown in Fig. 3.2.

The C_{18}, C_{19}, and C_{21} steroids have been identified in tissues and also often in the blood of all the main groups of vertebrate. The compounds present are, however, not identical in all of these, while the evidence of their precise identity in some (for instance, cyclostomes) has been noted as "tentative" or "only suggestive" (Idler, 1972). Steroids of a hormonal nature, nevertheless, undoubtedly have a wide phyletic distribution among vertebrates.

The sex hormones show a remarkable uniformity; testosterone, progesterone, and estradiol-17β are common throughout the vertebrates. This possibly reflects the "conservative" nature of the sexual process and the early evolution of a mechanism of such efficiency that little subsequent endocrine modification of the hormonal excitants could be advantageous. It can be seen in Fig. 3.2 that, when the structures of the steroid hormones are compared, the chemical differences appear surprisingly minor. Nevertheless, each molecule exerts distinct effects. A high degree of specificity based on such simple structural differences probably allows little room for subsequent successful evolutionary "experiments". A number of novel gonadal steroids have been identified in nonmammals, where they may exert special effects. The testes of some teleost fish and urodele amphibians can secrete 11-ketotestosterone (11-KT), which is formed from 11 β-hydroxytestosterone under the influence of 11β-hydroxysteroid dehydrogenase. This steroid is not present in the Australian lungfish *Neoceratodus forsteri* (Joss, Edwards, and Kime, 1996). In teleosts, 11-ketotestosterone has been found to have 10 times the androgenic activity of testosterone. Unlike testosterone it cannot be converted to an estrogen. Several special steroids have also been identified in teleosts where they have a role in inducing the maturation of the oocytes. They include 17,20β-dihydroxy-4-pregnen-3-one (17,20βP) (Nagahama *et al.*, 1994) and 17,20β, 21-trihydroxy-4-pregnen-3-one(20βS) (Trant and Thomas, 1989).

Among the adrenocorticosteroids, different molecules have emerged and these often have a distinct systematic distribution (Fig. 3.3). Such corticosteroid hormones exist in the cyclostome fishes. The steroid 1α-hydroxycorticosterone is widespread in the adrenal tissues and blood of the Chondrichthyes. This hormone is, however, only present in the Elasmobranchii (sharks and rays) and not in the Holocephali (chimaeroids), which instead have cortisol (Idler and Truscott, 1972). Among the Actinopterygii (including the Holostei, Chondrostei, and Teleostei), cortisol is the predominant corticosteroid in the blood, but corticosterone is also present. Cortisone, aldosterone, and corticosterone have also been identified in teleosts but the quantities appear to be small compared with cortisol. The criteria for such identifications are, however, sometimes in doubt. The Dipnoi (lungfishes) possess cortisol, like other bony fishes, and a fascinating discovery has been the additional identification of aldosterone in the South American lungfish *Lepidosiren paradoxa* (Idler, Sangalang, and Truscott, 1972) and the Australian lungfish *Neoceratodus forsteri* (Joss, Arnold-Reid, and Balment, 1994). Aldosterone has been identified in the blood of representatives of all the tetrapod groups so that its presence in the Dipnoi, but not apparently in most other fish (though it has been found in a few species), is consistent with their suggested phylogenetic relationships to tetrapods. A second major corticosteroid, corticosterone, is also present in amphibians, reptiles, and birds. This hormone is also the major corticosteroid in some mammals, whereas in most

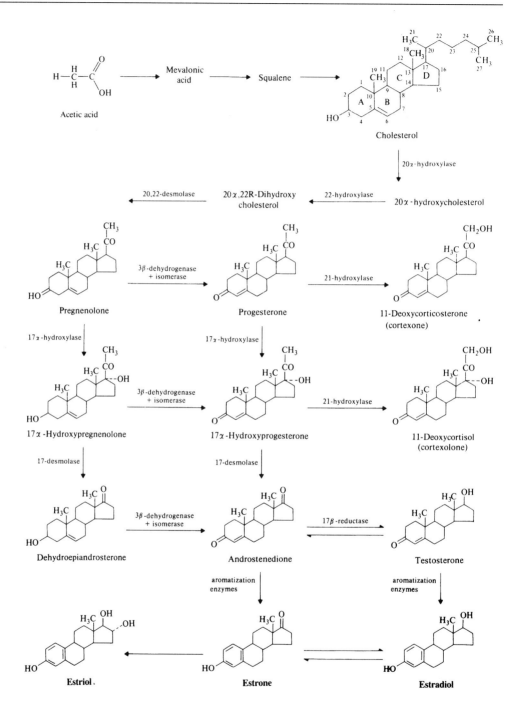

Fig. 3.2. Interrelationships and formation of the steroid hormones.

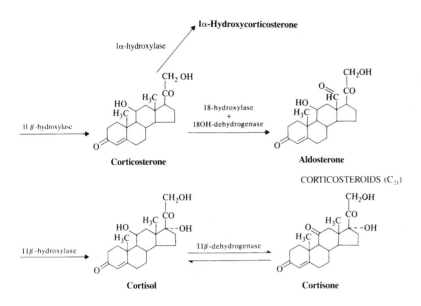

1α-Hydroxycorticosterone

1α-hydroxylase

11 β -hydroxylase

Corticosterone

18-hydroxylase
+
18OH-dehydrogenase

Aldosterone

CORTICOSTEROIDS (C$_{21}$)

11β -hydroxylase

Cortisol

11β-dehydrogenase

Cortisone

ANDROGENS (C$_{19}$)

5α -reductase

5α-Dihydrotestosterone

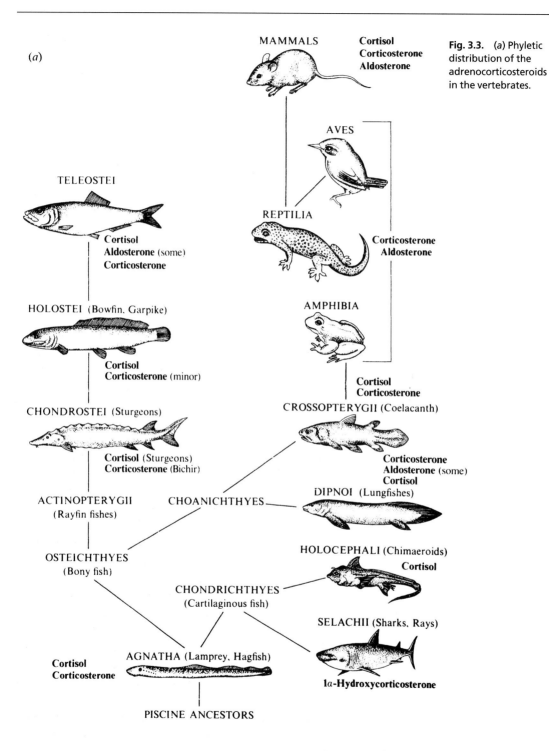

(a)

TELEOSTEI

Cortisol
Aldosterone (some)
Corticosterone

HOLOSTEI (Bowfin, Garpike)

Cortisol
Corticosterone (minor)

CHONDROSTEI (Sturgeons)

Cortisol (Sturgeons)
Corticosterone (Bichir)

ACTINOPTERYGII
(Rayfin fishes)

CHOANICHTHYES

OSTEICHTHYES
(Bony fish)

CHONDRICHTHYES
(Cartilaginous fish)

Cortisol
Corticosterone

AGNATHA (Lamprey, Hagfish)

PISCINE ANCESTORS

MAMMALS

Cortisol
Corticosterone
Aldosterone

AVES

REPTILIA

Corticosterone
Aldosterone

AMPHIBIA

Cortisol
Corticosterone

CROSSOPTERYGII (Coelacanth)

Corticosterone
Aldosterone (some)
Cortisol

DIPNOI (Lungfishes)

HOLOCEPHALI (Chimaeroids)

Cortisol

SELACHII (Sharks, Rays)

1a-Hydroxycorticosterone

Fig. 3.3. (a) Phyletic distribution of the adrenocorticosteroids in the vertebrates.

(b) The principal structures of the corticosteroid hormones.

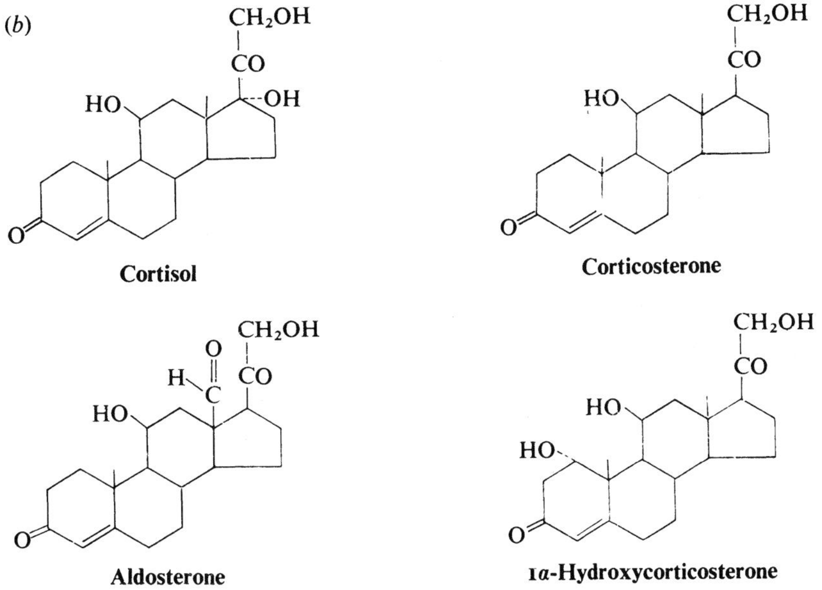

(b)

Cortisol

Corticosterone

Aldosterone

1α-Hydroxycorticosterone

others cortisol is predominant. The zona reticularis of humans and some other primates has been found to secrete large amounts of dehydroepiandrosterone (DHEA) (Hornsby, 1995). This steroid, which is also produced by the gonads of mammals, can function as a major precursor of testosterone and estradiol (Fig, 3.2) in peripheral tissues. It may also have other roles.

The ratio of cortisol to corticosterone varies among the mammals. Rats, rabbits, and mice secrete little or no cortisol from their adrenal cortexes; corticosterone (aldosterone is also present) predominates. Other mammals secrete a mixture of cortisol and corticosterone, usually with the former predominant. It was at one time suggested that the ratio cortisol : corticosterone may be a characteristic of a species and, therefore, determined genetically. It has, however, been found that this ratio can vary considerably, even in a single animal, depending on the physiological conditions. Nevertheless, the inability of the rat to form cortisol reflects the inactivity of an enzyme, 17α-hydroxylase, and it seems likely that this, at least, is genetic. Most mammals, including placentals and marsupials, secrete more cortisol than corticosterone. An interesting exception is the echidna *Tachyglossus aculeatus*, a monotreme in which corticosterone predominates (Weiss and McDonald, 1965). This pattern is more like that in reptiles and birds than that in most other mammals, but it is not seen in another monotreme, the platypus (Weiss, 1980).

The corticosteroids, in contrast to the sex steroids, display different chemical structures, which probably reflects evolutionary changes. Sex is a relatively uniform process, but the roles of corticosteroids show some variation. This may be reflected in the different structures of these steroids. The role of

1α-hydroxycorticosterone in the Elasmobranchii is uncertain. In other vertebrates, cortisol and corticosterone influence intermediary metabolism, which is a basic function in all vertebrates. Aldosterone and, to a lesser extent, corticosterone exert a prominent effect on sodium and potassium metabolism in tetrapods. These animals have special osmotic problems not faced by their piscine ancestors, so that it is conceivable that the solutions to them were accompanied not only by the evolution of special effector mechanisms but also by hormones to fit them.

A steroidal compound similar to that of the plant steroid ouabain has been identified in the circulation of several mammals (Hinson, Dawnay, and Raven, 1995). This observation is of special interest as ouabain also belongs to a group of drugs (digitalis cardiac glycosides) that is used to promote the contractility of the failing human heart. It is an inhibitor of the enzyme Na–K-activated ATPase (Na–K ATPase). The plasma steroid has been called endogenous digitalis-like factor (EDLF) and it may be secreted by the adrenal gland (Dorris and Stocco, 1989; Boulanger *et al.*, 1993). The synthesis of EDLF at this site is still, however, considered to be questionable (Hinson *et al.*, 1995). It is possible that it has a dietary origin, but EDLF may be a newly discovered adrenal steroid hormone.

Vitamin D may have been "one of the first hormones to have evolved on earth" (Holick, 1989). It is formed, under the influence of sunlight, by small plants and invertebrates (phytoplankton and zooplankton) that mainly live in the ocean. Vitamin D_3 is a steroid compound that in vertebrates functions as a prohormone. In amphibians and other tetrapods it is formed in the skin from 7-dehydrocholesterol. This photochemical reaction occurs under the influence of ultraviolet light. Whether or not this reaction can occur in fishes is controversial, but unless they are near the surface they are screened from such light anyway. The mature hormone 1,25-dihydroxycholecalciferol (1,25-dihydroxyvitamin D_3; 1,25 $(OH)_2D_3$) is formed by the addition of two hydroxyl groups; one addition is made in the liver and the other in the kidney (Chapter 6). Fish can accumulate vitamin D_3 from their diet and thus provide an entry for it into the vertebrate food chain (Rao and Raghuramulu, 1996). (A physiological role for vitamin D in fish is, however, controversial). Dietary sources of vitamin D may be important, especially in domestic animals and humans that live in higher latitudes and away from sunlight. Dietary vitamin D can exist in several forms including vitamin D_2, which can substitute for the prohormone in many species. However, in some animals, such as poultry and New World monkeys, it is not very effective. Vitamin D, and even the mature hormone 1,25-$(OH)_2D_3$, may be present in some higher plants, where, as described earlier (Chapter 1), it can have toxic effects in foraging animals. The vertebrates may have originally borrowed vitamin D from plants and used it as a prohormone, or even as a hormone, and it seems that such a link has never been fully severed.

The steroid and thyroid hormone receptor superfamily

The proteins of this receptor superfamily are encoded by a series of genes, homologs of which were originally identified in RNA tumor viruses. The viral oncogenes had been originally transduced, or "captured," from their cellular gene equivalents. The latter are thus proto-oncogenes and encode a variety of regulatory proteins, including growth factors and their receptors, intracellular transducer proteins, and nuclear transcription factors. Such transcription factors include the steroid and thyroid hormone receptors. The product of the v-*erbA* gene of avian erythroblastosis exhibits structural homologs to the C-terminal amino acids of steroid hormone receptors. The protein of the cellular homolog of the v-*erbA* gene (c-*erbA*) was found to be able to bind thyroid hormone and corresponded to a truncated form of the thyroid hormone receptor. These observations indicated that such nuclear receptors belonged to the same superfamily of receptors.

Receptors for the host of steroid hormones, thyroid hormone and $1,25(OH)_2D_3$ (see the next section) display many homologies in their structural organization and amino acid sequences, as well as in their ways of functioning. They have, therefore, been classified as members of a distinct superfamily (Green and Chambon, 1986; Evans, 1988; Parker, 1988; Tsai and O'Malley, 1994). This relationship became apparent following the cloning of the receptors and the deduction of their amino acid sequences. The steroid hormone receptors include those for estrogens (ER), progesterone (PR), adrenocortical mineralocorticoids (MR) and glucocorticoids (GR). The thyroid hormone receptor (T_3R) and vitamin D_3 are also related. Many nonhormonal receptors are also present in this superfamily, including those for the vitamin A ligands all-*trans*-retinoic acid (RAR) and 9-*cis*-retinoic acid (RXR) (Mangelsdorf and Evans, 1995). The various types of nonsteroid hormone receptor can interact with each other and the steroid hormone receptors to initiate many responses. Such receptors are the product of separate genes and vary considerably in their size, containing from about 450 to 1000 amino acid residues. The principal regions of amino acid homology define protein domains, which have various specific functions including ligand (hormone) binding (domain E), DNA binding (domain C) and transactivation (A/B) (Fig. 3.4). Domain C contains 60 to 70 amino acid residues and domain E about 250. Amino acid homologies in domain C of GR compared with PR, ER, and T_3R are about 90%, 52%, and 47%, respectively. The homologies within domain E are somewhat less strong; GR compared with PR, ER, and T_3R is 55%, 36%, and 17%, respectively. Homologies between identical domains from the same receptor but from different species are usually greater than 90%. The entire 418 residue amino acid sequence of VD_3R from chickens and quail are identical but differ somewhat from the human and rat receptors (Elaroussi, Prahl, and DeLuca,

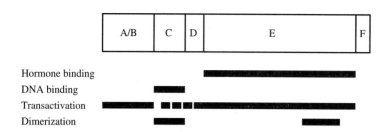

Fig. 3.4. The functional domains of the steroid hormone receptors. Some of their roles are indicated by the solid lines below. For more details see the text. (Based on Tsai and O'Malley, 1994.)

1994). However, comparisons of the avian hormone-binding domain and DNA-binding domain with those from the mammals showed a 98.5% homology for the former and 87.5% for the latter

The receptors for the steroid hormone/thyroid hormone superfamily function as factors that control the transcription of target genes and so influence various metabolic and developmental processes (p. 448). Their ability to function in this way is usually dependent on their binding to the specific hormones under discussion. Before ligand binding, these receptors are present in the nucleus or even bound to DNA. They may also be present in the cytoplasm but their former proposed ubiquitous localization in this region of the cell prior to ligand binding is no longer generally considered to be tenable. Following binding to a hormone, these receptors undergo an activation, possibly involving a conformational change, and may form dimers that bind to specific hormone-response elements (HREs) in chromatin. These elements are specified by specific nucleotide sequences. Following the hormone – ligand interaction, an inhibition of transcription is relieved and transcription is stimulated (Chapter 4). These hormones interact with their receptors, which form homodimers for the steroid hormones, or monomers or heterodimers for VD$_3$R, T$_3$R and RXR (Hawa *et al.*, 1994; Williams, 1994; Mangelsdorf and Evans, 1995).

Evolution of such receptors may have begun in simple organisms, perhaps a 1000 million years ago (O'Malley, 1989). Intracellular proteins or "regulators" may have controlled metabolic and growth processes following their binding to ligands derived from inside the cell or even from the external environment. Such regulators or receptors may have been perpetuated and have multiplied following gene duplication. The structural and functional relationships seen in the receptors of contemporary species may reflect such an ancestry.

The GR and MR present in tetrapod vertebrates are thought to have evolved from an ancestral gene coding for a single type of corticosteroid receptor (DiBattista, Mehdi, and Sandor, 1984; Ducouret *et al.*, 1995). Teleost fish only possess a single corticosteroid receptor, which utilizes cortisol to mediate both mineralocorticoid and glucocorticoid types of effect (Sandor, DiBattista, and Mehdi, 1984). It is found in tissues such as liver and skeletal muscle as well as in the gills and intestine. It was, therefore, an event of much awaited interest when a corticosteroid hormone receptor was cloned from tissue of the rainbow trout, *Oncorhynchus mykiss* (Ducouret *et al.*, 1995). The deduced 758 residue amino acid sequence of this protein indicated that it was related to other such steroid hormone receptors. The DNA-binding domain exhibited a 97% homology to the amino acid sequence of the human glucocorticoid receptor. The ligand-binding domains in the two species had a 70% homology. In spite of these homologies, corticosteroid receptors display many differences in their precise structures, even including their DNA-binding domains. Comparisons of the teleost receptor with the GRs and MRs in tetrapod vertebrates indicates that it could not have been a common ancestral prototype but appears to have diverged from a GR-like molecule before the origin of the tetrapod GRs. The tetrapod MRs apparently evolved separately to the teleost cortisol receptor.

Hormones that are amino acid derivatives

Catecholamines

The adrenal medulla and other chromaffin tissues secrete two hormones, epinephrine and norepinephrine (Fig. 3.5). These are amine derivatives of catechol; hence their name catecholamines. Such compounds are found in all vertebrates, where they also act as neurotransmitters in the sympathetic nervous system and brain. They are also present in many invertebrates and even in the ciliated protozoan *Tetrahymena*, where they influence metabolism in a manner reminiscent of that in more sophisticated metazoan animals (Blum, 1967). Dopamine is formed in some nerve cells and may have an endocrine function.

The catecholamines are made from a single amino acid, tyrosine, and could be considered to have the simplest structure of all the vertebrate hormones. They clearly have a long phylogeny and could be primordial hormones.

Using phenylalanine, then tyrosine, as substrates, norepinephrine (Fig. 3.5*a*) is formed in chromaffin tissue. This may, under the influence of a methyltransferase enzyme, PNMT, have a methyl group added to become epinephrine. These two hormones have differing actions though a crossover of their effects occurs. Their actions were originally classified as α-adrenergic and β-adrenergic but α_1-, α_2-, β_1-, and β_2-subtype effects are now known, each reflecting the presence of a specific receptor. The α_1-adrenergic effects include the constriction of vascular and some other types of smooth muscle, a decreased release of renin, increased hepatic glycogenolysis in some species, and sweat gland secretion. The α_2 adrenergic responses include constriction of vascular smooth muscle and a decreased release of insulin. The β_1-adrenergic responses principally involve promotion of the contraction of the heart and an increased release of renin. The β_2-adrenergic effects are reflected by the

(a)

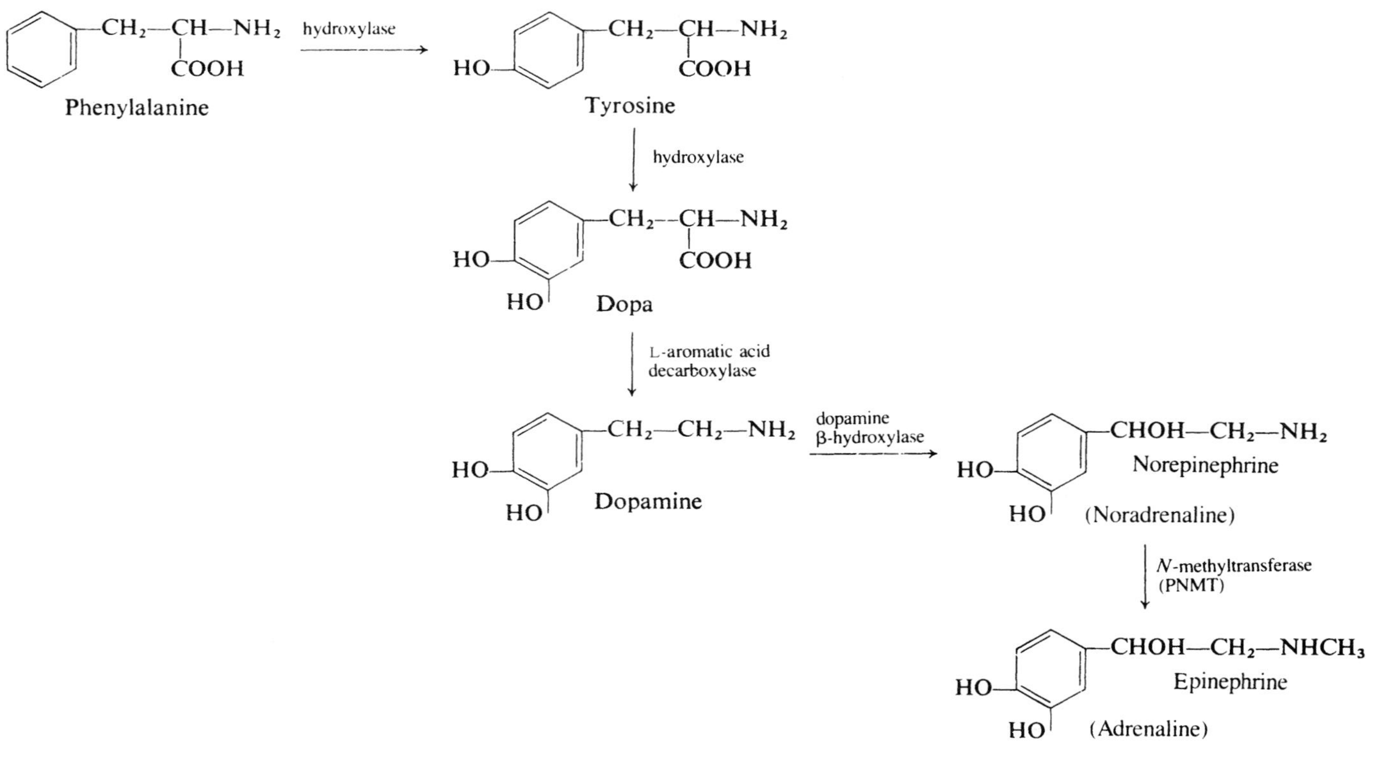

Phenylalanine — CH_2—CH—NH_2 with COOH

hydroxylase →

Tyrosine — CH_2—CH—NH_2 with COOH, HO—

hydroxylase →

Dopa — CH_2—CH—NH_2 with COOH, HO— HO—

L-aromatic acid decarboxylase →

Dopamine — CH_2—CH_2—NH_2, HO— HO—

dopamine β-hydroxylase →

Norepinephrine (Noradrenaline) — CHOH—CH_2—NH_2, HO— HO—

N-methyltransferase (PNMT) →

Epinephrine (Adrenaline) — CHOH—CH_2—$NHCH_3$, HO— HO—

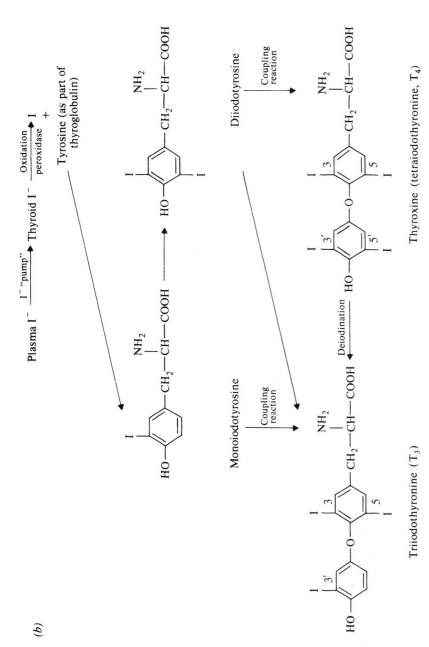

(b)

Fig. 3.5. Chemical structure and biological synthesis of (a) the catecholamine hormones; (b) the thyroid hormones.

Table 3.1. *Norepinephrine as a percentage of the total catecholamines in the adrenals of various species of vertebrates*

	Norepinephrine as % total catecholamines
Whale	83
Domestic fowl	80
Dogfish	68
Turtle	60
Pigeon	55
Frog	55
Toad	55
Pig	49
Sheep	33
Ox	26
Humans	17
Rat	9
Rabbit	2
Guinea pig	2

Source: Based on West, 1995.

relaxation of many smooth muscles, including the bronchi and some blood vessels, as well as increased glycogenolysis. Lipolysis and fatty acid oxidation in the fat cells of some species appears to be mediated by the recently discovered β_3-adrenergic receptor. Both epinephrine and norepinephrine can exert such effects, but their relative activities at each site vary. While epinephrine usually has an equal or greater systemic effect than norepinephrine, the latter's neural roles often predominate. Epinephrine and norepinephrine can, however, exert true endocrine actions. As shown in Table 3.1, both catecholamines are present in the adrenal medulla but there are large species differences in the ratios of the amounts that are present. Norepinephrine can contribute 80% of the total stored catecholamines in a whale or the domestic fowl, or only 2% in the rabbit.

Catecholamine receptors have been isolated and found to belong to a related family. The different types (α_1, α_2, β_1, etc.) are each coded by a separate gene and consist of about 400 to 500 amino acid residues, exhibiting homologies of 30 to 40% (Caron and Lefkowitz, 1993). These proteins are glycosylated at their N-terminus, which lies in the extracellular space. Seven transmembrane domains cross and recross the plasma membrane, while the C-terminus lies in the cytoplasm. The transmembrane domains are highly conserved (for instance, 70 to 80% between different types of β-adrenoreceptors) and appear to provide a site on the external face of the cell at which the hormone can bind. These receptors are all linked to GDP- and GTP-binding proteins (G proteins), which help mediate their effects. These adrenergic receptors belong to a larger superfamily of receptors that are also linked to G proteins and contain seven transmembrane domains. It includes receptors for calcitonin, glucagon, secretin, and vasoactive intestinal peptide (Segre and Goldring, 1993), as well as those for vasopressin and oxytocin (Birnbaumer *et al.*, 1992; Lolait *et al*, 1992). Homologous amino acid sequences between such diverse receptors in this superfamily are not as great but combined with their similar structural organization suggest that they may have shared a common ancestor long ago.

Thyroid hormones

The thyroid hormones are unique as they contain, as part of their structure, the halogen iodine. The formation of thyroxine (3,5,3',5'-tetraiodo-L-thyronine, T_4) and triodothyronine (3,5,3'-triiodothyronine, T_3) by the thyroid gland is shown in Fig. 3.5b. There are quantitative differences in the effects of T_4 and T_3; those of the former are much slower in onset but longer in duration than those of the latter. In order to act, T_4 must be converted to T_3. T_4 can be bound more strongly to plasma proteins in a complex, which may contribute to the difference in the time course of its effects from that of T_3. In mammals, T_4 is probably secreted at about five times the rate of T_3.

While the thyroid gland of teleost fish secretes T_4, a significant release of T_3 has been in doubt (Eales, 1990). However, in rainbow trout the proportion of secreted T_3 rises during fasting to about 25% of the total thyroid hormone (Sefkow *et al.*, 1996). The levels of T_3 in the thyroid gland of amphibians and birds is very low (Kuhn, 1990). However, it is questionable whether T_4 can itself interact with the thyroid hormone receptors, which have a high affinity for T_3. The T_4 may, therefore, be only functioning as a prohormone. Most T_3 is formed peripherally from T_4 after its release from the thyroid gland. This process can involve several monodeiodinase enzymes (Edmonds, 1987; McNabb and Freeman, 1990). A 5'-monodeiodinase (5'-MD) removes iodine from the 5'-position to produce T_3. A Type I 5'-MD is present in the liver, kidneys, brain, and pituitary. Type II 5'-MD is present in the brain, pituitary, placenta, and brown fat. A third type occurs in skin, brain, and placenta and removes iodine from the 5-position (type III 5-MD). This latter deiodinase converts T_3 to inactive 3,3'-diiodothyronine and T_4 to 3,3',5'-triiodothyronine, which is also known as reverse T_3 (rT_3). The last lacks known thyroid activity and at high, pharmacological concentrations can act as a T_3 antagonist. In mammals, it may be present in concentrations similar to those of T_3 but a possible physiological role remains an enigma. These monodeiodinase enzymes not only help regulate the levels of T_3 in the circulation but also convert T_4 to T_3 in its effector tissues. The 5'-MD has been identified in several mammals, as well as birds, amphibians, reptiles, and teleost fish (McNabb and Freeman, 1990; Santini *et al.*, 1992; Kuhn *et al.*, 1993; Wong,

Lam, and Chiu, 1993). A 5-MD has also been identified in amphibians (Galton, 1988) and birds (Kuhn *et al.*, 1993). The liver of the rainbow trout possesses two such enzymes with the characteristics of mammalian types I and II 5'-MD (Orozco, Silva, and Valverde-R, 1997). The 5'-MD found in the liver of the teleost *Fundulus heteroclitus* has remarkable structural similarities to mammalian type II 5'-MD, indicating its conservation during the process of evolution (Valverde-R *et al.*, 1997).

Thyroid hormones have been identified in the blood of representatives of several groups of lower vertebrates, including cyclostomes, teleosts, amphibians, and reptiles (Higgs and Eales, 1973; Chiu *et al.*, 1975; Packard, Packard, and Gorbman, 1976). Biologically active iodothyronine compounds occur throughout the vertebrates and have also been identified in a number of protochordates (Chapter 2). Iodine readily combines with proteins containing tyrosine (*in vitro*) so that the natural occurrence of such compounds, especially in iodine-rich solutions like sea water, is perhaps not surprising. Their transformation to iodothyronine compounds, however, seems to depend on specialized metabolic pathways and conditions such as those that occur in the thyroid gland, which has a unique ability to trap and oxidize iodide. Some of these mechanisms have been demonstrated in ascidians (Urochordata) (Chapter 2). This process can be imitated *in vitro* provided the appropriate amounts of iodine and tyrosine-containing proteins are incubated together.

The spontaneous occurrence of T_4 compounds in nature, even before the origin of the thyroid gland, is not inconceivable. Whether such compounds did arise and acquire a usefulness as hormonal excitants is sheer conjecture. If this did occur, subsequent specializations may have led to the hormones' more efficient formation in the thyroid gland.

The thyroid and catecholamine hormones are clearly very "conservative" with respect to evolutionary changes in their chemical structure. This would seem to be, at least partly, because of their small size, which limits the possibility for change in their molecules, and the fact that they are made from simple precursors that are abundant in nature. The catecholamines and thyroid hormones (and to a slightly lesser extent the steroid hormones) provide us with an illustration of the dictum that, "it is not the hormones that have evolved but the uses to which they have been put." As will become particularly apparent in the succeeding sections, this is not always true.

Melatonin

Melatonin (*N*-acetyl-5-methoxytryptamine, MT) is a hormone that is formed in a rhythmical manner by an ancient gland, the pineal (Binkley, 1988; Reiter, 1991a). It is also found in the retina, in some species, in the Harderian gland and in the ubiquitous MT-containing cells that are present in the gut

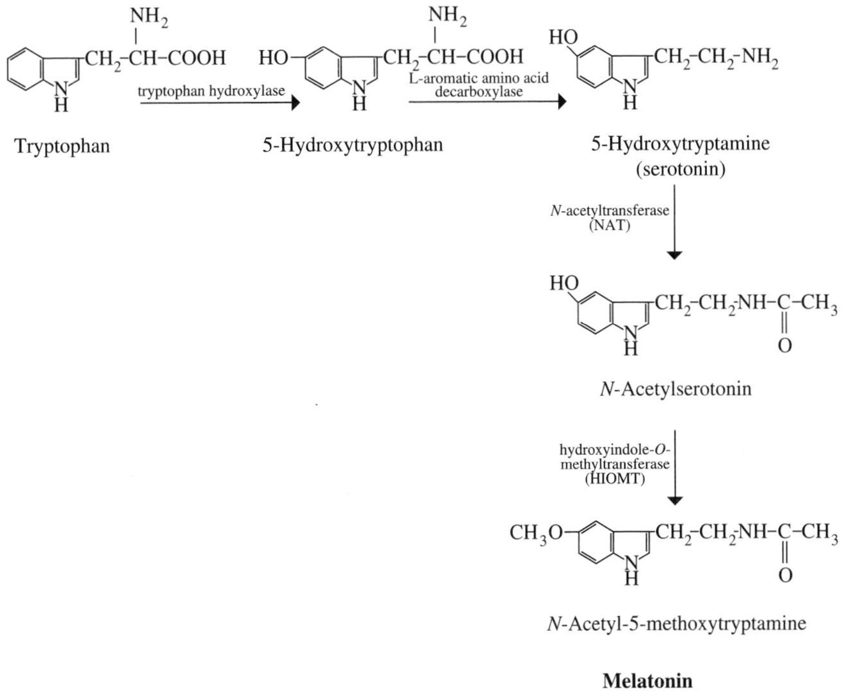

Fig. 3.6. The chemical structure of melatonin and the pathway of its biological synthesis.

Tryptophan → (tryptophan hydroxylase) → 5-Hydroxytryptophan → (L-aromatic amino acid decarboxylase) → 5-Hydroxytryptamine (serotonin) → (N-acetyltransferase (NAT)) → N-Acetylserotonin → (hydroxyindole-O-methyltransferase (HIOMT)) → N-Acetyl-5-methoxytryptamine

Melatonin

and many other tissues (Kvetnoy, Sandvik, and Waldum, 1997). Synthesis occurs in the pinealocytes and increases at night, as a result of darkness. This rhythm is coordinated by the receipt of light by the retina, which inhibits a neural control mechanism present in the hypothalamic suprachiasmatic nucleus. A "clock" mechanism associated with these nerve cells is activated in darkness and in mammals signals the pineal gland through a sympathetic postganglionic nerve pathway. The diurnal release of melatonin can provide information to various tissues of the body, including other endocrine glands, regarding the time and length of the day, and hence the season of the year. It can coordinate various physiological activities, including reproduction (Reiter and Sorrentino, 1970). Melatonin also induces a palor of the skin in amphibian tadpoles and lampreys during the night hours. This response results from an aggregation of melanin in the melanophores and it is the reason that melatonin was so named. Melatonin is synthesized from tryptophan, which is hydroxylated to 5-hydroxy-tryptophan. Decarboxylation follows to form 5-hydroxytryptamine (5-HT) or serotonin (see Fig. 3.6). Serotonin is an important neurotransmitter in the brain and is also present at other sites in the body including the enterochromaffin cells of the gut. It has a widespread distribution in animals and plants. In some diseases it is released from the enterochromaffin cells, with unpleasant results. It is not considered to be a hormone but may nevertheless be considered as an ancient forebear of melatonin. It is converted to a hormone by the action of two enzymes:

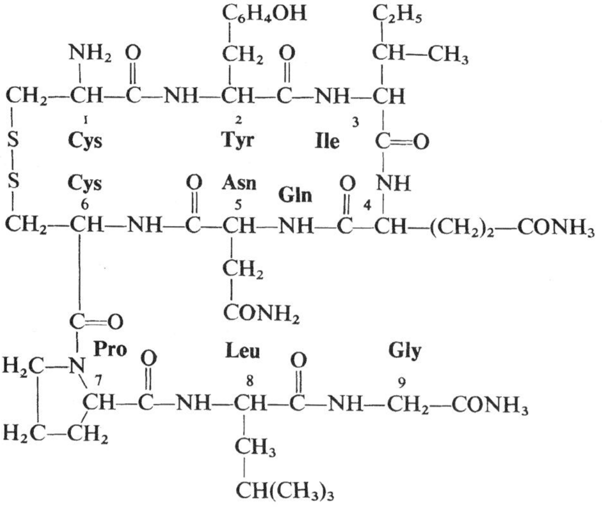

$$C_6H_4OH \quad C_2H_5$$

$$\underset{\text{Cys}}{\underset{1}{\underset{S}{\underset{|}{\underset{S}{\underset{|}{CH_2-CH-C-NH-CH-C-NH-CH}}}}}}$$

NH$_2$ O   CH$_2$ O   CH—CH$_3$

CH$_2$—CH—C—NH—CH—C—NH—CH

Cys (1)   Tyr (2)   Ile (3)   C=O

Cys (6)   Asn (5)   Gln (4)   NH

CH$_2$—CH—NH—C—CH—NH—C—CH—(CH$_2$)$_2$—CONH$_3$

CH$_2$

CONH$_2$

C=O

Pro (7)   Leu (8)   Gly (9)

H$_2$C—N

CH—C—NH—CH—C—NH—CH$_2$—CONH$_3$

H$_2$C—CH$_2$

CH$_3$

CH(CH$_3$)$_3$

Fig. 3.7. The structure of oxytocin showing the conventional numbering of the amino acids.

N-acetyltransferase (NAT) which converts serotonin to *N*-acetyl-serotonin and, subsequently by hydroxyindole-*O*-methyltransferase (HIOMT) which methylates it to *N*-acetyl-5-methoxytryptamine, or melatonin. *N*-Acetyltransferase is an inducible enzyme and this process can occur in response to the animal being in darkness (Chapter 4). Proposed family associations of hormones are currently popular. Melatonin would appear to possess an affiliation with serotonin, which is a venerable neurotransmitter, possibly with an endocrine history. Melatonin has been identified in several plants (Migitaka *et al.*, 1996). It is, for instance, present in high concentrations in members of the rice family. There has been speculation about the possible effects of the dietary use of such plants.

The peptide hormones of the neurohypophysis

Two chemically related hormones are usually secreted by the neurohypophysis. These are peptides containing nine amino acid residues. They are arranged in a five-membered ring, joined by a disulfide bridge (contributed by two half-cystine residues) and a side chain with three amino acids (Fig. 3.7 and Table 3.2). In most mammals, the two hormones are arginine-vasopressin (AVP, also called antidiuretic hormone or ADH) and oxytocin. These differ by two amino acid substitutions; vasopressin has phenylalanine and arginine at positions 3 and 8 in the molecule, where oxytocin has isoleucine and leucine. This change confers considerable differences in biological activity; vasopressin enhances water reabsorption across the renal tubule and so reduces urine flow, while oxytocin can contract the uterus and initiate "milk letdown' from the mammary glands. There is little crossover in their actions.

Homologous hormones have been identified in the neurohypophyses of representatives of all the systematic groups of vertebrates and they may also be expressed in the mammalian and avian ovaries and testes (Wathes and Swann, 1982; Kasson, Adashi, and Hsueh, 1986; Saito, Kinzler and Koike, 1990; Nicholson *et al.*, 1991). Considerable differences in chemical structure exist, however, so that, so far, 12 such peptides have been identified in nature. Amino acid substitutions occur at the 2, 3, 4, and 8 positions in the molecule (Table 3.2). The occurrence of these natural analogs has a well-defined systematic distribution (Fig. 3.8). For example, arginine-vasopressin is confined to mammals whereas arginine-vasotocin (a combination of the ring of oxytocin and the side chain of vasopressin) is present in all other vertebrates. The second oxytocin-like (or neutral) peptide in nonmammals exists in eight variant forms: mesotocin (isoleucine instead of leucine at position 8) is present in birds, reptiles, amphibians, and lungfishes; isotocin (isoleucine at 8, serine instead of glutamine at 4) is found in all the myriad of bony fishes except chondrosteans and lungfishes. The chondrichthyeans exhibit more variability, vasotocin and oxytocin being present in the Holocephali, and vasotocin as well as glumitocin, valitocin, aspargtocin, asvatocin, and phasvatocin are distributed among the Selachii (Fig. 3.9). The rays (Batoidea) possess vasotocin and glumitocin. The physiological roles of vasopressin and oxytocin in mammals, and vasotocin in tetrapods, are reasonably well understood, but the functions of the other peptides remain unknown, particularly in fish. They are nevertheless present and, from our knowledge of extant species, apparently have persisted for about 500 million years since the first cyclostomes evolved (see Fig. 3.9).

Such polymorphism of the hormones is genetically determined. It can be examined more closely among mammals where a variant of arginine-vasopressin occurs. Some pig-like mammals (Suiformes, including the true pigs, Suidae, the peccaries, Tayassuidae, and the hippopotamus Hippopotamidae) possess a vasopressin with lysine instead of arginine present in the 8 position (but see Rouille *et al.*, 1988). This probably arose as a result of a single-step mutation from arginine-vasopressin. Its present distribution suggests that this transformation occurred in an ancestor of the Suiformes before the hippopotami broke away from the pig–peccary stock (Ferguson and Heller, 1965). The change occurred in the Eocene Epoch, about 60 million years ago. The neurohypophyses of domestic pigs contain only lysine-vasopressin (and oxytocin), but among other Suiformes, such as peccaries, warthogs, and hippopotami, both arginine- and lysine-vasopressin may, or may not, be present in the same individual. It is possible that the homozygotes contain one such peptide, the heterozygotes both. The evolutionary persistence of lysine-vasopressin seems to reflect the fact that its biological potency is only a little less than that of its arginine-containing relative so that it is not appreciably disadvantageous. In addition, an adaptive increase in sensitivity of the kidney to lysine-vasopressin may occur (Stewart, 1973). It is, of course, possible that

Table 3.2. *Amino acid sequences of neurohypophysial peptides. Comparison with two analogous peptides from molluscs*

Common structure (variations in positions 2,3, 4, and 8 indicated by (X))	1	2	3	4	5	6	7	8	9
	Cys	(X)	(X)	(X)	Asn	Cys	Pro	(X)	Gly(NH$_2$)

	Amino acids in position			
	2	3	4	8
Basic peptides				
Lysine-conopressin (invertebrates)[a]	Phe	Ile	Arg	Lys
Arginine-conopressin (invertebrates)[b]	Ile	Ile	Arg	Arg
Arginine-vasopressin (AVP)	Tyr	Phe	Gln	Arg
Lysine-vasopressin (LVP)	Tyr	Phe	Gln	Lys
Phenypressin[c]	Phe	Phe	Gln	Arg
Arginine-vasotocin (AVT)	Tyr	Ile	Gln	Arg
Neutral (oxytocin-like) peptides				
Oxytocin		Ile	Gln	Leu
Mesotocin		Ile	Gln	Ile
Isotocin (ichthyotocin)		Ile	Ser	Ile
Glumitocin		Ile	Ser	Gln
Valitocin		Ile	Gln	Val
Aspargtocin		Ile	Asn	Leu
Asvatocin[d]		Ile	Asn	Val
Phasvatocin[d]		Phe	Asn	Val

[a] Present in the water snail *Lymnaea stagnalis* (van Kesteren *et al.*, 1992a,b).
[b] present in the mollusc *Conus striatus* (van Kesteren *et al.*, 1992a,b).
[c] Chauvet *et al.* (1980).
[d] Chauvet *et al.* (1994).
Source: Modified from Heller, 1974.

Fig. 3.8. The phyletic
distribution of the
neurohypophysial hormones
among the vertebrates.

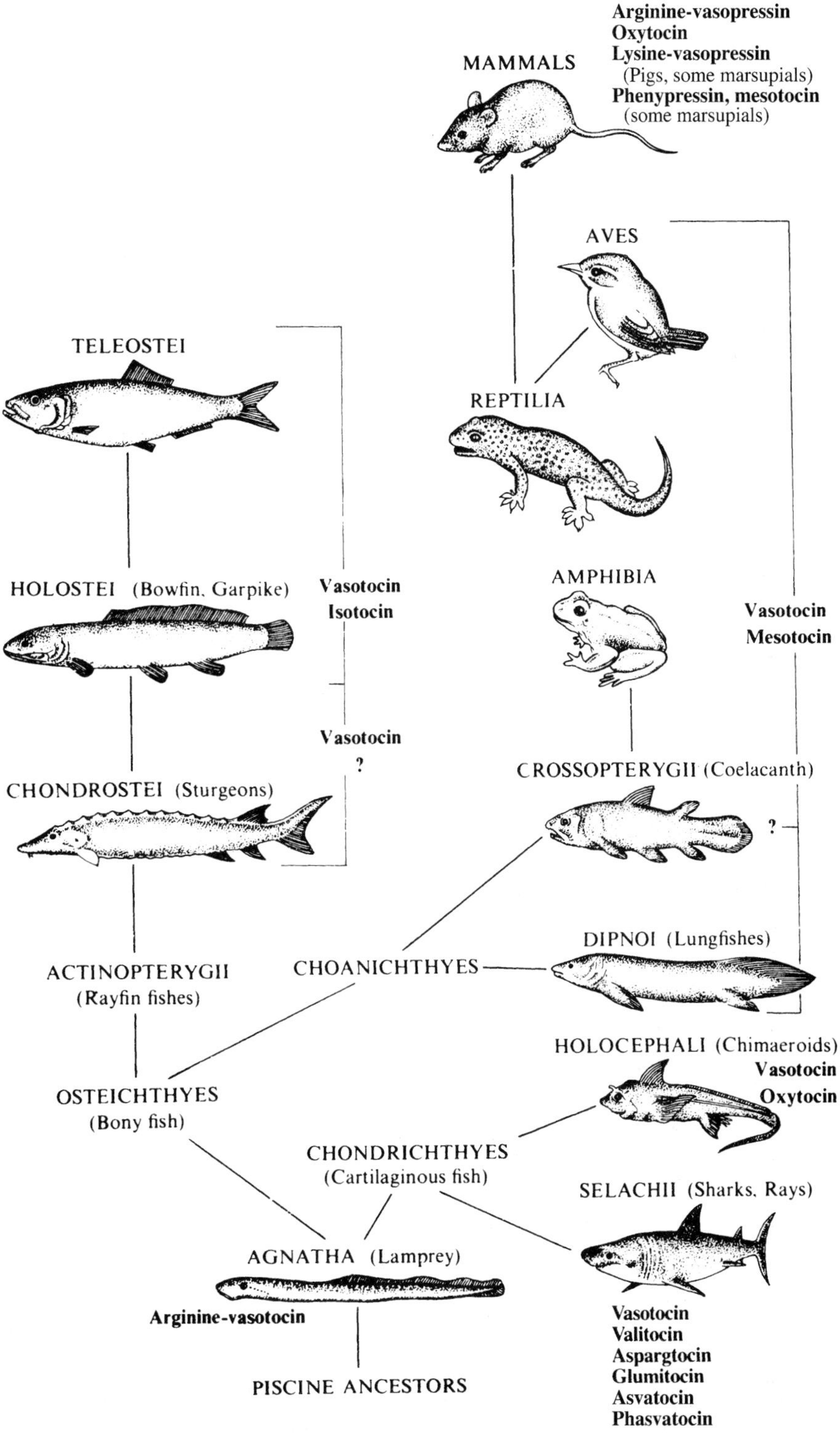

MAMMALS

Arginine-vasopressin
Oxytocin
Lysine-vasopressin
(Pigs, some marsupials)
Phenypressin, mesotocin
(some marsupials)

AVES

TELEOSTEI

REPTILIA

HOLOSTEI (Bowfin, Garpike)

Vasotocin
Isotocin

AMPHIBIA

Vasotocin
Mesotocin

Vasotocin
?

CHONDROSTEI (Sturgeons)

CROSSOPTERYGII (Coelacanth)

?

DIPNOI (Lungfishes)

ACTINOPTERYGII
(Rayfin fishes)

CHOANICHTHYES

HOLOCEPHALI (Chimaeroids)
Vasotocin
Oxytocin

OSTEICHTHYES
(Bony fish)

CHONDRICHTHYES
(Cartilaginous fish)

SELACHII (Sharks, Rays)

AGNATHA (Lamprey)

Arginine-vasotocin

Vasotocin
Valitocin
Aspargtocin
Glumitocin
Asvatocin
Phasvatocin

PISCINE ANCESTORS

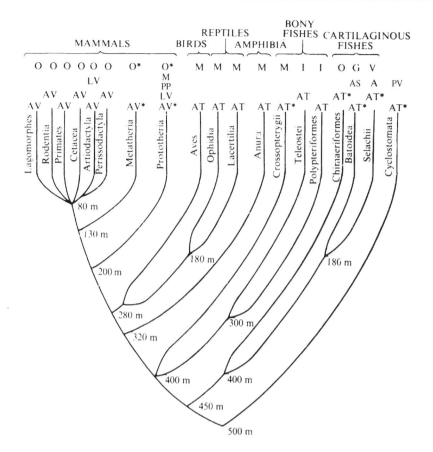

Fig. 3.9. Evolution of the neurohypophysial hormones. The letters represent the hormones that have been identified in extant species from each group of vertebrates. A, aspargtocin; AS, asvatocin; AT, arginine-vasotocin; AV, arginine-vasopressin; G, glumitocin; LV, lysine-vasopressin; M, mesotocin; O, oxytocin; PP, phenypressin; PV, phasvatocin. Million years since divergence is indicated by m, i.e. 500m is 500 million years since divergence. (Modified from Acher, Chauvet, and Chauvet, 1972, with additional information from Chauvet *et al.*, 1980; 1994.)

the presence of this hormone also confers adaptive advantages that we do not know about.

The marsupials (or Metatheria) have been separated from Eutheria for about 130 million years. It has been shown (Chauvet *et al.*, 1980) that they may possess two vasopressin-like hormones: lysine-vasopressin, which is the predominant one, and phenypressin (Phe^2-Arg^8-vasopressin). It was suggested that this duality of vasopressins in a single animal may have resulted from duplication of the vasopressin gene with subsequent single-step mutations. These two peptides appear to be quite widespread among Australian marsupials and have even been observed in a South American species (an opossum). Both of these peptides have an antidiuretic activity that is only a little less than that of arginine-vasopressin so that their acquisition may not be expected to be disadvantageous in this respect. The Marsupials may also possess two oxytocin peptides: oxytocin itself and, in kangaroos, mesotocin (Acher, 1990). When tested in eutherians, the latter peptide has a similar spectrum of action to oxytocin but it is slightly less effective.

The neurohypophysial hormones are stored in granules where they are associated with a polypeptide called neurophysin (Acher, 1978; 1990). This polypeptide exists in two forms in mammals, one of which is combined with vasopressin and contains 95 amino acid residues, and the other combines with oxytocin and has 93 residues. The two neurophysins are apparently identical within each species except for amino acid substitutions at positions 2, 3, 6, and 7. In vasopressin-neurophysin (NP II) these substitutions consist of methionine (M), serine (S), glutamic acid (E), and leucine (L) and hence it is also called MSEL-neurophysin. The oxytocin-neurophysin (NP I) is the VLDV form. Interspecific differences between the neurophysins are relatively minor. In 1982 Land and his collaborators (1982) isolated the precursor (preprohormone) for arginine-vasopressin in cattle. It was found to consist of 166 residues and, starting at the N-terminal, contains a signal peptide followed by vasopressin, then neurophysin and a 39 amino acid residue glycoprotein tail. These sections are separated by small groups of amino acids, such as glycine–lysine–arginine, which are called "processing sequences" and provide cleavage sites for the separation of the different segments. Mutations resulting in changes in the amino acid sequences of neurophysins may result in a failure to process the precursor to produce the mature hormone (Sonnemans et al., 1996). The Brattleboro strain of laboratory rats secrete large volumes of urine (diabetes insipidus) owing to a lack of arginine-vasopressin. The neurophysin in these rats exhibits such a mutation. The neurophysins appear to "chaperone" the precursors during the formation of their final secretory products. The terminal glycopeptide of the precursor was called "copeptin" and it has been suggested that it may function as a prolactin-releasing hormone (p. 101). In 1984, Ivell and Richter described the structural organization of the vasopressin and oxytocin genes in rats. They each consist of three exons interspersed by two introns. This pattern appears to be common to nearly all vertebrates. The oxytocin gene, however, lacks the nucleotide coding for the C-terminal glycopeptide. Genes coding for the neurohypophysial peptides have been described in nonmammals, including the vasotocin and mesotocin genes in a frog (Nojiri et al., 1987) and the vasotocin and isotocin genes in several species of teleost fish (Heierhorst et al., 1989; 1990; Morley et al., 1990; Urano, Kubokawa, and Hiraoka, 1994). The vasotocin genes, like that of vasopressin, contain three exons and two introns and also have an extended copeptin-like tail, which in teleosts and cyclostomes appears to lack the potential for glycosylation. The mesotocin gene is similar to that of oxytocin, also lacking the C-terminal glycopeptide sequence (Nojiri et al., 1987). The isotocin gene in the white sucker is unlike other genes for the neurohypophysial hormones as it lacks introns and consists of one exon (Urano et al., 1994). In the chum salmon it is also different, consisting of four exons and three introns (Kuno et al., 1996). These observations suggest that the isotocin gene has followed a different lineage to that of the other neurohypophysial hormones. The precursor of teleost isotocin has an extended copeptin-like C-terminus but as in the vasotocin precursor its glycosylation may not be possible. A duplication of the vasotocin and isotocin genes has been observed in several species of teleosts, especially salmonids, where it apparently reflects their tetraploidy. As a result, they may each possess two separately coded vasotocins and two isotocins. In each fish, the paired genes are not necessarily expressed equally (Hiraoka *et al.*, 1993). There is also conjecture as to whether the products of such duplicated genes may have different functions. Such hormone gene duplication occurs elsewhere among the vertebrates and it is not necessarily related to polyploidy.

On the basis of information about the structures of the neurohypophysial peptides, their genes and their precursors, tentative suggestions have been made about their evolution. Arginine-vasotocin is present in all the major groups of vertebrates, even the cyclostome fishes where it is the sole such peptide. It has, therefore, been proposed (Urano, Hyodo, and Suzuki, 1992; Urano *et al.*, 1994) that vasotocin is the archetypal "mother" neuro-hypophysial hormone that may have originated about 500 million years ago in the earliest vertebrates. The precursor molecules for vasotocin in the two groups of extant cyclostomes (the hagfishes and the lampreys) exhibit considerable differences. They only have an homology between their amino acid sequences of 47% (Suzuki *et al.*, 1995). The vasotocin precursor in the lamprey *Lampetra japonica* is more similar to that of mammalian vasopressin than to the vasotocin precursors in the hagfish, *Eptatretus*. The lamprey hormone would, therefore, appear to be in the main vertebrate phyletic line. (It was noted that the hagfish vasotocin precursors may be more similar to that of a molluscan peptide conopressin, see below.) Evolution of other related peptides may reflect a process of gene duplication followed by mutations, some of which were perpetuated. Such events could have provided the origin of isotocin, along with vasotocin in the holostean and teleost fish, and the plethora of oxytocin-like peptides in the elasmobranchs (Chauvet *et al.*, 1994). Mesotocin possibly evolved from isotocin, appearing initially in the lungfishes (Dipnoi) and has been perpetuated in the amphibians, reptiles, and birds. However, the organization of its precursor in the Australian lungfish *Neoceratodus forsteri* is similar to that of the tetrapod neurohypophysial hormones rather than that of isotocin (Hyodo, Ishii, and Joss, 1996). Mesotocin may, therefore, have been directly derived from the vasotocin gene (Urano *et al.*, 1994).

The earliest mammals, represented by the echidna (Prototheria), possess arginine-vasopressin and oxytocin, possibly reflecting the origins of the provasopressin-like and prooxytocin-like lines of evolution. The former is probably descended from arginine-vasotocin and the latter from mesotocin. Lysine-vasopressin, which is present in some members of the Suiformes (pigs) and some marsupials (kangaroos), may have evolved from arginine-vasopressin, an event that could have been preceded by gene duplication. Both

Fig. 3.10. A comparison of the domains for the prohormones for conopressin from a mollusc (*Lymnaea stagnalis*) and arginine-vasopressin-neurophysin II from mammals. The homologies between the amino acid sequences of paired domains (%) are given. AVP, arginine-vasopressin; CHO, glycosylation site; CP, conopressin; SP, signal peptide; Y, putative glycosylation site. (Based on van Kesteren *et al.*, 1992a.)

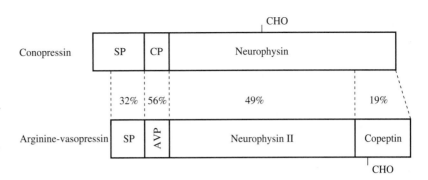

hormones have been observed together in some species. Mesotocin is also present in some marsupials and it has been postulated that it may have persisted separately or even been "reinvented." The construction of such evolutionary trees from information in contemporary species is obviously quite speculative and open to alternative suggestions.

Early studies, usually based on immunocytochemical interactions, suggested that some invertebrates, including molluscs and insects, may possess peptides that are similar to the vertebrate neurohypophysial hormones (Mizuno and Takeda, 1988). Modern molecular genetic techniques have confirmed such observations by the isolation and decoding of nucleotide sequences for such peptides and their precursors (van Kesteren *et al.*, 1992a,b). Thus, the central nervous system of a mollusc, the fresh water snail *Lymnaea stagnalis*, contains a nonapeptide, lysine-conopressin, that has a similar structure to the vertebrate hormones (Table 3.2). The precursor (preproconopressin) has also been described and has a remarkably similar structure to preprovasopressin (Fig. 3.10). Both of the neurophysins present in these molecules have 14 cysteine residues and there is a 49% correspondence in their amino acid sequences (41% to human oxytocin-neurophysin). A cephalopod mollusc, *Octopus vulgaris*, also possesses an oxytocin-like peptide called cephalotocin. Several other such peptides have been described in other invertebrates. These fascinating observations suggest that the neurohypophysial hormones may have had even earlier origins than previously envisaged. They could have arisen about 600 million years ago, before the divergence of the vertebrates.

The evolution of receptors for hormones is usually assumed to have occurred concurrently with the hormones themselves, but there is little information about this possibility. A start has been made for the neurohypophysial hormones by the isolation and cloning of the genetic material encoding such receptors in rats and humans. A model of the neurohypophysial hormone receptor has been constructed (Fig. 3.11). There are several types of such receptor. Vasopressin V_{1a} receptors mediate contraction of vascular smooth muscle and glycogenolysis in the liver; V_2 receptors are

Table 3.3. *Biologically active peptides and putative neurotransmitters in the median eminence of the rat*

Substance	Content (pmole/mg protein)
GnRH	19.0
TRH	110.0
Somatostatin	189.0
Vasopressin	717.0
Oxytocin	416.0
Norepinephrine	118.0
Dopamine	523.0
Epinephrine	3.4
5-Hydroxytryptamine	87.0
Histamine	160.0

Source: Based on Brownstein, 1977.

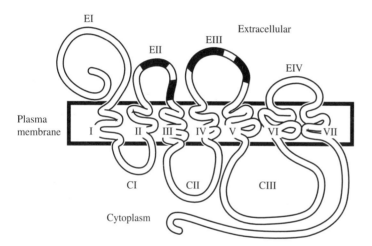

Fig. 3.11. Topography of the arginine-vasopressin V_2 receptor. The structure has been deduced from the nucleotide sequences of clones for receptor proteins from human and rat kidney. In the rat, the protein contains 370 amino acid residues. The receptor contains seven helical transmembrane-spanning domains (I–VII) and is a member of the G protein-coupled receptor superfamily. There are four extracellular domains (EI–EIV) and three cytoplasmic domains (CI–CIII). A conservation of amino acid sequences in EII and EIII (compared with related receptors for other neurohypophysial hormones) suggests that the interaction with vasopressin occurs in this region of the receptor. In other receptors of this superfamily the interactions with G proteins appear to occur between trans-membrane domains V and VI in the region of CIII. (Based on Sharif and Hanley, 1992; utilizing information from Birnbaumer *et al.*, 1992; Lolait *et al.*, 1992.)

present in the kidney and produce an increase in water permeability of the renal tubules. Oxytocin receptors mediate contraction of uterine muscle. Human and rat V_2 receptors have been cloned (Birnbaumer *et al.*, 1992; Lolait *et al.*, 1992) as have rat V_{1a} receptors (Morel *et al.*, 1992) and human oxytocin receptors (Kimura *et al.*, 1992). The sequences of the nearly 400 amino acid residues in these receptors are quite similar and they have been classified as members of the G protein-coupled receptor superfamily.

An exciting step into the unravelling of possible evolutionary changes that may have occurred in such receptors has recently been made following the cloning of vasotocin receptors from three species of teleost fish and an amphibian (Mahlmann *et al.*, 1994). The vasotocin receptor from the white sucker, *Catastomus commersonii*, contains 435 residues and is a typical mem-

Table 3.4. *Hypothalamic hormones believed to control the release of pituitary hormones*

Hypothalamic hormone (or factor)	Abbreviation
Corticotropin-releasing hormone	CRH, CRF
Thyrotropin-releasing hormone	TRH
Gonadotropin-releasing hormone	GnRH, LHRH, LH/FSH-RH
Growth hormone-releasing hormone	GH-RH
Growth hormone release-inhibiting hormone, somatostatin	GH-R-IH, somatostatin
Prolactin release-inhibiting hormone, dopamine	P-R-IH, dopamine
Prolactin-releasing factor, TRH, copeptin?	PRF
Melanocyte-stimulating hormone release-inhibiting hormone, dopamine, melanostatin (?)	MSH-R-IH

The roles of copeptin, TRH and melanostatin are not established in all species.
Based with numerous modifications on Schally, Arimura, and Kastin, 1973.
(Copyright © 1973 by the American Association for the Advancement of Science.)

ber of the G protein-coupled superfamily. Its amino acid sequence shows a 61% identity to the rat V_{1a} receptor, 42% to the rat V_2 receptor and 48% to the human oxytocin receptor. The teleost vasotocin receptor was expressed and studied in toad oocytes. It was shown to interact with vasotocin, arginine-vasopressin, oxytocin, mesotocin, and aspargtocin, but not with teleost iso-tocin. This observation appears to be consistent with separate roles for vasotocin and isotocin in teleosts. (Pharmacological studies on the pituitary of trout indicate that the affinity of isotocin for such receptors is about 800 times less than that of vasotocin (Pierson, Guibbolini, and Lahlou, 1996).) Two conopressins from molluscs also interacted with the teleost vasotocin receptor, suggesting an ancient familiarity between the two types of molecule.

The neurohormones of the hypothalamus

The hypothalamus at the base of the brain contains a host of biologically active materials, some of which act as neurotransmitters whereas others are released into the portal blood vessels that supply the adenohypophysis. These compounds are usually either amines or peptides, and they tend to accumulate in the median eminence, which is adjacent to the portal vessels. The concentrations of some of these substances in this region of the hypothalamus are shown in Table 3.3. There is physiological, pharmacological, and even chemical evidence that possibly seven or more such substances can be secreted and influence the release of hormones from the adenohypophysis (Table 3.4). Most of these hypophysiotropic hormones (or factors) appear to be peptides

(Vale, Rivier, and Brown, 1977) that are formed by special neurons in the hypothalamus. They may either increase (and thus have the suffix -*releasing hormone*, -RH) or decrease (-*release-inhibiting hormone*, -R-IH) the release of an adenohypophysial hormone. Two such hormones with opposite actions may regulate the secretion of a single pituitary hormone.

The amino acid sequences of five such hypophysiotropic hormones are known: thyrotropin-releasing hormone (TRH) which is a tripeptide; GnRH or luteinizing hormone (LH)-releasing hormone (LHRH or LH/FSH-RH), a decapeptide; somatostatin (growth hormone) (growth hormone-release-inhibiting hormone (GH-R-IH)), a tetradecapeptide; and two larger molecules corticotropin-releasing hormone (RH) and growth hormone-releasing hormone (GH-RH). These last two are 41 and 45 residue chains, respectively, in mammals. These hormones may exhibit actions other than that which their name implies.

Thyrotropin-releasing hormone

TRH is a tripeptide, pyroGlu–His–Pro–NH$_2$, a size that provides little opportunity for evolutionary experiments. Substituted analogs have little biological activity. One natural analog pyroGlu–Glu–Pro–NH$_2$ has been found in the pituitary glands and other tissues of several mammals and the domestic fowl (Harvey *et al.*, 1993). It may act as a natural paracrine antagonist of the effects of TRH. Using an immunoassay, TRH has been identified in the hypothalamus of a variety of nonmammals, including the domestic fowl, a reptile, an amphibian, a teleost fish, and even from the brain of a larval cyclostome and the head region of a protochordate amphioxus (Jackson and Reichlin, 1974). It is also interesting that TRH has been identified in other parts of the brain, apart from the hypothalamus, in both mammals and nonmammals, suggesting that it may have a more widespread physiological role. It has even been identified in frog skin, where it is present in very high concentrations (Jackson and Reichlin, 1977). Rather remarkably, TRH has also been found in alfalfa plants (Jackson, 1981b). The effects of TRH in stimulating the release of thyrotropin are often unpredictable in nonmammals, which could reflect differences in the structures of the homologous hormones or physiological differences in function. TRH may have other roles to play. For instance, it has been shown to promote the release of growth hormone in a wide range of vertebrates, especially mammals and birds and, *in vitro*, in amphibians and fish (Harvey, 1990). It also has a well-known prolactin-releasing effect in mammals (Neill and Nagy, 1994).

Gonadotropin-releasing hormone

GnRH (LHRH) has been identified throughout the vertebrates, from cyclo-

Fig. 3.12. The amino acid sequences of vertebrate GnRH. The structure of mammalian GnRH (mGnRH) is given at the top of the figure; (–) indicates an amino acid identical to that in mGnRH. It should be noted (see the text) that each peptide described is not necessarily confined to the phyletic group that its name indicates. Mammalian GnRH is, for instance, also found in some amphibians and teleosts, and chicken GnRH II is present throughout the vertebrates. (Based on information provided in Powell *et al.*, 1994.)

	1	2	3	4	5	6	7	8	9	10
Mammal, mGnRH	pGlu	His	Trp	Ser	Tyr	Gly	Leu	Arg	Pro	Gly-NH$_2$
Chicken I, cGnRH I	-	-	-	-	-	-	-	Gln	-	-
Chicken II, cGnRH II	-	-	-	-	His	-	Trp	Tyr	-	-
Salmon, sGnRH	-	-	-	-	-	-	Trp	Leu	-	-
Catfish, cGnRH	-	-	-	-	His	-	-	Asn	-	-
Dogfish, dfGnRH	-	-	-	-	His	Asp	Trp	Lys	-	-
Lamprey I, lGnRH I	-	-	Tyr	-	Leu	Glu	Trp	Lys	-	-
Lamprey III, lGnRH III	-	-	-	-	His	Asp	Trp	Lys	-	-

stome fishes to mammals (King and Millar, 1992; Sherwood, Lovejoy, and Coe, 1993; Millar and King, 1994). It promotes the release of the gonadotropins LH and FSH and so is often called LHRH, but it may have other paracrine or even neural roles in such organs as the gonads, placenta and brain. Nine variants of this decapeptide have been identified in vertebrates (Fig. 3.12). They are quite similar, their lowest homology to each other, involving lamprey GnRH, being about 50%. They are considered to all be members of a single hormone family. The amino acids at position 1, 2, 4, 9, and 10 are invariant between species, apparently reflecting their essential roles in the hormone's mechanism of action. Eutherian mammals (with a single exception) appear to possess only a single GnRH, mGnRH, but others, including the marsupials, have two or sometimes even three such peptides (Sower *et al.*, 1993; Powell *et al.*, 1994). Among the teleosts, three such GnRHs have been identified in the South American sabalo (*Prochilodus lineatus*) and the Eastern Pacific grass rockfish (*Sebastes rastrelliger*) (Somoza *et al.*, 1994; Powell *et al.*, 1996). Usually such GnRHs are found at different sites, such as the hypothalamus, pituitary, different regions of the brain, and the placenta. They may each have different roles at such sites. However, two can be present together in the same neuron and even in the same storage granules, as seen in an African catfish, *Clarias gariepinus* (Schultz *et al.*, 1993). Chickens have two different GnRHs: cGNRH I and cGnRH II. The cGnRH II has also been found in representatives of vertebrates extending from elasmobranch fishes to marsupials, but not in eutherians. (Such hormones are named according to the species in which they were first identified. Thus, cGnRH I and II were originally characterized from chicken hypothalami.) The lamprey GnRHs are quite different to those in other vertebrates; lGnRH I has five amino acid substitutions compared with mGnRH (Fig. 3.12) and it has little effect in other species. It has been suggested (Fig. 3.13) that a gene duplication occurred prior to the divergence of the elasmobranchs, resulting in two main lines of evolution: one forming the cGnRH II line, and the other mGnRH and some similar molecules including, cGnRH I and salmon GnRH (sGnRH). Phylogenetic analyses based on the amino acid sequences

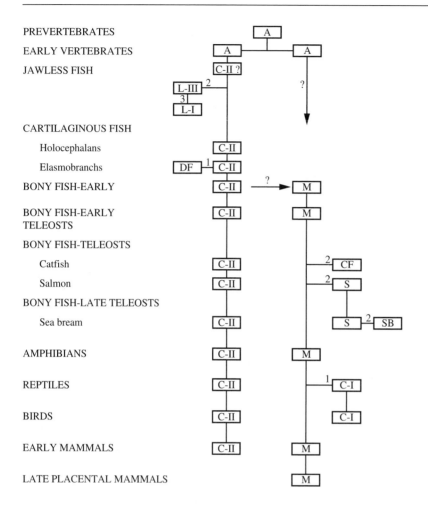

Fig. 3.13. A hypothetical evolutionary pathway for GnRH based on the phyletic distribution of the amino acid sequences of the hormone in extant species. The GnRHs are coded as follows: A, ancestral form; C-I, chicken; C-II, chicken II; CF, catfish; DF, dogfish; L-I, lamprey; L-III, lamprey III, S, salmon; SB, seabream. The numbers indicate the differences in amino acids between the pairs indicated. (From Sherwood et al., 1994.)

of such small peptides are considered to be of limited reliability (Dores *et al.*, 1996). However, such an analysis based on the sequences of the prepro-GnRHs from 16 species supports an ancestral role for the cGnRH II gene (Grober *et al.*, 1995).

An African teleost fish, *Haplochromis burtoni*, possesses sGnRH and cGnRH II. Each is encoded by a distinct gene. The two precursors were found (White *et al.*, 1994) to lack a distinct homology, suggesting a separate evolution or divergence even before the origin of the teleost fish. The ancestral gene may be traced back to even more lowly origins. A yeast has been shown to possess a mating pheromone called α-factor, which is a tridecapeptide. It exhibits an 80% homology in its amino acid sequence to mGnRH and can even stimulate the release of mammalian LH (Loumaye, Thorner, and Catt, 1982).

The receptors for GnRH are members of the G protein-coupled superfamily (Millar *et al.*, 1996). The mouse, rat, sheep, and human receptors have been cloned and consist of a polypeptide chain with extracellular and cyto-plasmic regions that characteristically, crosses the plasma membrane seven times. There is an homology of about 85% between the sequences of the 328 amino acid residues present in each receptor. Pharmacological observations indicate that the GnRH receptors in mammals are very selective with respect to the particular GnRHs that can trigger their responses (Millar *et al.*, 1996). In contrast, in nonmammals, these receptors are quite promiscuous in their behavior and they readily interact with a variety of GnRHs. However, at this time no direct information is available about the structures of GnRH recep-tors in nonmammals.

Somatostatin

Somatostatin was originally identified in the mammalian hypothalamus where it was found to inhibit the release of growth hormone. It was, therefore, called growth hormone- release-inhibiting hormone (GH-R-IH) but it is now usually referred to as somatostatin. It occurs in a variety of tissues including the gut and pancreas, where it has a role in inhibiting gastric secretions and the release of insulin and glucagon. Somatostatin has been identified in the brain of representatives of all the main groups of vertebrates, including cyclostome fish (King and Millar, 1979a), it has even been found, using immunocytochemical procedures, in two species of pond snails (Grimm-Jørgensen, 1983) and a plant (spinach) (LeRoith *et al.*, 1985; 1986). Somatostatin is clearly an ancient hormone but the chemical structures of the somatostatins that occur in different vertebrates indicate that little change has occurred in its basic configuration (Andrews *et al.*, 1988; Conlon, 1990; Nishii *et al.*, 1995). However, several variants exist in which N-terminal fragments of different length are retained from the precursor as parts of the secreted hormone. Considerable phyletic variation may occur in the amino acid sequences of these N-terminal appendages. The prototype somatostatin is a cyclic tetradecapeptide called somatostatin-14 (Fig. 3.14). Its structure includes a disulfide bridge linking two cysteine residues at positions 4 and 14. The integrity of this section of the molecule is vital for its activity and it is strongly conserved.

A variant of somatostatin with a widespread distribution and which occurs in the endocrine pancreas (D-cells) and intestinal epithelial cells is somato-statin-28. It is identical to somatostatin-14 but it inherits a 14 residue extension of its N-terminal from the precursor molecule. It is possible that somatostatin-28 is the circulating hormone while somatostatin-14 fulfills the molecule's paracrine roles (Martin and Faulkner, 1996). Other such variants have been described and include somatostatin-25, which occurs in eels and salmon, and somatostatin-34 and somatostatin-37 in cyclostomes (Conlon, 1990). Such differences reflect genetic variations in the processing of the

	1	2	3	4	5	6	7	8	9	10	11	12	13	14
Mammalia Pig, sheep	Ala	Gly	Cys	Lys	Asn	Phe	Phe	Trp	Lys	Thr	Phe	Thr	Ser	Cys
Amphibia Frog	-	Pro	-	-	-	-	-	-	-	-	-	-	Met	-
Teleostei Anglerfish II	-	-	-	-	-	-	Tyr	-	-	Gly	-	-	-	-
Chondrostei Sturgeon	-	Pro	-	-	-	-	-	-	-	-	-	-	-	-
Holocephali Ratfish	-	-	-	-	Ser	-	-	-	-	-	-	-	-	-
Cyclostomata Lamprey	-	-	-	-	-	-	-	-	-	-	-	Ser	-	-

Fig. 3.14.   The amino acid sequences of somatostatin-14 from different vertebrates. (–) indicates identical amino acids. A second somatostatin-14 has been identified in anglerfish and lampreys that is identical to the mammalian hormone. (Based on information presented by Hobart *et al.* (1980) and Nishii *et al.* (1995.)

precursor molecules. The basic activity of such variants appears to reside in their somatostatin-14 segments, but the appendages may also contribute. For instance, the stability of the molecule in the circulation may be enhanced.

Few amino acid substitutions, compared with the structure of mammalian somatostatin-14, have been identified among the vertebrates (Fig. 3.14). Differences have been observed at position 12 in a cyclostome (a lamprey), position 2 in a chondrostean (the Russian sturgeon), position 5 in a chondrichthyean (the ratfish) and position 13 in an amphibian (a frog) (Nishii *et al.*, 1995). In the anglerfish, two somatostatin-14 hormones with separate precursors are encoded by two genes (Goodman *et al.*, 1980; Hobart *et al.*, 1980). The somatostatin-I in this teleost fish has the same amino acid sequence as mammalian somatostatin. The somatostatin-II has two amino acid substitutions, at positions 7 and 10, and it has been identified in other teleosts but not elsewhere. Lampreys also possess two somatostatin-14 peptides, one of which is identical to the mammalian hormone. Considering what appears to be a long phylogenetic history, the structure of somatostatin-14 has been remarkably well conserved, probably reflecting little flexibility in the process of its interactions with its receptors.

Growth hormone-releasing hormone

The presence of a GH-RH in the hypothalamus was suspected for over 20 years but it evaded isolation. It was finally found, not in the hypothalamus, but in a human pancreatic tumor where its gene was also being expressed (Guillemin *et al.*, 1982; Rivier *et al.*, 1982). Three variants have since been identified in the mammalian hypothalamus: a 44, 40, and 37 residue polypeptide known, respectively as GH-RH-(1–44)NH$_2$, GH-RH-(1–40)OH, and GH-RH-(1–37)OH. These peptides have similar biological activities, the differences reflecting post-translational processing (Mayo *et al.*, 1983). GR-RH belongs to the secretin–glucagon family of hormones (see below).

Such growth hormone-releasing factors have been identified in several species of mammals, birds, reptiles, amphibians, and teleost and chondrostean fish (Harvey, 1993; Sherwood *et al.*, 1994). Pig and human GH-RH differ by only three amino acid substitutions but mouse and human differ by 18 such changes (Sherwood *et al.*, 1994). Teleost and chondrostean GH-RH-like peptides contain 45 amino acid residues. The salmon and carp peptides differ from each other by four residues and the salmon and sturgeon by six. Homologies of the amino acid sequences between the human and fish peptides are about 40%. The genes for human GH-RH and salmon GH-RH-like peptide have been described (Sherwood *et al.*, 1994). The human gene only encodes GH-RH but the salmon gene also encodes a peptide containing 38 amino acid residues that has been given the acronym PACAP. It has been identified in the mammalian pituitary (Miyata *et al.*, 1989), including the hypophysial portal system, and, *in vitro*, activates adenylate cyclase activity in this tissue (hence *p*ituitary *a*denylate *c*yclase *a*ctivating *p*eptide) (Rawlings and Hezareh, 1996). The PACAP has no defined physiological role but *in vitro* it can promote the release of and modulate the responses to growth hormone, corticotropin, and LH. It has also been found in the pituitary of the frog *Rana ridibunda*, where it only differs by one amino acid from human, sheep, and rat PACAP (Chartrel *et al.*, 1991). The PACAP structures in bony fish have a remarkable 89–92% homology in deduced amino acid sequences to the human peptide. However, human PACAP, in contrast to that in fish, is encoded by a separate gene to that of GH-RH. It is possible that, like in contemporary fish, the ancestral gene for GH-RH also encoded PACAP and following subsequent gene duplication there was a separation of the encoding exons onto different genes. However, several ancestral scenarios are possible (Sherwood *et al.*, 1994).

Corticotropin-releasing hormone

The elucidation of the structure of CRH was initially quite elusive. However, following the collection and extraction of the hypothalami from many thousands of sheep it was isolated and found to be a peptide containing 41 amino acid residues (Vale *et al.*, 1981). The ovine hormone differs from human and rat CRH (which are identical) by seven amino acid substitutions (Esch *et al.*, 1984). Ovine and bovine CRH differ by a single amino acid residue. Corticotropin-releasing factors (CRF) with biological activity and immunocytochemical interactions similar to mammalian CRH have been identified in many vertebrates (Lederis, 1987). The amino acid sequence of a CRF has been predicted from its encoding nucleotides in an amphibian (the toad *Xenopus laevis*), and only differs from human CRH by three amino acid substitutions (Stenzel-Poor *et al.*, 1992). The structure of a CRF has also been described in a teleost fish, the white sucker (Lederis, 1987). It only differs from the rat and human CRH by two amino acid substitutions, which is a remarkable example of conservation of a molecule since the evolutionary divergence of the teleosts about 300 million years ago. CRH has distinct similarities to another neuropeptide, urotensin I, which is present in teleost fish (p. 103). It also bears homologies to a peptide called sauvagine, which has been isolated from the skin of some frogs. Another related peptide, which also interacts with CRH receptors, has been identified in the brain of rats (Vaughan *et al.*, 1995). Immunological evidence suggests that it may also be present in other tissues. It contains 40 amino acid residues, the sequence of which bears a 45% homology to rat CRH, a 63% homology to teleost urotensin I and a 35% homology to amphibian sauvagine. It has been called *urocortin.* Urotensin I, CRH (or CRF) and urocortin have been classified as fellow members of a peptide family. They also exhibit similarities to a series of peptides found in insects (the diuretic peptides, DP I and DP II) (Lederis *et al.*, 1994). An evolutionary relationship between such peptides appears to be tenuous.

Arginine-vasopressin can also initiate the release of corticotropin and this activity resulted in some problems during the early identification of CRH as a distinct hormone. Each peptide appears to affect different aspects of the release of corticotropin and may act on different corticotrope cells (Jia *et al.*, 1991). Vasopressin may mediate stress-related release of corticotropin while CRH controls its basal and diurnal rhythmical secretion. In view of these functional relationships of the two hormones it is interesting to observe that the sequences for the precursor of ovine CRH and that of vasopressin-neurophysin II exhibit sufficient similarities to suggest a possible evolutionary relationship (Furutani *et al.*, 1983).

A corticotropin release-inhibiting factor (CRIF) has been identified in the mammalian hypothalamus (Redei, Hilderbrand, and Aird, 1995). Its physiological role has not been defined but it is a peptide containing 22 amino acid residues that shares its precursor with TRH. It is preproTRH-(178–199).

Dopamine

Dopamine is a catecholamine and a nerve transmitter that in the hypothalamus is released into the hypophysial portal vessels and travels to the pituitary where it inhibits the release of prolactin. It thus functions at this site as a prolactin release-inhibiting hormone (MacLeod and Lehmeyer, 1974; Neill and Nagy, 1994). The release of prolactin also appears to be influenced by a releasing factor (a PRF), but despite the identification of several putative candidates its identity is still equivocal. TRH can promote the secretion of prolactin and is often favored. Various other peptides have been found to promote release of prolactin under experimental conditions. Currently there is interest in the 39 amino acid residue tail fragment from the precursor for vasopressin-neurohypophysin II (called copeptin, see above) (Nagy *et al.*, 1988) and a substance present in the pars intermedia, probably α-MSH (Frawley, 1994; Zheng *et al.*, 1997). Dopamine may also influence the release of other hypophysial hormones. In the goldfish, it can inhibit the release of GnRH (de Leeuw *et al.*, 1989) and it also promotes the release of growth hormone in these fish (Wong, Chang, and Peter, 1992). As teleosts lack a hypophysial portal system, such effects appear to reflect its role as a neuro-transmitter, when its action would be a paracrine-like one. Dopamine also inhibits the release of MSH from the pars intermedia, where its role is also a neural one. Apart from prolactin, dopamine has been observed experimen-tally to influence the release of other pituitary hormones in mammals, but a consensus as to the physiological significance of such observations has not been reached.

Paracrine and autocrine modulation of the hypothalamo–hypophysial system

Apart from the established hormones of the hypothalamus, a variety of other biologically active molecules may influence the activities of the pituitary gland. Such modulation may affect the thresholds for the release of the hypophysiotropic hormones and the sensitivity of such target cells as the somatotropes and lactotropes. These modulators are produced at local sites and may exert paracrine and autocrine effects. Two of the most notable such modulators are *nitric oxide* (NO) and *galanin*. However, a number of pep-tides, which are described in subsequent sections, can also interact with and influence the release of the hypothalamo–hypophysial hormones. These include the endorphins, which can decrease release of vasopressin and oxy-tocin, and, possibly, GnRH. Angiotensin II has the opposite effect, promo-ting the release of vasopressin and oxytocin, while neuropeptide Y (NPY) can increase the release of gonadotropins. Melanocyte-concentrating hormone (MCH) can decrease the release of MSH in amphibians. It should be emphasized that many such effects have only been demonstrated under pharmacological conditions.

Nitric oxide

There is currently a burgeoning interest in the actions of nitric oxide in the endocrine system (Costa *et al.*, 1996). These effects are somewhat unexpected as it is better known as a gas. However, in the body, nitric oxide may be present as a free radical or as a nitrosothiol compound (Grossman, 1994). Nitric oxide is a powerful vasodilator, which provided a clue to its discovery as a natural product of the vascular endothelium (Palmer, Ferrige, and Mon-cada, 1987). It is produced under the influence of nitric oxide synthase

(NOS) from L-arginine and can activate several cytoplasmic enzymes, including guanylate cyclase. Nitric oxide synthase has been identified in many tissues, including the hypothalamus and pituitary gland. For instance, nitric oxide is produced by the pituitary gonadotrope cells (Ceccatelli *et al.*, 1993). At this site, it blocks the action of GnRH and thus the release of LH. This action is an autocrine one, but nitric oxide also plays a paracrine role in the pituitary as it is formed by the folliculo-stellate cells, which are neighbours of the somatotropes. It can, therefore, also inhibit the release of growth hormone by modulating the action of GH-RH. Nitric acid synthase activity has also been identified in the hypothalamic supraoptic nuclei and it accumulates in the neurohypophysis (Kadowaki *et al.*, 1994). At this site, the nitric oxide appears to inhibit release of vasopressin and oxytocin. This nitrinergic system may have even more widespread effects on endocrine glands, the repercussions of which are only starting to be understood.

Galanin

Galanin is a linear polypeptide containing 29 amino acid residues (30 in humans) that was originally isolated from pig intestine (Tatemoto *et al.*, 1983). It was found to contract, *in vitro*, various rat smooth muscle preparations and produced a hypoglycemia in dogs. Galanin was named from its N-terminal glycine and C-terminal alanine amino acids. It is formed in nerve cells in various tissues in addition to the gut, including brain, adrenal medulla, gonads, pancreas, hypothalamus, and pituitary (Shine, 1994). Its physiological role is not yet precisely defined. Galanin has been colocalized in nerve cells with either dopamine or GH-RH (Meister and Hokfelt, 1988), and it has been observed in somatotropes, thyrotropes, lactotropes, and corticotropes. Currently it seems likely that it promotes the secretion of growth hormone by enhancing the release of GH-RH (Ottlecz, Snyder, and McCann, 1988), and decreasing that of somatostatin (Liposits *et al.*, 1993). However, many other possibilities remain to be investigated. Galanin may act in other tissues, such as the endocrine pancreas where it decreases the release of insulin, and the ovary, where it promotes the action of LH and synthesis of steroid hormones (Fox *et al.*, 1994).

Galanin has been identified in several mammals: humans, pigs, rats, rabbits, cattle, and dogs (Boyle *et al.*, 1994). It has also been found in the domestic fowl, alligator, a frog, trout, and bowfin (Chartrel *et al.*, 1995). The first 15 residues at the N-terminal are invariant in these species and may constitute the molecule's active core. In the remainder of the molecule, changes have been observed at eight positions including an additional amino acid (to make a total of 30) in dogs and humans. Overall homologies in the amino acid sequence are about 80% for human versus the rat and pig, and 79% for pig versus trout and domestic fowl. The levels of homology of the molecules in different species, especially at positions 1 to 15, are consistent with an important physiological role for galanin, but at present this has only started to be unravelled.

The caudal neurosecretory peptides: the urotensins

The teleostean urophysis and chondrichthyean Dahlgren cells have been described earlier (Chapter 2). The neurosecretory activities of these cells is related to the presence of a variety of peptides (Bern *et al.*, 1985; Lederis *et al.*, 1985). They are called urotensins, which reflects their initial characterization by their abilities to alter the contractility of various smooth muscle preparations. Urotensin I can decrease blood pressure of rats and some other species. Urotensin II contracts the trout urinary bladder and can increase blood pressure in eels and dogfish (Hazon, Bjenning, and Conlon, 1993) but decreases it in rats (Conlon *et al.*, 1996).

Urotensin I from carp has a precursor molecule containing 145 amino acid residues from which it is cleaved as a 41 residue peptide (Ishida, Ichikawa, and Deguchi, 1986). Urotensin I and its precursor have structural homologies to CRH. Thus, the precursor of urotensin I and ovine CRH have a 51% identity, suggesting that they could share a common ancestry. Urotensin I can promote the release of corticotropin in both mammals and fish. It has also been identified in the brain of teleosts but not that of tetrapod vertebrates (Lederis *et al.*, 1994). However, it is present in elasmobranchs. The giant Dahlgren nerve cells in the spinal cord of these fish have long been suspected as being homologous to the teleostean urophysis (Chapter 2). A 41 residue peptide has been identified in such tissue from the dogfish *Scyliorhinus canicula* (Waugh *et al.*, 1995a). It has a homology in its amino acid sequence of 51% compared with urotensin I from a teleost (the white sucker) and 56% compared with human CRH.

The structural homologies between CRH or CRF and urotensin I and their precursors are consistent with their membership of a common protein family or superfamily (Lederis *et al.*, 1994). Therefore, they could have shared a common ancestral gene. Such a gene may have been CRF-like, urotensin I-like, or CRH–urotensin I-like. Subsequently, each peptide could have been separately proliferated following gene duplication. Comparison of the structural organization of the extant CRF gene and urotensin-I gene could facilitate such conjecture. Based on the 34% homology of the precursors of CRF and urotensin I in the white sucker, it has been estimated that divergence of such genes could have occurred 660 million years ago (Lederis *et al.*, 1994). Such a projection would be consistent with a common ancestry in an invertebrate.

Urotensin II exists in multiple forms and more than one of these may exist in a single species, probably reflecting the presence of duplicated genes

	1	2	3	4	5	6	7	8	9	10	11	12
Cyclostomata Lampreys	Asn	Asn	Phe	Ser	Asp	Cys	Phe	Trp	Lys	Tyr	Cys	Val
Elasmobranchii Dogfish, Ray	-	-	-	-	-	-	-	-	-	-	-	-
Chondrostei Sturgeon, Paddlefish	Gly	Ser	Thr	-	Glu	-	-	-	-	-	-	-
Teleostei Sucker (II)	Gly	Ser	Asn	Thr	Glu	-	-	-	-	-	-	-
Trout	Gly	Gly	Asn	Ser	Glu	-	-	-	-	-	-	-
Carp (Beta)	Gly	Gly	Asn	Thr	Glu	-	-	-	-	-	-	-
Flounder	Ala	Gly	Thr	Thr	Glu	-	-	-	-	-	-	-
Goby	Ala	Gly	Thr	Ala	-	-	-	-	-	-	-	-
Amphibia Frog	Ala	Gly	Thr	* / -	Glu	-	-	-	-	-	-	-

Fig. 3.15. Amino acid sequences of urotenson II from different species of fish and the brain of a frog. (–) indicates an identical amino acid. The amphibian peptide contains 13 residues so that a "gap" has been inserted (*) for the purpose of alignment of the molecules, between positions 3 and 4. It is occupied by leucine. (Based on information given by Waugh and Conlon (1993) Waugh *et al.* (1995b).)

(Ohsako *et al.*, 1986). It consists of 12 amino acid residues incorporating a disulfide ring linking positions 6 and 11 (Fig. 3.15). At least 14 different natural analogs have been described in teleosts, elasmobranchs, a chondrostean, and the brain of cyclostomes (McMaster *et al.*, 1992; Waugh *et al.*, 1995b). Elasmobranch and cyclostome forms of urotensin II are identical and differ from the chondostean peptide by four amino acid substitutions. It was especially interesting to observe that one such variant was present in the brain of a frog (Conlon *et al.*, 1992). In fish, urotensin II has also been identified in the brain of several species (Waugh and Conlon, 1993). Four variants of urotensin II have been found in the urophysis of the carp and two in the white sucker. This duplication appears to reflect the presence of multiple genes, probably caused by polyploidization, which is not uncommon in teleosts. Amino acid substitutions have been identified at all positions in the molecule except with the disulfide ring, positions 6 to 11. Cleavage of the disulfide bond destroys the molecule's activity and the conservation of this section of the peptide appears to reflect its biological importance. The physiological role of urotensin II is uncertain but it may be concerned with piscine osmoregulation (Chapter 8). Its persistence in many species, in several variant forms, suggests that it has a physiological role, but this remains to be defined.

It has been noted that urotensin II bears some structural similarities to somatostatin-14. However, there does not appear to be any crossover in their biological activities and the precursors of the peptides show no evidence of a common evolutionary origin (Ohsako *et al.*, 1986). The precursors of urotensin I and urotensin II are also unrelated.

Fig. 3.16. The amino acid sequence of MCH from teleost fish and mammals. (–) indicates identical amino acids. (Based on information in Baker (1994).)

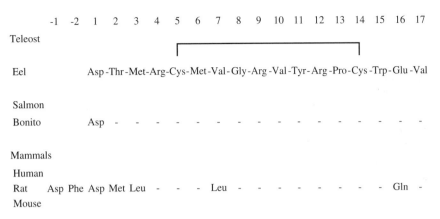

Immunocytochemical and radioimmunoassay techniques have identified urotensin I- and II-like peptides in the cerebral ganglia of the sea hare *Aplysia californica* (Gonzalez *et al.*, 1992). The presence of such substances in this mollusc could reflect an ancestral origin of the urotensins in an invertebrate.

Melanin-concentrating hormone

Melanin-concentrating hormone (MCH) was named because of its ability to bring about the aggregation of melanin in the melanophores in the scales of teleost fish (Rance and Baker, 1979). The fish thus become pale in color (Chapter 7). This hormone is released from the neurohypophysis of teleost fish, where it is present as a product of neurosecretion from nerve cells that extend up into the hypothalamus. Salmon MCH has been purified (Kawauchi *et al.*, 1983) and found to be a cyclic heptadecapeptide (Fig. 3.16). A similar peptide has been identified in the hypothalamus of rats (Vaughan *et al.*, 1989). This homolog of MCH is a nonadecaptide that apart from the additional two amino acid residues present at its N-terminus, has six amino acid substitutions compared with the salmon peptide. The MCH forms in rats, mice, and humans are identical. Salmon and bonito MCH only differs from that in eels by a single amino acid substitution, the former two being the same. Such evolutionary persistence suggests that MCH has functions other than just contributing to the color change mechanism in teleost fish. Neurons containing MCH-like immunoreactivity are widespread in the brain of vertebrates ranging phyletically from the cyclostome fish to humans (Baker and Rance, 1983). Such immunoreactivity has also been identified in the spleen, intestine, and testis. In salmon, there are two genes encoding MCH (Kawauchi, 1989) and the precursors not only contain MCH but two other peptides in tandem. It seems likely that such a neuropeptide assemblage has a role in neurotransmission or neuromodulation, but information about such possibilities is fragmentary. MCH can inhibit secretion of α-MSH and CRH in fish but apparently not in mammals. It is released in response to stress.

When injected into the brains of rats and mice it promotes feeding behavior (Qu *et al.*, 1996; Rossi *et al.*, 1997). MCH is an interesting example of how a biologically active substance was discovered first as a hormone in a lower vertebrate but was later found to be present higher up the phyletic scale. Its precise role there, however, has yet to be discovered.

The renin–angiotensin system

It has been known since the turn of the century that saline extracts of the mammalian kidney, when injected into mammals, produce a large increase in the blood pressure. This effect results from the interaction of an enzyme present in the kidney called renin, which, as described in the last chapter, is formed by the juxtaglomerular cells. Renin interacts with an α_2-globulin in the blood plasma to form angiotensin I, which is converted by a "converting" enzyme to angiotensin II, which is the hormone that actively constricts the peripheral blood vessels.

The substrate angiotensinogen contains 453 amino acid residues and is mainly formed in the liver. It is a member of a superfamily of proteins that includes antithrombin III, α_1-antitrypsin, and chicken ovalbumin (Doolittle, 1983). These proteins have an 18–23% homology of their amino acid sequences and as several members are serine protease inhibitors it is called the serpin superfamily. The utilization of one of these proteins as a hormone precursor is an interesting and, for this group, an apparently unique circumstance.

Angiotensin has other roles apart from its effect on vascular smooth muscle. For instance, it has a steroidogenic action on adrenocortical tissue. This effect involves aldosterone and corticosterone synthesis in tetrapods and lungfish (Joss *et al.*, 1994) as well as that of cortisol in bony fish and 1α-hydroxycorticosterone in elasmobranchs (O'Toole *et al.*, 1990). All of the components of the renin–angiotensin system have been identified in the brain where it may exhibit various effects of a paracrine nature (Mosimann, Imboden, and Felix, 1996). These actions involve an increased release of pituitary hormones, including vasopressin, LH, and prolactin (Steele, 1992). It may also act in the central nervous system to promote drinking, which appears to be a physiological response in mammals and has also been observed in many other vertebrates, even fishes (Nishimura, 1987).

Renin has a wide phyletic distribution among the tetrapods and also among teleost and elasmobranch fishes (Wilson, 1984; Uva *et al.*, 1992), but apparently, it is not present in the cyclostomes. Renin is a glycoprotein (Peach, 1977), which in mammals consists of 340 amino acid residues. It is stored with prorenin in the juxtaglomerular cells in the kidney. Both renin and prorenin are released into the plasma. Renin is an enzyme that could be considered also to have a hormonal status. Several isorenins have been

Fig. 3.17. The structure and synthesis of the angiotensins, shown in relation to that of their precursor angiotensinogen. A decapeptide (angiotensin I) is cleaved from the latter under the influence of renin, and this peptide is subsequently further broken down (or "activated") under the influence of "converting enzyme", to form an octapeptide (angiotensin II), and then a peptidase to form the heptapeptide angiotensin III. (From Peart, 1977.)

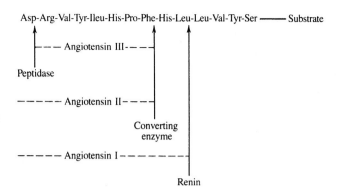

identified in other tissues, including the brain, uterus, and submaxillary gland, and, in teleosts, the corpuscles of Stannius.

Angiotensin I is a decapeptide (Fig. 3.17) that is the cleavage product resulting from the action of renin on angiotensinogen. It has little activity in this form but it is activated by a "converting enzyme" that removes the two C-terminal amino acids to produce the octapeptide angiotensin II. This peptide can be partially inactivated by an angiotensinase that removes the N-terminal aspartic acid, resulting in the heptapeptide called angiotensin III or angiotensin-(2–8). The last only has about 25% of the vasoconstrictor activity of angiotensin II but retains its parent's ability to stimulate the synthesis of aldosterone in mammals (Blair-West *et al.*, 1971). Other products of the metabolism of angiotensin II are angiotensin-(3–8) and angiotensin-(1–7). The latter heptapeptide has little effect on the blood pressure of rats but, in contrast to angiotensin II and III, *increases* urinary salt excretion apparently through blockade of the reabsorption of sodium from the renal tubule (Handa, Ferrario, and Strandhoy, 1996). The hexapeptide angiotensin-(3–8) is also called angiotensin IV (Wright *et al.*, 1995). Receptors for this peptide have been identified in the brain and also in peripheral tissues. It may be concerned with the acquisition of memory and the control of regional blood flow. However, the possible physiological roles of angiotensin-(1–7) and angiotensin IV remain to be defined.

Renin activity has been identified in the kidney tissue of species ranging from elasmobranchs to mammals (Nolly and Fasciola, 1973; Uva *et al.*, 1992). However, variations in the cross-reactions between renin from different species and angiotensinogen suggests that there are differences in its chemical structure. Teleost renin will interact with angiotensinogens of all tetrapods except mammals. Elasmobranch renin, however, also reacts with mammalian angiotensinogen (Henderson *et al.*, 1981). Mammalian renin, by comparison, will not react with substrates from nonmammals. Amphibian renin interacts with angiotensinogen from birds and reptiles, while reptilian renin reacts with those from birds and amphibians.

Amino acid position	1	2	3	4	5	6	7	8	9	10
Mammals										
Human	Asp	Arg	Val	Try	Ile	His	Pro	Phe	His	Leu
Cattle	-	-	-	-	Val	-	-	-	-	-
Bird										
Domestic fowl	-	-	-	-	Val	-	-	-	Ser	-
Reptiles										
Turtles	-	-	-	-	Val	-	-	-	His	-
Alligator	-	-	-	-	Val	-	-	-	Ala	-
Amphibian										
Bullfrog	-	-	-	-	Val	-	-	-	Asn	-
Teleosts										
Salmon	Asn	-	-	-	Val	-	-	-	Asn	-
Eel	Asn	-	-	-	Val	-	-	-	Gly	-
Goosefish	Asn	-	-	-	Val	-	-	-	His	-
Elasmobranch										
Dogfish	Asn	-	Pro	-	-	-	-	-	Gln	-

Fig. 3.18. The amino acid sequences of angiotensin I from different vertebrates. (–) indicates identical amino acids. (Based on information given by Takei *et al.* (1993).)

While there is no direct information about the structures of renins in nonmammals, the amino acid sequences of angiotensin I from at least nine species ranging from elasmobranchs to birds have been described (Fig. 3.18). Among mammals, the structures of angiotensin I are identical in humans, rats, pigs, and horses. Cattle angiotensin I differs from these peptides by a single amino acid substitution: valine replaces isoleucine at position 5. In the nonmammals, such substitutions have been observed at positions 1, 3, 5, and 9. Positions, 2, 4, 6, and 8 are invariant and they are considered to be important for the binding of the hormone to its receptor. In fish, asparagine replaces the aspartic acid which is present at position 1 in tetrapods. Such angiotensins are often referred to as fish angiotensin and tetrapod angiotensin. Seven different amino acid substitutions have been observed at position 9, which although not a part of the active octapeptide is involved in the action of the converting enzyme. This enzyme would, therefore, not appear to be very fastidious. The most notable variation, which only occurs in the dogfish, is the substitution of proline for valine at position 3. It has been predicted that this change could alter the tertiary structure of the molecule and have important effects on its biological activity (Takei *et al.*, 1993). The significance of this change, however, remains to be studied. Angiotensin II, nevertheless, can be considered to be a well-conserved hormone.

Angiotensin receptors have been identified in various tissues (including blood vessels, brain, muscle, adrenal, intestine, and the gills of teleost fish) in vertebrates ranging phyletically from teleost fish to mammals. In the latter, there are at least four subtypes (AT_{1A}, AT_{1B}, AT_2, and AT_4) (Wright *et al.*, 1995; Mosimann *et al.*, 1996). The AT_1 and AT_2 receptors contain about 360 amino acid residues and possess the characteristic serpentine structure of the

G protein receptor superfamily, which cross the plasma membrane seven times. The AT_4 receptors are larger and may consist of two or three subunits. In rats, the AT_{1A} and AT_{1B} receptors exhibit an identity of 95% in their amino acid sequences and they have an homology of 33% to the AT_2 receptors. Such structural variations are reflected in their different abilities to interact with angiotensins I, II, III, and IV, and with several artificial antagonist drugs. Angiotensin receptors have been identified in a bird (turkeys), an amphibian (*Xenopus laevis*), teleost fish (trout and eels), and an elasmobranch fish (the Japanese dogfish) (Sandberg *et al.*, 1991; Cobb and Brown, 1992; Murphy *et al.*, 1993; Marsigliante *et al.*, 1996; Tierney, Takei, and Hazon, 1997). In the eel and turkey, they are of the AT_1 type but they exhibit substantial differences from the mammalian receptors in the profiles of their response to various agonist and antagonist molecules. Clearly such receptors have evolved structural differences that may parallel changes in the structures of the endogenous angiotensins present and, possibly, reflect their diverse physiological roles. It is apparently unknown if multiple subtypes of the angiotensin receptors are present in "lower" vertebrates or if they have only evolved in mammals.

Parathyroid hormone, calcitonin, and stanniocalcin

Parathyroid hormone (PTH) and calcitonin are, respectively, the peptide hormones originating in the parathyroids and ultimobranchial bodies (or in mammals the thyroid C-cells). Stanniocalcin is found in the corpuscles of Stannius of teleostean and holostean fish. PTH increases calcium levels in the plasma and calcitonin decreases them.

Parathyroid hormone

PTH is present in all tetrapods but not fishes. However a PTH-like activity, which interacts with antibodies to the bovine hormone, has been detected in the brain, pituitary, and plasma of trout and goldfish (Harvey, Zeng, and Pang, 1987). Variations in its structure undoubtedly exist but have not been chemically elucidated in nonmammals except birds. The mammalian PTH contains 84 amino acid residues.

Human PTH differs (Fig. 3.19) from both the porcine and bovine hormones by substitutions of amino acids at 11 positions, whereas the latter two species differ from each other at seven positions. All the changes, except at position 43, may be accounted for genetically with a single base change. When tested in the rat (*in vivo*, blood calcium levels) the porcine and bovine PTHs do not exhibit different biological activities, though human PTH is only about one-third as active. It is likely that the active hormone at the effector site represents only a portion of the whole polypeptide molecule and that the hormone, once released, is converted at some peripheral site into an

Parathyroid Hormone

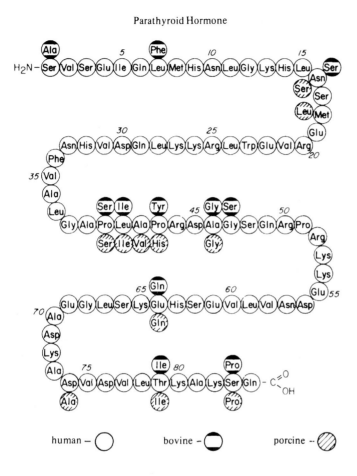

Fig. 3.19. Amino acid sequences of human, bovine, and porcine parathyroid hormone. (Reprinted with permission from H. T. Keutmann *et al.* Copyright © 1978 by the American Chemical Society.)

active fragment. The complete molecule is not essential for the exertion of a biological effect as it has been shown that a portion, the amino acids 1–34 at the N-terminal of the bovine hormone, has a similar activity to the complete molecule with its 84 amino acid residues (Tregear *et al.*, 1973). This observation suggests that considerable polymorphism of the PTH molecule is possible.

Chicken PTH is quite different to the mammalian hormones and contains 88 amino acid residues. There is only a 60% homology to the human N-terminal 1–34 amino acid fragment and this relationship is even less when the whole molecules are compared (Khosla *et al.*, 1988).

PTH is formed from a precursor (preproparathyroid hormone) that contains 115 residues. During synthesis of the hormone, the signal segment, containing 25 residues is removed from the N-terminus followed by the six residues in the "pro" part of the precursor. Internal structural homologies within the preproparathyroid molecule suggest that it may have arisen following the duplication and fusion of an ancestral gene (Cohn, Smardo, and

Morrissey, 1979). This hypothesis is given credence by the observation that segments of PTH that contain as few as 27 amino acid residues are still biologically active. Such a small molecule may reflect the general nature of the ancestral hormone.

A homolog of PTH has been identified in mammals and also, surprisingly, in some fish. It is called parathyroid hormone-related protein (PTHrP) and it has been found in a variety of tissues (Fraser, 1989; Philbrick *et al.*, 1996). In humans, PTHrP can be secreted in large amounts by some tumors, when, like PTH, it produces a hypercalcemia by an action on bone cells and the kidney. Its mechanism of action also involves an activation of adenylate cyclase. (Human PTHrP has, therefore, also been called human adenylate cyclase-stimulating protein or hACSP.) PTHrP is secreted by the mammary glands and appears in the milk of humans, cows, and rats (Budayr *et al.*, 1989). In the young, following its absorption from the gut, it may influence calcium metabolism. It has been identified immunologically in the pituitary and plasma of a teleost fish, the sea bream *Sparus auratus* (Danks *et al.*, 1993), and an elasmobranch, the dogfish *Scyliorhinus canicula* (Ingleton *et al.*, 1995). Several isoforms have been identified in the sea bream. In the dogfish, PTHrP was also found in the rectal gland and kidneys but not the gut or gills. Whether this peptide is the same as that identified in the pituitary of the goldfish (Kaneko and Pang, 1987) is not clear. A hypercalcemic role for such a hormone in marine fish, which normally experience a surfeit of calcium, is considered doubtful. Elasmobranchs also lack its principal tetrapod effector organ, a bony skeleton. In mammals, the structure of PTHrP has been described; it consists of 141 amino acid residues. The 1–34 residue N-terminus only has a 30% homology to that in PTH, though the homology is 62% for the first 13 residues. Both polypeptides can interact with the same receptor in a bone-derived cell line (Jüppner *et al.*, 1988). PTH and PTHrP may be members of a single protein family that arose following the duplication of a primeval gene, possibly in early fishes (Martin, Moseley, and Gillespie, 1991). The PTHrP in extant fish could even be the prototypical forebear of tetrapod PTH. Its ancestral function could also have involved the regulation of calcium metabolism, especially in fresh water species. However, it may have had other roles, including paracrine ones, which its ubiquitous tissue distribution suggests that it may have in extant species.

Calcitonin

Calcitonin activity has been measured in all vertebrates except the cyclostomes (Table 3.5). The hormones contain 32 residues. Chemical analysis indicates that considerable differences in their sequence occur that result in quantitative differences of biological activity. The amino acid sequence of the calcitonins in four mammals and a teleost fish (salmon) is shown in Fig. 3.20.

Table 3.5. *Calcitonin concentration in glands from various vertebrates*

| Class and species | Units (MRC)/g fresh gland wt | | | Unit/kg body wt |
	Thyroid	Ultimo-branchial	Internal parathyroid	
Mammalia				
Homo sapiens				
Normal thyroid	0.4	—	0.1–0.5	0.16
Medullary cell carcinoma of thyroid	17	—	—	—
Rat, *Rattus rattus*	5–15	—	—	0.2–0.6
Hog, *Sus scrofa*	2–5	—	—	0.4–0.8
Dog, *Canis familiaris*	1–4	—	1.5–3.3	0.25–0.50
Rabbit, *Oryctolagus cuniculus*				
Lower pole	1.5–2	—	2.1–2.5	—
Upper pole	a	—	—	—
Aves				
Domestic fowl, *Gallus domesticus*	a	30–120	—	0.5–0.8
Turkey, *Meleagris gallopavo*	a	60–100	—	0.5–0.9
Reptilia				
Turtle, *Pseudemys concinna suwaniensis*	a	3–9	—	0.002–0.006
Amphibia				
Bullfrog, *Rana catesbeiana*	—	0.5–0.8	—	0.001–0.002
Teleosti				
Chum salmon, *Onchorhynchus keta*	—	25–40	—	0.4–0.6
Gray cod, *Gadus macrocephalus*	—	10–20	—	0.2–0.4
Elasmobranchii				
Dogfish shark, *Squalus suckleyi*	a	25–35	—	0.25–0.40

[a] No detectable hypocalcemic activity.
Source: Copp, 1969.

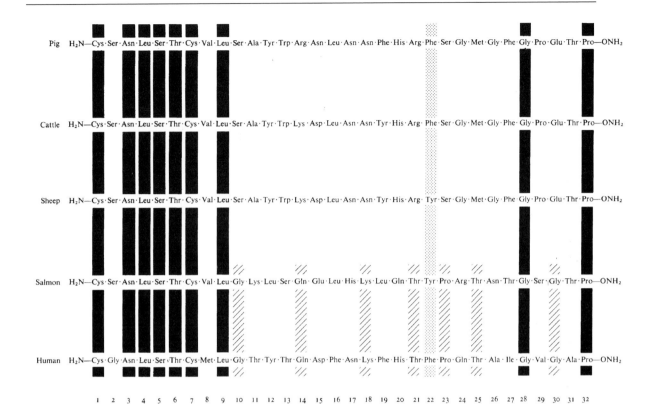

Fig. 3.20. A comparison of the amino acid sequence of calcitonin in four mammals and the salmon. The solid bars indicate that the amino acids are homologous in all species. It can be seen that extensive differences exist, especially in the central part of the molecules. Cross-hatched bars indicate homologies between salmon and human; stippled bar indicates comparable hydrophobic residues. (From Potts *et al.*, 1972.)

Only nine amino acid positions are commonly shared by all five species. The differences, however, can nearly all be accounted for by single-base changes in the genetic code (Potts *et al.*, 1972). The salmon calcitonins are of special interest. Three variants have been identified among four different species of salmon, with amino acid substitutions at four or five positions. All species have a common hormone, calcitonin I, but others, calcitonin II or III, may also be present. Salmon calcitonin is much more active when tested in mammals (20–100-fold) than is the natural (homologous) hormone (Deftos, 1997). This is probably a unique situation and is caused by two factors; a slow rate of destruction of the piscine hormone and a greater affinity for the receptor in the kidney and in bone (Marx, Woodward, and Aurbach, 1972).

The amino acid sequence of an avian (chicken) calcitonin has been deduced from its encoding nucleotide sequence (Lasmoles *et al.*, 1985). Somewhat surprisingly it was found to have a structure more similar to that of fish calcitonin than to that of mammals. It has an homology of 84% with salmon calcitonin but only about 45% with the mammalian hormones. Of special interest has been the purification of calcitonin from an elasmobranch fish, the stingray *Dasyatis akatei* (Takei *et al.*, 1991). This event was not only of note as it involved a species near the bottom of the vertebrate phyletic scale but also because it was a species that lacks a bony skeleton, which is the principal site of action of calcitonin in most other vertebrates. The elasmobranch calcitonin has an homology in its amino acid sequence of 78% to the avian hormone but it was only 66% to that of salmon. The homology to mammalian calcitonins was 32%. The elasmobranch polypeptide has a unique substitution of valine for leucine at position 9 and, in common with the avian hormone but with no other species, serine replaces asparagine at position 2. Therefore, it only shares amino acids at seven positions with all other vertebrates. Nevertheless, the stingray calcitonin, like salmon calcitonin, has a more potent hypocalcemic action in rats than the mammalian hormones (Sasayama *et al.*, 1992). The seven universally conserved amino acid sites may represent the basic framework needed for the hormone to interact with its receptors. During their evolution, these receptors may have been more strongly conserved than their complementary hormones.

The calcitonin gene can provide another peptide apart from calcitonin. By a process involving alternate splicing of its mRNA, a peptide called calcitonin gene-related peptide (CGRP) can be formed (Rosenfeld *et al.*, 1983). This substance, which is encoded by an exon adjacent to that for calcitonin, contains 37 amino acid residues. It has been described in several mammals, as well as the domestic fowl, a frog, *Rana ridibunda* (Esneu *et al.*, 1994), and salmon (Jansz and Zandberg, 1992). CGRP has been identified in the hypothalamus and peripheral neurons. Its amino acid sequence has been highly conserved (homologies of human to that of chicken, frog, and salmon being, respectively, 92, 87, and 80%) which is consistent with it having a physiological role. CGRP has a powerful vasodilatatory action (Brain *et al.*, 1985) and it has been shown to influence the activities of the gut, inhibit glycogenic responses to insulin in skeletal muscle, and exhibit behavioral effects (Taché, Holzer, and Rosenfeld, 1992). In osteoblast-like bone cells it may contribute to the regulation of intracellular calcium ion concentrations (Kawase *et al.*, 1996). CGRP is generally considered to be a neuropeptide rather than a hormone and its presence in the plasma may reflect a "spillover" from vascular nerves. However, an endocrine role has not been excluded and could be occurring in some species such as teleost fish (Lamharzi and Fouchereau-Peron, 1996). CGRP has an homology of nearly 50% in its amino acid sequence with that of a peptide called *amylin* that is secreted along with insulin from pancreatic B-cells (Cooper, 1994). Amylin has an uncertain physiological role but it also exhibits vasodilatatory and anti-insulin effects. Adrenomedullin (p. 170) also shares some pharmacological and structural features with CGRP, including common amino acids at 11 homologous positions (Kitamura *et al.*, 1993a.) These three peptides are considered to be members of the same superfamily (Wimalawansa, 1996).

A calcitonin-like molecule has been identified, using immunohistochemical methods, in the nervous tissues and brain of birds, reptiles, and cyclostome fishes (Girgis *et al.*, 1980). It has also been found in the pituitary gland of mammals and teleost fishes (Deftos *et al.*, 1980). Even more remarkably, it is present in the nervous system of several protochordates, including amphioxus and the ascidian *Ciona intestinalis*. The calcitonin gene may be very ancient and its expressed products may have had different roles. Originally it may have had a role in regulating the activity of the nervous system. Calcitonin receptors have been identified in the brain of rats, which is consistent with this possibility (Rizzo and Goltzman, 1981). Its functioning as a calcemic hormone may have been a subsequent development, possibly associated with the migration of the neuroectodermal calcitonin-forming cells (C-cells; Chapter 2) to the region of the thyroid gland.

Stanniocalcin

The corpuscles of Stannius of teleostean and holostean fish (Chapter 2) contain a hormone called stanniocalcin, which exerts a hypocalcemic effect in teleosts (Butkus *et al.*, 1987; Wagner *et al.*, 1988; Verbost *et al.*, 1993; Wagner, 1994). It is secreted into the plasma (Mayer-Gostan, Flik, and Pang, 1992). Stanniocalcin was formerly called hypocalcin or, sometimes, teleocalcin. The latter name has now been reserved for another biologically active molecule that also occurs in the corpuscles of Stannius. Stanniocalcin is a glycoprotein. Estimates of its size vary from 32 kDa in eels (Butkus *et al.*, 1987) to 60 kDa in rainbow trout (Lafeber *et al.*, 1988). The differences appear to reflect the formation of disulfide bridge-linked dimers and variations in the carbohydrate content of the extracted hormones. In the Australian eel, *Anguilla australis*, stanniocalcin contains 231 residues (Butkus *et al.*, 1987) and in the coho salmon, *Oncorhynchus kisutch*, 233 residues (Wagner, 1994). The proteins have no other identified family relatives among the vertebrates. The overall homology in the amino acid sequence between the eel and salmon stanniocalcin is about 65% but the differences are mainly present in the C-terminal region. The N-terminal 1–20 amino acid sequences of stanniocalcin from eels, salmon, and trout are compared in Fig. 3.21. It can be seen that there are considerable interspecific homologies between these parts of the molecules. They also exhibit considerable hypocalcemic activity (Milliken *et al.*, 1990; Verbost and Fenwick, 1995). However, a C-terminal fragment was also found to have this action, it was not as effective in eels though it was the only part of the molecule found to decrease the influx of calcium across the gills (Verbost and Fenwick, 1995). The different regions of the hormone may interact with different receptors.

A second, smaller (3 kDa) glycopeptide has been identified in the corpuscles of Stannius of trout and carp (Verbost *et al.*, 1993). It appears to be identical to the teleocalcin discovered by Ma and Copp in 1978. It is not a fragment of stanniocalcin. Teleocalcin does not influence calcium movements across the gills but it may have other effects on calcium metabolism.

Stanniocalcin has, somewhat belatedly, been identified in the plasma of elasmobranch fish, an amphibian, and mammals (Chapter 2). Immunocytochemical observations and its mRNA indicate its presence in various tissues of such vertebrates including renal tubular cells. The cDNA for human stanniocalcin shows that it is a protein containing 247 amino acid residues and that their sequence bears an homology of 73% to that of teleost stanniocalcin (Olsen *et al.*, 1996). Such a similarity in the structures of these glycoproteins in two such phyletically disparate vertebrates suggests that stanniocalcin retains a physiological role in mammals. However, unlike in bony fish, a discrete endocrine gland where its gene is expressed appears to be lacking. Its synthesis may occur in dispersed groups of cells in various tissues in tetrapod vertebrates. Stanniocalcin may still have an endocrine role in such species but an autocrine or paracrine one is also possible. Such effects may even influence responses involving mineral metabolism, as seen in fish (Olsen *et al.*, 1996).

The hormones of the islets of Langerhans and the gastrointestinal tract

The insulin family

Insulin is the prototype of this family of hormones. In most vertebrates it is produced by the B-cells of the islets of Langerhans (Chapter 2). It has a major role in controlling the intermediary metabolism of glucose, fats, and proteins. Insulin molecules have been identified throughout the vertebrates and insulin-like substances are present in invertebrates, including protochordates such as amphioxus. They have also been found in the nervous systems of molluscs and insects. Other members of this family, which also have related structures and genes, are the insulin-like growth factors, IGF-I and IGF-II, and an ovarian hormone called relaxin. The IGFs are present in the plasma and they are synthesized in the liver and some other tissues. They can modulate the growth of various tissues. Relaxin helps prepare viviparous, and possibly ovoviviparous, vertebrates for the delivery of the young. A large amount of chemical and genetic information about these polypeptides have allowed some fascinating speculation as to their common origins.

The insulin molecule consists of two main parts, an A-chain with 21 amino acid residues and a B-chain, usually with 31. These are joined by two disulfide bridges contributed by four cysteine residues (Table 3.6). Among the species so far examined (but excluding the hagfish) amino acid substitutions have been recorded at 35 of the 51 positions in the insulin molecules. The A-chain is identical in the insulin present in humans, pigs, rabbits, dogs, and sperm whales, and the B-chain is the same in pig, horse, ox, sheep, goat, sperm whale, and sei whale. The intact insulin in the pig, dog, sperm whale, and fin whale are identical. Most of the differences that occur in mammals are localized at three positions (8, 9, 10) in the A-chain and in one position (3) in the B-chain.

The amino acid sequences of insulins from a wide range of vertebrates are now available as a result of their direct chemical analysis and the decoding of the nucleotide sequences of their genes. These observations include two species of cyclostome fish, several chondrichthyeans, and various other vertebrates from all the major groups. As expected, closely related species usually show little variability but there are some exceptions. Among the cyclostomes, the insulins from two Holarctic lampreys (order Petromyzontiformes, family Petromyzontidae), *Petromyzon marinus* and *Lampetra fluviatilis* are identical (Conlon *et al.*, 1995a). However, they only have an homology of 65% in their amino acid sequences with the insulin of a hagfish *Myxine glutinosa* (order Myxiniformes) (Peterson *et al.*, 1975; Plisetskaya *et al.*, 1988). The magnitude of this difference is similar to that all three species exhibit from human insulin (62 to 75%). Insulin from a southern hemisphere lamprey, *Geotria australis* (order Petromyzontiformes, family Geotriidae), differs from the Holarctic lampreys by 17 amino acid substitutions (a 67% homology) (Conlon *et al.*, 1995b), apparently reflecting a long period of separate evolution. The chondrichthyean fishes show more consistent differences in the relationships of their insulins to those of tetrapods. The ratfish and the rabbitfish (Holocephali) have identical insulins and they have an amino acid sequence homology to human insulin of 61% (Conlon *et al.*, 1988). In a dogfish (Elasmobranchii) there is a 68% homology to the human hormone (Bajaj *et al.*, 1983). The insulin of the American eel *Anguilla rostrata* has a 74% homology to human insulin (Conlon *et al.*, 1991a). The toad *Xenopus laevis* has two insulins (I and II) that bear a 94% homology to each other (Shuldiner *et al.*, 1989). The *Xenopus* insulin II has an 84% homology to human insulin but only a 62% identity to hagfish insulin. Reptile and bird insulins display similarities to each other that appear to reflect their phyletic relationships. Thus, insulin from the domestic fowl and turkeys (order Galliformes) and ostriches are identical to those of two chelonian reptiles: the turtles *Pseudemys scripta* and *P. dorbigni* (Chevalier *et al.*, 1996; Cascone *et al.*, 1991) (an 86% homology in amino acid sequence to that of human insulin). Nevertheless, this insulin differs from that of ducks and geese (order Anseriformes) by three amino acid substitutions, and from that of the alligator and rattlesnake by, respectively, three and seven substitutions (Kimmel *et al.*, 1976; Lance *et al.*, 1984). The homology of the last to human insulin is about 73%. Among mammals, the insulins are usually quite similar to each other. Even that from a marsupial, the kangaroo *Macropus giganteus*, only differs from human insulin by four amino acid substitutions (Treacy *et al.*, 1989). However, even among mammals large differences in the structures of the insulins may exist, as seen in some South American rodents. In the

Table 3.6. *Amino acid sequence in vertebrate insulins*
(a) Insulin A-chains

Types of insulin	Amino acid sequences									
	1	2	3	4	5	6	7	8	9	10
Human[a]	Gly	Ile	Val	Glu	Gln	Cys	Cys	Thr	Ser	Ile
Sei whale	Gly	Ile	Val	Glu	Gln	Cys	Cys	*Ala*	Ser	*Thr*
Horse	Gly	Ile	Val	Glu	Gln	Cys	Cys	Thr	*Gly*	Ile
Beef	Gly	Ile	Val	Glu	Gln	Cys	Cys	*Ala*	Ser	*Val*
Sheep, goat	Gly	Ile	Val	Glu	Gln	Cys	Cys	*Ala*	*Gly*	*Val*
Elephant	Gly	Ile	Val	Glu	Gln	Cys	Cys	Thr	*Gly*	*Val*
Rat, mouse (I and II)	Gly	Ile	Val	*Asp*	Gln	Cys	Cys	Thr	Ser	Ile
Guinea pig	Gly	Ile	Val	*Asp*	Gln	Cys	Cys	Thr	*Gly*	*Thr*
Chicken, turkey	Gly	Ile	Val	Glu	Gln	Cys	Cys	*His*	*Asn*	*Thr*
Cod	Gly	Ile	Val	*Asp*	Gln	Cys	Cys	*His*	*Arg*	*Pro*
Tuna (II)	Gly	Ile	Val	Glu	Gln	Cys	Cys	*His*	*Lys*	*Pro*
Angler fish	Gly	Ile	Val	Glu	Gln	Cys	Cys	*His*	*Arg*	*Pro*
Toadfish (I)	*Gly*	*Ile*	*Val*	*Glu*	*Gln*	*Cys*	Cys	*His*	*Arg*	*Pro*
Toadfish (II)	Gly	Ile	Val	Glu	Gln	Cys	Cys	*His*	*Arg*	*Pro*
Hagfish	Gly	Ile	Val	Glu	Gln	Cys	Cys	*His*	*Lys*	*Arg*

Types of insulin	Amino acid sequences										
	11	12	13	14	15	16	17	18	19	20	21
Human[a]	Cys	Ser	Leu	Tyr	Gln	Leu	Glu	Asn	Tyr	Cys	Asn
Sei whale	Cys	Ser	Leu	Tyr	Gln	Leu	Glu	Asn	Tyr	Cys	Asn
Horse	Cys	Ser	Leu	Tyr	Gln	Leu	Glu	Asn	Tyr	Cys	Asn
Beef	Cys	Ser	Leu	Tyr	Gln	Leu	Glu	Asn	Tyr	Cys	Asn
Sheep, goat	Cys	Ser	Leu	Tyr	Gln	Leu	Glu	Asn	Tyr	Cys	Asn
Elephant	Cys	Ser	Leu	Tyr	Gln	Leu	Glu	Asn	Tyr	Cys	Asn
Rat, mouse (I and II)	Cys	Ser	Leu	Tyr	Gln	Leu	Glu	Asn	Tyr	Cys	Asn
Guinea pig	Cys	*Thr*	*Arg*	*His*	Gln	Leu	Glu	*Ser*	Tyr	Cys	Asn
Chicken, turkey	Cys	Ser	Leu	Tyr	Gln	Leu	Glu	Asn	Tyr	Cys	Asn
Cod	Cys	*Asp*	*Ile*	*Phe*	*Asp*	Leu	*Gln*	Asn	Tyr	Cys	Asn
Tuna (II)	Cys	*Asn*	*Ile*	*Phe*	*Asp*	Leu	*Gln*	Asn	Tyr	Cys	Asn
Angler fish	Cys	*Asp*	*Ile*	*Phe*	*Asp*	Leu	*Gln*	*Ser*	Tyr	Cys	Asn
Toadfish (I)	Cys	*Asp*	*Ile*	*Phe*	*Asp*	Leu	*Gln*	*Ser*	Tyr	Cys	Asn
Toadfish (II)	Cys	*Asp*	*Lys*	*Phe*	*Asp*	Leu	*Gln*	*Ser*	Tyr	Cys	Asn
Hagfish	Cys	*Ser*	*Ile*	*Tyr*	*Asp*	Leu	*Gln*	Asn	Tyr	Cys	Asn

Table 3.6. *(cont.)*

(b) Insulin B-chains

Type of insulin	Amino acid sequences																
	−1	1	2	3	4	5	6	7	8	9	10	11	12	13	14	15
Pig[b]		Phe	Val	Asn	Gln	His	Leu	Cys	Gly	Ser	His	Leu	Val	Glu	Ala	Leu
Human, elephant		Phe	Val	Asn	Gln	His	Leu	Cys	Gly	Ser	His	Leu	Val	Glu	Ala	Leu
Rabbit		Phe	Val	Asn	Gln	His	Leu	Cys	Gly	Ser	His	Leu	Val	Glu	Ala	Leu
Rat, mouse (I)		Phe	Val	*Lys*	Gln	His	Leu	Cys	Gly	*Pro*	His	Leu	Val	Glu	Ala	Leu
Rat, mouse (II)		Phe	Val	*Lys*	Gln	His	Leu	Cys	Gly	Ser	His	Leu	Val	Glu	Ala	Leu
Guinea pig		Phe	Val	*Ser*	*Arg*	His	Leu	Cys	Gly	Ser	*Asn*	Leu	Val	Glu	*Thr*	Leu
Chicken		*Ala*	*Ala*	Asn	Gln	His	Leu	Cys	Gly	Ser	His	Leu	Val	Glu	Ala	Leu
Cod	Met	*Ala*	*Pro*	*Pro*	Gln	His	Leu	Cys	Gly	Ser	His	Leu	Val	*Asp*	Ala	Leu
Tuna (II)	*Val*	*Ala*	*Pro*	*Pro*	Gln	His	Leu	Cys	Gly	Ser	His	Leu	Val	*Asp*	Ala	Leu
Angler Fish	*Val*	*Ala*	*Pro*	*Ala*	Gln	His	Leu	Cys	Gly	Ser	His	Leu	Val	*Asp*	Ala	Leu
Toadfish (I)	Met	*Ala*	*Pro*	*Pro*	Gln	His	Leu	Cys	Gly	Ser	His	Leu	Val	*Asp*	Ala	Leu
Toadfish (II)	Met	*Ala*	*Pro*	*Pro*	Gln	His	Leu	Cys	Gly	Ser	His	Leu	Val	*Asp*	Ala	Leu
Hagfish		*Arg*	*Thr*	*Thr*	*Gly*	His	Leu	Cys	Gly	*Lys*	*Asp*	Leu	Val	Asn	Ala	Leu

Types of insulin	Amino acid sequences															
	16	17	18	19	20	21	22	23	24	25	26	27	28	29	30	
Pig[b]	Tyr	Leu	Val	Cys	Gly	Glu	Arg	Gly	Phe	Phe	Tyr	Thr	Pro	Lys	Ala	
Human, elephant	Tyr	Leu	Val	Cys	Gly	Glu	Arg	Gly	Phe	Phe	Tyr	Thr	Pro	Lys	*Thr*	
Rabbit	Tyr	Leu	Val	Cys	Gly	Glu	Arg	Gly	Phe	Phe	Tyr	Thr	Pro	Lys	*Ser*	
Rat, mouse (I)	Tyr	Leu	Val	Cys	Gly	Glu	Arg	Gly	Phe	Phe	Tyr	Thr	Pro	Lys	*Ser*	
Rat, mouse (II)	Tyr	Leu	Val	Cys	Gly	Glu	Arg	Gly	Phe	Phe	Tyr	Thr	Pro	Met	*Ser*	
Guinea pig	Tyr	*Ser*	Val	Cys	*(Gln*	*Asp*	*Asp)*	Gly	Phe	Phe	Tyr	*Ile*	Pro	Lys	*Asp*	
Chicken	Tyr	Leu	Val	Cys	Gly	Glu	Arg	Gly	Phe	Phe	Tyr	*Ser*	Pro	Lys	Ala	
Cod	Tyr	Leu	Val	Cys	Gly	*Asp*	Arg	Gly	Phe	Phe	Tyr	*Asn*	Pro	Lys		
Tuna (II)	Tyr	Leu	Val	Cys	Gly	*Asp*	Arg	Gly	Phe	Phe	Tyr	*Asn*	Pro	Lys		
Angler Fish	Tyr	Leu	Val	Cys	Gly	*Asp*	Arg	Gly	Phe	Phe	Tyr	*Asn*	Pro	Lys		
Toadfish (I)	Tyr	Leu	Val	Cys	Gly	*Asp*	Arg	Gly	Phe	Phe	Tyr	*Asn*	Pro	Lys		
Toadfish (II)	Tyr	Leu	Val	Cys	Gly	*Asp*	Arg	Gly	Phe	Phe	Tyr	*Asn*	*Ser*			
Hagfish	Tyr	*Ile*	*Ala*	Cys	Gly	*Val*	*Arg*	Gly	Phe	Phe	Tyr	*Asp*	Pro	*Thr*	Lys	Met

Note: The italicized amino acids indicate the principal differences.

[a] Sequence is identical in humans, rabbit, dog, pig, and sperm whale.

[b] Sequence is identical in pig, horse, ox, dog, sheep, sperm whale, and sei whale.

Source: Humbel, Bosshard, and Zahn, 1972. The hagfish sequence is from Peterson *et al.*, 1975.

	1	2	3	4	5	6	7	8	9	10	11	12	13	14	15	16	17	18	19	20
Salmon	Phe	Ser	Pro	Asn	Ser	Pro	Ser	Asp	Val	Ala	Arg	Cys	Leu	Asn	Gly	Ala	Leu	Asp	Val	Gly
Eel	-	-	Ala	Ser	-	-	-	-	-	-	-	-	-	-	-	-	Gln	-	-	
Trout	-	Ser	-	-	-	-	-	-	-	-	-	-	-	-	-	-	Ala	-	-	

Fig. 3.21. N-terminal 1–20 amino acid sequences of salmon, eel, and trout stanniocalcins. The intact hormone in the eel contains 231 amino acid residues while that of the salmon contains 233. (Based on information provided by Lafeber (1988) and Millikan et al. (1990).)

IGF-I

10
Gly-Pro-Glu-Thr-Leu-Cys-Gly-Ala-Glu-Leu-Val-Asp-Ala-Leu-Gln-Phe-Val-Cys-Gly-
20
Asp-Arg-Gly-Phe-Tyr-Phe-Asn-Lys-Pro-Thr-Gly
30
-Tyr-Gly-Ser-Ser-

Area of homology

IGF-II

10
Ala-Tyr-Arg-Pro-Ser-Glu-Thr-Leu-Cys-Gly-Gly-Glu-Leu-Val-Asp-Thr-Leu-Gln-Phe-
20
Val-Cys-Gly-Asp-Arg-Gly-Phe-Tyr-Phe-Ser-Arg-Pro-
30

IGF-I   (cont.)

40
-Ser-Arg-Arg-Ala-Pro-Gln-Thr-Gly-Ile-Val-Asp-Glu-Cys-Cys-Phe-Arg-Ser-Cys-Asp-Leu-Arg-Arg-
50
60
-Leu-Glu-Met-Tyr-Cys-Ala-Pro-Leu-Lys-Pro-Ala-Lys-Ser-Ala
70

Fig. 3.22. Amino acid sequences of IGF-I and IGF-II extracted from human plasma. Apart from the similarities between each of these polypeptides, they also exhibit considerable homology to the amino acid sequences in the insulin B-chain (positions 1–29) and insulin A-chain (positions 12–62). The half-cystine and glycine residues of insulin are all conserved which suggests that the three-dimensional structure of IGF-I is probably similar to that of insulin. (Based on Rinderknecht and Humbel (1976a: b: 1978).)

domestic guinea pig, *Cavia porcellus*, the casiragua, *Proechimys quairae*, and the coypu, *Myocastor coypus*, there are about 18 such substitutions compared with human insulin (Treacy *et al.*, 1989). There is, nevertheless, an underlying pattern in the similarities of the vertebrate insulins and the closeness of their phyletic relationships, but there are also some startling exceptions. However, phylogenetic trees based on the known structures of vertebrate insulins and proinsulins have been prepared (Hedges *et al.*, 1990; Dores *et al.*, 1996). Groupings that correspond to their phyletic relationships are generally apparent but there are exceptions, such as a clade containing rat, human, and lamprey insulin. A larger sample of insulins may be necessary to resolve such anomalies.

Despite the chemical differences in the structure of the vertebrate insulins, there is surprisingly little demonstrable variation in their specific biological activities when they are tested on mammalian preparations. This observation presumably reflects the conservation of essential chemical sites.

Insulin-like growth factors I and II were first identified in human plasma (Froesch *et al.*, 1963). Antibodies to insulin were found to be only partially effective in inhibiting its metabolic actions and hence the activity was called nonsuppressible insulin-like activity of NSILA. In 1957, Salmon and Daughaday identified a substance in rat plasma that increased the incorporation of sulfate into cartilage. It was called "sulfation factor" and, later, somatomedin C. Somatomedin C and NSILA were subsequently found to be identical and it was then called insulin-like growth factor-I (Klapper, Svoboda, and van Wyk, 1983). Two such IGFs were identified in human plasma and found to be polypeptides with molecular weights equivalent to about 75 kDa. Human IGF-I contains 70 residues and IGF-II contains 67. IGF-I which is mainly synthesized in the liver under the influence of growth hormone, appears to play a vital role in promoting postnatal growth. IGF-II may be more important in regulating growth in the fetus. (IGF-I may also act in the fetus.) The mechanism controlling the synthesis of IGF-II is uncertain but it does not appear to involve growth hormone. The IGFs bear a strong structural similarity to each other (Fig. 3.22) and share about 66 of their amino acid positions. They also have internal domains of amino acid sequences that bear remarkable similarities to the A- and B-chains of insulin. Human insulin and IGF-I have 25 common amino acid positions including most of those considered to be invariant in the insulins. These similarities extend to their precursor molecules and suggest a common lineage. Using radioimmunoassays, IGF activity has been identified in the plasma and tissues of many vertebrates (Poffenbarger, Burns, and Bennett-Novak, 1976, Shapiro and Pimstone, 1977; Drakenberg *et al.*, 1989; Bautista, Mohan and Baylink, 1990; Anderson *et al.*, 1993; Guillette, Cox, and Crain, 1996) including several teleost and chondrichthyean fish. Determination of their amino acid sequences (see, for instance, Francis *et al.*, 1989 a,b; Chan *et al.*,

1992) indicate that with the exception of a cyclostome (the hagfish) the structures are remarkably conserved. Thus, IGF-I in humans, cattle, and pigs is identical while sheep IGF-I only differs from these by a single amino acid substitution. Rat and mouse IGF-I differ, respectively, from the human polypeptide at three and four amino acid positions. In the domestic fowl, IGF-I differs from human IGF-I by eight substitutions, a toad (*Xenopus*) by 11, and Atlantic salmon by 14. IGF-I has not been detected in the plasma of cyclostome fish but an IGF-I gene has been identified in the hagfish *Myxine glutinosa* (Nagamatsu *et al.*, 1991). Its amino acid sequence has been decoded from its nucleotides and differs from human IGF-I at 38 positions. Apart from the hagfish, IGF-I appears to be more highly conserved than the corresponding insulin molecules. They, therefore, may have evolved more slowly, reflecting more stringent structural requirements for their activity.

Relaxin   Relaxin is a hormone that was originally identified in the serum of pregnant guinea pigs (Hisaw, 1926). Subsequent interest in what was then only considered to be an exotic "biological activity" was sporadic for a period of about 50 years (Sherwood, 1994). Relaxin is a polypeptide that is secreted by the corpora lutea and placenta during pregnancy. It can aid parturition by relaxing the ligaments of the pubic symphysis, softening the cervix, and inhibiting uterine contractions. Relaxin has also been identified in the reproductive system of the human male, where it is present in seminal fluid.

Relaxin consists of two chains of amino acids (A and B) linked by two disulfide bridges in the same manner as seen in insulin (Fig. 3.23). The cysteine residues have been conserved and there are a number of other similarities in their amino acid sequences giving a homology of about 25%. The tertiary structures of relaxin and insulin appear to be quite similar (Bedarkar *et al.*, 1977).

Relaxins have been identified in many species of mammals. Apart from guinea pigs, these include humans, monkeys, pigs, horses, whales, dogs, and rats. They have also been found in viviparous and ovoviviparous elasmobranch fish (sharks and rays) (Schwabe, 1994; Sherwood, 1994). A relaxin has even been identified in the ovary of an ascidian (Protochordata) (Georges *et al.*, 1990) and an oviparous species, the domestic fowl (Brackett *et al.*, 1997).

The chemical structures of several relaxins have been determined (Stewart *et al.*, 1991; Schwabe, 1994) and 12 of about 50 amino acid residues are invariant. Human and gorilla relaxins have identical A-chains while the B-chains have an amino acid sequence homology of 93%. Two elasmobranchs, the sand tiger shark and the dogfish, show a 71% homology in their A-chains and 79% in the B-chains. The similarity of the A-chain of another elasmobranch, a skate, to the human hormone is 29% while that of the B-chain is 39%. There is much more variability in the amino acid sequences of the relaxins than there is in their corresponding insulins. Rats and pigs only

Relaxin

A-chain: H - Arg - Met - Thr - Leu - Ser - Glu - Lys - Cys - Cys - Glu - Val - Gly - Cys - Ile - Arg - Lys - Asp - Ile - Ala - Arg - Leu - Cys - OH

B-chain: <Glu - Ser - Thr - Asn - Asp - Phe - Ile - Lys - Ala - Cys - Gly - Arg - Glu - Leu - Val - Arg - Leu - Trp - Val - Glu - Ile - Cys - Gly - Val - Trp - Ser - OH

Insulin

A-chain: H - Gly - Ile - Val - Glu - Gln - Cys - Cys - Thr - Ser - Ile - Cys - Ser - Leu - Tyr - Glu - Leu - Glu - Asn - Tyr - Cys - Asn - OH

B-chain: H - Phe - Val - Asn - Gln - His - Leu - Cys - Gly - Ser - His - Leu - Val - Glu - Ala - Leu - Tyr - Leu - Val - Cys - Gly - Glu - Arg - Gly - Phe - Phe - Tyr - Thr - Pro - Lys - Ala - OH

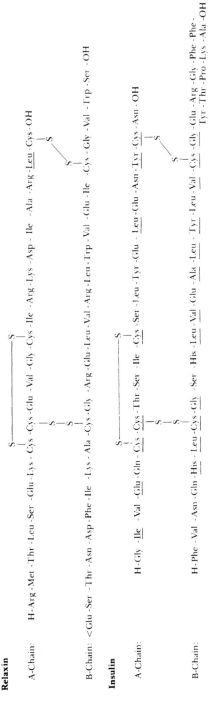

Fig. 3.23. Primary structures of porcine relaxin and insulin with their cysteine residues aligned. In the insulin structure, the underlined residues signify those that are either homologous or conservatively replaced with reference to the respective chains in relaxin. It can be seen that the disposition of the disulfide bridges is the same in both relaxin and insulin. (From Schwabe and McDonald, 1977. © 1977 by the American Association for the Advancement of Science.)

show a 40% homology in the amino acid sequences of their relaxins but a 92% one between their insulins. The evolution of the relaxins appears to be the least restrained in the insulin family.

Origins of the insulin family: insulin-like molecules in invertebrates   Immunocytochemical observations using antibodies to mammalian insulins have identified insulin-like immunoreactivity in several insects, molluscs, and protochordates (Ebberink, Smit, and van Minnen, 1989). Some of these substances have been further studied in biological extracts and using molecular genetic procedures. A mollusc, the fresh water snail *Lymnaea stagnalis*, has insulin-like activity that is localized in neuroendocrine cells of the cerebral ganglia. The secretions appear to contribute to the growth of the snails and their protein and glycogen metabolism: effects reminiscent of the actions of insulin in vertebrates. Gene cloning procedures in these snails have identified three polypeptide precursors (preprohormones). These molecules incorporate the main components seen in vertebrate insulins: an A- and B-chain and connecting C peptide (Smit *et al.*, 1991; 1993). They are called *molluscan insulin-related peptides* (MIP) I, II and III. The amino acid sequences of these putative hormones are quite diverse and only bear a 40 to 50% homology to each other. The A-chain of MIP III (Smit *et al.*, 1993) has an homology to the A-chains of human insulin, relaxin, and IGF-I of, respectively 33, 29, and 42%. Homologies between the B-chains are somewhat less than these values. The organization of the precursors of MIPs is remarkably similar to that of human insulin (Fig. 3.24). The genes for MIP and human insulin are also organized in a similar way, consisting of three exons and two introns with exon II mainly encoding the signal sequence and B-chain and exon III the C peptide and the A-chain. It has been suggested (Smit *et al.*, 1993) that this insulin-like molecule may have evolved about 600 million years ago in an ancestral group common to both the vertebrates and invertebrates.

Details regarding the divergence of such an ancestral insulin to produce vertebrate insulin, relaxins, and the IGFs would be of particular interest. As described earlier, a relaxin may be present in the ovaries of an ascidian (Protochordata) and an insulin-like substance has also been found in the gut of another member of this group, a sea tulip (Galloway and Cutfield, 1988). An IGF-I gene has been identified in a cyclostome, a hagfish (Chan *et al.*, 1992), but IGF has not been found in invertebrates. It was, therefore, of special interest to find that a "hybrid" gene encoding an insulin-like peptide that incorporated some IGF characteristics was present in a protochordate, amphioxus *Branchiostoma californiensis* (Chan, Cao, and Steiner, 1990). This gene is of the insulin type having the typical three exons interspersed by two introns. The encoded precursor contains the usual signal A-, B-, and C-peptide domains but, like the precursor of IGF, it also contains D- and E-like domains at its C-terminal. It, therefore, may be the type of gene that could

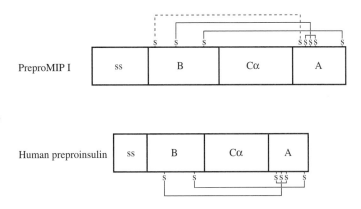

Fig. 3.24. A comparison of the structural organization of the domains of human preproinsulin with MIP I from the water snail *Lymnaea stagnalis*. SS, signal sequence; A, A-chain domain; B, B-chain domain; C, connecting peptide domain. The bars represent the regions of dibasic amino acids where proteolytic cleavage can occur. S–S, homologous disulfide bridges in preproinsulin and MIP I. (Based on Smit *et al.* (1993).)

have linked the evolution of the insulins to the IGFs. A scheme for the evolution of insulin and the IGF-I and IGF-II genes has been proposed (Chan *et al.*, 1992). It involves duplications of such a hybrid ancestral gene to give separate insulin, IGF-I and IGF-II genes. The relaxin gene may have had a separate origin, possibly somewhat later in the elasmobranchs.

The insulin receptor superfamily It seems likely that the related hormones of the insulin family may also have receptors that have structural similarities to each other. Insulin and IGF-I receptors have been extensively studied in mammals and are considered to belong to the same family (Prager and Melmed, 1993). They both contain a cytoplasmic tyrosine kinase and are glycoproteins. They each consist of two pairs of subunits, α and β, that are linked by disulfide bridges to form a heterotetramer (Kahn *et al.*, 1993). The receptor for each hormone is coded by a single but separate gene. The ligand (hormone)-binding recognition site is present on the α-subunit and faces the extracellular fluid. The β-subunit is the site of the cytoplasmic tyrosine kinase and it also provides the pathway for transmembrane signaling. The α-subunits have molecular masses of about 130 kDa and the β ones 120 kDa. When the entire amino acid sequences of the insulin and IGF-I receptors are compared, they exhibit a homology of 50%. For such a molecule, containing about 1400 amino acid residues, this similarity is considered to be remarkable. The IGF-II receptor is quite different to the insulin and IGF-I receptors. It forms a single polypeptide chain that crosses the plasma membrane and lacks tyrosine kinase activity. The affinities of each type of receptor is greatest for its namesake hormone but the insulin receptor can also interact, with decreasing affinities, with IGF-I and IGF-II. The IGF-I, receptor can, likewise, also interact with IGF-II and insulin. The IGF-II receptor combines with IGF-II and IGF-I, but not with insulin.

Insulin and IGF-I receptors have also been identified in birds, where they appear to have similar structures to those in mammals (Goddard *et al.*, 1993; Simon and Taouis, 1993). A novel insulin receptor has been described in an elasmobranch, the stingray *Dasyatis americana* (Stuart, 1988). It has many features in common with the mammalian receptor but instead of being assembled as a heterotetramer it forms a homodimer. Like the mammalian receptor it contains a cytoplasmic tyrosine kinase domain, a transmembrane segment, and an extracellular ligand-binding domain. The last interacts with mammalian insulin and IGF-I with similar affinities but less readily with IGF-II. Unlike the mammalian receptor it does not form a heterotetramer and only consists of two subunits, which are identical and not linked by disulfide bridges. Each is like a mammalian proreceptor prior to its processing to α- and β-subunits. There was no evidence of IGF-I or IGF-II receptors in this species. It was suggested that this insulin receptor could reflect the structural scenario prior to the evolutionary divergence of the vertebrate insulin and IGF-I receptors.

An insulin IGF-I type receptor has been studied in a protochordate (amphioxus) (Chan *et al.*, 1992). It has similarities to both mammalian insulin and IGF-I receptors, which is consistent with the presence of an insulin–IGF-like hormone in this animal (see above). Like the prototype mammalian receptors, it has a tyrosine kinase domain. It has been suggested that duplication of the gene for such a receptor, and subsequent mutations and intron rearrangement, may have resulted in the separate evolution of genes for insulin, IGF-I and IGF-II receptors. Insufficient information is available about the relaxin receptor to speculate about its place in such a scheme.

A somewhat unexpected, and indeed startling, discovery was the identification of an insulin-like receptor in the fruit fly *Drosophila melanogaster* (Nishida *et al.*, 1986; Petruzzelli *et al.*, 1986; Roth, 1990; LeRoith, 1995). Like the mammalian insulin receptor, it is a heterotetramer with a β-subunit and tyrosine kinase domain and an α-subunit containing a ligand-binding domain. It binds mammalian insulin and, with less affinity, also IGF-I and IGF-II. The tyrosine kinase domain has an amino acid sequence with a 64% homology to that of the mammalian insulin receptor, while the ligand-binding domains exhibit a 25% identity. The overall homology between the two receptors is about 35%. Although there are some prominent structural differences, the insect and mammalian insulin receptors would appear to share a common ancestral gene. Mutations of this gene in *Drosophila* are associated with deficiencies in growth that can be lethal (Chen, Jack, and Garofalo, 1996). Phylogenetic relationships between insects and vertebrates are, to say the least, tenuous, and one cannot help wonder how they each come to have such a similar gene. It has been suggested that roles for this gene in the regulation of growth may have arisen prior to the evolutionary divergence of the vertebrates and insects, about 600 million years ago (Chen *et al.*, 1996).

Glucagon

This hormone is a linear peptide containing 29 amino acid residues. It is a member of the secretin–glucagon superfamily (see below). The glucagon gene is expressed at two sites: the A-cells of the endocrine pancreas and the L-cells in the gut. The precursor is processed differently in each type of cell (Mojsov *et al.*, 1986; Bataille, 1989; Fehmann, Göke, and Göke, 1995). Glucagon-29 (proglucagon 33–61) is formed in the pancreatic A-cells. In the L-cells in the intestine a proglucagon fragment called glicentin (proglucagon 1–69) is formed. This polypeptide is further processed to oxyntomodulin (proglucagon 33–69), which may have an endocrine role as an inhibitor of gastric secretion. The C-terminal portion of the precursor in mammalian L-cells gives rise to two more peptides: glucagon-like peptides I and II. (GLP-I is proglucagon 78–107 and GLP-II is proglucagon 126–158). In the chicken and a teleost (the anglerfish) only one GLP is formed. In mammals, the amino acid sequences of GLP-I and GLP-II are highly conserved, which suggest that they have a physiological function. GLP-I has been shown to stimulate the secretion of insulin and inhibit that of glucagon in mammals and may be involved in the regulation of hormone release associated with feeding. It has also been identified in the hypothalamus. Infusion of GLP-I into the brain fluids of rats inhibits their feeding behavior, suggesting that it may contribute to the sensation of satiety (Turton *et al.*, 1996).

Glucagon has prominent effects on intermediary metabolism and promotes glycogenolysis, lipolysis, and gluconeogenesis (Chapter 5). These actions tend to oppose those of insulin. Glucagon has been identified in all the major groups of the tetrapods and also in teleostean, holostean, chondrichthyean, and cyclostome fishes. The chemical structures of several of these hormones is shown in Fig. 3.25. Structurally, glucagon is generally a well-conserved hormone. Among mammals it displays virtually no variations, with the interesting exception of some South American rodents. These animals include the domestic guinea pig and some of its hystricognathan relatives, where as many as five amino acid substitutions may occur in a single species (Iturriza, Verzi, and Di Maggio, 1995). In birds and amphibians only one or two amino acid substitutions have been observed compared with the sequence in mammals. More variation is, however, apparent among the fishes. Glucagon in the sea lamprey *Petromyzon marinus* differs from the human hormone by eight amino acid substitutions, yet there is still a 72% homology (Fig. 3.25). The sea lamprey glucagon differs even more from that of another lowly fish, the ratfish (*Hydrolagus colliei*: Holocephali), with which it only has an homology of 59%. (The homology between glucagon from the sea lamprey and that from the river lamprey, *Lampetra fluviatilis*, is 79% (Conlon *et al.*, 1995a).) Among teleosts, positions 1–13 appear to be identical in all species so far examined, reflecting a substantial degree of homology

Positions: 1 … 5 … 10 … 15 … 20 … 25 … 29

Group	Species	Sequence (positions 1–29)
	Human	His Ser Gln Gly Thr Phe Thr Ser Asp Tyr Ser Lys Tyr Leu Asp Ser Arg Arg Ala Gln Asp Phe Val Gln Trp Leu Met Asn Thr
Birds	Chicken[1]	– Ser –
	Duck[2]	– – – – – – – – – – – – – – – Thr – – – – – – – – – – – Ser –
Amphibia	Bullfrog[3]	– Ser – – –
Teleostei	Tuna[4]	– – Glu – – – Ser Asn – – – – – – Glu Thr – – – – – – – – – – Lys – Ser
	Coho salmon[5]	– – Glu – – – Ser Asn – – – – – – Glu Glu Glu – Met – – – – – – – – – Ser
Chondrostei	Alligator gar[6]	– – – – – – – Asn – – – – – – – Thr – – – – – – – – – – – Ser –
Elasmobranchii	Dogfish[7]	– – Glu – – – – – – – – – – – Met Asn – – – Lys – – – – – – – – –
	Ray[8]	– – Glu – – – – – – – – – – – – Asn – – – Lys – – – – – – – – –
Holocephali	Ratfish[9]	– Thr Asp – Ser – – – – – – – – – – Asn – – Thr Lys – – – – – – Leu Ser Thr
Cyclostomes	Lamprey[10]	– – Glu – – – – – – – – – – – Glu Asn Lys Gln – – – – – – Arg – – – Ala

Fig. 3.25. The structures of vertebrate glucagons. (–) indicates identical amino acids. 1. Pollock and Kimmel (1975). 2. Sundby et al. (1972). 3. Pollock et al. (1988a). 4. Navarro et al. (1991). 5. Plisetskya et al. (1986). 6. Pollock et al. (1988b). 7. Conlon, O'Toole, and Thim (1987). 8. Conlon and Thim (1985). 9. Conlon et al. (1989). 10. Conlon, Nielsen, and Youson (1993).

within this group. For instance, the identity in amino acid sequences between salmon and tuna is 84%. The glucagon of a more primitive bony fish, the alligator gar (Holostei), and that of a ray (*Torpedo*, Elasmobranchii) differ from human glucagon by only three amino acid substitutions. At least 13 of the 29 residues in glucagon appear to be invariant among all vertebrates studied, but in tetrapods there are fewer changes. Little evolution of the glucagon molecule appears to have occurred among the tetrapods, possibly reflecting restrictions related to its biological activity, which are apparently not as stringent among the fishes. Glucagon receptors appear to be quite fastidious, especially in tetrapod vertebrates. Early studies on teleost glucagon showed that while it was effective in such fish it has little action in mammals, presumably reflecting differences in the structures of their receptors.

The pancreatic polypeptide family

In 1968 Kimmel, Pollock and Hazelwood (see Hazelwood, 1990) extracted a peptide from the pancreas of chickens that had several biological activities including an ability to reduce the exocrine secretion from the pancreas and the mobility of the gut. The active molecule was found to be a peptide containing 36 amino acid residues. This exercise in comparative endocrinology heralded the discovery of a group of related peptides that phyletically extend from cyclostome fish to mammals, and possibly even includes invertebrates. They have been called the pancreatic polypeptide family and are related because of the homologies in their amino acid sequences and a common feature in their tertiary structure, the pancreatic polypeptide fold (PP-fold). All have 36 amino acid residues with a tyrosine amide at the C-terminus and often also an N-terminal tyrosine. They consist of three main groups.

The pancreatic polypeptides   The pancreatic polypeptides or PPs are found in the F-cells of the endocrine pancreas (sometimes also called PP-cells) which apart from in birds, have been identified in many mammals, as well as reptiles and amphibians, but apparently not in any of the fishes. The PPs are released into the plasma and exhibit their effects on pancreatic secretion and gut motility described above.

Peptide YY   Peptide YY (PYY), which is an abbreviation for peptide tyrosine–tyrosine (Y being the single letter code for tyrosine) was discovered in extracts of pig intestine by Tatemoto and Mutt in 1980 (see Tatemoto, Carlquist, and Mutt, 1982). It has similar effects to PP but it also has a powerful vasoconstrictor action. This peptide has the normal number of 36 amino acid residues but has tyrosine residues at both the N- and C-termini (hence PYY). It is formed in the L-cells of the intestine and it is also found in neural tissue. Peptide YY has been identified in species extending phyletically from mammals to fishes. A PYY-like peptide has been identified in the skin of a South American tree frog, *Phyllomedusa bicolor* (Mor *et al.*, 1994). It has a 94% homology to the gut PYY of another frog, *Rana ridibunda*, and it has been called skin peptide tyrosine–tyrosine or SPYY. The peptide can inhibit the release of MSH from the frog pars imtermedia and so it has been called *melanostatin*.

Neuropeptide Y   Neuropeptide Y (NPY, neuropeptide tyrosine) was first found in the brain of pigs (Tatemoto *et al.*, 1982). Its effects on the gut are less prominent than those of its relatives but it has the same powerful vasoconstrictor actions as PYY. It is usually found in neural tissue and may be colocalized with catecholamines and other peptides. In some fish a NPY-related peptide has been found in the pancreas (Conlon *et al.*, 1991b). NPY has the same general 36 residue structure as other members of its family but it only has a single C-terminal tyrosine, hence it is called neuropeptide tyrosine. Like PYY it has been identified in many species extending from the fishes to mammals. NPY has several putative functions apart from its vascular actions. It can potentiate the release of LH by increasing the release and effects of GnRH (Freeman, 1993). It can also promote feeding behavior (orexigenic effect) (Sahu and Kalra, 1993) and, like SPYY, it inhibits the release of MSH from the frog pars intermedia (Valentijn *et al.*, 1994). It can also potentiate the release of vasopressin from the rat pituitary (Larsen *et al.*, 1994). Such a plethora of effects, and there are others, need not all reflect physiological functions, but some might.

NPY and PYY are structurally more similar to each other than to PP. In pigs, NPY and PYY share 25 out of 36 amino acid positions while NPY and PP only have 18 residues in common. NPY is a remarkably well conserved molecule. For instance, it only varies in two amino acid positions among the mammals (Blomquist *et al.*, 1992) and in an elasmobranch fish, the ray *Torpedo marmorata*, it has a 93% homology in its amino acid sequence to that of mammalian NPYs. (A similar comparison of insulins would give a value of 71% and of relaxins 41%.) PP and PYY show greater structural variability than NPY: the homologies of the amino acid sequences of NPYs are human to dogfish, 92%; human to trout, 78%; and human to frog, 97% (Jensen and Conlon, 1992). A similar comparison of the PYYs yield values of about 75%. Among the PPs, the differences are even greater, with such homologies ranging from 44 to 58% (see Blomquist *et al.*, 1992). Chicken PP, for instance, differs from human PP by 17 amino acid substitutions.

The information available regarding the phyletic distribution of the peptides of the PP family invite phylogenetic speculation (Blomquist *et al.*, 1992; Jensen and Conlon, 1992). NPY is the most highly conserved member of the family and it is present in all vertebrates from elasmobranchs to mammals.

(The cyclostomes will be discussed below.) PYY, while less well conserved, has a similar distribution. PP, by comparison, has a very variable structure and first occurs in the amphibians (Pollock *et al.*, 1988a). The evolution of the group can be considered as being, possibly, diphyletic with NPY and PPY being the ancestral peptides. (An earlier origin from a single NPY-like peptide is also considered to be feasible.) As a result of gene duplication, PP appeared later, following the divergence of the fishes and the tetrapods. Subsequent evolution of NPY would be constrained by its preordained biological roles, but this limitation may not have applied to PP, which shows a remarkable number of variations in its amino acid structure.

A member of the PP family has been isolated from the sea lamprey *Petromyzon marinus* (Conlon *et al.*, 1991b). Its amino acid sequence exhibits similar homologies to porcine NPY (64%) as to porcine PYY (61%). It lacks an N-terminal tyrosine (methionine is present at this site) and so it has been called peptide methionine–tyrosine or PMY. Differences in its amino acid sequence from other members of the family suggest that its tertiary PP-fold has been modified. Its place in the evolutionary tree of the pancreatic polypeptide family is not clear. It will be of interest to see if it is present in other cylostomes and if it is the sole PP family member in these lowly vertebrates.

Receptors for the polypeptides of the PP family have been identified at numerous sites including blood vessels, intestine, peripheral nerves, the brain, and the hypothalamus. At least six, possibly seven different types (Y1 to Y4, two types of Y5 and an unnamed "novel receptor") have been described based on their different interactions with the various PP polypeptides and their synthesized analogs (O'Shea *et al.*, 1997). In vertebrates, four of these receptors have been cloned: Y1, Y2, Y4, and Y5 (Larhammar *et al.*, 1992; Bard *et al.*, 1995; Gerald *et al.*, 1995; 1996). They all belong to the G protein-coupled superfamily of receptors and possess the characteristic seven transmembrane domains. However, their structures vary considerably, with the cDNAs encoding, respectively 384, 381, 375, and 456 amino acid residues. The most favored natural ligands for Y1 and Y2 are NPY and PYY; for Y3, NPY; for Y4, PP; and for Y5, NPY, PYY, and PP. The homology between the predicted amino acid sequences of the Y1 and Y2 receptors is 31%, Y1 and Y4, 42%, while that between Y5 and the others is less than 35%. Such diversity within such a group of receptors is unusual and suggests that they diverged long ago. A Y2-like receptor (called PR4 receptor) has been cloned from an insect, *Drosophila melanogaster* (Li *et al.*, 1992). Its cDNA encodes a predicted 449 amino acid residues and it also appears to be a member of the G protein-linked receptor superfamily. Although it interacts with mammalian PP polypeptides in a manner suggesting that it is Y2-like, a natural excitant has apparently not yet been identified in this insect. Its situation and possible role in *Drosophila* are, therefore, somewhat mysterious.

The secretin–glucagon superfamily

Secretin was the first hormone to be discovered. It is now known to be a member of a superfamily of related polypeptides that include glucagon, vasoactive intestinal peptide (VIP), gastric inhibitory polypeptide (GIP) and GH-RH. The pancreatic polypeptide family are also members of this large superfamily, which is called the secretin–glucagon superfamily (also the secretin–glucagon–VIP superfamily). It also includes a variety of other peptides, such as pituitary adenylate cyclase-activating peptide (PACAP), which have undefined roles (Dockray, 1989; Fahrenkrug, 1989; Sherwood *et al.*, 1994). They have widespread phyletic distribution among vertebrates but clear affinities with invertebrate peptides do not appear to have been identified.

Secretin is a hormone that is secreted by the S-cells of the upper part of the intestinal tract. It has a number of actions, the most notable being an ability to stimulate the secretion of digestive juices by the acinar cells of the pancreas. Secretin is a peptide containing 27 amino acid residues. Porcine secretin (Fig. 3.26) has 15 amino acids at identical positions to those in glucagon, nine to VIP and 10 to GIP. GH-RH is a more recently discovered member of the superfamily and is not as closely related as it only shares six of its N-terminal amino acid positions (it has 44 such residues) with secretin. However, it has nine such residues in common with VIP and eight with glucagon. The ancestors of this peptide superfamily are considered to have undergone considerable gene duplication to produce a multiplicity of excitants. Each particular hormone may exhibit considerable conservation of its structure in different species, as already described for glucagon. VIP is even more conserved.

VIP was so named as it was originally identified in intestinal extracts. However, it is apparently confined to nerve cells from which it can be released and so it is classified as a neuropeptide. It is present in the nerves supplying many tissues and on its release it relaxes vascular and other types of smooth muscle and stimulates secretion by a number of exocrine and endocrine glands (Fahrenkrug, 1989). Pig, ox, rat, and human VIP have identical sequences of 28 amino acid residues. The VIP from a marsupial, the North American opossum, differs from these eutherian peptides at five positions (Eng *et al.*, 1992). The VIP from a bird (the domestic fowl), a reptile (an alligator) and an amphibian (a frog) are identical and only differ from the principal eutherian VIP by four amino acid substitutions (Wang and Conlon, 1995). Among the fishes, trout (Teleostei) and bowfin (Holostei) VIP are identical and differ from the nonmammalian tetrapod VIPs by one amino acid substitution. The VIP in the spiny dogfish (Elasmobranchii) is less similar and displays changes at four or five positions compared with all other vertebrates. Such remarkable phyletic conservation is consistent with a con-

Fig. 3.26. The secretin–glucagon family of hormones: the amino acid sequences of porcine VIP, secretin, glucagon, and GIP. (Only the first 29 of the 43 residues in the GIP are shown.) Amino acids that are identical to those in VIP are shown by the blank spaces while those common to secretin, glucagon, and GIP are in boxes. (From Dockray, 1978.)

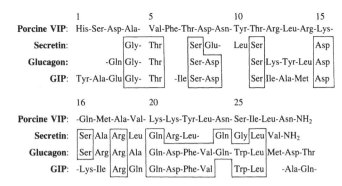

Fig. 3.26. The secretin–glucagon family of hormones: the amino acid sequences of porcine VIP, secretin, glucagon, and GIP. (Only the first 29 of the 43 residues in the GIP are shown.) Amino acids that are identical to those in VIP are shown by the blank spaces while those common to secretin, glucagon, and GIP are in boxes. (From Dockray, 1978.)

tinuing physiological role among the vertebrates, with little tolerated change in the relationship of its structure to its biological activity. GIP (Brown *et al.*, 1989) was first identified in extracts of the pig duodenum and jejunum where it is present in K-cells. It has two principal roles: it inhibits gastric acid and pepsin secretion and stimulates the synthesis and release of insulin. (It is also called glucose-dependent insulinotropic polypeptide.) GIP is released into the circulation in response to the presence of fat in the intestine. It contains 43 amino acid residues (Fig. 3.26). The pig and human hormones differ at two amino acid positions.

Immunocytochemical observations have identified substances like secretin, GIP and VIP in the endocrine cells of the gut of a wide phyletic range of species (Rawdon and Andrew, 1990). Secretin-like immunoreactivity or biological activity has been found in the gut of most vertebrate classes, including (apart from mammals) birds, reptiles, and teleost fish (Dockray, 1975; 1989). There is some uncertainty as to whether it is present in cyclostomes but it is present in protochordates. VIP-like immunoreactivity has also been identified in protochordates, as well as in the main groups of vertebrates. GIP-like substances have not, it appears, been found in cyclostomes but are present in elasmobranchs.

The possible evolution of the secretin–glucagon superfamily has been suggested on the basis of the structures of its precursors (Sherwood *et al.*, 1994). The primordial gene may have contained a single exon coding for a single peptide. A duplication of this exon could have occurred resulting in a gene that encoded two such peptides. Following a duplication of this gene, these peptides may have been the progenitors of two major lines of hormones: one of these gave rise to glucagon, VIP, and PACAP, and the other to secretin, GIP, and GH-RH.

The cholecystokinin–gastrin family

The cholecystokinin–gastrin family comprises linear peptides that are clearly related by the homologies of their amino acid sequences (Fig. 3.27). At the

Gastrin	$\overset{\text{SO}_3\text{H}}{\mid}$ Glu-Glu-Glu-Ala-Tyr-Gly-Trp-Met-Asp-Phe-NH_2
CCK	$\overset{\text{SO}_3\text{H}}{\mid}$ Asp-Arg-Asp-Tyr-Met - - - - -
Caerulein	$\overset{\text{SO}_3\text{H}}{\mid}$ Glu-Gln-Asp-Tyr-Thr - - - - -
Cionin	$\overset{\text{SO}_3\text{H}\ \ \text{SO}_3\text{H}}{\mid\quad\ \mid}$ Asn-Tyr-Tyr - - - - -
Leucosulfakinin	$\overset{\text{SO}_3\text{H}}{\mid}$ Glu-Glu-Phe-Glu-Asp-Tyr - His - Arg -
Drosulfakinin	$\overset{\text{SO}_3\text{H}}{\mid}$ Phe-Asp-Asp-Tyr - His - Arg -

Fig. 3.27. The C-terminal amino acid decapeptide sequences of gastrin compared with that of cholecystokinin (CCK) and of related peptides found in frog skin (caerulein) and the neural ganglion of a protochordate (cionin). The structures of two members of groups of insect neuropeptides leucosulfakinin I and drosulfakinin I are also shown (see text). (–) indicates identical amino acid residues. (Based on information in Johnsen and Rehfeld (1990).)

C-terminus of such peptides there is a heptapeptide sequence that is essential for the activities of both types of hormone and appears to constitute an active core. Sulfated tyrosine residues at positions 6 and 7 from the C-terminals in, respectively, gastrin and cholecystokinin (CCK) enhance their activities. The remaining sections of the molecules have quite variable structures and may influence such properties as their duration in the circulation and confer species and organ specificities.

Gastrin is a hormone that is formed by the G-cells in the pyloric region of the stomach and the duodenum. Its principal action is to increase the secretion of gastric acid, but it displays a number of other effects on glandular secretion and smooth muscle contractility in the gut and on the proliferation of gastric mucosal cells (Ohning *et al.*, 1996).

CCK is formed by the I- or CCK-cells in the upper parts of the intestine. It has two prominent actions: contraction of the gall bladder and increased secretion of an enzyme-rich fluid from the pancreas. The latter effect initially led to its being named pancreozymin, which was once thought to be a separate hormone from CCK (Jorpes and Mutt, 1966). CCK may also promote the release of insulin and contribute to the sensation of satiety following feeding (Chapter 5). CCK has been identified immunocytochemically in the brain and peripheral nerves of many vertebrates, including fishes (Jönsson, 1989). The peptide, therefore, appears to have a dual hormonal and neural role.

Gastrin exists in two main forms in the body: "big" gastrin (or gastrin I), which contains 34 amino acid residues, and "little" gastrin (gastrin II), which has 17 such residues. Both forms appear to be secreted by the G-cells in mammals. Little gastrin is more potent than big gastrin, but it is destroyed much more rapidly in the circulation. Such truncation of the hormone molecules of this family is common and results from their selective proteolytic cleavage during post-translational processing of the precursors. The little gastrins (G-17) from the pig, ox, dog, cat, and rat have amino acid substitutions at five positions (Dockray and Gregory, 1989). The "big" gastrins are more variable in their structures. Chicken gastrin (36 residues) shows little homology to mammalian gastrin except at the C-terminus and even the sulfated tyrosine is different, being in position 7 from the C-terminus, which is otherwise characteristic of mammalian CCK (Dimaline, Young, and Gregory, 1986). While gastrin–CCK immunoreactivity has been observed in species throughout the vertebrates, the specificity of such antibodies usually does not distinguish between the two hormones. Indeed it is arguable as to whether a distinct gastrin is present in species phyletically lower than the reptiles (Vigna, Thorndyke, and Williams, 1986).

Cholecystokinin exists in a variety of forms. CCK-33 (with 33 residues) is historically the prototype (Mutt, 1994) but in humans a CCK-58 has also been isolated. A truncated CCK-8 (the C-terminal octapeptide) is also present. They are all biologically active, both as hormones and as neuropeptides. The physiological actions of CCK are usually distinct from those of gastrin, though in pharmacological doses it can also increase gastric acid secretion. This effect appears to reflect the chemical similarities of the two hormones. The octapeptide sequence, with the sulfated tyrosine at position 7 from the C-terminal, instead of the 6 position as in gastrin, appears to be important for the differences in activity. Apart from mammals, CCK-like activity has been identified in the intestines of cyclostomes, chondrichthyeans, teleosts, amphibians, reptiles, and birds (Barrington and Dockray, 1970; 1972; Nilsson, 1970; Dockray, 1979). CCK-like activity has also been identified in the stomach of dogfish, but they apparently lack gastrin (Vigna, 1979). The gut of lampreys also lacks gastrin activity but CCK is present (Holmquist et al., 1979). Immunological evidence shows that gastrin and CCK cannot be identified in different types of cell in amphibians and teleosts; they react with antisera to either hormone (Larsson and Rehfeld, 1977).

It has been proposed that CCK and gastrin evolved from a common ancestral CCK-like molecule like that found in some amphibians (Larsson and Rehfeld, 1977). The skin of a colorful tree frog, *Hyla caerulea*, contains a decapeptide (Fig. 3.27) with an identical C-terminal pentapeptide sequence to that present in gastrin and CCK. It has a sulfated tyrosine at position 7 from the C-terminus and hence it is considered to be CCK-like. It is called *caerulein* and it was suggested (Larsson and Rehfeld, 1977) that CCK and gastrin may have evolved from such an ancestral molecule. Duplication of such a gene (the caerulein gene has been identified (Lasak *et al.*, 1987)) may have occurred, allowing the separate evolution of the two hormones.

The possible origins of the CCK-gastrin family may be traced back further to invertebrates. Clues provided by immunocytochemical observations led to the identification of two groups of neuropeptides from insects, the Madeira cockroach, *Leucophora maderae* (Nachman *et al.*, 1986), and the fruit fly,

Drosophila melanogaster (Nichols, Schneuwly and Dixon, 1988). The *leuco-sulfakinins* from the head region of the cockroach and the *drosulfakinins* from the fruit fly have a common C-terminal heptapeptide sequence that includes a sulfated tyrosine at position 6 from this terminus (Fig. 3.27). They exhibit structural homologies to gastrin and CCK but lack the exact C-terminal tetrapeptide sequence that is a prerequisite for CCK–gastrin family membership. They have, nevertheless, been considered as candidate ancestral-type (or ancestral-like) molecules. Closer to the vertebrates, the neural ganglion of a protochordate, the sea squirt *Ciona intestinalis*, has been shown to contain an octapeptide (Fig. 3.27) that includes the family C-terminal tetrapeptide sequence. In addition, at positions 6 and 7 from the C-terminus there are two sulfated tyrosine residues (Johnsen and Rehfeld, 1990). This neuropeptide, which is called *cionin*, therefore has profound similarities to both CCK and gastrin. It has been suggested that such a hybrid would "suit" a common ancestor for CCK and gastrin. A caerulein-like molecule such as that found in the skin of amphibians provides another, phyletically closer, model. A divergence of CCK and gastrin could have occurred in reptilian-like ancestors of the homeotherms (Vigna *et al.*, 1986). The structures of CCK-like molecules among the fishes is at present unknown and such information could be especially illuminating.

Receptors for CCK and gastrin have been identified in the brain and various peripheral tissues (Wank, 1995). They were originally classified on the basis of their abilities to bind different CCK/gastrin peptides and selective agonist and antagonist drugs. Those in the mammalian brain were found to be different to those found in the rat pancreas (Innis and Snyder, 1980). The principal peripheral alimentary receptors were called CCK-A receptors and those in the brain CCK-B receptors. The CCK-A subtype was also identified at some sites in the brain while the CCK-B subtype is also present in gastric parietal cells, some smooth muscle, and the pig pancreas (Morisset *et al.*, 1996). The CCK-A receptors readily interact with sulfated forms of CCK but not gastrin, while the CCK-B ones interact with both sulfated and nonsulfated CCKs and gastrins. Both types of these receptor have been cloned in humans and rats. The CCK-A receptors contain 428 to 444 residues and the CCK-B ones about 450 residues. They belong to the G protein-coupled superfamily of receptors containing seven transmembrane domains. Such receptors in different mammals display an identity of 85 to 93% in their amino acid sequences. The genes for human CCK-A and CCK-B receptors are organized similarly, with five exons and four introns. A third type of receptor, the "gastrin receptor," has been identified pharmacologically but its cloning has revealed that it is identical to the CCK-B receptor. It is now also called the CCK-B/gastrin receptor (Lee *et al.*, 1993). In 1986, Vigna and his collaborators, on the basis of measurements of differential binding of CCKs and gastrins to such receptors, suggested that cold-blooded vertebrates (fish, amphibians, and reptiles) only had a single type of CCK receptor. This receptor bound sulfated forms of gastrins and CCKs with a similar specificity. In contrast, in birds and mammals the two types of receptor present displayed a selectivity towards such peptides. It was, therefore, suggested that the CCK/gastrin receptors in warm-blooded vertebrates evolved from a single type of such receptor, probably in a reptile-like ancestor. A possible prototype for such a receptor has been cloned from an amphibian, the toad *Xenopus laevis* (Pratt *et al.*, 1994). This receptor exhibits a different pharmacological binding profile to mammalian CCK-A and CCK-B receptors with which it has a predicted amino acid sequence identity of, respectively, 55% and 56%. It has been dubbed the CCK-X receptor and it possibly reflects an ancestral type that gave rise to both CCK-A and CCK-B receptors by a process of involving gene duplication. Pharmacological binding studies have also identified a comparable single "primitive" CCK/gastrin receptor in the brain of a teleost fish, the goldfish (Himick, Vigna, and Peter, 1996).

Hormones of the opioid prohormone family

A single prohormone, coded by one gene, may with appropriate post-translational processing provide several hormones. The best-known example of such a multihormonal precursor is provided by a related group of prohormones that share several features, including the presence of a series of peptides with opioid (analgesic, pain killing) types of effect. These peptides have been called enkephalins and range in size from chains of 5 to 30 amino acid residues. Their amino acid sequences are incorporated in three prohormones: proenkephalin A, proenkephalin B (also called prodynorphin) and proopiomelanocortin (POMC). The last is also the parent molecule for several other hormones including corticotropin (adrenocorticotropin, ACTH), α-MSH (α-melanotropin) and β-MSH. All three of these prohormones are polypeptides and contain about 260 residues; they have a molecular mass of about 30 kDa. The arrangement of the components of their genes are remarkably similar, suggesting that they have an ancient relationship. Proenkephalins A and B have homologies of their amino acid sequences of 50 to 60% and probably arose from the duplication of the gene of a common ancestor (Horikawa *et al.*, 1983; Lewis and Erickson, 1986).

The enkephalins

The enkephalins are a family of peptides that are principally present in the brain but they are also formed in some other tissues. The prototypes are pentapeptides: methionine-enkephalin (or met-enkephalin), which has the amino acid sequence Tyr–Gly–Gly–Phe–Met, and leucine-enkephalin (Leu-enkephalin), which is identical to it except that leucine replaces the methionine. These peptides have analgesic effects and appear to act as neurotransmitters. Proenkephalin A is processed to produce four molecules of Met-enkephalin and one of Leu-enkephalin, while proenkephalin B provides three molecules of Leu-enkephalin. The POMC precursor contains one molecule of Met-enkephalin, which is incorporated in a 31 residue chain called β-endorphin (named from "endogenous morphine"). This peptide is an even more effective analgesic than the pentapeptides. There are several endorphins, some of which may have endocrine roles. They are present in substantial concentrations in the pituitary and hypothalamus and can be released into the circulation. Endorphins are able to influence the release of other pituitary hormones (Bicknell, 1985) possibly by exerting local, paracrine, effects.

The enkephalins are ancient molecules that have also been identified among the invertebrates. In vertebrates they are distributed phyletically from cyclostome fish to mammals (see King and Millar, 1980; Dores and Gorbman, 1990; McDonald, Joss, and Dores, 1991). Proenkephalin A appears to be present throughout the vertebrates (Lewis and Erickson, 1986; Lindberg and White, 1986) but variations in its amino acid sequence may occur. For instance, in amphibians proenkephalin A provides Met-enkephalin, but Leu-enkephalin is missing (Martens and Herbert, 1984). It appears, however, that this change is unique to amphibians and does not reflect a phylogenetic pattern (McDonald *et al.*, 1991).

The POMC peptides

The POMC prohormone (Smith and Funder, 1988) is formed at a variety of sites but it occurs mainly in the corticotrope cells of the pars distalis, the melanotropes of the pars intermedia, and the hypothalamus. Its enzymic processing can result in a variety of peptides, some of which are hormones. The particular peptides and hormones that are formed depend on the type of cell involved and they are different in the corticotropes and melanotropes. These events are summarized in Figs. 3.28 and 3.29. Two major peptides that can be directly cleaved from POMC are ACTH-(1–39) and β-lipotropin-(1–91). Beta-lipotropin (Fig. 3.30) was once considered to be a hormone but it is now merely viewed as a prohormone. It has incorporated in its amino acid sequence β-MSH (β-lipotropin-(41–58)), β-endorphin (β-lipotropin-(61–91)), and Met-enkephalin (β-lipotropin-(61–65)). The ACTH-(1–39) can be further processed to α-MSH (ACTH-(1–13)) and corticotropin-like intermediate lobe peptide (CLIP, ACTH-(18–39)). ACTH-(22–39) has also been shown to display biological activity by promoting the release of insulin (Eagle *et al.*, 1996). (It has, therefore, been dubbed beta-cell tropin or BCT.) In the corticotrope cells, the principal secreted product is corticotropin. In the melanotropes they are α-MSH, CLIP (via ACTH), and β-MSH (via β-

Fig. 3.28. A diagrammatic representation of the bovine proopiocortin molecule that is the precursor of pituitary ACTH and β-lipotropin. The sequences of amino acids can also be seen to contain several other peptides, including α- and β-MSH, β-endorphin, and Met-enkephalin. The active hormone fragments are progressively cleaved, in a cascade, from the precursor. Based on Nakanishi *et al.* (1979). (Modified from Krieger and Liotta (1979). Copyright © 1979 by the American Association for the Advancement of Science.)

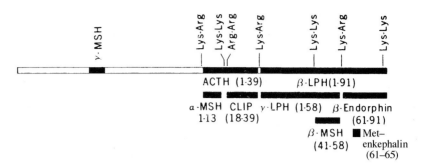

lipotropin). Beta-endorphin may be formed in both types of cell from β-lipotropin.

POMC-like molecules have been identified in many groups of the vertebrates from cyclostome fish to mammals. Differences in its structure have been observed. In salmon the γ-MSH that was first identified in the N-terminal region of mammalian POMC (Fig. 3.28) is absent and glycosylation, which also may occur in the mammalian precursor, does not occur (Kawauchi, 1983). In a toad, *Xenopus laevis*, the POMC is more similar to that in mammals, including the presence of the γ-MSH sequence and a glycosylation site (Martens, Civelli, and Herbert, 1985). The amino acid sequence of the amphibian POMC is 55% homologous to that of humans, which is greater than its similarity to that of the salmon precursor. Elasmobranch (the dogfish *Squalus acanthias*) POMC also has the γ-MSH sequence (Denning-Kendall, Sumpter, and Lowry, 1982) so that its absence in salmon appears to reflect a special evolutionary change. A cyclostome, the sea lamprey *Petromyzon marinus*, has two POMC-like precursors that are encoded by separate genes (Takahashi *et al.*, 1995). One of these contains domains composed of amino acids that correspond to ACTH, an MSH-like peptide, and β-endorphin, but it lacks β- and γ-MSH-like sequences (Heinig *et al.*, 1995). However, at the N-terminus the gene encodes a novel putative hormone called *nasohypophysial factor* (NHF), which has been identified in the sea lamprey (Sower *et al.*, 1995). This precursor has been called

Fig. 3.29. A diagram summarizing the hypothesis that α- and β-MSH are formed from the same precursor molecule as β-lipotropin and ACTH but only in cells in the pars intermedia. In the latter, enzymes are present that can cleave the parent molecule at certain positions (see Fig. 3.28) that results in the separation of the MSH forms. (Modified from Lowry and Scott (1975).)

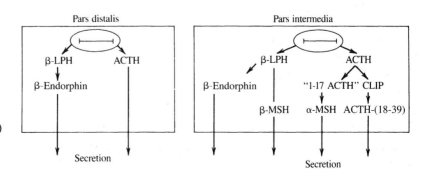

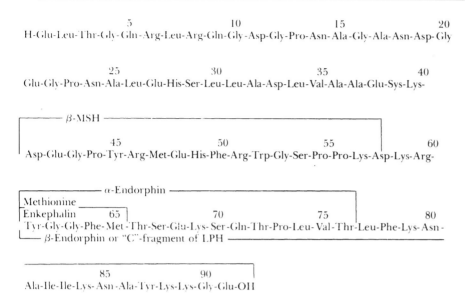

Fig. 3.30. Amino acid sequence of human β-lipotropin. The sections of the molecule that correspond to some of the known active fragments have been superimposed. (From Li and Chung, 1976.)

proopiocortin (POC) and its gene is expressed in the pars distalis. The amino acid sequences of the peptides that may be derived from it differ markedly from those in other vertebrates. A second gene encodes a precursor that contains the domains for two novel MSHs called MSH-B and MSH-A (p. 143) and a different β-endorphin to that present in POC. This precursor is called proopiomelanotropin (POM) and it is formed in the pars intermedia. (Immunocytochemical evidence (Nozaki *et al.*, 1995) suggests that MSH-A is also present in the pars distalis so that the POM gene may also be expressed in this lobe of the pituitary.) The two precursors exhibit amino acid sequences with a homology of 32% and similarities in their structural organization that suggest their evolution from a common ancestral gene (Takahashi *et al.*, 1995). The ancestral origins of the POMC protein can therefore, it seems, be traced back about 500 million years. Interspecific variability in the amino acid sequences of POMC will, inevitably, be reflected by the various hormones it gives rise to. Some of these are described below.

ACTH is a linear peptide containing 39 amino acid residues. This hormone acts on the zona fasciculata cells of the adrenal cortex to promote the synthesis of corticosterone and cortisol. It has been identified in the pars distalis of vertebrates ranging from cyclostomes to mammals. In the cyclostomes, an ACTH-like activity has been detected in the pituitary glands of lampreys using immunological and biological assays (Scott, Besser, and Ratcliffe, 1971; Eastman and Portanova, 1982; Baker and Buckingham, 1983). It has also been identified immunocytochemically at this site (Dores, Finger, and Gold, 1984). A detailed description of an ACTH from the sea lamprey *Petromyzon marinus* has been provided by A. Takahashi (quoted by

Active core of α and β-MSH / β-MSH   "Immunogenic tail" region of mammalian corticotropin

Human

H - Ser - Tyr - Ser - Met - Glu - His - Phe - Arg - Trp - Gly - Lys - Lys - Arg - Arg - Pro - Val - Lys - Val - Tyr - Pro - Asn - Gly - Ala - Glu - Asp - Glu - Ser - Ala - Glu - Ala - Phe - Pro - Leu - Glu - Phe - OH

  1   2   3   4   5   6   7   8   9 10 11 12 13 14 15 16 17 18 19 20 21 22 23 24 25 26 27 28 29 30 31 32 33 34 35 36 37 38 39

Substitutions relative to the human sequence (position: residue):

Pig	31: Leu
Ox	33: Gln
Sheep	25: Asp; 33: Gln
Turkey	13: Arg, 14: Arg, 15: Lys; 20: Ile; 27: Ser, 28: Val; 30: Glu, 31: Glu, 32: Ala, 33: Ser, 34: Tyr; 37: Val
*Xenopus***	15: Arg; 20: Ile; 27: Val; 30: Glu; 33: Ser; 34: Ser, 35: Tyr; 38: Met, 39: Leu
Salmon*	15: Arg; 24: Thr; 27: Val; 30: Glu, 31: Glu; 33: Ser; 34: Gly; 37: Ser; 39: Met
Dogfish / *Squalus acanthias*	14: Met; 15: Arg; 20: Ile; 27: Ser, 28: Phe; 32: Val; 34: Asn, 35: Met, 36: Gly, 37: Pro; 39: Leu

Beta-MSH (in mammals β-lipotropin-(41–58))

Pig, sheep: Asp-Glu-Gly-Pro-Tyr-Lys-Met-Glu-[His-Phe-Arg-Trp-Gly-Ser-Pro-Pro-Lys-Asp] (boxed: His-Phe-Arg-Trp-Gly-Ser-Pro-Pro-Lys-Asp)

Species	Substitutions
Xenopus	Asn; Asn; Arg; Arg
Salmon	Ser; Arg; Gly; Thr, Ala, Ile
Squalus acanthias	Gly, Asp, Asp; Phe, Gly; Leu; Ser, Val
Scyliorhinus canicula	Gly, Ile, Asp; Gly; Met, Asp, Lys; Ala

Fig. 3.31. The amino acid sequences of ACTH in various vertebrates. In the lower section of the figure are shown the structures of β-MSH. The boxed section indicates the presence of identical amino acids. (* From Dores, Adamczyk and Joss, 1990; ** from Martens *et al.* 1985. Based on Lowry and Scott, 1975.)

Nozaki *et al.*, 1995). Its N-terminal section displays clear homologies to the ACTHs in other vertebrates. There are, however, many differences including an extension of five amino acids at the N-terminus. The amino acid sequence of ACTH-(1–24) appears to be invariant in mammals, and this sequence exhibits full biological activity. Substitutions do occur in this region of the hormone in nonmammals but they are infrequent (Fig. 3.31). The principal variations in the structures of ACTH occur in the C-terminal region of the molecule (ACTH-(25–39)) which, as they can result in immunological differences, is called its immunological tail. Such changes occur among all species, including mammals, but tend to be greater as the phyletic distance increases: in dogfish, *Squalus acanthias*, corticotropin differs from human corticotropin by 11 amino acid substitutions, three of which are within positions 1–24. About 19 amino acid positions in corticotropin appear to be invariant, of which 18 are in ACTH-(1–24). Every position in the tail region appears to have been substituted in one or other species except, possibly, position 38.

The MSH types are peptides that in cold-blooded vertebrates may induce a darkening of the skin by promoting the dispersal of the pigment melanin in melanophores (Chapter 7). In mammals, they can increase the formation of melanin in the melanocytes and so increase the pigmentation of the skin. Two main types of MSH may be released into the circulation: α-MSH, which is a linear sequence of 13 residues, and β-MSH, which, depending on the species, may contain 16 to 22 amino acid residues (Fig. 3.31). The MSHs in all species so far examined, with two exceptions, contain a common pentapeptide "core" sequence (–His–Phe–Arg–Tyr–Gly–) that may be important for their mechanism of action. (In the dogfish and Russian sturgeon the glycine in this segment is replaced by serine.) Gamma-MSH was initially identified from its nucleotide sequence in mammalian POMC. It is formed in the pars intermedia and contains 26 amino acid residues (Martens *et al.*, 1985). The pentapeptide core sequence in γ-MSH has one substitution, aspartic acid replaces glycine. The amino acid sequences of human, rat, and *Xenopus* γ-MSH has been deduced from their nucleotide coding sequences (Martens *et al.*, 1985). Substitutions occur at 14 of the 26 positions. The physiological significance of γ-MSH is unknown. As described above, it seems to be absent in cyclostome and teleost fish.

Alpha-MSH (ACTH-(1–13)) undergoes a post-translational α-amidation at the C-terminus (ACTH-(1–13)amide) that protects it from breakdown by carboxypeptidase enzymes. It can also undergo acetylation at its N-terminus resulting in two possible variants: *N*-acetyl-ACTH-(1–13)amide and *N,O*-diacetyl-ACTH-(1–13)amide. These modifications result in an increase in potency. Acetylation occurs under the influence of an acetyltransferase enzyme in the hormone-secretory granules in the pars intermedia (where β-endorphin is similarly processed). Such acetylated α-MSH hormones have been identified in most vertebrates (Dores, Kaneko, and Sandoval, 1993a; Keller *et al.*, 1994). Phyletic differences, however, appear to exist in the relative predominance of the mono- and the diacetylated forms, leading to speculation as to the phylogeny of this post-translational processing. The monoacetylated α-MSHs are more frequent among some groups of fish: the Holocephali, Chondrostei, and Branchiopterygii. However, in others (the Holostei, Teleostei, and Dipnoi) diacetylated MSH is predominant. It is not clear whether mono- or diacetylation may be the ancestral condition (Keller *et al.*, 1994).

The amino acid sequences of α-MSH (ACTH-(1–13)) are remarkably conservative, possibly reflecting constraints on the structure of its precursor (ACTH-(1–39)). Only two changes have been described in this hormone: at single positions in a toad (*Xenopus*) and a dogfish (*Squalus acanthias*) hormone (Fig. 3.31). Beta-MSH is much more variable; only seven of its amino acid positions are invariant. Even two species of dogfish only display a homology of about 65% in their β-MSH amino acid sequences. An amphibian (*Xenopus*) β-MSH has a homology to the mammalian hormones of about 80% while the similarity for dogfish and salmon compared with mammals is, respectively, about 50% and 60%. In the Russian sturgeon, identity to mammalian β-MSH is greater than in the other fish, about 70% (Nishii *et al.*, 1995). Despite this variability in overall structure, the pentapeptide core is well conserved in β-MSH, only one change being observed, serine being substituted for glycine in a dogfish and the Russian sturgeon.

An unequivocal demonstration of the presence of MSH in the cyclostome fishes has been elusive. Immunocytochemical procedures and bioassays suggested that while such peptides may be present in the pituitary glands of lampreys, their precise chemical structures probably differ substantially from those in other vertebrates (Baker and Buckingham, 1983; Dores *et al.*, 1984; Nozaki and Gorbman, 1984). Recently two MSH-like molecules have been identified in the pituitary of the sea lamprey *Petromyzon marinus* (A. Takahashi, quoted by Nozaki *et al.*, 1995). They possess the characteristic pentapeptide core but otherwise differ considerably from the MSHs in other vertebrates. They have been called MSH-A (19 amino acid residues) and MSH-B (20 residues). The position of MSH-A on its precursor molecule (POM) tentatively suggests that it may correspond to β-MSH in higher vertebrates (Takahashi *et al.*, 1995). Immunocytochemical observations using antisera to these peptides indicate that MSH-A is present in both the pars distalis and pars intermedia but MSH-B is only present in the latter (Nozaki *et al.*, 1995).

A considerable number of changes in the amino acid sequences of the peptides in the POMC family have been described. It is tempting to try to ascribe to each of these an optimal physiological functioning of the particular hormones or the mechanisms of their biosynthetic processing from the parent precursor. However, it should also be recalled that two or more hormones or peptides may share sequences in the precursor. Therefore, α-MSH in mammals has an identical sequence to ACTH-(1–13) and β-MSH to γ-lipotropin-(41–58). There is little variation in α-MSH and the active region (1–24) of ACTH. Thus, the requirements of one hormone may place restraints on evolutionary changes in another with which it shares a common sequence in the precursor. The greater variability in the structure of β-MSH may partly reflect a lack of evolutionary restraint on γ-lipotropin, which has no known physiological role. The evolution of the hormones in the POMC family may be expected to exhibit a "togetherness." The evolution of such family hormones may reflect changes not only in the gene coding for the precursor but also in the mechanisms of the post-translational processing. Such an event probably led to the perpetuation of the separate corticotrope and melanotrope cells, each of which produces distinctive products from the same precursor.

The gonadotropin–thyrotropin family: luteinizing hormone, follicle-stimulating hormone, and thyroid-stimulating hormone

The anterior lobe of the pituitary gland is the site of synthesis and release of hormones that control the activities of the testes and ovaries, as well as the thyroid gland. These hormones (Papkoff, 1972; Licht *et al.*, 1977; Bousfield *et al.*, 1994) are proteins with a molecular weight of about 30 000 and they contain about 12–20% carbohydrate. They exhibit strong similarities in their general tertiary structure and amino acid sequences. The carbohydrate content of each may vary and this contributes to the presence of isoforms of the hormones that have differences in their activities. All of these hormones consist of two nonidentical subunits (α and β), which are not covalently bound. In mammals two have been named according to their particular actions on the gonads (Chapter 9): *follicle-stimulating hormone* (FSH) and *luteinizing hormone* (LH). The latter also stimulates hormone production from interstitial cells in the testis. These hormones are included in the broad category of gonadotropic hormones. Other such hormones with a similar structure and lineage have been identified in members of all classes of the vertebrates, where they seem to perform similar functions. Two structurally related variants of the pituitary gonadotropins are found in mammals, where they are formed during pregnancy in the placenta. They are human *chorionic gonadotropin* (hCG) (other primates also appear to possess an analogous hormone) and *pregnant mare's serum gonadotropin* (PMSG), which has been renamed *equine chorionic gonadotropin* (eCG). The pituitary gland also secretes TSH, which has remarkable similarities to the gonadotropins.

The pituitary and placental gonadotropins and thyrotropin each contain an α-subunit and a β-subunit. The α-subunits are identical in all such hormones within a single species. The β-subunits, however, have differing structures that determine the type of activity the intact hormone exhibits. These subunits are called FSH-β, LH-β, TSH-β, and in teleost fish, gonadotropic hormones GTH-Iβ and GTH-IIβ. The subunits can be chemically separated and later recombined, either with each other or with subunits from other hormones. Each subunit by itself lacks biological activity; however, if TSH-α is combined with LH-β the product regains activity but this corresponds to LH. Conversely, if LH-α is combined with TSH-β the molecule has TSH activity. Therefore, the β-subunit determines which type of action the molecule has, while the α-subunit is necessary for its expression. Such recombinations of subunits from different sources also suggest that it is the β-subunit that determines the species specificity of a particular hormone. The ability of the hybrid turtle LH-α–ovine LH-β to increase secretion of testosterone from turtle testis was only 12% of that seen by the homologous combination (turtle LH-α–turtle LH-β) whereas that of ovine LH-α–turtle LH-β was 64% of that elicited by the native hormone (Licht, Farmer, and Papkoff, 1978).

The precise sequence of the amino acids and carbohydrates have been determined in the subunits of the mammalian and some nonmammalian glycoprotein hormones. In humans (Fig. 3.32) it can be seen that LH-α, TSH-α, and hCG-α are identical in their structures. The β-subunits show variation but they are still in many respects similar (Fig. 3.33). The homologies in the sequences of FSH-β and hCG-β are about 40%, and LH-β and hCG-β are over 80%. There are also differences in the chain lengths, FSH-β contains 118 residues, LH-β 115, TSH-β 112, and hCG-β 145. The similarities between hCG and both LH and FSH may be reflected in the observation that hCG has both LH-like and FSH-like activity (Siris $et\ al.$, 1978), though the former predominates.

Each subunit of the glycoprotein hormones is coded by a separate gene (Daniels-McQueen $et\ al.$, 1978; Kourides and Weintraub, 1979; Godine, Chin and Habener, 1980; Fiddes and Goodman, 1980). Duplications of such genes occur, however, and in the instance of hCG multiple genes may code for the β-subunit. The similarities in the structures of the products of all the genes have led to the suggestion (Fontaine and Burzawa-Gerard, 1977) that they may have evolved from an ancestor to form an α-subunit and a family of β-subunit genes coding for the specific structure in each type of hormone (Fig. 3.34). The structural organization of the genes for the α- and β-subunits of the gonadotropins have been described in several mammals and two teleost fish, carp and chinook salmon (Xiong, Suzuki, and Hew, 1994). The α-subunit genes vary considerably in size but they all contain four exons separated by three introns. The β-subunit genes are different and, apart from being smaller, contain only three exons interspersed by two introns. This structural arrangement has also been observed in the TSH-β-subunit gene.

hLH-α:
Val-Gln-Asp-Cys-Pro-Glu-Cys-Thr-Leu-Gln-Glu-Asn-Pro-Phe-Phe-Ser-Gln-Pro-Gly-

hCG-α:
Ala-Pro-Asp-Val-Gln-Asp-Cys-Pro-Glu-Thr-Leu-Gln-Glu-Asp-Pro-Phe-Phe-Ser-Gln-Pro-Gly-

Alignment
position: 1   10

Ala-Pro-Ile-Leu-Gln-Cys-Met-Gly-Cys-Cys-Phe-Ser-Arg-Ala-Tyr-Pro-

Ala-Pro-Ile-Leu-Gln-Cys-Met-Gly-Cys-Cys-Phe-Ser-Arg-Ala-Tyr-Pro-

20   30
Thr-Pro-Leu-Arg-Ser-Lys-Lys-Thr-Met-Leu-Val-Gln-Lys-Asn-Val-
                                          CHO

Thr-Pro-Leu-Arg-Ser-Lys-Lys-Thr-Met-Leu-Val-Gln-Lys-Asn-Val-

         40   50
Thr-Ser-Glx-Ser-Thr-Cys-Cys-Val-Ala-Lys-Ser-Tyr-Asn-Arg-Val-Thr-Val-
Thr-Ser-Glu-Ser-Thr-Cys-Cys-Val-Ala-Lys-Ser-Tyr-Asn-Arg-Val-Thr-Val-

                 60
Met-Gly-Gly-Phe-Lys-Val-Glx-Asn-His-Thr-Ala-Cys-His-Ser-Cys-Thr-Cys-Tyr-Tyr-His-Lys-Ser
                 CHO

Met-Gly-Gly-Phe-Lys-Val-Glu-Asn-His-Thr-Ala-Cys-His-Cys-Ser-Thr-Cys-Tyr-Tyr-His-Lys-Ser-

     70   80   90

Fig. 3.32. The amino acid sequences of the α-subunits of human LH (hLH) and hCG. The α-subunit of human TSH has the same structure as human LH-α. (Based on Sairam, Papkoff, and Li, 1972; Morgan, Birken and Canfield, 1975.)

The pituitary glands of nonmammals have all (except cyclostomes) been shown to possess a biological activity that indicates the presence of a hormone or hormones similar to the mammalian gonadotropins. Members of all the classes of tetrapod vertebrate have been shown to possess two such hormones (Licht, 1983) though in some species only one has been identified. A survey of several families of snakes (Ophidia, Reptilia), for instance, has failed to uncover a second gonadotropin (Licht *et al.*, 1979). The one that is present has neither a distinct FSH nor LH activity. Earlier investigations on teleost fish also revealed only a single gonadotropin (Burzawa-Gerard and Fontaine, 1972), and the same situation seems to exist in a chondrostean, the sturgeon (Burzawa-Gerard, Goncharov, and Fontaine, 1975) and an elasmobranch, the dogfish (Sumpter *et al.*, 1978). However, several species of teleosts, including carp and some salmon, have since been shown to possess two gonadotropins with the same types of α- and β-subunits as seen in other vertebrates (Idler and Ng, 1979; Trinh *et al.*, 1986; Itoh, Suzuki, and Kawauchi, 1988; Swanson *et al.*, 1991; van der Kraak *et al.*, 1992). The similarities of their biological activities to those of LH and FSH are not clear, so they have been called gonadotropins I and II (GTH-I and GTH-II). GTH-II appears to be quite LH-like but the relationship of GTH-I and FSH is a phylogenetically distant one (see later). It remains possible that some

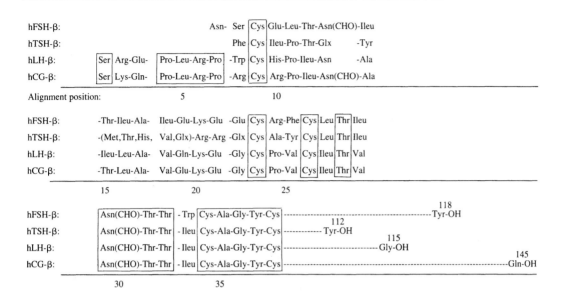

Fig. 3.33. Comparison of the N-terminal amino acid sequences of hFSHβ, hLH-β, hTSH-β, and hCG-β. The boxed-in sections indicate regions of the molecules where the amino acid sequences are identical. The remaining parts also show considerable homologies. (Based on Saxena and Rathnam, 1976.)

teleosts possess only one GTH but isoforms could contribute to some structural diversity in such a situation (Banerjee *et al.*, 1993).

The phyletic distribution of gonadotropins suggests that in ancestral forms there may have been a single gonadotropic hormone that fulfilled all the required functions. A contemporary gonadotropin that can clearly exhibit such a dual activity is pregnant mare's serum gonadotropin. It has been suggested that this ability may reside in a special C-terminal region of its β-subunit (Moore and Ward, 1980). Subsequently, another such hormone may have evolved that facilitated a more specialized control over gonadal function.

"There are no clear cut, well documented cases of species specificity of gonadotrophic hormones" (Nalbandov, 1969). Different species invariably show *some* response to heterologous gonadotropins. There are, however, considerable variations in the biological potency of such hormones, confirming that polymorphic variations exist. Mammalian gonadotropins exhibit some activity in all vertebrates. In teleost fish, mammalian LH and hCG are sometimes effective but FSH is inactive (Burzawa-Gerard and Fontaine, 1972). Teleost gonadotropins, however, while being very active in teleosts, have little effect in mammals. Amphibians, reptiles, and birds show considerable responsiveness to gonadotropins from teleost, chondrichthyean, and dipnoan fishes, as well as to those from mammals. The avian hormones are more effective in lizards than in mammals (Licht and Stockell Hartree, 1971; Burzawa-Gerard and Fontaine, 1972; Donaldson *et al.*, 1972; Scanes *et al.*, 1972). While a reptilian gonadotropin preparation, from the snapping turtle, is ineffective *in vivo* on mammalian test preparations it stimulates the ovarian granulosa cells of a monkey *in vitro* and is also active in birds, amphibians,

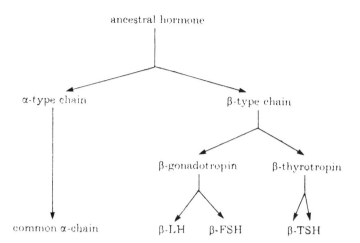

Fig. 3.34. A scheme showing the possible pathway of the evolution of the pituitary glycoprotein hormones: LH, FSH and TSH. Each of these hormones consists of two subunits, α and β. The α-subunit is remarkably similar in all these hormones but differences exist in the β-subunits, which confer on the hormone its particular selective, physiological effect and species specificity. (The α-subunit, however, is vital for the response to occur.) It is suggested that each subunit originally arose as a result of gene duplication. The structure of the α-subunit was then largely conserved while that of the β-subunit underwent further dup-lications and point mutations to provide the materials needed to assemble the three hormones. (From Acher, 1980.)

and other reptiles (Channing *et al.*, 1974; Licht and Papkoff, 1974). No gonadotropin preparations would appear to be completely species specific, but they exhibit considerable differences in their activity when tested on preparations from other phyletic groups. Such differences are assumed to reflect variations in their molecular structures and those of their receptors.

Immunological cross-reactions are also indicative of chemical relationships between hormones. A survey had been made (Licht and Bona Gallo, 1978) comparing the abilities of FSH from various tetrapods to interfere with the binding of human FSH to antibodies to ovine FSH. FSH from the various eutherian mammals was usually most effective in preventing the binding of human FSH; birds and reptiles showed less cross-reactivity. The amphibians showed a very poor reactivity and the curves were also not parallel to those of FSH from all the other species. The snakes were an interesting exception as they showed no activity in this system. One marsupial, the kangaroo *Macropus eugenii*, showed less cross-reactivity than all of the eutherians, falling among the birds and reptiles. One cannot trace the phylogenetic history of hormones in this way but it emphasizes the variations between them. It should be emphasized that similarities and differences in biological and immunological responses of hormones need not parallel each other, as the associated changes in the molecules may have evolved independently for each type of activity. The structure of the receptors, which could also change, may influence the hormone's biological activity.

The amino acid sequences of many of the α- and β-subunits of the gonadotropins have been determined in species ranging from fish to mammals. A comparison of the homologies between some of these subunits is shown in Fig. 3.35. It can be seen that in closely related species, such as cattle and pigs, chickens and quail, and salmon and carp, the structures of each type of subunit are usually quite similar to each other. The homologies between

Fig. 3.35. Homology (%) matrices of the amino acid sequences of the apoproteins of the precursors of the gonadotropin α-subunits (a) and the LH-β subunits (b) from various vertebrates. (Based on Ishii *et al.*, 1993.)

(a) Gonadotropin α-subunits

Cattle	97	85	85	72	66
	Pigs	84	84	74	64
		Quail	100	71	66
		Domestic fowl		71	66
			Carp		74
			Salmon		

(b) LH β-subunits

Cattle	85	44	43	44	39	42
	Pigs	45	44	44	40	43
		Quail	92	39	45	47
		Domestic fowl		41	45	48
			Bullfrog		48	50
			Salmon			77
			Carp			

the α-subunits, and between the β-subunits are, respectively, in each of these listed pairs of species 97% and 85%, 100% and 92% and 74% and 77%. Phyletically more distant species display greater differences, the homologies often then only being 40 to 50%. The amino acid sequences of the subunits in GTH-I and GTH-II of the striped bass, *Morone saxatilis*, have been deduced from their encoding nucleotide sequences and these have been compared with other teleosts (Hassin, Elizur, and Zohar, 1995). Homologies varied from 40 to 97% with the greatest differences being exhibited by the GTH-I β-subunit. The GTH-I β-subunit and GTH-II β-subunit showed sequence identities to human FSH β-subunit of 39 and 43% respectively, and to human LH β-subunit of 36 and 46%, respectively. This degree of conservation across the phyletic scale is remarkable.

In 1940, Gorbman found that the goldfish thyroid tissue was stimulated by pituitary extracts from a teleost fish, two amphibians, a bird, and a mammal. This suggested that TSH had a wide phyletic distribution. Subsequent measurements using a greater variety of species to compare the activity of such glandular extracts (in addition to the goldfish, a salamander, lizard, and guinea pig were used) indicated that the hormones present in the various species were not identical, though they exerted the same general biological effects. Although mammalian TSH preparations are active in teleosts, teleost TSH has little activity in mammals (Fontaine, 1969a,b). As the phyletic scale is ascended, it is found that TSH preparations from a lungfish (*Protopterus*), amphibians, reptiles, and birds can exert well-defined effects on the thyroid of both a mammal (mouse) and teleost fish (trout). The thyroid of chondrichthyean fish (the stingray *Dasyatis sabina*), although responding to its own, homologous TSH, shows no response to the mammalian or even teleost hormones (Jackson and Sage, 1973). Chondrichthyean TSH, however, stimulates the mammalian thyroid (Dodd and Dodd, 1969). Mammalian TSH increases thyroidal activity in the Pacific hagfish *Eptatretus stouti* (Cyclostomate), though TSH activity (when tested in mammals) has not been demonstrated in the pituitary of the Atlantic hagfish, *Myxine glutinosa* (Dodd and Dodd, 1969; Kerkof, Boschwitz, and Gorbman, 1973). One obviously cannot construct an ordered story from these observations, but they serve to show that TSH, like other hormones, has suffered changes in its structure during evolution.

In mammals (humans, horses, cattle, and pigs) there is a 75% to over 90% identity between the amino acid sequences of TSH-α (Chin *et al.*, 1981). The TSH-β of humans, mice, pigs, and cattle show a similar, 85–90% identity (Gurr, Catterall, and Kourides, 1983). Such structural information is not, apparently, available in nonmammals, with the exception of the rainbow trout. The TSH-β-subunit of this teleost has an impressive 51% homology of its amino acid sequence to that of human TSH-β (Ito *et al.*, 1993). (The sequence identity to GTH-Iβ and GTH-IIβ in the trout was, respectively, 31% and 33%). Purified preparations of such thyrotropins have been made in various species including turtles (MacKenzie, Licht, and Papkoff, 1981), ostriches, bullfrogs (MacKenzie and Licht, 1984) and several teleost fish (Ng, Idler, and Eales, 1982; Swanson, Dickoff, and Gorbman, 1987; Byamungu, Darras, and Kuhn, 1991; Banerjee *et al.*, 1994). These hormones exhibit a thyrotropic action in homologous and generally, also heterologous species (see, for instance, MacKenzie and Licht, 1984; Ng, Idler, and Eales, 1991). They appear to have the usual two subunit structure and similar amino acid compositions (Banerjee *et al.*, 1994).

Gonadotropins can exert thyrotropic actions in a variety of species (heterothyrotropic effect; Fontaine, 1969a). For instance, hCG has a thyrotropic effect in humans and bullfrog LH is quite potent in this respect in chickens (MacKenzie and Licht, 1984). Such observations provide support for the proposal (Fontaine, 1969a) that the gonadotropins and thyrotropin may have shared a common ancestral gene (Fig. 3.34). A single gene could have encoded a hormone with both actions and following its duplication a divergence of more specific subunits and hormones could have occurred. Phylogenetic trees based on similarities in the amino acid sequences of the β-subunits of the gonadotropin/thyrotropin family have been constructed (Hassin, Elizur, and Zohar, 1995; Wako and Ishii, 1995). The GTH-I appears to have diverged very early in vertebrate evolution. Subsequently two further groups emerged: FSH/TSH and GTH-II/LH/hCG. Thus FSH appears to be more closely related to TSH than it is to LH or GTH-II. The last apparently originated much later than GTH-I, which appears to occupy an extant primeval position on this evolutionary tree.

The transforming growth factor-β superfamily

The inhibins, activins, and antimullerian hormone (AMH) are gonadal hormones and are members of a superfamily that is called after their prototypical member transforming growth factor-β (TGF-β). It includes a variety of growth factors that have been identified in several species including, apart from mammals, an amphibian (*Xenopus*) and even an insect (*Drosophila*). Such growth factors can inhibit or promote the growth of cells, such as those in epithelia, bone, and blood, as well as in embryos and tumors. They are bonded by the homologies of their chemical structures and structural organization. They appear to share a common ancestry. These hormones and growth factors are glycoproteins and exist as homo- or heterodimers linked by disulfide bridges. They are formed from larger precursor molecules containing about 400 amino acid residues and their biological activities usually residing in the C-terminal regions (Fig. 3.36). It is here that the closest homologies in their structures are observed. Usually the C-terminal domain contains 110 to 135 residues including seven highly conserved cysteines. As this section of the molecule contains its biologically active sequences, such evolutionary conservatism appears to be apt. This domain can be cleaved from its precursor at strategic sites consisting of basic amino acids, such as arginine and lysine. The inhibins, activins, and antimullerian hormone have homologies in their amino acid sequences of 20–40% with TGF-β.

Inhibins

Inhibin was discovered and named in 1932 (McCullagh, 1932) and was subsequently almost forgotten for about 40 years. Aqueous extracts of the testes of bulls prevented the hypertrophy of the pituitary gland that normally follows castration in rats. It was demonstrated, much later, that inhibins (two have been identified) block the increased release of FSH, but not that of LH, that occurs in such rats (see Lincoln, McNeilly, and Sharpe, 1989; Vale, Bilezikjian, and Rivier, 1994). Inhibins have also been extracted from the

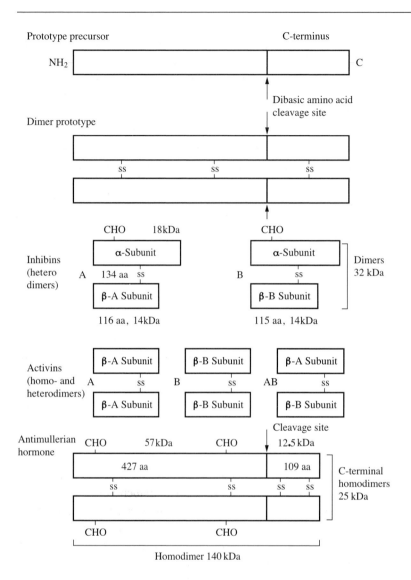

Fig. 3.36. Diagramatic representation of the precursors of the various subunits of hormones of the TGF-β superfamily. The information on inhibins and activins are based on the human hormones. aa, amino acid residues; CHO, carbohydrate; S–S, disulfide bridges. (Based on Mason *et al.*, 1985; Lee and Donahoe, 1993.)

fluids of the ovarian follicles of sows and cows, and the rete testis of rams. They are produced by the granulosa cells of the ovarian follicle, by the corpus luteum, and by the Sertoli cells of the seminiferous tubules. The inhibins (Mason *et al.*, 1985) are glycoproteins comprising two subunits that are assembled as dimers which are linked by one or two disulfide bridges. The subunits are formed from the C-terminal segments of the precursors (Fig. 3.36). There are three types of subunits: α (a glycosylated molecule with a molecular mass of 18 kDa), β-A (nonglycosylated, 14 kDa) and β-B (nonglycosylated, 14 kDa). The precursor for each is coded by a separate gene but there are strong similarities between each of these genes in a single species suggesting that they arose by gene duplication. The homologies of the amino acid sequences between the β-A- and β-B-subunits are about 70%. The structural identities of the α- and β-subunits are about 30%. Inhibin A is composed of an α-subunit and a β-A-subunit, while inhibin B consists of α- and β-B-subunits. Such an arrangement of linked subunits is reminiscent of that which occurs in the pituitary gonadotropins (FSH and LH) and thyro-tropin. Interspecific homologies between the amino acid sequences of the α-subunits of inhibins in mammals (humans, pigs, cattle, and rats) are about 80%. Such similarities are even greater for the β-subunits, which differ at fewer than three amino acid sites in such species (see Vale *et al.*, 1988). In the domestic fowl, the α-subunit has an amino acid sequence that is about 60% homologous to that of mammals (Johnson and Wang, 1993). Such relation-ships are impressive even within a superfamily.

Activins

Some extracts of gonadal fluids contain substances that stimulate the release of FSH from cultured pituitary cells and they are, therefore, called activins. This activity was found to result from differently assembled subunits from the same precursors that form the inhibins. However, activins only utilize the β-subunits. Three such possible combinations exist and all of these have been identified in tissue extracts. They are the combination of two β-A-subunits to form a homodimer called activin A, two β-B-subunits called activin B, and the heterodimer composed of the β-A- and β-B-subunits that is activin AB (Fig. 3.36). All of these dimers exhibit FSH-releasing activity *in vitro* but their roles *in vivo* are so far undefined.

There is little information available about the presence of inhibins and activins in cold-blooded vertebrates. An immunocytochemical study using antibodies to mammalian inhibin and activin α- and β-subunits has indicated their presence in the ovaries and testes of goldfish (Ge *et al.*, 1993). Under *in vitro* conditions, mammalian inhibins and activins *both* stimulate the release of gonadotropin from goldfish pituitaries (Ge *et al.*, 1992).

The physiological roles of both the inhibins and the activins are not completely understood at this time. Their actions do not appear to be to be confined to the reproductive system. They have been identified in many nongonadal tissues including brain, adrenals, spleen, and placenta (see, for instance, Bilezikjian and Vale, 1992). It is considered likely that they may exert local regulatory paracrine or autocrine effects at such sites, including the testes and ovaries (see Chen, 1993).

Follistatins

Follistatins inhibit the release of FSH, *in vitro*, apparently by binding to the activins (Nakamura *et al.*, 1990). They were originally isolated from ovarian follicular fluids (see Lincoln *et al.*, 1989; Vale *et al.*, 1994) and are glyco-proteins that are rich in cysteine residues. The two that have been identified contain 315 and 288 residues. They are *not* members of the TGF-β superfamily but are the products of a single unrelated gene. Follistatins have been identified in many tissues and may act locally to modulate the effects of activins.

Antimullerian hormone

The fetal testis of mammals secretes a hormone that inhibits the differentiation of the embryonic Müllerian ducts (Lee and Donahoe, 1993; Josso, 1994). It is called either antimullerian hormone (AMH) or mullerian-inhibiting substance (MIS). In the fetal genetic female embryo, the Müllerian ducts are the precursors of the female reproductive tract, and this hormone is absent at this stage of their life. Its action in the male only occurs during a brief period of fetal life but it then assures the suppression of the development of the female reproductive ducts. The male pattern of development then proceeds. Following birth, antimullerian hormone is present in the young of both sexes but its levels decline at puberty. It is produced by the Sertoli cells of the seminiferous tubules and the granulosa cells of the ovarian follicles. Antimullerian hormone is a homodimeric glycoprotein with a molecular mass of 140 kDa. Its subunits are linked by disulfide bridges (Fig. 3.36). The structure of antimullerian hormone indicates that it is a member of the TGF-β superfamily. It is not clear whether the entire 140 kDa molecule is the active hormone (MacLaughlin *et al.*, 1992). In other members of the TGF-β superfamily, the C-terminal domain is separated and it is this section of the molecule that constitutes the active hormone. A mutant form of antimullerian hormone in which such cleavage is not possible has been found to be inactive. The C-terminal 24 kDa dimer of the molecules can be separated *in vitro* and retains activity. The remaining N-terminal portion is inactive. However, what happens *in vivo* is still not agreed upon. The amino acids in the 25 kDa active domain are highly conserved and in humans, cattle, and rats have homologies in their amino acid sequences of about 95%.

The growth hormone–prolactin family

Growth hormone (somatotropin) and prolactin are the prototypes of an extensive family of protein hormones with molecular masses of about 20 to 26 kDa and which contain about 200 amino acid residues. It also includes somatolactin and several placental lactogens. Growth hormone, prolactin, and somatolactin mainly occur in the pituitary gland, the last in the pars intermedia the other two in the pars distalis. Somatolactin appears to be confined to teleost fishes but growth hormone and prolactin occur throughout the vertebrates, with the possible exception of the cyclostome fishes. As their name indicates, the placental lactogens are mammalian hormones that occur in the placenta. They are structurally and phylogenetically a quite diverse group that includes human placental lactogen (somatomammotropin) and are also present in other primates. At least two other types of placental lactogen occur: in rodents and in ungulates, such as cows, ewes, does, and goats (Talamentes, 1975; Talamentes *et al.*, 1980). These hormones display a variety of effects on growth, development, intermediary metabolism, and reproductive functions (such as lactation and pregnancy), as well as in osmoregulation, molting, and, in amphibians, metamorphosis.

Several members of this family are usually present in a single species where they have different roles that may vary at different stages of the species life cycle. Each hormone is encoded by a separate gene that usually displays homologies which are consistent with a shared evolution. In addition, each hormone may have variants, or isoforms, that result from alternate splicing of their mRNA, modifications such as glycosylation and phosphorylation, and the formation of aggregates. Such variation may provide the raw material for mediating the many different physiological effects exhibited by this remarkable hormone family (see, for instance, Sinha, 1995).

Growth hormones and prolactins have been identified, using biological and immunological assays, in all the major groups of vertebrates apparently, with the exception of the cyclostomes. The amino acid sequences of many of these hormones have been determined thus allowing speculation as to the reasons for many of their common effects and their shared ancestry. The structure of human growth hormone is shown in Fig. 3.37. It is a single-chain polypeptide containing 188 residues and two cross-linking disulfide bridges. The amino acid sequence of growth hormone, prolactin, and placental lactogen have been compared and considerable homologies exist. There are 160 (out of 190) identical residues in human growth hormone and placental lactogen although, of the remainder, only seven positions are occupied by what are considered "nonhomologous" amino acids (Li, 1972). Such similarities in mammalian hormones led to the suggestion (Bewley and Li, 1970; Niall *et al.*, 1971) that the three hormones may have arisen from a common ancestral molecule. Various segments of each hormone molecule also bear considerable similarities to each other (internal homologies). The ancestral molecule may have been a smaller peptide of 25–50 residues that, by a process of genetic reduplication in a "tandem" manner, led to an increase in the chain length of the hormones (but see p. 163).

The chemical similarities in the molecules are reflected in their biological activities. Apart from the dual effects of placental lactogen on growth and lactation, prolactin exhibits growth hormone-like activity whereas growth hormone has (though more limited) prolactin-like actions. Such common properties of these hormones are not simply caused by similarities in their

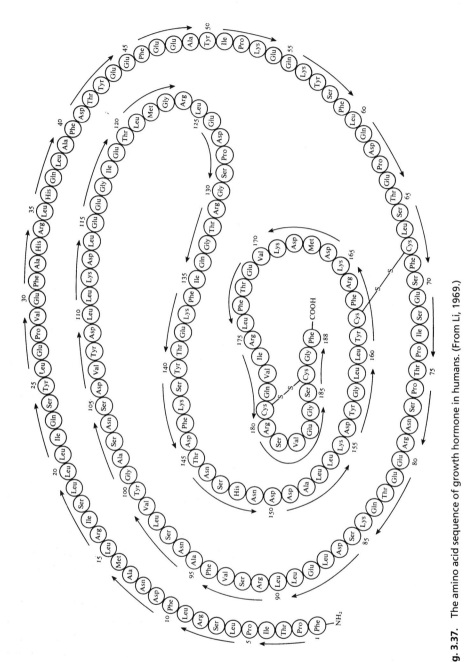

Fig. 3.37. The amino acid sequence of growth hormone in humans. (From Li, 1969.)

Fig. 3.38. Molecular phylogenies of growth hormone based on the homologies of their amino acid sequences. The numerals indicate the number of amino acid changes per 100 residues that may be expected between the various points of divergence indicated. (Modified from Noso, Lance, and Kawauchi, 1995.)

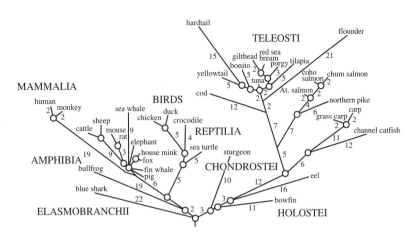

amino acid sequences but also by the common sites of intramolecular disulfide bridges and the molecule's three-dimensional structure. Crystallographic studies on porcine growth hormone have revealed the presence of a clump of four helices comprising groups of 13 to 32 amino acid residues (Abdel-Meguid *et al.*, 1987). The components of these structures have been identified in growth hormones from other mammals and bony fishes and are thought to extend to other members of the growth hormone/prolactin family (Rubin and Dores, 1994; Goodman, Frick, and Souza, 1996; Rubin *et al.*, 1996). These characteristic α-helical motifs appear to contribute to an optimal three-dimensional structure that is necessary for the interactions of such hormones with their receptors.

The amino acid sequences of many growth hormones that are distributed throughout the vertebrates from elasmobranch fish to mammals have been described. The homologies of 39 such hormones have been summarized by Noso, Lance, and Kawauchi (1995) and some of this information is included in Table 3.7. A molecular phylogenetic tree has also been constructed (Fig. 3.38). The observations based on chemical structures of the hormones generally confirm earlier pioneering studies that used immunological interactions between pituitary extracts and specific antibodies to the hormones to predict their phylogenetic relationships (Tashjian, Levine, and Wilhelmi, 1965; Hayashida, 1970; 1971). Amino acids at only 35 positions in growth hormone appear to be invariant among all vertebrates (Goodman *et al.*, 1996). Nonprimate mammals share 135 such positions. Only 102 amino acid sites are known to be common to primate (humans and rhesus monkeys) and nonprimate mammals. Evolution of growth hormone in the primates, therefore, appears to have been rapid since they diverged from ancestral mammals about 40 million years ago. About 60 amino acid changes appear to have occurred during that time.

Prolactins have been isolated and their amino acid sequences determined in most of the major groups of vertebrates. They usually contain about 200

Table 3.7. *Homologies of the amino acid sequences of growth hormone and prolactin among vertebrates*

	Percentage identity	
	Growth hormone	Prolactin
Mammalia		
Human	100	100
Cattle	67	76
Bird		
Domestic fowl	59	72
Reptilia		
Sea turtle	48	75
Amphibia		
Bullfrog	46	65
Dipnoi		
Lungfish	64[a]	58
Teleostei		
Salmon	34	35
Holostei		
Bowfin	43	-
Chondrostei		
Sturgeon	51	36
Elasmobranchii		
Blue shark	54	-

Homologies between growth hormones and prolactins from the same species are about 20–26%.
[a] Based on a partial sequence in the African lungfish from H. Kawauchi (unpublished data quoted by Rubin and Dores (1994)). The percentage homology is versus the rat, which has a homology of about 65% with human growth hormone.
Data from Kawauchi, Yasuda, and Rand-Weaver, 1990; Noso, Lance, and Kawauchi, 1995; Noso, Nicoll, and Kawauchi, 1993a; Noso *et al.*, 1993b.

amino acid residues but there are fewer (about 190) in some fish. Greater differences in chemical structure exist among the prolactins than the growth hormones. Especially notable are the prolactins in teleost fish, which, apart from being smaller, only contain two disulfide bridges instead of three as seen in most other vertebrates. Some homologies between the prolactins are summarized in Table 3.7. The sturgeon (Chondrostei), which is considered to be a phyletic precursor of the teleosts, has a prolactin with three disulfide bridges (Noso *et al.*, 1993b). The bowfin (Holostei), which represents a link between the Chondrostei and Teleostei, however, has two such disulfide bridges, like the teleosts (Dores *et al.*, 1993a). Therefore, the teleostean-type prolactin may have arisen in an ancestor it shared with the holosteans. The lungfishes (Dipnoi) have three disulfide bridges in their prolactin (Noso,

Nicholl, and Kawauchi, 1993a) like tetrapod vertebrates and the chondro-stean fish. The prolactins of the teleosts and holostean fish appear to reflect a separate line of evolution of this hormone.

Some species of vertebrates have been shown to possess two prolactins. These include the toad *Xenopus laevis* (Yamashita *et al.*, 1993). The two proteins are not identical in this species but they have an identity of their amino acid sequences of 90%. This duplication appears to reflect the doub-ling of the entire genome of this species. The teleost tilapia *Oreochromis niloticus* also has two prolactins (Specker *et al.*, 1993) but these have only a 69% homology to each other. One of these hormones has 188 residues and the other 177. In this species, a gene duplication followed by an evolutionary divergence appears to have occurred. In tilapia, both prolactins are produced in a single type of cell and they are apparently both present in the same storage granules.

Prolactin increases milk secretion in mammals, and this response can be used to measure the hormone's activity even in the low concentrations that appear in the plasma (Frantz, Kleinberg, and Noel, 1972). Pituitary gland extracts from birds, reptiles, and amphibians all promote this response but not those from fishes. Pigeons secrete a milk-like paste (pigeons-milk) from their crop-sac with which they feed their young. This response is stimulated by prolactin from tetrapods *and* lungfishes, but that from other fishes is ineffective. Prolactin, when injected into certain newts (*Notophalmus (Diemictylus) viridescens*) at a particular stage in their life cycle, causes them to seek water preparatory to breeding. This is called the "eft (or newt) water-drive response" and can be initiated by prolactin from all the principal groups of vertebrates except the cyclostomes, which seem to lack a prolactin hormone (Bern and Nicoll, 1968). This response cannot be mimicked by any other pituitary hormone and has been used to demonstrate the presence of an analogous prolactin-like secretion throughout the vertebrates. Further evi-dence of the occurrence of this hormone in fishes has followed the discovery that certain teleost fishes, when in fresh water, usually die following removal of the pituitary gland; death results from excessive losses of sodium. When injected with mammalian prolactin, they retain sodium and survive. Teleosts' pituitaries contain a hormone that also has the latter effect and which, as a reflection of its difference from mammalian prolactin, has also been called "paralactin." The phyletic distribution of all these effects follows a precise pattern that is shown in Fig. 3.39.

The foregoing observations suggested two things: first, that the prolactin hormone is not identical in all vertebrates and has been subject to evolution-ary change and, second, that it seems likely that it has assumed diverse biological roles.

Neither a growth hormone nor a prolactin have been directly identified in the pituitary gland of cyclostome fish. This apparent absence is phylogeneti-

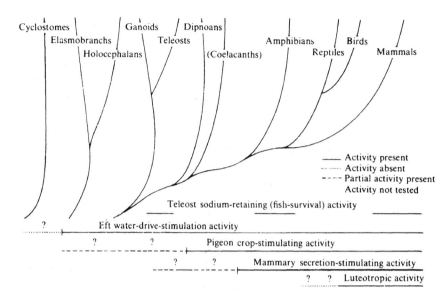

Fig. 3.39. Distribution of some of the biological activities that can be initiated by mammalian prolactin and prolactin-like hormones from other vertebrates. (From H. A. Bern, personal communication in Bentley, 1971.)

cally discomforting to many. It could, apart from a lack of experimental persistence, reflect very low concentrations of the hormones or a perplexing dissimilarity of their structures from such hormones in other vertebrates. An immunocytochemical study using antisera to mammalian growth hormone and prolactin showed interactions in two different types of cell in the pars distalis of the sea lamprey *Petromyzon marinus* (Wright, 1984). Modern gene technology could, perhaps, help resolve this mystery.

Somatolactin is the most recent hormone to be identified in the growth hormone–prolactin family. Histological clues were provided by a novel type of cell present in the teleost pars intermedia that stained with periodic acid–Schiff (PAS) reagent. A putative hormone was isolated from the pituitary glands of flounder (Ono *et al.*, 1990) and it was localized immunocytochemically in such "PIPAS" cells. The hormone is present in the plasma of teleosts where it may mediate such functions as are involved in reproduction (Rand-Weaver *et al.*, 1992). In contrast to prolactin and growth hormone, it is a glycoprotein but it has, among its sequence of 207 residues, homologies of 23 to 28% with members of the growth hormone–prolactin family. To celebrate this relationship it has been called somatolactin. In addition to that in the flounder, the structure of somatolactin has been described in Atlantic cod and chum salmon, to which homologies of 73 to 81% in amino acid sequence have been observed. These similarities are greater than those for the corresponding growth hormones (58 to 62%) (Takayama *et al.*, 1991). Somatolactin has been identified in at least 10 species of teleosts. The possibility that the somatolactin gene may be present and expressed in tetrapod vertebrates has been pursued and it has been stated that "tetrapod SL is now being characterized" (Ono and Kawauchi, 1994). The role of somatolactin in bony fish is uncertain but, apart from correlations with reproductive processes, its presence in the plasma has been observed to respond to changes in external osmotic concentration, environmental calcium levels, and acidity (see, for instance, Kakizawa, Kaneko, and Hirano, 1996). Its release has recently been shown to be promoted in red drum, *Sciaenops ocellatus*, adapted to a dark background colour, but a physiological role in mediating color change responses is undefined (Zhu and Thomas, 1996). Nevertheless, the injection of somatolactin into the red drum promotes skin palor, which reflects an aggregation of melanin in the melanophores (Zhu and Thomas, 1997).

Immunocytochemical observations indicate that somatolactin is present in the pars intermedia of two holostean fish: the bowfin *Amia calva* and the longnose gar *Lepidosteus osseus* (Dores *et al.*, 1996). However, it was not found in a chondrostean fish, the lake sturgeon *Acipenser fulvescens*. The Holostei are considered to lie phylogenetically between Chondrostei and Teleostei. Somatolactin, therefore, possibly originated in a common ancestor of the Holostei and Teleostei over 200 million years ago.

Several members of the growth hormone–prolactin family are secreted by the mammalian placenta. They are quite diverse hormones and are called placental lactogens. They have various effects on growth, intermediary metabolism, and lactation. Such hormonal activities were originally extracted from the human placenta and were subsequently found also to be present in Rodentia, Lagomorpha, and Artiodactyla (see Talamantes, 1975; Talamantes *et al.*, 1980). The hormones have been isolated and studied in detail in humans, rats, mice, hamsters, cows, and ewes. They can be divided into at least three main subgroups based on their structures, genetic origins, and effects.

Human placental lactogen (hPL, also called human somatomammotropin), like prolactin, promotes lactation but also has a prominent growth hormone-like effect. In pregnancy, it has a glucose-sparing action that may be its most important physiological role (Conley and Mason, 1994). The amino acid sequence of human placental lactogen is 85% homologous to human growth hormone but only 21% to human prolactin. It is, therefore, thought to have arisen as a result of the gene duplication of a growth hormone gene. Rodent placental lactogen has principally lactogenic rather than somatotropic activity and is, therefore, thought to have originated following the duplication of a rodent prolactin gene. The ruminants (Artiodactyla) possess placental lactogens that exhibit both lactogenic and somatotropic effects, though their amino acid sequences are more similar to prolactin than growth hormone. Ovine placental lactogen has an homology of its amino acid sequence to ovine and bovine prolactin of 48% (Warren *et al.*, 1990). It also has a 67% identity to bovine placental lactogen but only 29% to rat and mouse placental lactogen. The ruminant placental lactogens may have originated from prolactin genes but have a different lineage to the rodent hormones.

The placental lactogens have been found to be expressed by genes that are often present in duplicate. Other related genes code for proteins called "prolactin-related proteins," which have quite diverse structures (see, for instance, Wallis, 1992; Anthony *et al.*, 1995). The pituitary growth hormones and prolactins have much more conservative structures than the placental lactogens. The latter appear to have evolved quite rapidly following their appearance in the mammals. Their potentially important roles and evolution, which are apparently associated with the appearance of mammalian viviparity, are at present not well defined.

The cytokine receptor superfamily: growth hormone and prolactin receptors

The hormones of the growth hormone–prolactin family exhibit precise relationships between their chemical structures and their biological effects (Nicoll, Mayer, and Russell, 1986). Nevertheless absolute specificity of action of an individual hormone for a particular species is rare. There is usually a substantial crossover in their activities when they are administered to other (heterologous) species but the potencies of their effects may then differ. Such observations are presumed to reflect the similarities and differences of their receptors in such species. A parallel evolution of receptors and hormones would appear to have occurred but each retains common characteristics. In mammals, the receptors for growth hormone have been described. In rabbits and humans these receptors are different but they retain an 84% homology of their amino acid sequences (Leung *et al.*, 1987). They form a single chain with an extracellular, transmembrane, and intracellular domains (Chapter 4). The extracellular hormone-binding domains of human growth hormone receptors form dimers when they interact with a single molecule of their ligand (De Vos, Ultsch, and Kossiakoff, 1992). Prolactin receptors are often smaller than growth hormone receptors, as observed in rats (Boutin *et al.*, 1988), but their size varies. The receptors in this family have been shown to have a structural kinship to receptors for a number of cytokines such as erythropoietin, granulocyte-macrophage colony-stimulating growth factor, and interleukins (Bazan, 1990; Patthy, 1990; Nakagawa *et al.*, 1994). Similarities have been observed in modules of such receptors that contain about 90 amino acid residues from their extracellular hormone-binding domains, and in the structural organization of their genes. These relationships resulted in the suggestion that they all belong to a cytokine superfamily of receptors. Such a structural similarity may reflect the retention of an important primeval building block domain in the receptor's genes. It could have been dispersed during evolution by exon duplication and shuffling from a primeval encoding gene or exon (Nakagawa *et al.*, 1994). During subsequent evolution, the binding domain may have been conserved (Russell and Nicoll, 1990) and differences between the receptors could have been the result of nearby structural changes that "hindered" interactions of hormones with the binding sites. Such changes could produce a "lock and key" situation, which has been used by pharmacologists to describe such relationships between receptors and their ligands. Information about the structures of growth hormone and prolactin receptors and their genes in nonmammalian vertebrates is sorely needed to test this interesting hypothesis.

Evolution of the growth hormone-prolactin family

This family of hormones, which also includes somatolactin and the placental lactogens, is thought to have evolved from a single primordial protein. Comparison of the amino acid sequences of the hormones and those of their precursors indicate an ancestral relationship. Common features in the organization of the exons and introns in their genes are also consistent with such a shared origin.

Internal homologies of amino acid sequences of some of the mammalian hormones in this family suggested to Niall and his collaborators (1971) that the primordial gene may have been quite small. Evolution may have initially occurred as a result of tandem-like duplications within the gene resulting in its elongation. This interesting idea, however, was not found to be consistent with subsequent observations of the amino acid sequences of such hormones in nonmammals, especially the fishes (Nicoll *et al.*, 1986; Agellon *et al.*, 1988).

The genes encoding the hormones of contemporary family members vary considerably in their size but generally exhibit a common five exon and four intron pattern in their organization (Chen *et al.*, 1994; Ono and Kawauchi, 1994). It is considered possible that a similar ancestral gene may have been present in an invertebrate with divergence occurring about 500 million years ago (Kawauchi, Yasuda, and Rand-Weaver, 1990). Growth hormone in some teleost fish (salmonids and tilapia) has an additional intron inserted in exon V and this event may have occurred subsequent to the divergence. The diaspora of the different hormones in this family presumably took place following gene duplications and in the case of the placental lactogens occurred relatively recently, less than 80 million years ago, during the evolution of the mammals. A scheme for the evolution of these hormones is shown in Fig. 3.40. Many contemporary teleost fish use both growth hormone and prolactin in the regulation of their water and salt metabolism (Chapter 8). It has been suggested that the ancestral hormone may have been utilized for a similar purpose (Chen *et al.*, 1994).

Leptin

In the summer of 1949, a plump type of the yellow house mouse, *Mus musculus*, was identified at the Jackson Memorial Laboratory in Maine (Ingalls, Dickie, and Snell, 1950). The young mice ate voraciously and attained a body weight about four times that of their normal siblings. This difference was associated with the mutation of a gene called *obese* (*ob*), which is an autosomal recessive character. The mutant *ob/ob* mice feed apparently without adequate satiation and they also exhibit a decreased fat catabolism. A variety of such rodent models for obesity have since been identified (Friedman and Leibel, 1992). Another such fat mutant mouse, which develops diabetes mellitus, has a mutant *diabetes* gene and is termed a *db/db* mouse. The pattern of its metabolic disorder is similar to that of the *ob/ob* mice but a different gene and mechanism is involved (see below). In 1994, a group at the Rockefeller University cloned the normal *ob* gene in mice as well as its homolog in humans (Zhang *et al.*, 1994). The genes mature protein (Ob) product contains 167 amino acid residues and it has been identified in the circulation. The principal hormonal activity may reside in a 35 amino acid residue segment near the N-terminus, abutting the signal peptide of the gene product (Samson *et al.*, 1996). The mouse and human gene products display an homology of 84% in their amino acid sequences. Homologous genes were detected in other mammals, as well as in the domestic fowl and a teleost fish (an eel). No known similarity to other known proteins was identified but its three-dimensional structure suggests a relationship to cytokines such as interleukin-2 and growth hormone (Madej, Boguski, and Bryant, 1995). When administered to normal and *ob/ob* mice, but not *db/db* mice, the normal protein product of the *ob* gene results in decreased feeding and a loss of body weight (Campfield *et al.*, 1995; Pelleymounter *et al.*, 1995). The Ob protein has been called *leptin* from the greek *leptos*, meaning thin. Leptin is released from white fat cells (not brown fat cells (Cinti *et al.*, 1997)) following feeding and the response appears to be mediated by elevated plasma insulin concentrations and possibly also by glucocorticoids (de Vos *et al.*, 1995; MacDonald *et al.*, 1995; Saladin *et al.*, 1995). Receptors for leptin (Ob-R) have been identified in several tissues including the choroid plexus and hypothalamus (Tartaglia *et al.*, 1995). In the latter, it is associated with the paraventricular nucleus (Woods and Stock, 1996). Isoforms of this receptor have been found in several embryonic tissues and in reproductive organs (Cioffi *et al.*, 1996). In the *db/db* mice, a receptor that spans the plasma membrane once has been identified and it originates at the *db* gene locus (Baumann *et al.*, 1996; Chen *et al.*, 1996). However, in this mutant form of the mouse, the receptor displays an abnormal shortening of its intracellular signaling domain so that it cannot respond to administered leptin. Indeed leptin in these mice is already present in an excess in the circulation. In the mutant *ob/ob* mice, the leptin molecule is defective so that adequate satiety is not signalled during feeding. In some obese humans, the leptin receptors appear, like in the *db/db* mice, to be abnormal and unable to respond to normal or elevated plasma leptin levels (Rohner-Jeanrenaud and Jeanrenaud, 1996). However, it is possible that there are structural changes in the leptin molecule in such humans that render it ineffective (McGregor *et al.*, 1996). It is also possible that leptin may fail to gain access across the choroid plexus to appropriate sites in the brain (Schwartz *et al.*, 1996). Leptin is a relative newcomer on the endocrine stage and it remains possible that it exhibits other physiological roles, such as ones involved in reproduction and development (Cioffi *et al.*, 1996) and the physiology of starvation (Chapter 5).

The natriuretic peptide family

The heart is a somewhat unexpected endocrine gland. Information that it had such a physiological role initially arose from histological observations on changes in the numbers of granule-like structures in atrial muscle cells (de Bold, 1985). The administration of sodium chloride to rats reduced the numbers of these granules while sodium restriction increased them (de Bold, 1979). These organelles have the appearance of storage sites that are utilized by various secretions such as neurotransmitters and hormones. When extracts of such atrial muscle or the separated granules were injected into mammals, they were found to produce a brief decline in blood pressure accompanied by a diuresis and natriuresis. The active material was purified and found to be a peptide (Fig. 3.41) containing 28 amino acid residues, 17 of which form a ring connected by a disulfide bridge (Flynn, de Bold, and de Bold, 1983). The precursor of this peptide in humans and rats was found to consist of 126 residues with the 18-membered active peptide as its C-terminus (Kangawa *et al.*, 1984). This peptide is called either atrial natriuretic peptide (ANP), atriopeptin, or atrial natriuretic factor (ANF). Several variants have been identified, not only in the heart but also in the brain and hypothalamus. They have also been found in many nonmammals, including the cyclostome fishes and even some invertebrates (Reinecke, 1989; Takei and Balment, 1993; Vesely *et al.*, 1993). On the basis of their structure and the organization of their separate encoding genes, they have been classified into three (or possibly four) groups (Rosenzweig and Seidman, 1991; Samson, 1992).

Type A natriuretic peptide or ANP-28 (the numeral refers to the number of amino acid residues it contains) is principally found in atrial heart muscle. Its chemical structure has been highly conserved in mammals (human ANP differs from rat, mouse, and rabbit ANP by a single amino acid substitution). ANP has also been identified in fish (the eel *Anguilla japonica*, Takei *et al.*, 1989), amphibians (the frog *Rana catesbeiana*, Sakata, Kangawa and Matsuo, 1988) and reptiles (the turtle *Pseudemys scripta*, Reinhart and Zehr, 1994).

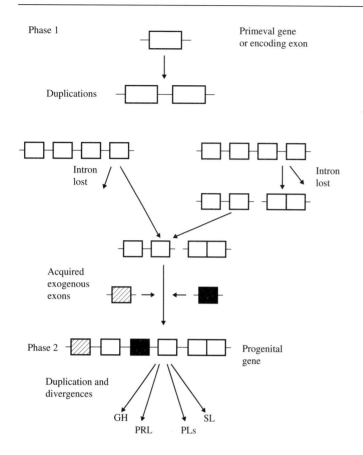

Fig. 3.40. A model for the evolution of the growth hormone–prolactin family. The extant genes for the hormones in this family usually consist of five exons and four introns, which code areas of homologous amino acid sequences. Such a gene is considered to be the progenital (precursor) gene for the family that, by a process of duplication (phase II), gave rise to growth hormone, prolactin, placental lactogens, and somatolactin. The insertion of an additional intron into exon V in the growth hormone gene in some teleost fish is viewed as occurring after the divergence of this ancestral gene.

In mammals, the identification of regions of internal homology in the genes for these hormones resulted in the suggestion that the original primeval gene was quite small but was enlarged as a result of tandem duplications of exons, the loss of an intron to form exon V, and the acquisition of two novel exons, II and III. This process is shown in phase I. Such internal homologous domains are, however, not apparent in the teleost hormones so that the details suggested in phase I are equivocal. GH, growth hormone; PRL, prolactin: PLs, placental lactogens; SL, somatolactin. (Based on Agellon *et al.*, 1988 and Ono and Kawauchi, 1994.)

The amino acid sequence of the eel ANP was found to have a 59% homology to the mammalian peptide and 45% to that of the frog. The amino acids in the 17-membered ring structure are the most highly conserved ones.

Type B natriuretic peptide (BNP)) is present in the heart and brain. It has the same ring structure as ANP but has extensions of its N-terminal amino acid chain (Sudoh *et al.*, 1988). In humans and pigs (Fig. 3.41) it contains 32 residues (BNP-32), but the rat has 45 such residues (BNP-45) and cattle 35 (BNP-35). Apart from maintaining the characteristic ring structure, BNP has many amino acid substitutions, which are seen in different species. Chicken BNP-29 (Miyata *et al.*, 1988) is quite similar to porcine BNP but differs substantially from that in the rat. Type B is the least conserved group of the natriuretic peptides.

Type C natriuretic peptides (CNPs) are mainly found in the brain (Sudoh *et al.*, 1990) but they have also been identified in the heart of some fish. In pigs, CNP contains 22 amino acid residues and it characteristically lacks the C-terminal amino acid tail that is present in Types A and B (Fig. 3.41). The CNP-22 peptide has been found in the brain of two species of teleosts, three elasmobranchs, the domestic fowl, and the bullfrog (Suzuki *et al.*, 1994;

Type A (ANP-28)   Ser-Leu-Arg-Arg-Ser-Ser-Cys-Phe-Gly-Gly-Arg-Met*-Asp-Arg-Ile-Gly-Ala-Gln-Ser-Gly-Leu-Gly-Cys-Asn-Ser-Phe-Arg-Tyr(28)

Type B (BNP-32) Ser-Pro-Lys-Thr-Met-Arg-Asp-Ser-Gly-Cys-Phe-Gly-Arg-Arg-Leu-Asp-Arg-Ile-Gly-Ser-Leu-Ser-Gly-Leu-Gly-Cys-Asn-Val-Leu-Arg-Arg-Tyr(32)

Type C (CNP-32)   Gly-Leu-Ser-Lys-Gly-Cys-Phe-Gly-Leu-Lys-Leu-Asp-Arg-Ile-Gly-Ser-Met-Ser-Gly-Leu-Gly-Cys(22)

Fig. 3.41. Amino acid sequences of type A, type B and type C atrial natriuretic peptides. The boxed residues within the ring structure indicate identical residues in all three types. The underlined residues indicated homologies in two types. * Replaced by isoleucine in rat, mouse, and rabbit. Type B may have an extended N-terminus in several species to give human BNP-32 (shown), rodent BNP-45, porcine BNP-26, and bovine BNP-35. (Based on Atlas and Maack, 1992.)

Takei, Ueki, and Nishizawa, 1994c). The structures of the CNPs are highly conserved. For instance, the amino acids present in the 17-membered ring in pig and killifish CNP are identical (Takei and Balment, 1993). Rat, pig, and human CNP are completely identical (Suga *et al.*, 1992). Elasmobranch CNP has an 85% identity of its amino acid sequence to that in mammals (Suzuki, Takahashi, and Takei, 1992).

A fourth type of natriuretic peptide has been found in the ventricles of the heart of a teleost fish, the eel *Anguilla japonica*, and the rainbow trout, *Oncorhynchus mykiss* (Takei *et al.*, 1994b). It may also be present in quail. This peptide, which has been called *ventricular natriuretic peptide* (VNP), contains 36 amino acid residues including the 17-membered ring. A less potent VNP-25 also exists in eels (Takei *et al.*, 1994a). VNP has an elongated tail containing 14 residues at its C-terminus. Trout and eel VNP have a 77% homology of their amino acid sequences and they are more similar to ANP than to BNP. The VNPs differ from the other natriuretic peptides on the basis of their structures and genes and may belong to a unique group.

There is experimental evidence suggesting that there may be a local production of a natriuretic peptide by the mammalian kidney tubule (Goetz, 1991). It has been called *urodilatin*. A physiological role for this substance is, however, uncertain (Sonnenberg, 1994). In humans the infusion of urodilatin results in a natriuresis and diuresis while dietary supplements of sodium enhance its excretion of the urine (Meyer *et al.*, 1996).

Several cleavage products of the precursor of ANP, proANP-(1–126), in addition to ANP-28 (proANP-(99–126)), exhibit biological activity in mammals (Martin *et al.*, 1990; Vesely *et al.*, 1994). These peptides all lower blood pressure and may promote diuresis, natriuresis, and a kaliuresis. They have, to emphasize their special individual actions, been called long-acting sodium stimulator peptide (proANP-(1–30)), vessel dilator peptide (proANP-(31–67)) and kaliuretic stimulator peptide (proANP-(79–98)). The promotion of renal potassium excretion by the kaliuretic peptide is an especially interesting effect that is mediated by an inhibition of the enzyme Na–K ATPase (Chiou and Vesely, 1995). Their normal physiological roles remain to be established.

The most prominent observed effects (Atlas and Maack, 1992) of the natriuretic peptides involve dilatation of peripheral blood vessels and an enhanced renal excretion of water and sodium (Chapter 8). They may, therefore, be involved in homeostatic adjustments of the volumes of the body fluids. Their release occurs in response to distension of the right atrium. The natriuretic peptides have also been observed to exhibit effects that suggest they may be acting as neurotransmitters and neuromodulators, as they can decrease the release of vasopressin, renin, aldosterone, and CRH. Three types of plasma membrane receptor for these peptides have been identified. The effects of two of these are linked to an activation of guanylate cyclase (Chinkers *et al.*, 1989) and they are therefore called GC-A (or ANP-A) and

GC-B (or ANP-B). These receptors are considered to be prototypes for guanylate cyclase-linked hormone receptors. The GC-A receptor reacts most strongly with ANP and BNP and GC-B with CNP (Suga *et al.*, 1992). Specific binding sites for the natriuretic peptides that are also considered to be receptors have been identified in kidney aorta, adrenal cortex, and the brain of mammals (Atlas and Maack, 1992; Levin, 1993). They have also been found in the kidney and aorta of a cyclostome fish (the hagfish *Myxine glutinosa*) (Kloas *et al.*, 1988). They are called "clearance" or C-type receptors.

The natriuretic peptide family appears to be a persistent and unusually conservative group of hormones with an ancient lineage (Takei, 1994). They may have evolved in fishes living in a marine environment where their actions, involving the regulation of excessive accumulations of sodium chloride, may have been opportune. Despite considerable available information, a phylogenetic tree of their possible evolution has not, it seems, yet been proffered.

The guanylin peptide family

In 1978 M. Field and his collaborators identified a heat-stable toxin that was produced by the intestinal bacterium *Escherichia coli*. It causes a watery diarrhea owing to its ability to promote the secretion of chloride by intestinal mucosal cells. This enterotoxin (Forte and Hamra, 1996) was found to be a peptide consisting of 16 amino acid residues including six cysteines that form three disulfide bridges. Its mechanism of action involves an activation of guanylate cyclase in the apical plasma membranes of the intestinal epithelial cells. Receptors for this bacterial toxin were initially identified in several tissues of the opossum *Didelphys virginiana*, including the intestine and renal tubule. A search for possible natural endogenous excitants that interacted with these receptors resulted in the identification of two peptides, which also activated guanylate cyclase and were called *guanylin* and *uroguanylin*. Guanylin was first found in the rat jejunum (Currie *et al.*, 1992) and uroguanylin in the urine of the opossum (Hamra *et al.*, 1993). Both peptides have been identified in other mammals, including humans, where they are present in the blood (Kuhn *et al.*, 1993; Kita *et al.*, 1994). Their principal source in the body appears to be the intestine, especially the colon. The guanylins, like the enterotoxin produced by *E. coli*, are peptides containing 15 to 16 residues and six highly conserved cysteine residues. Differences in the sequences of their N-terminal amino acids appear to determine their abilities to interact preferentially with guanylin receptors under alkaline (guanylin) or acidic (uroguanylin) conditions following their secretion into the lumen of the intestine (Hamra *et al.*, 1997). Guanylin and uroguanylin have homologies in their amino acid sequences with the *E. coli* toxin of, respectively, 47% and 67% (Forte and Hamra, 1996). Human and opossum guanylin have homologies to each other of 73%. The abilities of guanylins to promote salt excretion by the intestine and the kidney, and their presence in the circulation, has resulted in the suggestion that they may function as hormones contributing to the regulation of sodium chloride excretion (Chapter 8). The bacterium *E. coli* apparently possesses a toxin that mimics the mammalian hormone, and which in excess can produce diarrhea.

Adrenomedullin

Adrenomedullin (Kitamura *et al.*, 1993a,b; 1995; Schell, Vari, and Samson, 1996), is a polypeptide that was initially identified in extracts of tumors of human adrenomedullary tissue (pheochromocytomas). It was biologically characterized in rats by its ability to increase cyclic AMP levels in blood platelets and by a potent vasodilatatory effect. This polypeptide contains 50 to 52 amino acid residues. It includes a disulfide bridge which forms a six-membered intramolecular ring that is essential for the expression of its vasodilatatory actions. As described above, its C-terminal region, including the ring structure, exhibits structural homologies in its amino acid sequence to calcitonin gene-related peptide and amylin. Such peptides and polypeptides have even been sometimes shown to interact with the same receptors (see, for instance, Zimmermann, Fischer, and Muff, 1995; Mazzocchi *et al.*, 1996). Porcine adrenomedullin (containing 52 residues) differs from the human polypeptide by a single amino acid substitution. The rat polypeptide has 50 residues and there are six substitutions compared with human adrenomedullin. Using bioassays, immunocytochemical procedures, and RNA identification, adrenomedullin has been identified in the adrenal medulla, heart ventricles, kidneys, lungs, vascular endothelial tissue, hypothalamus, and adenohypophysis. It appears to be present in highest concentrations in the adrenal medulla. Adrenomedullin is also present in the blood and its levels there are increased in diseases associated with an expansion of the volume of the body fluids, such as occurs in congestive heart failure (Kobayashi *et al.*, 1996). Apart from its vasodilatatory effect it promotes a diuresis and natriuresis (see for instance, Vari, Adkins, and Samson, 1996). Conversely, it may inhibit drinking and salt appetite (Samson and Murphy, 1997). Immunocytochemical observations indicate that an adrenomedullin-like polypeptide may be present in neural tissue in an echinoderm (a starfish), suggesting that it has a long evolutionary history (Martinez, Unsworth, and Cuttitta, 1996).

The evolution of hormones and their receptors: summary

Speculation about the origins and evolution of hormones and their receptors can be provided from knowledge of their contemporary structures and phyletic distribution in extant species. The genes encoding many such molecules have been identified and their structural organization studied. The comparative anatomy of such genes and that of the hormone's precursors they encode has provided important new information for such conjecture. It now seems unlikely that hormones and their receptors each coevolved from a common ancestral gene. Receptors and the genes that encode them have little in common with their hormonal ligands and their genes. They appear to have different lineages. The complementary interactions between the hormones and their receptors usually only involve limited regions of such molecules. In the instance of the receptors, their ligand- and DNA-binding sites consist of well-conserved domains containing only a small proportion of their amino acids. The amino acid sequences at these binding sites are, however, usually highly conserved. Nevertheless, a parallel evolution involving ligand binding and recognition sites on receptors and hormones appears to have been inevitable. Such mutual dependence suggests that each will impose constraints on the evolution of the other. Hormones and their receptors appear to have had separate origins, becoming acquainted with each other in some primordial organism, possibly a eukaryotic microorganism.

Hormones may have originated in such microorganisms (O'Malley, 1989) as part of a mechanism for controlling transcription. A transcription factor or prototype receptor could have been regulated by utilizing an internal (autocrine) or external (paracrine) ligand. An interaction of such a transcription factor and "hormone" may have facilitated the control of metabolism and growth of the cell. Other types of hormone-dependent homeostatic mechanism, not involving transcription factors, could have also arisen concomitantly, such as ligand-activated ion channels and enzymes situated in the plasma membrane. Subsequently, simple multicellular organisms would have required more complex coordination for cell-to-cell communication, where paracrine-type messages would have been appropriate. In more highly evolved animals with a circulatory system, classic hormonal control systems may have emerged (Csaba, 1986).

It is conceivable that hormones may have been able to act without receptors. Lactoferrin is a polypeptide that is released from neutrophils in response to invasion of the blood by microorganisms. Lactoferrin enters its target natural killer (NK) cells where it binds directly to DNA and initiates transcription. This event is remarkable as it does not involve any intervention from other proteins, such as receptors or transcription factors (He and Furmanski, 1995; see also Baeuerle, 1995). Such a mechanism for the control of transcription may provide a model for cell communication prior to the appearance of classical receptors. The lactoferrin could even be considered to have a hormone-like role but with no requirement for a receptor to mediate its effects. On the basis of this analogy, there could be life for a hormone without a receptor. Nevertheless, the intervention of receptors may provide nuances in the control processes that facilitate the responses to the hormones.

The evolution of particular hormones and receptors each appear to have occurred within related families and superfamilies. Such relationships suggest that each group had a common ancestral gene and that their numbers multiplied as a result of gene duplication and subsequently diversified by point mutations and genetic rearrangements. Alternate splicing of transcripts from a single gene may contribute to the diversification of these families of proteins. Such families of hormones and receptors exhibit homologies of their structures including their tertiary configuration.

The first molecules to function as hormones could have been quite simple ones that were utilized because they were available and fulfilled certain chemical and physical requirements, such as size, electrical charge, and solubility in water and lipids. Amino acids (see, for instance, O'Malley, 1989) and their derivatives provide a modern pattern for such excitants. Glycine, GABA, glutamate and 5-HT function as neurotransmitters, while catecholamines, which are derived from tyrosine, can fulfill both neural and hormonal roles. Such excitants are present throughout the animal kingdom, even in unicellular species. Melatonin synthesized from tryptophan and T_3 from tyrosine are two other examples of such relatively simple extant hormones. Sterols, such as cholesterol, are quite abundant constituents of animal cells and provide the template for several very conservative hormones. Indeed one such product, vitamin D_3, can even be acquired as a prohormone in the diet to be subsequently converted to the active hormone.

The emergence of more complex polypeptide and protein hormones required a greater genetic involvement. It is, however, unlikely that a primeval polypeptide hormone was specifically invented but rather that a gene product with another function was coopted as it fulfilled certain prerequisites. It has been suggested that such molecules could have been provided by proteins such as digestive enzymes (Adelson, 1971; Weinstein, 1972), plasma proteins like prealbumin (Jörnvall et al., 1981), or, as described above, cellular polypeptides such as lactoferrin. Large protein molecules could have been subsequently broken down to produce smaller polypeptides that may also have served as hormones. Another strategy for promoting a multiplicity of hormones involves their assembly from different modules or subunits each of which may be under separate genetic control. In extant species, such subunits usually have similarities in their structures that indicate their ancestral origins from a common gene. Such an assembly process is seen among contemporary hormones such as the gonadotropins, thyrotropin, and the inhibins, activins, and antimullerian hormone.

A variety of hormone receptors have been cloned and their primary structures deduced from the DNA sequence. Such information has been used, along with X-ray crystallography, to predict their tertiary structures and the sites of various domains that are associated with hormone binding, DNA binding and binding to transducing proteins (G proteins) and enzymes. They are usually large molecules consisting of several hundred amino acid residues. Receptors have been grouped into several superfamilies that may contain several more closely related families (Segre and Goldring, 1993). Each superfamily is considered to have originated from a single ancestral gene. Such receptor superfamilies can be classified into several groups.

1. Receptors coupled to G proteins. This superfamily of receptors (Martens, 1992) utilizes many types of hormone including the catecholamines, vasopressin, oxytocin, the secretin–glucagon family, calcitonin, calcitonin-gene related peptide, parathyroid hormone, the gonadotropins, corticotropin, angiotensin II, CRH, somatostatin, GnRH, TRH, neuropeptide Y, adrenomedullin, and the endorphins. Such receptors are also involved in the sensing of odors and, possibly, taste. It includes several hundred genes. All of these receptors can be coupled to G proteins, which transduce their effects to adenylate cyclase or phospholipase C. The polypeptide chain of the receptor characteristically crosses the plasma membrane in a serpentine manner seven times (Fig. 3.11). The amino acid sequences of such receptors can be highly homologous. Thus the dopamine D_2 receptor, which controls the release of α-MSH from the pars intermedia in the toad *Xenopus laevis*, has a 75% overall homology in its amino acid sequence to the dopamine D_2 receptors from humans, cattle, rats, and mice (Martens *et al.*, 1991). This similarity is remarkable considering that this amphibian diverged on its own path of evolution about 350 million years ago. Some special regions of the receptor are even more similar than the overall homologies for the molecules indicate.

2. The tyrosine kinase family of receptors have enzymic activity inherent to their cytoplasmic domains (RTKs). They are lodged in the plasma membrane but their amino acid chain only crosses it once. The constituent units of these receptors are usually assembled as tetramers and mediate the actions of insulin and some growth factors, including IGF-I.

3. Guanylate cyclase in the plasma membrane is directly activated by the natriuretic peptides and guanylins, which appear to bind to its N-terminal extracellular domain.

4. The growth hormone–prolactin–cytokine–hematopoietic superfamily are receptors that are present in the plasma membrane and their amino acid chain only crosses it once. They apparently have no intrinsic enzymic activity. Their mechanism of action involves associated cytosolic protein kinases.

5. The steroid–thyroid hormone superfamily of receptors are usually

present in the nuclear region of the cell but sometimes also in the cytoplasm. After binding to a steroid hormone (estrogens, progestins, androgens, corticosteroids), $1,25(OH)_2D_3$ or T_3, they fulfill their roles by initiating the transcription of DNA on the chromatin.

At present there appear to be only five well-defined superfamilies of hormone receptors though some others will probably emerge. They can account for the actions of virtually all the known vertebrate, and some invertebrate, hormones. Their evolution is considered to be quite conservative compared with that of their ligands, the hormones (Csaba, 1986). Each superfamily appears to retain their basic tertiary structures and organization in relation to the transduction of their effects. Some domains, such as those that the hormones bind to, are well conserved, especially among receptors mediating the actions of homologous hormones in different species.

We can only speculate about the origins of receptors that are utilized by hormones. They are not a prerogative of the endocrine system and they probably preceded its origin. Such basically important molecules that transduce messages in cells could have subsequently interacted with aspiring hormones to produce a fruitful effect. Prior to such an event, these receptors may have reacted with other types of ligand or may even have acted alone. Their transducing mechanisms including G proteins, adenylate cyclase, and phospholipase C may already have existed at that time (Janssens, 1987). Genes for receptor-like molecules have been found but their binding ligands and even, usually their physiological activities have not been identified. Their encoding nucleotides, however, indicate that they belong to contemporary receptor families (O'Malley *et al.*, 1991; Soontjens, Rafter, and Gustafsson, 1996). They have been called "orphan receptors." Some such molecules may be unidentified hormone receptors.

Evolutionary change occurs progressively in receptors for hormones (Fontaine, Leloup-Hatey, and Dufour, 1991). Such changes may alter their selectivity and binding affinities for their hormonal ligands. Hormones from species lower down on the phyletic scale usually bind less readily to receptors from species higher up this scale than hormones from the latter bind to receptors from species in the lower parts of this spectrum. This observation has been referred to as the "one way rule," though actually it is not inviolate. However, it suggests that the receptors may become more fastidious as their evolution progresses. This selectivity, and its molecular basis, is illustrated by the interactions of different growth hormones and their receptors (Souza *et al.*, 1995). Growth hormones from humans and rhesus monkeys (order Primates) are effective when administered to other mammals, and even fishes. However, humans and rhesus monkeys display a complete "species specificity" and do not even respond to growth hor mones from other orders of mammals. The growth hormones from humans and monkeys have an aspartic acid residue in place of histidine at position 171 on the hormone and this structure is apparently unique to theses species. Complementary to this change is the presence of an electropositive amino acid, arginine, (instead of the electroneutral leucine in other mammals) at position 43 on the recognition, hormone-binding domain of the human growth hormone receptor. The structures of other growth hormones are not compatible with an interaction at such a site. Receptor evolution may also involve their ability to gain expression in novel tissues and organs, thus providing a framework for potentially new roles for hormones.

The evolution of hormones and their receptors appears to have been a quite parsimonious affair. There are numerous extant hormones and receptors but they belong to relatively few superfamilies that, apparently, emerged quite early in evolution. Their subsequent divergence may have initially involved only a few ancestral genes. The conservatism of these origins may be more surprising than the subsequent diversity.

Conclusions

An examination of the chemical, biological, and immunological behavior, as well as the chemical structure, of homologous hormones and their receptors from different species suggests that many of these may have been subject to an orderly evolutionary change. This possibility is also indicated by the similarities that persist between each hormone from closely, in contrast to distantly, related species and within the principal systematic groups of the vertebrates, and even invertebrates. In some instances, a genetic background for such changes has been described. It is nevertheless noteworthy that some hormones display little or no difference in their structure even when they are present in such distantly related species as a lamprey and a human. On some occasions, it even appears that a completely "new" hormone that lacked a homolog in its ancestors has evolved.

While the evolution of "new" hormones has potential importance in novel processes of coordination in the body, the functional significance of alterations in the structure of "old" hormones is less clear. Such changes may take place at chemical sites on the hormone molecule that apparently are not essential for its action and so have little or no effect on its functioning. However, if more important sites are involved, the hormone's activity may be altered and this could have important results in the animal. These changes may include a virtual absence of its effect, differences in the quantities of the hormone required to mediate the response, or an alteration in its relative ability to influence different processes within the same animal (specificity).

As we shall see in succeeding chapters, the roles of hormones in the body have often changed completely during the course of evolution. Such a modification of a hormone function is not necessarily accompanied by an alteration in its chemical structure, but it often is; we are uncertain, however, as to how important this change may be in the transition.

4 The life history of hormones

The use of hormones for the purpose of coordination involves a complex series of physiological events. Such a life history begins with the formation of the excitant by the endocrine glands and concludes with the response of a target, or effector tissue, and the hormone's ultimate destruction or its excretion from the body. The events that determine the action of a hormone are shown in Fig. 4.1. This basic pattern persists throughout the vertebrates, though, as will be described, certain differences exist.

The formation of hormones

Although the formation of all hormones is determined at the genetic level, it can be either a relatively direct translational procedure or, alternatively, occur as a result of the prior formation of enzymes that mediate synthesis. Enzymes are also involved in the former process, however.

The synthesis and processing of hormones

Hormones, like other proteins, are synthesized by the process of DNA transcription to produce RNA, followed by translation into a primary protein that may be subsequently modified to produce a mature hormone (see, for instance, Andrews, Brayton, and Dixon, 1987; Conlon, 1989). Such processes occur in the endocrine glands but may exhibit differences depending on the particular type of cell where the gene is being expressed. Thus, the products of the proopiomelanocortin (POMC) genes are different in the pars distalis and pars intermedia in the pituitary. The glucagon gene also produces different polypeptides when it is expressed in the pancreatic A-cells and the intestinal L-cells. Different species may also process hormone precursor proteins differently and contribute to the diversity of the endocrine system. The primary RNA transcript of the genomic DNA is spliced to remove introns, cleaved, and polyadenylated, resulting in the formation of mature mRNA. Such splicing can, however, follow more than one pattern ("alternate splicing") to produce multiple mRNA species encoding different polypept-

PROCESS   MOLECULAR MECHANISM

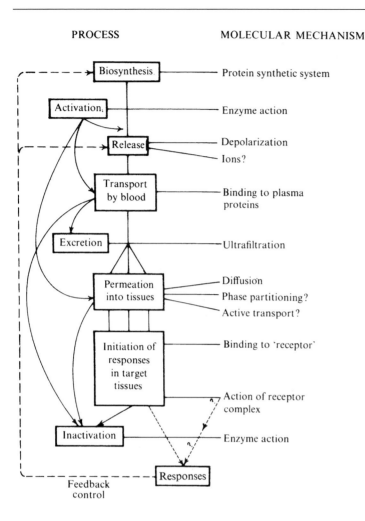

Fig. 4.1. Diagrammatic summary of the life history of a hormone commencing with its biosynthesis and concluding with the response and its inactivation. (From Rudinger, 1968. Reprinted by permission of the Royal Society.)

ides. For instance, the calcitonin gene can produce calcitonin and calcitonin gene-related peptide (Chapter 3). The translation of the nucleotide sequence of the mRNA occurs on the ribosomes. The N-terminal signal sequences of amino acids targets the nascent polypeptides to the endoplasmic reticulum during the translation process in preparation for the hormone's secretion. Cleavage of the signal peptide in the endoplasmic reticulum converts the preprohormone to the prohormone. The preprohormone may, in the instance of low-molecular-weight peptides, contain "spacer regions" consisting of amino acid chains that ensure that it has a minimal length consistent with an ability to emerge from the ribosome and interact with the endoplasmic reticulum (Wolin and Walter, 1993). The prohormone may be further processed by tryptic-like endopeptidases and carboxypeptidases. Such cleavages usually occur at dibasic sites, such as Arg–Arg and Arg–Lys, or monobasic ones such as Arg–Pro and Lys–Ser. An assortment of polypeptides may

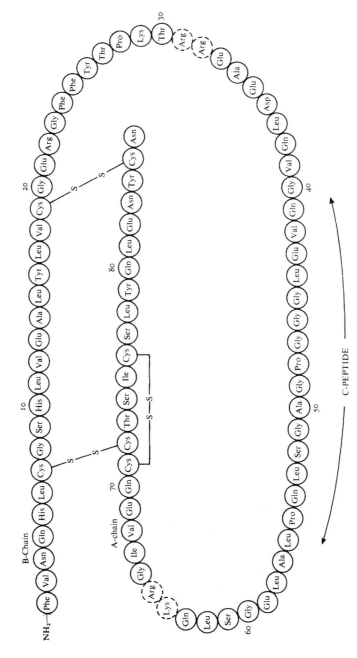

Fig. 4.2. The proposed amino acid sequence of proinsulin from humans. The basic residues, indicated by the broken circle, have been assigned as they are known to occur in bovine and porcine proinsulin. (From Oyer *et al.*, 1971.)

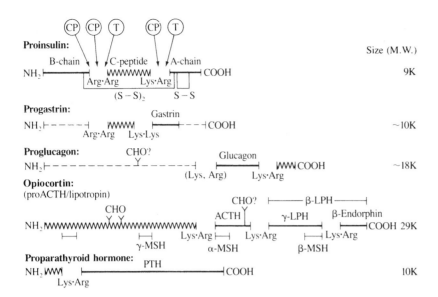

Fig. 4.3. Structures of some prohormones. The heavy lines indicate the biologically active fragments, or hormones. The dashed lines are the regions of amino acid sequences of uncertain biological significance. The parent molecules are broken down as a result of the activities of carboxypeptidase B-like (CP) and trypsin-like enzymes (T). The reactions occur in the regions of the arginine and lysine residues, as indicated for proinsulin. (Based on Steiner *et al.*, 1980).

result, some of these being known hormones. A single prohormone may provide a single hormone, several copies of the same hormone (for instance six copies of TRH arise from a single precursor) or multiple types of hormone. Thus, POMC contains the amino acid sequences of several opioid endorphins, three MSHs and corticotropin. Further post-translational changes may be made to peptides and polypeptides during their processing to hormones. These modifications include their assembly into an appropriate tertiary structure; this may involve the alignment of cysteine residues to form disulfide bridges. Glycosylation, sulfation and phosphorylation at specific sites may occur, which can influence the hormone's duration of action and selectivity for specific receptors. Alpha-amidation and N-acetylation at terminal amino sites can also influence hormone function.

Some polypeptide products of such processing of hormone precursors have no known function but nevertheless possess biological activities that suggest a physiological role. Thus a prolactin-releasing substance called copeptin has been identified in the prohormone sequence for arginine–vasopressin–neurophysin (Chapter 3). Various products resulting from the processing of the glucagon precursor are also in such a putative category. These polypeptides often have amino acid sequences that are well conserved between different species, which supports the likelihood that they may have a contemporary endocrine role. They could also represent relics of former hormones.

Several endoproteases that contribute to the processing of prohormones such as POMC, proinsulin, and prorenin have been identified in mammals (Steiner *et al.*, 1992; Marcinkiewicz *et al.*, 1993). They belong to an ancient family of proteases that apparently originated in prokaryotic organisms. The prototype is subtilisin, which was first identified in the hay bacillus *Bacillus subtilis*. These enzymes are referred to as subtilisin-related preprotein convertases (SPCs and PCs). Several such enzymes have been identified in mammals, where they have a ubiquitous distribution including endocrine glands such as pancreatic B-cells and the pituitary. Two members, SPC 2 and SPC 3 have been characterized in the protochordate amphioxus (*Branchiostoma californiensis*) (Oliva, Steiner, and Chan, 1995). Amphioxus SPC 2 has a 71% homology in its amino acid sequence to human SPC 2 while the SPC 3s have a 55% homology. Such a remarkable conservation of molecular structures appears to reflect an essential function in proprotein and prohormone processing.

The process of hormone formation from mRNA is well illustrated in the production of insulin (Steiner *et al.*, 1974; Chan, Keim, and Steiner, 1976; Lomedico *et al.*, 1977), which provides the historical prototype. Such processes can also be seen in the formation of other hormones. In the insulin system, the initial product is a protein with a molecular weight of about 11 500 (compared with about 6000 for insulin). It is a linear peptide with a special 23 amino acid residue segment at its N-terminus (Fig. 4.2). This "signal peptide" attaches the molecule, and the ribosome on which it is being formed, to the endoplasmic reticulum. The molecule is called preproinsulin and it only has a brief existence as it passes into the Golgi apparatus where its 86 amino acid residue C-terminal segment is incorporated into storage granules. This proinsulin is then broken down under the influence of carboxypeptidase-like and tryptic-like enzymes that fragment the molecule in the region of certain arginine and lysine residues (Fig. 4.3). The A-chain and B-chain are released but as a result of the folding of the prohormone they remain aligned and joined to each other by two disulfide bridges. The "C" peptide fragment is released with the hormone. The structures of several other prohormones are shown in Fig. 4.3: these are converted to hormones in a similar manner to that in proinsulin.

Many polypeptide hormones are stored in granules present in the endocrine cells. These structures are bounded by membranes and are 0.1–0.4 μm in diameter. They appear to originate in the Golgi apparatus of the cell. The precursor, or prohormone, becomes associated with the granules and it seems likely that conversion to the hormone takes place here. The granules can travel to the peripheral regions of the cell and, in response to releasing stimuli, combine with the plasma membrane and discharge their contents into the region of blood vessels. A summary of this process as it is thought to occur for insulin is shown in Fig. 4.4.

Such granules apparently furnish sites for the formation and storage of many hormones. If released into the cytoplasm of the cell the hormones may be destroyed, as has been observed for the catecholamines when they are exposed to the mitochondrial enzyme monoamine oxidase (MAO). In addition, storage granules may afford convenient vehicles in which hormones can

Beta Granule Formation

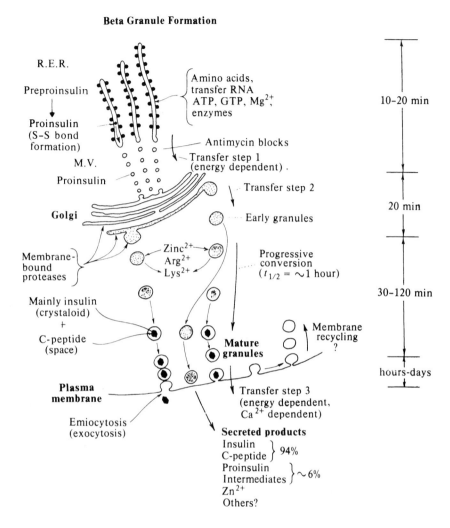

Fig. 4.4. Schematic summary of the insulin biosynthetic mechanism in the pancreatic B-cells. The time scale on the right side of the figure indicates the time required for each of the major stages in the biosynthetic process. R.E.R, rough endoplasmic reticulum; M.V., microvesicles. (From Steiner *et al.*, 1974.)

be transported for considerable distances along nerve cells.

Some neurons form hormones by a process called *neurosecretion*. These neurons are like ordinary nerve cells and consist of a cell body with an extended axon and they can also be depolarized and so convey electrical information. The axon, instead of terminating at another neuron or an effector tissue, such as a gland or muscle, lies near a capillary into which it can discharge certain of its products (Fig. 4.5). These products may be hormones, the formation of which is initiated some distance away in the cell body. The hormones, parceled up in their granules, travel along the nerves to the peripheral sites in the axon where they can be released into the blood.

Hormones that are formed as a result of neurosecretion include those of the neurohypophysis and hypothalamus including vasopressin, oxytocin, and the

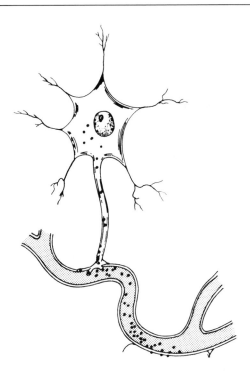

Fig. 4.5. A neurosecretory cell. The hormonal products are transported from the cell body down to the axon from which they can be released into capillaries. In contrast, ordinary nerve cells have axons that abut onto other neurons (instead of capillaries). (From R. Guillemin and R. Burgus, *The Hormones of the Hypothalamus*. Copyright © 1972 by Scientific American, Inc. All rights reserved.)

various releasing hormones that control the adenohypophysis. In mammals, vasopressin and oxytocin are formed in the supraoptic and paraventricular nuclei that are situated at the base of the brain. These hormonal products pass down the axons in granules to the neural lobe. Inside the granules they are attached to protein molecules of neurophysin, the synthesis of which is closely associated to that of the hormone (Chapter 3). The mRNA for vasopressin has also been identified in the neurohypophysis of dehydrated rats, but it is not present following cutting of the nerve tracts (Levy and Lightman, 1989). Therefore, like the processed hormone itself, it may also travel down the axons and be translated in the pituicyte cells of the neural lobe. (Alternatively the vasopressin gene may be expressed in these cells as a result of neural stimuli.) Amphibians and fishes have a single preoptic nucleus where both the neurohypophysial hormones appear to originate. The putative hormones of the urophysis in fishes (Chapters 2 and 3) are also formed by a process of neurosecretion.

The formation of hormones by enzymically controlled synthesis

Thyroid hormones   The endocrine secretions of the thyroid, the adrenal medulla, the pineal, the gonads, and the adrenal cortex are the result of biosynthetic processes controlled by enzymes. Although the enzymes themselves are the result of genetic translational processes, the hormones are synthesized in chemical reactions controlled by the enzymes. This synthesis may involve many reactions, some of which have been summarized in Fig. 3.2 (p. 70, steroid hormones), Fig. 3.5 (p. 79, thyroid and catecholamine hormones), and Fig. 3.6 (p. 83, melatonin). Thyroid hormones contain iodine and the thyroid gland has a special ability to concentrate inorganic iodide from the blood. This ability to transport iodide actively against an electrochemical gradient is shared by some other tissues, including the intestine and salivary glands. This ability may be controlled by a single gene: in humans a congenital inability to accumulate iodide in the thyroid is accompanied by a parallel deficiency at the other iodide transport sites. The accumulated iodide is oxidized to iodine, which combines with tyrosine to form the precursor of thyroxine and triiodothyronine. The latter reaction occurs with specific tyrosine residues present in a large dimeric glycoprotein called thyroglobulin. It is stored extracellularly in the follicles of the thyroid gland. This prohormone has been cloned and in cattle its encoded amino acid sequence contains 2750 residues (Mercken *et al.*, 1985). It has an homologous structure in other mammals, such as humans, rabbits and rats, and also in a reptile, the turtle *Pseudemys scripta* (Roe *et al.*, 1989). Four specific tyrosine hormogenic sites (A, B, C, and D) have been identified in mammalian thyroglobulin (Dunn *et al.*, 1987) and five in the turtle. In the rabbit, sites A and B are the principal ones involved in the synthesis of thyroxine and site C with triiodothyronine. In the turtle, little triiodothyronine is present in the thyroid gland and site B is relatively more important than it is in the rabbit. The biosynthetic process for the thyroid hormones appears to be common to all vertebrates and was apparently attained early in their evolution. Nevertheless, most of our information has been derived from studies of mammals.

Thyroglobulins have been identified in thyroid tissues of species from most groups of vertebrates, even including larval cyclostomes (lampreys) where it is present in the subpharyngeal gland or endostyle (Suzuki and Kondo, 1973; Suzuki *et al.*, 1975). These proteins exhibit many similarities with respect to their molecular size (though a few differences have been observed), as determined by centrifugation in sucrose gradients, but their amino acid constitutions may differ. In hagfish, however, a smaller protein may have assumed the role of thyroglobulin. Thyroglobulins also exhibit different immunological behavior. Antibodies to specific thyroglobulins have been prepared and these react, *in vitro*, with the homologous protein, which can be radioactively labeled. Thryoglobulins from different species may compete with this labeled protein for binding to its antibodies. The relative ability to do this suggests the degree of immunological similarity to the homologous thyroglobulin. Considerable interspecific differences have been observed in such radioimmunoassays (Torresani *et al.*, 1973). Sheep thyroglobulin readily displaces its labeled form from anti-sheep thyroglobulin antibodies, but thyroglobulins from other mammals, such as pigs and rabbits, are much less effective. Thyroglobulin from a python and a crocodile also compete with the homologous labeled protein for such binding, but this is also much less than that for the sheep protein. Bird thyroglobulin, from ducks, has no ability to bind with the sheep antibodies. Although it is tempting to construct phylogenetic trees with such information, the paucity of species examined makes such predictions of doubtful significance. The measurements nevertheless illustrate the diversity that can occur among thyroglobulins from different species.

Catecholamines   Epinephrine (adrenaline) and norepinephrine (noradrenaline) are formed in chromaffin tissues. These hormones are present not only in the adrenal gland but also are associated with nervous tissue in other parts of the body. Norepinephrine is also formed in certain nerve endings in the sympathetic nervous tissue and the brain. The original precursor of these catecholamines is tyrosine (Fig. 3.5), which by a series of enzymically controlled reactions is converted to 3,4-dihydroxyphenylalanine, or dopa, and thence to dopamine. These reactions occur in the cell's cytoplasm. The dopamine is accumulated by storage granules in which it is converted, under the influence of dopamine β-hydroxylase, to norepinephrine. Norepinephrine can be N-methylated to epinephrine under the influence of the enzyme phenylethanolamine-N-methyltransferase (PNMT), which, in mammals (Chapter 3), can be induced in the presence of high concentrations of corticosteroids. There is evidence to suggest that norepinephrine and epinephrine are stored in different granules and even different cells in the adrenal medulla. This could be determined by regional differences in the access of corticosteroids to the medullary tissue influencing the local levels of PNMT (Pohorecky and Wurtman, 1971).

Melatonin   The synthesis of melatonin is of special interest as it provides a model for the circadian control of the expression of a gene. The enzymically mediated four-step process for the synthesis of melatonin in the pinealocytes is described in Fig. 3.6 (p. 83). Tryptophan is the initial substrate; it is first hydroxylated and the product is then decarboxylated to form 5-hydroxytryptamine (serotonin). This intermediate substrate is acetylated by N-acetyltransferase (NAT) to produce N-acetylserotonin, which is 5-methylated by hydroxy-O-indolemethyltransferase (HIOMT) resulting in the formation of melatonin (N-acetyl-5-methoxytryptamine). Synthesis of melatonin is increased by 10 to 20 times when the animal is in darkness and under these conditions the NAT gene increases its transcription of NAT mRNA. The controlling mechanism is stimulated in mammals following the receipt of nerve impulses from the suprachiasmatic nucleus, which during the hours of darkness no longer receives inhibitory nerve impulses from the retina. The

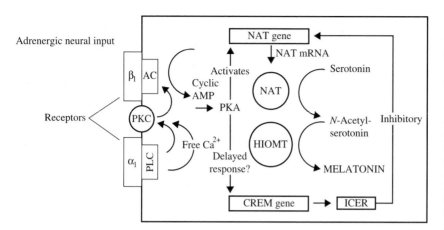

Fig. 4.6. A model for the circadian control of melatonin synthesis by the pinealocyte, involving an input from an adrenergic nerve and the expression of the N-acetyltransferase gene (NAT) and the cyclic AMP-responsive early modulator gene (CREM). For details see the text. AC, adenylate cyclase; CRE, cyclic AMP-response element; HIOMT, hydroxy-O-indolemethyltransferase; ICER, inducible cyclic AMP early repressor; PKA, phosphokinase A; PKC, phosphokinase C: PLC, phospholipase C. (Based on Stehle *et al.*, 1993; Takahashi, 1993.)

neural pathway from the suprachiasmatic nucleus to the pineal appears to be a tortuous one, utilizing postganglionic sympathetic nerves that release norepinephrine. This neurotransmitter interacts with β_1-adrenoceptors on the pinealocytes resulting in the activation of adenylate cyclase and the formation of cyclic AMP (Fig. 4.6). The response is potentiated by a parallel interaction of norepinephrine with α_1-adrenoceptors, which results in the activation of phospholipase C and the phosphatidylinositol metabolism pathway. As a result, there is a synergism that increases the activation of adenylate cyclase. The cyclic AMP activates protein kinase A, which induces transcription by phosphorylating the cyclic AMP-response element (CRE) on the NAT gene. Synthesis of melatonin is thus increased. After a time lag of several hours of darkness, the response starts to decline, even though the cyclic AMP levels are still high. This effect is mediated by a gene called CREM (for *cyclic AMP responsive early modulator* (Foulkes and Sassone-Corsi, 1992; Stehle *et al.*, 1993; Takahashi, 1993). This gene expresses a repressor of nuclear transcription processes that are mediated by cyclic AMP. It is called ICER (for *inducible cyclic AMP early repressor*). The levels of its mRNA exhibit a circadian rhythm that in the rat pineal attains a maximum about 9 hours after the onset of darkness. The ICER then represses the activity of the NAT regulator gene during the period of approaching daylight. The formation of the ICER is promoted by β-adrenergic stimuli. It is possible that other processes involved in the synthesis of melatonin may be influenced by the circadian control mechanism; for instance the number of β_1-adrenoceptors and the tryptophan hydroxylase concentrations, but their roles remain to be investigated further (Reiter, 1991a).

Steroid hormones   The basic pattern for the synthesis of the steroid hormones (Fig. 3.2. p.70) (androgens, estrogens, progestins, and corticosteroids) is the same in all vertebrates (Sandor, 1969). However, some novel steroid hormones have appeared during vertebrate evolution, such as aldosterone,

1α-hydroxycorticosterone and 11-ketotestosterone, which are accompanied by appropriate synthesizing enzymes. The sites of such synthesis may also display some interspecific variability but they are usually the testes, ovaries, and adrenocortical tissues. Peripheral sites, such as brain, skin, and adipose tissue, may, however, make special contributions to synthesis. Activation processes, such as the conversion of testosterone to 5α-dihydrotestosterone and cortisone to cortisol, may involve their target tissues and the liver. The common substrate for steroid hormone synthesis is cholesterol, which may be formed *in situ* but it is usually obtained from sources outside the cell. Synthesis of each hormone involves a succession of enzymically mediated changes that occur on the endoplasmic reticulum and in the mitochondria (Fig. 4.7). Several of the major enzymes involved, including 20,22-desmolase (for cholesterol side-chain cleavage), 21-hydroxylase, 11β-hydroxylase, the aromatase complex, and 17α-hydroxylase belong to the cytochrome P450 enzyme superfamily. This extensive group contains at least 10 gene families (Nebert and Gonzalez, 1987). It is thought to have originated about 1.5 billion years ago, before multicellular organisms evolved. These enzymes are heme-containing proteins, usually associated with the endoplasmic reticulum but also with the mitochondria. Their primeval role, and main contemporary physiological contribution, is thought to be the detoxification of lipid substances that are accumulated from the environment. The cytochrome P450 enzymes are monooxygenases, which, in conjunction with NADP, molecular oxygen, and cytochrome P450 reductase, convert such substances to less lipid-soluble materials and so enhance their excretion. Four families of these enzymes appear to have been coopted for use in the synthesis of the vertebrate steroid hormones. Recognition of their membership of the cytochrome P450 superfamily has resulted in an alternative method for naming the steroidogenic enzymes. Therefore, the aromatase complex is labeled $P450_{arom}$, 11β-hydroxylase, $P450_{11\beta}$, 20-22-desmolase (cholesterol side-chain cleavage), $P450_{scc}$ and so on. The steroidogenic members of the cytochrome P450 superfamily are among its oldest members; $P450_{11\beta}$ and $P450_{scc}$ (which are mitochondrial proteins) originated about 1.1 billion years ago, and $P450_{arom}$ about 850 million years ago. (Renal 1α-hydroxylase, which converts 25-hydroxyvitamin D_3 to the active hormone $1,25(OH)_2D_3$, is also a member of the cytochrome P450 superfamily).

The genes for each of these steroidogenic enzymes appear to be under dual control (Parker and Schimmer, 1993). A nuclear receptor protein has been identified that may coordinate the common induction of all the enzymes involved in the synthesis of corticosteroids (a "gene battery"), but individual responses of each gene also occur. The latter involve the tropic effects of corticotropin and angiotensin II, which promote synthesis of cortisol and aldosterone, respectively. The effect of corticotropin is mediated by cyclic AMP and angiotensin II by stimulation of phosphatidylinositol metabolism.

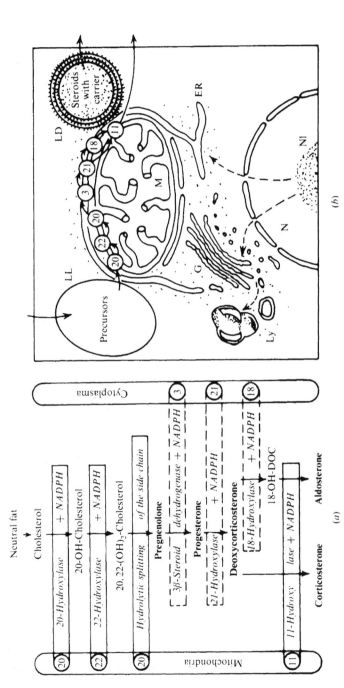

Fig. 4.7. The biosynthesis of corticosteroid hormones (a) in the adrenocortical tissue of a frog. This is illustrated with reference to the cell organelles (b). Corticosterone can also act as a precursor for aldosterone. In some species, 18-hydroxylase may be a mitochondrial rather than a microsomal enzyme. In vertebrates that form cortisol, another microsomal enzyme, 17α-hydroxylase, active on progesterone and leading to 17α-OH-progesterone, deoxycortisol, and cortisol is present. ER, endoplasmic reticulum; G, Golgi apparatus; N, nucleus; LD, electron-lucid lipid droplet; LL, electron-dense lipid droplet; Ly, lysosomes; M, mitochondrion; Nl, nucleolus. Solid arrows are pathways of steroid synthesis from precursors to steroid bound to a carrier; broken arrows indicate cellular responses activated by ACTH. (From Lofts and Bern, 1972.)

The entire process of steroid hormone synthesis may occur in a single cell (Fig. 4.7) but in the ovarian follicle two or even three types of cell may be involved. Therefore, in birds, it has been suggested (Porter *et al.*, 1989) that the granulosa cells, which synthesize progesterone, may transfer some of this product to the theca interna cells, which can form androgens from it, and the latter then goes to the theca externa cells where aromatization to estrogens takes place. A two-cell system has been identified in some mammals in which androgens are synthesized by the theca interna cells but, in contrast to the birds, they are subsequently aromatized to estrogens by the granulosa cells. Estrogen synthesis may also occur in nonovarian tissues such as the skin and adipose tissue in the human male (Hodgins, 1989) or in the brain of the male zebra finch. The latter species, in common with other song birds, can aromatise androgens in the brain in such quantities that they appear in the peripheral circulation (Schlinger and Arnold, 1993). The female birds also have aromatase activity in the brain so that it is possible that they may also synthesize significant amounts of peripheral estrogens in this manner.

The release of hormones from the endocrine glands

Nature of the stimuli

The role of the endocrine glands in the regulation of bodily functions is dependent on the release of their secretions on appropriate occasions. Secretion is initiated upon the receipt, by the gland, of a suitable stimulus, which may increase or decrease the discharge of its hormone. The message may arrive either by way of a nerve or be carried in the blood that perfuses the tissue. The primary event that initiates this stimulus may arise either from the external environment (exteroceptive stimulus) or from inside the body (interoceptive stimulus).

Exteroceptive stimuli that may affect the endocrine glands include the receipt of light, a change in temperature or of the osmotic concentration (of an aqueous environment), and the acquisition of food, water, and salts. Social situations such as the proximity of prey, a predator, a mate, or the young may evoke psychogenically mediated responses in the endocrine glands. Climatic events such as rain, temperature, and even, possibly, humidity and atmospheric pressure can also influence a hormone's release. The receipt of an endocrine response to such external stimuli help the animal to maintain an equitable relationship with the events that happen around it. Exteroceptive stimuli are especially useful in providing cues that are involved in reproduction.

Interoceptive stimuli are those that result from changes in the physiochemical conditions within the body. Ultimately they may reflect the external conditions: for instance, a lack of drinking water and a hot, dehydrating environment will lead to an increase in the osmotic concentration of the body fluids. Internal stimuli include changes in the concentration of salts, such as sodium, potassium, and calcium, in the body fluids, alteration of the hydrostatic pressure of the blood vascular system, oscillations of the levels of nutrients, like glucose, amino acids, and fatty acids, as well as changes in the body temperature. The physiological factors influencing release of hormones are summarized in Table 4.1.

Hormones can exert tropic effects on other endocrine glands and the secretions, once released, may travel back to the region where the ultimate tropic hormone originated and inhibit its further release. This last effect completes the cycle of events that closes the loop of a *negative-feedback system* that plays a vital role in regulating the endocrine system. The release of hormones from the median eminence, which in turn controls the formation and discharge of the tropic hormones of the adenohypophysis, is regulated in this manner. The hormones that exert the inhibitory effects in the median eminence may alter the thresholds for stimulation of the neurosecretory cells. In addition, such a negative feedback can also act directly on the adeno-hypophysis, as seen with the action of thyroid hormones, which inhibit the release of thyrotropic hormone in this way. The interrelations of the hor-mones of the hypothalamus, adenohypophysis, and more peripheral endo-crine glands are shown in Fig. 4.8.

The feedback mechanism involving the action of the peripheral endocrine secretions on the hypothalamus is called a *long-loop feedback*. There is also evidence suggesting that the adenohypophysial secretions may exert a similar action on the hypothalamus by what is termed a *short-loop feedback*. It should be noted that peripheral hormones do not necessarily initiate a negative-feedback inhibition on the hypothalamus. High estrogen levels can stimulate the release of GnRH, which initiates the events that result in ovulation. Such a *positive feedback* has also been shown in the hypothalamus of the goldfish, where thyroxine stimulates the release of a thyrotropin-inhibiting hormone (Peter, 1971). The sporadic information available from experiments on nonmammals suggests that feedback control working through the hy-pothalamus and the adenohypophysis is widespread. There is some doubt as to the importance of such effects in cyclostomes (Larsen and Rosenkilde, 1971; Fernholm, 1972).

A negative-feedback inhibition of hormone secretion also results from changes in the concentrations of the products of the hormone's actions. The retention of water or sodium, as a result of the actions of antidiuretic hormone and aldosterone, respectively, reduces the further release of these hormones. Comparable mechanisms exist involving glucose levels and the regulation of insulin, glucagon, and growth hormone, as well as calcium, parathyroid hormone, calcitonin, and stanniocalcin.

The endocrines, apart from influencing each other's release through tropic

Table 4.1. *Principal stimuli influencing the release of hormones*

Hormone	Releasing stimuli
Aldosterone	Low plasma Na^+ concentration, angiotensin II and III
Angiotensin	Renin
Calcitonin	Hypercalcemia
Cholecystokinin	Digestive products in the upper intestine
Cortisol and corticosterone	Corticotropin
Epinephrine	Neural stimuli (mediated by acetylcholine)
Estrogens	LH
FSH	External stimuli, such as light, low estrogen levels, GnRH
Gastrin	Feeding (vagal reflex; local reflex from food in stomach)
GIP	Fats in intestine
Glucagon	Hypoglycemia, gastrin, CCK, high amino acids and low fatty acids in plasma, exercise
Glucagon-like peptide	Feeding
Growth hormone	Sleep, exercise, apprehension, hypoglycemia
Hypophysiotropic hormones: CRH, TRH, etc.	Hypothalamic neuronal stimuli (dopamine and monoamine transmitters), inhibited by negative-feedback mechanisms carried by hormones and metabolites
Insulin	Hyperglycemia and amino acids in plasma, glucagon, growth hormone, in ruminants high levels of propionic and butyric acid, vagal stimulation, CCK and GLPI. Inhibition by epinephrine
LH	External stimuli, sexual excitement (male), estrogen "surge" (female), low progesterone or testosterone levels
MSH	Light on retina, low plasma corticosteroid levels, inhibition by neural stimuli
Melatonin	Darkness (adrenergic neural stimulation)
Natriuretic peptides	Atrial distension
Oxytocin	Suckling, parturition
Parathyroid hormone	Hypocalcemia
Progesterone	LH, chorionic gonadotropin, prolactin
Prolactin	Diurnal rhythm (sleep), suckling, parturition, plasma osmotic concentrations (low in fish, high in mammals?), estrogens
Renin	Low Na^+ in plasma, hemorrhage, reduced renal blood flow, nerve stimulation (β-adrenergic), increased osmotic concentration in renal blood supply
Secretin	Acid in upper intestine
Stanniocalcin	Hypercalcemia
Testosterone	LH
Thyroid hormones	TSH
TSH	Low thyroxine, temperature reduction
Vasopressin, ADH	Increased osmotic concentration of plasma
$1\alpha,25\text{-}(OH)_2$-vitamin D_3	Low Ca^+ and phosphate levels in plasma; parathyroid hormone

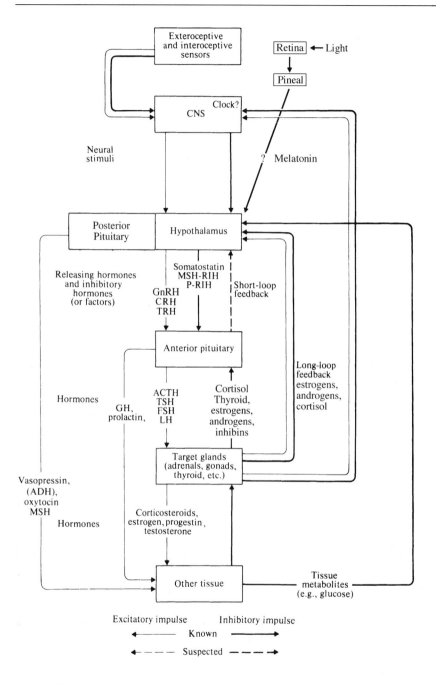

Fig. 4.8. Factors controlling the release of hormones from the anterior pituitary (adenohypophysis). Shown is the use of negative-feedback inhibition of secreted hormones and tissue metabolites in influencing further hormonal release. As can be seen, the hypothalamus (and its associated median eminence) plays a central role in this process. Stimulatory effects are shown by thin lines and inhibitory processes by thick lines. (Based on Krieger, 1971.)

and feedback mechanisms, may also interact with each other and so modify their secretory activity. Epinephrine can inhibit the release of antidiuretic hormone from the neurohypophysis and that of insulin from the islets of Langerhans. This inhibition is an α-adrenergic effect that contrasts with the β-adrenergic effect that increases the release of insulin. Glucagon promotes the release of insulin and growth hormone directly; such effects do not depend on changes in blood glucose levels.

Many less precise and less specific stimuli than those just described can initiate the discharge of hormones from endocrine glands. Such stimuli may contribute to the homeostatic process, though they may also confuse it. No endocrine gland can exist and be uninfluenced by events outside what we may like to think of as its homeostatic area of influence. Nonspecific stimuli, especially if strong enough, can elicit a discharge of many hormones. This is sometimes referred to as "stress" and is particularly likely to occur in experimental situations, which contributes to the confusion of the perpetrating scientists.

"Pulsatile" release of hormones

Hormones are not necessarily released in a continuous stream but may also be released in discrete bursts. This process may take place in a single large "surge" or in a pulsatile manner, one "pulse" succeeding another at regular intervals. The latter is also referred to as "episodic" release. This phenomenon did not become apparent until sensitive and convenient radioimmunoassay methods were developed for assaying hormones in small, serially collected samples of plasma. Many hormones may be released in this manner (Brabant, Prank, and Schöfl, 1992). In humans and other primates, GnRH is usually released about once an hour but, depending on the stage of the reproductive cycle, this event may vary from once every 3 to 4 hours to 4 or 5 times each hour. Prior to ovulation, when GnRH promotes a surge in LH release, the pulse frequency may, in sheep, speed up to 20 or more times an hour. In humans, insulin and glucagon may be released over 100 times each day, while aldosterone is secreted only six to ten times a day. These patterns probably reflect the acute need to regulate metabolites and minerals, such as follows feeding.

A pulsatile manner of release of a hormone provides the possibility of relaying a more detailed message to the target organ. Such an oscillating type of stimulus may be more effective than a single sustained tonic concentration. Indeed, many effectors may lose their sensitivity to a hormone and fail to respond altogether when exposed to a steady tonic concentration of a hormone. Such a refractory condition may reflect a loss (down-regulation) of receptors.

The rhythmical, pulsatile release of hormones can result in their attainment of stable basal, or tonic, concentrations in the plasma. When a hormone is released in such a manner it will at first be diluted, as a result of its redistribution in the body fluids, and it will then be removed from the circulation. The rate of this process is called its "clearance time." This occurs as a result of its uptake by tissues, its metabolic breakdown, and its excretion

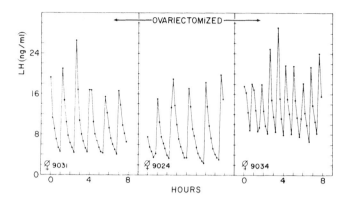

Fig. 4.9 The spontaneous pattern of release of luteinizing hormones (LH) in three ovariectomized ewes. Blood samples were collected every 12 minutes for 8 hours during the normal midanestrous period in May. It can be seen that the plasma concentrations of LH varied in a pulsatile manner from about 2 to 30 ng/ml. The time between the peak (or trough) levels reflects the times of release of the hormone from the pituitary gland and varies from about once every 50 to once every 100 minutes. The particular pattern was consistent in each sheep. The mean LH concentration in the plasma and the basal level, below which it did not drop, also varied, the highest being in the ewe that released LH most frequently. (From Karsch, 1980.)

in the urine and bile (see later in this chapter). Its levels will thus decline in the intervals between the pulses. The level to which they drop will depend on the factors that influence the hormone's clearance and the quantity that was originally released. If no further pulses occur, as in a surge of released hormone, the plasma concentrations will eventually be reduced to zero. A useful measure of this decline is the half-life ($t_{\frac{1}{2}}$) of the hormone in the circulation, which is the time it will take to reach 50% of its previous level. If a hormone is released more frequently than the time of its half-life, then it will tend to accumulate and its concentration in the plasma will rise. It will then reach a new equilibrium level after the equivalent of about four half-lives. The particular basal level attained will depend on the frequency of the pulses and the amount of hormone released on each such occasion. The basal level will appear graphically (Fig. 4.9) as an oscillating-type system with abrupt increases and declines centered around a steady median value. Such a mechanism provides a system whereby different tonic levels of hormones can be attained, and maintained, such as in daily rhythms of hormone concentrations in the plasma or in more long-term changes that accompany reproductive cycles. The factors that determine the frequency of the pulses and the amount of hormone released on each occasion are not well understood and explanations often seek hypothetical refuge in engineering jargon, such as "pulse generator".

The pulse generator for GnRH has been anatomically identified in rhesus monkeys. It consists of a neural network, including GnRH secretory cells, which is present in the region of the arcuate nucleus of the medial basal hypothalamus (Knobil, 1992). A cultured immortalized line of such cells has been found to continue to release GnRH in a pulsatile manner (Wetsel *et al.*, 1992). Such autonomy under *in vitro* conditions was similar to that observed *in vivo*. However, under the latter conditions the pulse frequency may be influenced by neural, and even hormonal, inputs from adjacent areas of the hypothalamus (Dyer and Robinson, 1989).

The testes of rams are small during the spring; these animals do not normally come into breeding condition until the autumn. When GnRH is injected into rams in the spring (Lincoln, 1979), in an episodic manner, for 60 seconds every 2 hours, the testes enlarge. These sheep then come into a reproductive condition similar to that which occurs normally in autumn. A continuous infusion of the hormone is ineffective under such conditions.

Rhythms of hormone release

The release of many hormones occurs in a regular rhythmical manner. Such events may occur in cycles, such as an approximately 24 hour daily period (diurnal and circadian rhythms), during the course of a period of reproduction, at a particular season of the year, or on an approximately annual basis (circannual rhythm) not directly related to external environmental events. Such timing may be especially important in coordinating the events of the reproductive cycle and insuring that this occurs during the times of the year most appropriate to the survival of the young. A predictably functioning release mechanism also insures that adequate hormone levels, necessary for the animal's optimal daily activities, are available. Such release of hormones is usually controlled by centers in the brain that are programmed by stimuli that include the length of the daily period of light, the external temperature, and changes in the seasons, as well as certain interoceptive (endogenous) stimuli. The evolution of such timing mechanisms appears to have been necessitated by the external physical changes resulting from the earth rotating on its axis every 24 hours and the occurrence of the vernal and autumnal equinoxes.

It has been proposed (Turek and van Cauter, 1994) that a basic timing mechanism exists in the brain which exhibits an endogenously generated rhythm over an approximately 24 hour period. It can receive and be synchronized by some external signals, such as light. In the presence of such signals it runs on an approximate 24 hour basis (circadian rhythm) but if shielded from such events (as has been shown in humans living in complete isolation) it runs on an approximately 26 hour basis (van Cauter, 1989). Such an alteration in timing is apparent from changes in such physiological rhythms as those of body temperature and the concentrations of hormones in the plasma. In mammals this master circadian clock has been identified, following surgical manipulations, near the suprachiasmatic nuclei (SCN) in the anterior hypothalamus (Miller, 1993). Surgical ablation of this area seriously disturbs or abolishes most bodily rhythmical activities including those of the endocrine system. Transplantation of homologous tissue from other animals to such sites restores the rhythms, but they have the pattern of the donor animal. This clock mechanism can receive neural inputs, such as from the retina. Such signaling based on the receipt of light provides information about day length and, thus, also the season of the year. "Clock," or circadian oscillator, genes have been identified in an insect, *Drosophila*, and recently in mice (Reppert and Weaver,1997). Mouse *clock* gene mRNA is expressed in several tissues but most prominently in the suprachiasmatic nuclei. Its product CLOCK belongs to a family of transcription factors. It may act to coordinate the activity of neurons in the suprachiasmatic nuclei or it could be an element in transcriptional regulation of the circadian oscillator mechanism. A possible interaction of such "clock" genes with hormones remains to be explored.

Rhythms of endocrine activity may also be related to seasonal changes in environmental temperature. Such effects appear to be more important in cold-blooded animals and may influence such processes as their feeding, migration, and reproduction (Chapter 9). Somatolactin, which is a hormone that appears to be confined to teleost fish (Chapter 3), shows a distinct circannual rhythm in its concentrations in the plasma (Rand-Weaver, Pottinger, and Sumpter, 1995). In trout, its levels are highest in summer and lowest in winter, a pattern that is related to the water temperature rather than the photoperiod.

The synthesis and secretion of melatonin by the pineal gland acts as a transducer for the circadian clock. As described earlier, melatonin is synthesized at an increased rate during the hours of darkness. In mammals this effect occurs as a result of neural adrenergic stimuli from the suprachiasmatic nucleus, which is programed following the receipt of light by the retina. The duration of the release of the melatonin mirrors the duration of the period of darkness and hence the length of the day and the season of the year. In many mammals, the pineal gland is an essential link for synchronizing day length to the initiation of reproduction in seasonal breeding animals. In conjunction with the clock in the suprachiasmatic nucleus, it acts as a type of calendar or almanac. The duration of the release of melatonin may also provide a signal for the rhythmic activities of other physiological activities apart from seasonal reproduction. The abilities of administered melatonin to readjust disturbed biological rhythms caused by jet lag has its supporters (Reiter, 1991a).

The physiology of the pineal gland displays variation even among mammals where a relationship of day length to synthesis of melatonin may not always be readily apparent. Diurnal rhythms in the activity of the pineal gland have been demonstrated among nonmammals (Binkley, 1988). Birds show a vigorous nightly increase in the production of melatonin (Gwinner *et al.*, 1993) and reptiles also display such nocturnal pineal activity (Firth, Kennaway, and Rozenbilds, 1979; Underwood and Hyde, 1989). Such rhythms of pineal gland activity have also been observed in amphibians (Rawding and Hutchison, 1992) and teleost fish (Gern, Owens and Ralph, 1978). Therefore, common diurnal rhythms of nightly melatonin synthesis appear to have a wide phyletic distribution among vertebrates and may be an ancient mechanism for timing physiological processes.

The levels of corticosteroids in the blood vary in a distinct pattern during the course of the day. In mammals, this is well-known in primates (including

Fig. 4.10. Diurnal variation in the levels of plasma cortisol in three sheep. It can be seen that the highest concentrations were recorded during the early hours of daylight. The concentrations of cortisol in the plasma were not uniform but showed continual oscillations, suggesting that this is released in sudden "pulses". (From McNatty, Cashmore, and Young, 1972.)

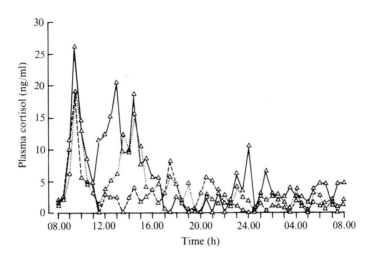

humans), dogs, rats, and mice. Release is related to the incidence of light and the corticosteroids are lowest in concentration during the night and reach a distinct peak after various periods of daylight. In humans and sheep (Fig. 4.10), this is seen in the morning hours, soon after dawn, though in laboratory rats, which are nocturnal, it is delayed until the early evening hours (Fig. 4.11). Comparable changes in the plasma corticosteroid concentrations have been observed in some teleost fishes (the channel catfish, *Ictalurus punctatus* (Boehlke *et al.*, 1966), and the gulf killifish, *Fundulus grandis* (Srivastava and Meier, 1972)), where peak concentrations occur about 8 hours after the onset of light. In another teleost (the eel *Anguilla rostrata*), diurnal variation of plasma cortisol levels does not seem to occur (Forrest *et al.*, 1973a). Prolactin and growth hormone in humans are released in greatest amounts during the period of sleep (Fig. 4.12) and do not appear to be directly dependent on light. Sheep exhibit a polyphasic pattern of sleep (cat naps) in contrast to the

Fig. 4.11. The circadian pattern in the concentration of corticosteroids in the plasma of rats. The effects of the administration of exogenous corticosteroids to young, developing rats on the subsequent circadian periodicity in the endogenous corticosteroid levels are shown. It can be seen that the administration of dexamethasone or hydrocortisone (cortisol) on days 2 to 4 after birth suppressed the rhythmical release. However, when dexamethasone was given on days 12 to 14 after birth no effect was seen: ●, control; ○, saline, 0.1 ml, days 2–4; ■, hydrocortisone acetate, 500 μg, day 3; ▲, dexamethasone phosphate, 1 μg, days 2–4; △, dexamethasone phosphate, 1 μg, days 12–14. (From Krieger, 1972. Copyright © 1972 by the American Association for the Advancement of Science.)

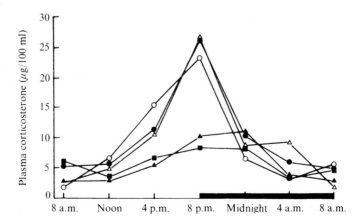

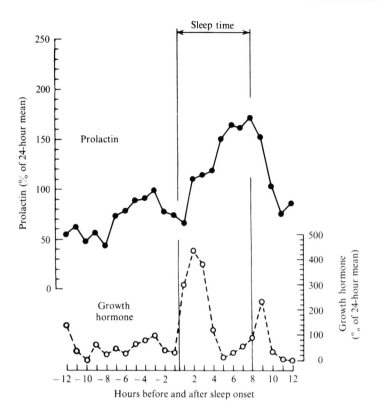

more usual monophasic one in humans. Growth hormone secretion, however, is still elevated in sheep during their short naps compared with their more vigilant periods (Laurentie *et al.*, 1989). Prolactin also exhibits rhythms related to the daily light cycle, as seen in birds (Hall, Harvey, and Chadwick, 1986) and some mammals (Chapter 9).

The activity of the endocrine glands and the release of their hormones often show profound changes that are associated with the season of the year. Such changes have most often been observed in relation to the breeding season, which, especially in animals from nonequatorial regions, usually only occurs at certain times of the year. The relationship of the size of the testes and cloacal gland of Japanese quail to the length of the day is shown in Fig. 4.13. These glands reach their maximum size in May, which corresponds to the onset of the breeding season. The morphological changes are related to the rising levels of gonadotropins and testosterone. The rates of secretion of these hormones start to increase in March, when the length of daylight is about 12 hours. Reproduction may be dictated or influenced by predictably favorable seasons or in less favored areas, like deserts, by the sudden appearance of rain. Apart from pituitary and gonadal sex hormones, cyclical changes in the activity of the thyroid gland and the adrenal cortex and medulla have been

Fig. 4.13. The relationship of the growth of the testis and cloacal gland of quail to the hours of daylight (the photoperiod, upper panel) throughout the year. The lower two panels show the related changes in the plasma concentrations of luteinizing hormone (LH), follicle-stimulating hormone (FSH), and testosterone. Each point is the mean from 12 birds: the vertical lines are standard error of the mean. (From Follett and Robinson, 1980.)

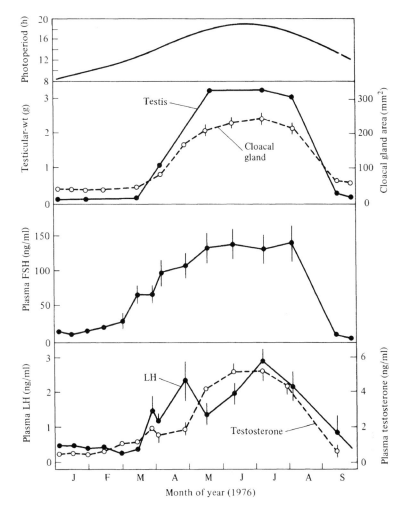

observed. The thyroid of the Japanese quail follows a pattern of activity that parallels that of the gonads (Follet and Riley, 1967). Thyroid activity (Leloup and Fontaine, 1960) is increased in fishes undergoing seasonal migrations. In the African lungfish (*Protopterus annectens*), thyroid activity is lowest during estivation at the time of seasonal drought. In toads and frogs, epinephrine attains its highest concentration in the plasma during autumn and winter (Donoso and Segura, 1965; Harri, 1972). Stores of growth hormone are highest in the pituitary of the perch (*Perca fluviatilis*) in June, a few weeks prior to the summer rapid-growth period (Swift and Pickford, 1965).

As described above, the cyclical and episodic release of hormones is usually controlled through the brain. This is especially apparent when light provides the cues for these changes. External temperature may also be involved in controlling reproductive and possibly other cycles, especially in poikilotherms

(Licht, 1972). It is likely that such heat stimuli act through the central nervous system, but they could also exert more direct effects on the pituitary and gonads and even influence the rates of a hormone's metabolism. Temperature impinges on receptors present in the skin but may also directly influence the brain.

Systematic differences in releasing stimuli

There are many instances among the vertebrates in which the physiological roles of analogous hormones exhibit systematic differences. Such a change in the use of a hormone necessarily results in an altered responsiveness to excitatory stimuli that prompt the endocrine gland to discharge its secretion. When certain teleost fish move from fresh water to sea water they lose water and accumulate salt so that the concentration of sodium in their body fluids rises. This initiates a release of corticosteroids. Tetrapods, however, usually release such analogous hormones in response to declines in the sodium concentration of the body fluids. Prolactin and oxytocin are released during suckling in mammals. Other vertebrates lack mammary glands so that this represents a unique and phyletically novel stimulus. In certain teleost fish prolactin is released when the fish migrate from the sea into fresh water. The prolactin is then released in response to the lower osmotic concentration in the fresh water. In homeotherms, thyroid hormones are discharged following a decline in the temperature of the blood flowing through certain areas in the hypothalamus but there is no indication that this happens in poikilotherms. Indeed, thyroid secretion usually increases in cold-blooded vertebrates exposed to elevated temperatures. Changes in glucose concentrations usually determine insulin release, but in ruminants (sheep and cattle) fatty acids are more important. In fishes, amino acids may be the major stimulant for secretion of insulin. Such phyletic differences in the propensity of endocrine glands to respond to certain stimuli are necessary for the evolution of a hormone's physiological role.

Conduction of stimuli to the endocrines

The conveyance of a stimulus to an endocrine gland may involve a complex series of events that take place along rather circuitous pathways. These are consistent with, and indeed are dictated by, the particular physiological requirements of the animal. The initial stimulus is usually translated into another form that may be a chemical compound or an electrical event or both. It travels to the endocrine gland, in such a modified form, by routes of varying complexity and may suffer further translation on the way. During this voyage, the stimulus may have modulated and interpolated with other information that is already available and other stimuli that also impinge on that particular communication pathway. This may take place in the brain and endocrine glands that are temporally proximal to the gland or tissue destined to receive, eventually, the final message. Such intermediary substations may involve neural areas in the brain, interconnecting endocrine glands, like the hypothalamus and the pineal, as well as the pituitary gland.

These events can be illustrated by summarily following the effects of light on reproduction. The stimulus is usually received by the eye (or in hypothalamic photoreceptors in many birds) where it is translated by the retinal receptors into electrical impulses that travel along nerves within the brain. These messages after further translation to other transmitter substances (such as acetylcholine, norepinephrine, and dopamine) eventually reach the hypothalamus and probably also, in some species, pass through the pineal gland, which may alter its release of melatonin. The hypothalamus modulates its release of GnRH, which crosses in the short portal blood vessels to the adenohypophysis. The receipt of this information will be further interpreted (in a process that involves the formation of cyclic AMP) in terms of an appropriate release of FSH and/or LH, which are carried in the blood to the ovaries or testes. Such an effect is illustrated by the levels of GnRH and the gonadotropins in the hypothalamus and pituitary of the Japanese quail. There is a fall in the concentration of GnRH that corresponds to the release of adenohypophysial gonadotropins. The gonadotropins may influence such events as ovulation and the formation and release of estrogens and progesterone as well as testosterone and inhibins. These hormones, in turn, are carried back to the hypothalamus and pituitary gland where they provide information about the current hormone levels that will be used to modify stimuli that subsequently pass through this tissue. The process is, in detail, undoubtedly even more complex than that which has been described.

In other instances, the mechanism of the hormone's release may be simpler. The discharge of ADH (vasopressin) from the neurohypophysis can occur in response to small increases in the osmotic concentration of plasma. This is thought to induce changes in osmoreceptors, in the region of the supraoptic nucleus, possibly by releasing small amounts of acetylcholine, which initiates a wave of depolarization along the axons of the supraopticohypophysial tract. This results in the release of ADH from the storage granules at the terminus of the nerve. Neurohypophysial hormones may also be released in other circumstances. This release may involve nonspecific stimuli, often termed "stress." that pass through higher centers in the brain to the nerve cells of the gland. Oxytocin is discharged in response to suckling in mammals; the initial receptor is in the nipple of the mammary gland from which the stimulus is transmitted along nerves to areas in the brain that initiate the release of the hormone from oxytocinergic neurons in the neurohypophysis.

Even simpler processes, not involving nerve pathways, may exist and determine the release of hormones in the body. The release of insulin can be demonstrated in isolated, perfused pieces of pancreatic tissues containing the islets of Langerhans. Elevated glucose and certain amino acid concentrations in the perfusate initiate a release of insulin. In a similar way, glucagon is discharged when the blood glucose concentration is depressed. The release of a hormone may, however, involve interacting excitants. The release of insulin (see below) is primarily controlled by plasma glucose concentrations but cholinergic and adrenergic nerve stimuli can also contribute to the response. Other hormones, such as glucagon, gastric inhibitory peptide and glucagon-like peptide can promote such release while somatostatin inhibits it.

The mechanism of release of the hormones

Upon receiving a stimulus, an endocrine gland may release its hormones from their storage sites. Many hormones, such as catecholamines, neuro-hypophysial peptides, and insulin, are spewed from their storage granules; steroid hormones are released from lipid droplets; and thyroid hormones are detached from thyroglobulins. In some instances, this is preceded by the formation of cyclic AMP as in the thyroid gland, the adrenal cortex, the ovary, and pancreatic B-cells. The phosphatidylinositol second messenger system may also be involved in such hormone release mechanisms, as seen in pancreatic B-cells, gonadotropes and somatotropes in the pituitary, and the zona glomerulosa cells of the adrenal cortex.

The release of hormones from intracellular storage granules has been studied in detail in the instances of the secretion of catecholamines from the adrenal medulla and ADH and oxytocin from the neurohypophysis (Douglas, 1972, 1974). Hormones that are stored in granules exist there as a nondiffusible complex with proteins and adenine nucleotides. Their release from the cell involves the process of emiocytosis (exocytosis or reverse pinocytosis) across the cell membranes. The nature of the events that result in such a release of hormones can be studied *in vitro* and is as follows:

1. As a result of nerve stimulation the cell membrane is depolarized and ions enter the cells.
2. An increase in the concentration of intracellular calcium ions occurs, as a result of its uptake across the cell membrane and probably also its mobilization within the cell. No release of hormone occurs *in vitro* following depolarization if the external media contain no calcium ions, which thus appears to be vital for excitation–release coupling.
3. Excitation–release coupling involves a calcium-mediated migration of the hormone storage granules toward the cell membrane, a process that may involve the cell microtubular system.
4. When contact between the cell membrane and the granules is made they

Fig. 4.14. A diagramatic summary of factors influencing the release of insulin by the B-cell of the endocrine pancreas. For details see the text. AC, adenylate cyclase; Ach, acetylcholine; CCK, cholecystokinin; DAG, diacylglycerol; ER, endoplasmic reticulum; GIP, gastric inhibitory peptide; GLP-I, glucagon-like peptide I; GRP, gastrin-releasing peptide; IP$_3$, inositol trisphosphate, NPY, neuropeptide Y; PIP$_2$, phosphatidylinositol-4-5-bisphosphate, PKA, phosphokinase A: PLC, phospholipase C; VIP, vasoactive intestinal peptide. (Based on information in Malaise (1990) Havel and Taborsky (1994) and Fehmann, Göke and Göke (1995.)

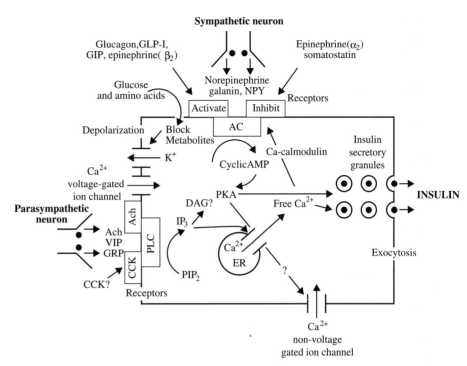

fuse and calcium ions may be important in the structural links. The entire contents of the granule (hormones, proteins, and adenine nucleotides) are then extruded and pass into the capillaries. The empty granule may then be reconstituted and return to the cell cytoplasm.

The process of the release of insulin from the pancreatic B-cells has been studied in detail (Wolf *et al.*, 1988; Malaise, 1990; Rushakoff *et al.*, 1990; Havel and Taborsky, 1994; Zawalich and Zawalich, 1996) and illustrates the many complexities of such a process (Fig. 4.14). Plasma concentrations of glucose play a pivotal role and when they are high promote the release of the hormone. A metabolite of glucose initiates a depolarization of the plasma membrane and an influx of calcium ions occurs into the cell. The resulting rise in cytoplasmic free calcium ions, together with its further release from intracellular sites, promotes the exocytosis of insulin from its storage granules and into the extracellular fluid. The process is modulated in several ways, principally involving the activities of the adenylate cyclase–cyclic AMP and phospholipase C(PLC)–phosphatidylinositol systems (p. 215). The latter is stimulated by acetylcholine released from parasympathetic nerves to the B-cells and the inositol trisphosphate (IP$_3$) formed mediates a mobilization of calcium ions from the endoplasmic reticulum. The activation of adenylate

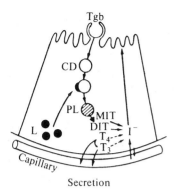

Secretion

cyclase and formation of cyclic AMP can be increased by several hormones: gastric inhibitory polypeptide and glucagon-like peptide 1 from the intestine, and glucagon from the pancreatic A-cells. The cyclic AMP activates protein kinase A (PKA), which enhances the exocytotic process. Adenylate cyclase is also under an inhibitory control by somatostatin from the pancreatic D-cells and intestine and norepinephrine and epinephrine released by the sympathetic nerve–adrenomedullary system (an α_2-adrenergic effect). Release of insulin can be reduced in this way. A variety of neuropeptides including vasoactive intestinal peptide, galanin, neuropeptide Y and gastrin-releasing peptide may be released from the autonomic nerve terminals and modulate the secretory response. The contemporary summary of the process of release of insulin provided in Fig. 4.14 is clearly only a tentative one.

Thyroid hormones are stored extracellularly, combined with thyroglobulin in the colloid of the thyroid gland follicles. Stimulation by TSH results in an activation of adenylate cyclase in the thyroid cells and the formation of cyclic AMP. In a manner that is not yet understood, this nucleotide is thought to stimulate pinocytosis during which fragments of follicular colloid are broken off and taken up by the thyroid cells; this process is thus called endocytosis. Lysosomes in the cell cytoplasm (Fig. 4.15) have been shown to contain four thiol endopeptidases and cathepsin D, which appear to act synergistically to release the thyroid hormones from thyroglobulin (Dunn and Dunn, 1982a,b). This process occurs in the phagolysosomes as they move towards the basal region of the cells. The thyroid hormones are thus freed to pass out of the cell.

The precise mechanism by which steroid hormones are released from the cell is not clear. The amounts stored in lipid droplets within the cell are relatively small compared with the large amounts of those hormones that can be accumulated in granules. The synthesis and release of steroid hormones are promoted by various substances, including tropic hormones. Corticotropin and LH activate adenylate cyclase so that cyclic AMP is formed, which plays a role in the initiation of the hormone synthesis and release; these two processes have not, however, been satisfactorily separated. It is possible that an increased hormone synthesis inevitably leads to its release by a type of "overflow mechanism," but this is usually considered to be an unsatisfactory explanation. Adenylate cyclase does not mediate the release of all steroid hormones: for example, aldosterone secretion, which occurs in response to angiotensin II and III, involves activation of phospholipase C and the phosphatidylinositol system.

The concentration of hormones in the blood

Hormones are normally present in the peripheral plasma at concentrations as low as 10^{-12} M and *in vitro* observations suggest that these levels may, in some instances, be similar to those necessary to stimulate the effector. However, for several reasons it is difficult to predict with certainty what the latter concentration will be. As we have seen, endocrines discharge their secretions in response to a great variety of stimuli; in some instances they may be sudden, or acute, occurrences whereas at other times they are of a more sustained, or chronic, nature. In the first case, the hormones may appear as a single surge or "spurt" of activity in the blood, such as seems to occur with the release of oxytocin in response to suckling. Thus, high concentrations of hormones may exist locally that briefly stimulate the effector and which are subsequently dissipated so that the remaining concentrations in the peripheral blood are not effective ones.

Differences in the patterns of release combined with seasonal and diurnal rhythms and environmental conditions make it difficult to generalize as to the concentrations of hormones that are normally present in the blood. ADH is normally present in the peripheral plasma of mammals at concentrations of 10^{-12} to 10^{-11} M. It is higher in small animals, such as the mouse (2×10^{-12} M), than in large ones such as humans (4×10^{-12} M). In the laboratory rat, ADH concentration increases about 25-fold during dehydration: from 10^{-11} to 2.5×10^{-10} M. As described earlier, the antidiuretic hormone in some mammals is lysine-vasopressin instead of arginine-vasopressin. A wallaby, *Setonix brachyurus*, has the former hormone and when adequate water is available the plasma concentration of this peptide is about 3.5×10^{-10} M but this level increases to 9×10^{-10} M in dry summer conditions when drinking water is not available on its island hideaway (Jones *et al.*, 1990). Some dehydrated amphibians (frogs and toads) have similar circulating levels of vasotocin: about 6×10^{-10} M (Bentley, 1969). The normal concentrations of vasotocin in the plasma have been found to be 10^{-11} to 10^{-12} M in a lizard, *Varanus gouldii* (Bradshaw and Rice, 1981), the domestic fowl (Arnason *et al.*, 1986), the galah *Cacatua roseicapilla* (a parrot) (Roberts, 1991), several teleost and elasmobranch fish (Warne *et al.*, 1994) and a cyclostome fish, the lamprey *Lampetra japonica* (Uchiyama *et al.*, 1994). Increases of three- to five-fold were observed in response to dehydration in all these species, with the exception of the fishes. The physiological role of vasotocin may be different in fish so that its concentrations may not parallel those of the tetrapods under such conditions. Other polypeptide hormones like glucagon and insulin are normally present in the blood of mammals at a concentration of about 10^{-9}.

The steroid hormones (Idler, 1972) attain much higher concentrations in the blood than do the polypeptides; therefore, plasma cortisol and corticosterone levels in teleostean fish are about 10^{-7} M, though in the holostean and chondrostean fishes it is 10 times less than this. In chondrichthyeans 1α-hydroxycorticosterone usually has a concentration of 10^{-8} M in plasma. Aldosterone, compared with other corticosteroids, is present at much lower concentrations, about 10^{-10} M in mammals and 10^{-9} M in birds, though it is 10^{-8} M in amphibians. The steroid sex hormones, owing to the cyclical nature of their release, are present in widely differing concentrations in the peripheral plasma: 10^{-7} – 10^{-10} M. Species differences in the plasma concentrations of steroid hormones have been observed in mammals. Glucocorticoid levels in the plasma of marsupials are generally lower than those of eutherian mammals (Bradley and Stoddart, 1992). New World primates usually have higher steroid hormone concentrations in the plasma than Old World primates (Klosterman, Mural, and Siiteri, 1986). The former species are sometimes referred to as "resistant" as in other species such concentrations of steroids can be toxic. The differences cannot be accounted for by increased binding of the steroids to plasma proteins (p. 209) and it has been suggested that the ligand-binding requirements of their receptors may be greater. The male Atlantic fin whale, *Balaenoptera physalus*, has testosterone levels that are about 10 times less than humans and dogs, and 2.5 times less than stallions (Kjeld, Sigurjonsson, and Arnason, 1992). Possibly these differences provide an example of how the size of a species may influence its hormone concentrations. In conclusion hormones act when present in extremely dilute solution. The levels vary somewhat with different species. The reasons for variation appear to be related to the animal's size, metabolic rate, and the hormone-binding capacity of the plasma. Receptors may also have different requirements for hormones, to which they may be bound with different strengths (affinities). It seems likely that differences in concentration may also be influenced by the body temperatures at which the animals usually function, because they affect the rate of the hormones' reactions with the effectors.

Differences in the potencies of homologous hormones, as reflected in their chemical structure, may also contribute to variations in plasma levels. A mutation that initiates the formation of a less active hormone analog could be compensated for physiologically by its release in greater quantity. An example of such an adaptation may be seen in mammals (Suiformes) that have lysine-instead of arginine-vasopressin.

Transport of hormones in the blood

Following their release from storage sites in the endocrine glands, hormones are carried, in the blood, to their various effector sites. This may be for very short distances, such as from the median eminence to the adenohypophysis, or involve much longer journeys, for example from the neurohypophysis to the kidney. Some hormones have a relatively long life in the circulation and may recirculate many times before they are finally destroyed or excreted.

The blood plasma is an aqueous solution that contains high concentrations of proteins made up of several distinct components. The hormone molecules may be dissolved in this solution or a substantial proportion can be bound to some of the proteins that are present. Plasma albumin provides a site for the binding of some hormones but additional specific proteins may be present. These binding proteins may have special properties for such sequestration. The binding of hormones in the blood is particularly important for the thyroid and steroid hormones, which are lipid soluble, but some protein hormones may also be bound.

The chemical nature of the binding of such hormones to plasma proteins is such that an equilibrium exists between the molecules that are bound and those that remain in solution. In other words, the binding is a reversible phenomenon and, therefore, involves relatively weak chemical forces such as hydrogen and ionic bonds and van der Waals forces. The relative strength of these bonds, however, may vary considerably (high and low affinity), depending on the particular hormone and the nature of the binding protein. The number of binding sites on such proteins may also differ and they may then be described as having a low or high capacity for sequestration of other molecules. There are several consequences of a hormone's binding to a protein in the plasma.

- Hormones are usually assumed to be unable to initiate their effects when so bound. The receptor is envisaged as interacting with hormone molecules present in the aqueous phase. Their removal from solution shifts the equilibrium and may thus result in a dissociation into solution of some of the hormone that is bound to the proteins. It is implicit that the receptor has an even stronger ability than the plasma proteins to bind the hormone, and it is even possible that more direct exchanges may occur between binding proteins and the receptors.

- The process of the hormone's inactivation, such as can occur in the liver, and its excretion, mainly in the urine and bile, is delayed while it is in the bound form.
- The bound hormone may constitute a circulating pool of the excitant that can extend or moderate the hormone's action.
- The distribution of the hormone within the body can be influenced by its binding to plasma proteins (Keller, Richardson, and Yates, 1969). A hormone–protein complex may have a special propensity to be blocked or to pass through the capillaries in certain vascular beds. It may thus help specifically to determine where the hormone is going.
- Protein binding may contribute to the specificity of a hormone's action in the cell (Funder, Feldman, and Edelman, 1973). In the instance of the corticosteroids, two type of response can occur: mineralocorticoid and glucocorticoid (referring to their effects on electrolytes and inter-mediary metabolism), for which there are two types of receptor in the cell. Although the aldosterone receptors have a higher affinity for aldosterone than corticosterone, the normal excess of the latter hor-mone (in the plasma of rats) would be sufficient to negate this difference so that corticosterone would be expected, inappropriately, to occupy the aldosterone receptors. Most of the corticosterone in the plasma unlike aldosterone, is bound to proteins in a way that obscures the effect of the aldosterone. There is, however, an additional effect that contributes to the specificity of the interaction of aldosterone with its receptors. Corticosterone, but not aldosterone, can be inactivated at local tissue sites by 11β-hydroxysteroid dehydrogenase (Funder, 1991). Such an effect occurs in mammals and is also seen in the domestic fowl (Grubb and Bentley, 1992).

High proportions of hormones often exist in a bound form in the plasma. In the instance of testosterone, it is usually greater than 90% of the total present though it is less in other cases. Cortisol binding in the plasma of teleost fishes varies from 30 to 55% of the total (Idler and Truscott, 1972). Thyroxine and triiodothyronine also are substantially associated with plasma proteins; therefore, in the plasma of kangaroos, more than 95% of the thyroxine and 90% of the triiodothyronine is so bound (Davis, Gregerman, and Poole, 1969).

The hormones may be bound to different protein components present in the blood plasma. In some instances, specialized proteins are present that have a high affinity for the hormones. These include a globulin that binds cortisol and corticosterone (corticosteroid-binding globulin (CBG) or transcortin). This protein is present in most vertebrates though quantitative differences in corticosteroid binding capacities of the plasma exist, indicating that interspecific variations occur (Seal and Doe, 1963). The plasma of New World primates generally has a low binding capacity and affinity for steroid hormones and this deficiency applies to the corticosteroids (Robinson, Hawkey, and Hammond, 1985; Klosterman *et al.*, 1986). Such species appear to lack a CBG that is homologous to that of Old World primates. Differences between the binding of cortisol and corticosterone to the CBGs of Old World monkeys, and those of apes and primates indicate that such variations can even occur between more closely related primates. The differences between the Old World species appear to reflect mutations that occurred after their evolutionary divergence. The echidna (Monotremata) lacks CBG, a deficiency that may reflect its low plasma levels of corticosteroids (Sernia, 1980). Cortisol binding is much less in the plasma of fishes than in most mammals (Idler and Truscott, 1972), and it is unlikely that the proteins are identical in the different species. Transcortin (CBG) also binds progesterone (but no aldosterone), and in mammals its levels increase during pregnancy as a result of the action of estrogen, which promotes its formation in the liver.

CBG may exert a protective role and guard animals against excessive toxic levels of corticosteroids in plasma. The males of a small species of shrew-like marsupials, *Antechinus stuartii*, all die about 2 to 3 weeks after the start of the breeding season. These little beasts become very aggressive at this time and their plasma corticosteroid levels rise sharply (Bradley, McDonald, and Lee, 1980). This change is paralleled by a decline in the CBG concentration, a change which appears to be androgen dependent as it can be prevented by castration or increased by the injection of testosterone. The death of these marsupials is associated with gastrointestinal hemorrhage and infection, apparently reflecting the high levels of free or loosely bound corticosteroids in the plasma. The females do not experience comparable oscillations in steroid or CBG and carry on, mateless, to bear their young. The binding capacity of CBG in marsupials is generally low compared with that of the eutherians (Bradley and Stoddart, 1992), which may compound the problems of the male *Antechinus stuartii*.

Plasma proteins that can bind sex steroids, androgens, estrogens, and progesterone with a high affinity have been identified in many vertebrates. They appear, however, to have undergone several evolutionary developments and the proteins involved are not identical. Cyclostome and elasmobranch fishes possess a plasma steroid-binding protein that binds not only androgens and estrogens but also corticosteroids, and so may serve a multiple role (Martin, 1975). Similar proteins have been identified in the plasma of teleost fish and amphibians (Ozon, 1972). A *sex hormone-binding globulin* (SHBG) is found in women (where its levels increase during pregnancy) and a protein that exhibits cross-immunoreactivity with it has also been found in other primates (Renoir, Mercier-Bodard, and Baulieu, 1980). This globulin binds both androgens and estrogens. In nonprimate eutherians, an analogous protein is present, but it only binds androgens and displays no cross-im-munoreactivity to the primate protein. Some marsupials also possess a SHBG (Sernia, Bradley, and McDonald, 1979) that only binds androgens. The appearance of a SHBG that can bind both androgens and estrogens may thus be a specialization that, among mammals, is confined to the primates. The marsupials belonging to the order Polyprotodonra (which includes the opos-sums) lack (except in 1 species out of 14 examined) a SHBG (Sernia *et al.*, 1979) and it is also absent in rodents. The occurrence of SHBG thus appears to be quite sporadic among the mammals, possibly reflecting differences in their reproductive requirements for the sex steroids.

Many, but not all, mammals possess an α-globulin that preferentially binds thyroxine, *thyroxine-binding globulin* or *TBG*, but which is absent in non-mammals (Farer *et al.*, 1962; Tanabe, Ishii, and Tamaki, 1969; Larsson, Pettersson, and Carlström, 1985). It also binds triiodothyronine but only about one-third as strongly as it does thyroxine. Although such plasma proteins strongly bind hormones, they are present in relatively small quanti-ties so that the total amount of hormone that they can associate with is limited (low binding capacity). The plasma albumins, however, have a high binding capacity though their affinity is low. They are phylogenetically the oldest thyroid-hormone binding proteins and may be the only such plasma proteins in fish, amphibians, reptiles, and many mammals (monotremes and various marsupials). They also play such a role in birds, eutherians, and marsupials when the binding capacity of the more specialized plasma proteins that they may possess is exceeded. Many species, however, initially utilize a specific plasma protein for binding thyroid hormones. It was first identified elec-trophoretically in the plasma and as it migrated behind the albumin it was called prealbumin, but it has now been dubbed *transthyretin*. It has an affinity and binding capacity for thyroid hormones that lies between TBG and albumin. Transthyretin has a molecular mass of about 60 kDa and consists of four similar subunits (Harms *et al.*, 1991; Richardson *et al.*, 1994). Its precise amino acid sequence varies in different species but a core section of the molecule that contains the thyroid hormone-binding site is well conserved. Transthyretin appears to be present in the plasma of all birds, eutherian mammals, and diprotodont marsupials (which includes kangaroos and most possums) (Richardson *et al.*, 1994). It has not, however, been identified in the plasma of monotremes, polyprotodont marsupials (mostly carnivorous spe-cies such as native cats, the Tasmanian devil and bandicoots), reptiles, amphibians, or fish. The transthyretin gene, however, is expressed in the choroid plexus of reptiles, birds, and all mammals but not in amphibians (Achen *et al.*, 1993). In order to appear in the plasma, the transthyretin gene must be expressed in the liver, a process which only occurs among the homeotherms. Its first phyletic expression in the choroid plexus of reptiles may reflect its origin in vertebrates about 350 million years ago. The role of transthyretin in the choroid plexus appears to be to facilitate the transport of thyroid hormones into the brain. Its subsequent appearance in the plasma of some homeotherms may have occurred independently in birds, eutherians, and diprotodont marsupials. There are, however, sporadic descriptions of a transthyretin-like plasma protein in several cold-blooded vertebrates including salmon (Larsson *et al.*, 1985), bullfrog tadpoles (Yamauchi *et al.*, 1993) and the slider turtle (Licht and Pavgi, 1992). Whether a transthyretin gene exists in vertebrates that are phyletically below the reptiles remains to be determined.

Protein hormones may also be bound to other proteins in the plasma. A specific, high-affinity protein that binds growth hormone and prolactin was first identified in rabbits and sheep (Ymer and Herington, 1985). Binding of growth hormone and prolactin to plasma proteins has since been observed in a variety of other mammals (Amit *et al.*, 1992), and domestic fowl and turkeys (Vasilatos-Younken *et al.*, 1991). In mammals, the nature of these binding proteins in different species varies. Types I to IV have been characterized on the basis of their relative affinities for growth hormone and prolactin. The amino acid sequence of the growth hormone-binding protein in the rabbit plasma is identical to the extracellular hormone-binding domain of the growth hormone receptor (Leung *et al.*, 1987). It appears to be encoded by the same gene as that of the receptor. Two mRNAs are produced, apparently as a result of alternate splicing of the primary transcript (Baumbach, Horner, and Logan, 1989). This is a fascinating strategy for producing a high-affinity binding protein. The insulin-like growth factors I and II can also be bound to plasma proteins (Baxter, 1993; Jones and Clemmons, 1995). Such proteins have been identified in humans and rodents, and also in teleost fish (Anderson *et al.*, 1993). In humans, six variants with molecular masses of 20 to 30 kDa have been identified. Their relative binding of insulin-like growth factors I and II differs. Such binding of protein hormone in the plasma may modulate their activities at the site of their receptors, prolong their actions, and even influence the targeting of effector tissues.

Peripheral activation of hormones

Some hormones are chemically altered, at sites peripheral to the endocrine gland from which they originated, in a manner than enhances their biological activity. This process is referred to as "activation" and may occur in the plasma or at tissue sites.

Angiotensin I is thus converted to its active forms, angiotensin II and III, by the action of enzymes. Not all the triiodothyronine present in the circulation is directly released from the thyroid gland. Most is formed from thyroxine in the liver and kidneys (Sterling, Brenner, and Saldanha, 1973). An integral step in the action of testosterone involves its conversion at its target site by 5α-reductase, to 5α-dihydrotestosterone. Cortisone can also be converted peripherally to the more active steroid cortisol. Cholecalciferol (vitamin D_3) undergoes several transformations before it can exert its effects. These changes involve the formation of 25-hydroxycholecalciferol in the liver and a further hydroxylation to 1α-dihydroxycholecalciferol in the kidney (Lawson *et al.*, 1971). It is also possible that some of the large protein hormones are fragmented peripherally into smaller pieces prior to their action. There is evidence that this may occur with parathyroid hormone, while large amounts of proinsulin are normally present in the plasma and this, possibly, may also be converted into the active hormone.

Termination of the actions of hormones

The durations of action of hormones vary: they may persist for many hours and even days, or only have a short-lived, transitory effect. This is usually in keeping with the nature of the homeostatic processes they mediate. Even a prolonged effect, however, will necessitate the renewed release of hormones because of the metabolic and excretory processes that inevitably result in their inactivation and elimination from the body.

Hormones persist in the circulation for different periods of time. The half-life of vasopressin in humans is about 15 minutes, that of cortisol is about 1 hour, and that of thyroxine is nearly a week. This reflects the speed of their degradation in the body, the rate at which they may be eliminated in the urine and bile, the protection afforded them as a result of binding to proteins, and the nature of their actions at the effector site. The effects of binding are well illustrated by comparing the rates of removal (clearance) of corticosteroid hormones from the blood. In humans, about 1600 liters of plasma are normally completely cleared of aldosterone each day. However, only about 180 liters are purged of cortisol in this time. The difference principally reflects the strong binding of cortisol to CBG in the plasma. Considerable inter-specific differences are apparent in the half-lives of hormones: arginine-vasopressin (AVP) has a half-life of about 1 minute in laboratory rats, compared with 15 to 20 minutes in humans. This probably reflects the effects of size and metabolic rate: small animals destroy and eliminate hormones more rapidly than large ones. The precise chemical structure of the hormone may also be important; lysine-vasopressin has a half-life that is nearly twice as long in the rat as arginine-vasopressin (Ginsburg, 1968). The considerable differences that have been observed in the potency of human, porcine, and salmon calcitonin substantially reflect the differences in their degradation rates (Habener *et al.*, 1971; DeLuise *et al.*, 1972). In rats, porcine calcitonin is destroyed by the liver whereas that from humans and the salmon is degraded by the kidney. Salmon calcitonin is much more resistant to inactivation (in rats, dogs, and humans) than the mammalian hormones. Body temperature will also be expected to have an effect on the rate of inactivation of hormones so that, in cold-blooded vertebrates, the inactivation and excretion process will be modified accordingly. The effects of differences in size, species, and temperature on hormone metabolism have not yet, however, been thoroughly evaluated.

The action of a hormone may be terminated in several ways.

1. In order to act it must attain a certain critical concentration in the neighborhood of its receptor site. If a hormone is released in a short burst as, for instance, usually occurs with oxytocin, the receptor will respond to a local high concentration that is sequestered like a small packet in the plasma. The response will then be terminated simply as a result of the subsequent dilution and redistribution of the hormone in the body fluids. A hormone may also be removed if it is bound or has accumulated at tissue sites. Epinephrine is readily taken up by the adrenergic nerve terminals. Less specific binding to tissues, such as that of neurohypophysial peptides to skeletal muscle, may also contribute to the removal of hormones from the circulation.

2. Small amounts of hormones may be eliminated unchanged in the urine and bile, but this usually amounts to less than 5% of the total released. The activity of hormones is generally reduced or destroyed as a result of their metabolism by enzymes in the tissues, particularly in the liver and kidneys. They are transformed in various ways and the by-products are usually then excreted in the urine and bile. The hormone's chemical structure may be altered in several ways. The catecholamines can be methylated by catechol-O-methyltransferase (COMT) or, to a lesser extent, they may be deaminated as a result of the action of monoamine oxidase (MAO). The action of the thyroid hormones is largely destroyed by the removal of iodine from the molecule by mono-de-iodinase enzymes. Protein hormones are broken up by proteolytic enzymes. Steroid hormones (and, to some extent, thyroxine) are combined chemically with glucuronic and sulfuric acids, in a process called "conjugation" that results in increased water solubility, which enhances their chances for excretion in the urine and bile. Prior to such a conjugation, considerable changes may be wrought in the chemical architecture of the steroids. Despite such chemical alterations, some hormones may subsequently still retain some of their biological activity. Water-soluble conjugates of steroid hormones, usually androgens and progestins, may be utilized as pheromones by some teleost fish (Scott and Vermeirssen, 1994; Sorenson et al., 1995). Following their excretion, probably in the urine and bile, they may be detected by olfactory receptors of the opposite sex and aid in the synchronization of their sexual behavior.

Mechanisms of hormone action

Hormones exert actions on every major group of tissues in the body. Their effects are numerous and include changes in the intermediary metabolism of fats, proteins, and carbohydrates, growth and development of the tissues, changes in the permeability of membranes, and the contraction or relaxation of muscles. The precise manner in which they effect such processes is only incompletely understood. The initiating event in the action of a hormone is its binding to its specific receptor. The combination of these two molecules results in a perturbation, such as a conformational change, that is usually followed by a chain, or cascade, of reactions that culminate in the hormone's physiological effect. The initial mechanisms can involve a series of biochemical events starting at the plasma membrane and then proceeding in the cytoplasm and sometimes the nucleus, where transcription of target genes may be affected. This process may involve changes in the activities of enzymes, such as protein kinases, the production of special metabolites, and changes in intracellular ion concentrations, especially calcium. An orderly cascade of such processes precedes the final definitive response.

Hormones utilize relatively few basic mechanisms to mediate their effects. Those hormones with receptors on the plasma membrane usually initiate their effects by activating an adjacent enzyme, such as adenylate cyclase, phospholipase C, guanylate cyclase or a tyrosine kinase. In some cases, the receptors themselves have protein kinase domains. Most other receptors are coupled to a special protein that can relay messages to the associated enzyme. These transduction proteins are members of a large and diverse family. They are heterotrimers with α-,β-, and γ-subunits and are called G proteins (Neer, 1995). The α-subunit binds GDP but when a hormone receptor interaction occurs the GDP dissociates from its binding site and is replaced by GTP. The α-subunit–GTP subsequently dissociates from the holoprotein leaving the dimer comprising β–γ-subunits. When mediating a link between the receptor and adenylate cyclase, different α-subunits may mediate a stimulatory effect on the enzyme, α_s (G_s proteins), or an inhibitory one, α_i (G_i proteins). The intricacies of the G proteins associated with phospholipase C have not yet been satisfactorily unravelled. The succeeding step in the actions of hormones utilizing such G protein-coupled receptors is the formation of a "second messenger" (the hormone is the first messenger) that triggers the succeeding series of biochemical events. There are several such molecules. The historical prototype is the nucleotide adenosine-3',5'-monophosphate (cyclic AMP, cAMP) which is formed from ATP by adenylate cyclase. A pair of lipid-derived second messengers are formed by the hydrolysis of the membrane lipid phosphatidylinositol-4,5-bisphosphate (PIP_2) under the influence of phospholipase C. They are inositol-1,4,5-trisphosphate (IP_3) and diacylglycerol (DAG). The calcium ion is also considered to be a second messenger as it can activate many enzymes, either alone or in combination with the protein calmodulin.

Such second messengers can activate protein kinases (Krebs, 1972; Hunter, 1995) that are integral to many responses to hormones. They are a large, diverse, related protein family that can phosphorylate serine–threonine and tyrosine residues on proteins. Phosphorylation can alter the activities of enzymes and transcription factors so as to produce the hormone's prescribed effects.

There are other types of membrane receptor mediating the effects of hormones but less is known about their mechanisms than for the G protein-coupled receptors. The insulin and IGF-I receptors appear to utilize their inherent tyrosine kinase activity but the pathways leading to the final effects of these hormones has not yet been fully delineated. Following an interaction with insulin, the insulin receptors undergo a dimerization. The receptor tyrosine kinase then autophosphorylates (Saltiel, 1996). The resulting increase in kinase activity can create binding sites and mediate the phosphorylation, and sometimes dephosphorylation, of at least five other cell proteins that are involved in the diverse responses to insulin. Even less appears to be known about the mechanisms utilized by the growth hormone–prolactin receptor family, but a dimerization and an activation of associated cytosolic protein kinases occurs following the interaction of the receptors with their ligands (Horseman, 1994; Finidori and Kelly, 1995).

A hormone may utilize more than one type of mechanism and second messenger. When the latter occurs, distinct subtypes of receptor may be involved, such as vasopressin V_1 and V_2, and the catecholamine adrenergic α- and β-receptors.

Hormone responses linked to adenylate cyclase

In the absence of an interaction between a hormone and its receptor, an associated adenylate cyclase enzyme remains in an inactive state (Rodbell *et al.*, 1975; Citri and Schramm, 1980). However, following the binding of the hormone to the receptor, and if the response is a stimulatory one, its adjoining G_s protein binds GTP and dissociates to release α_s–GTP. This subunit on binding to adenylate cyclase stimulates its activity so that cyclic AMP is produced from ATP (Fig. 4.16). If an inhibitory effect is involved, a G_i protein binds GTP and the subunit α_i–GTP will reduce the enzyme's activity and the levels of cyclic AMP. Owing to its intrinsic GTPase activity, the α–GTP is subsequently transformed to the inactive α–GDP and the complete G protein is then reconstituted. The cyclic AMP produced initiates the subsequent cascade of reactions by releasing protein kinase A from its inactive dimeric form bound to two regulatory units (R) to the free active form. The reaction is:

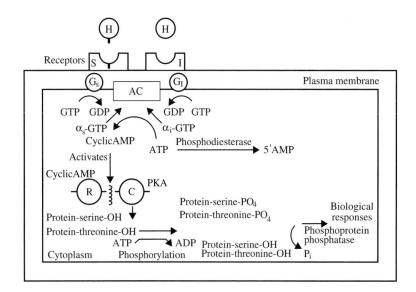

Fig. 4.16. A diagramatic summary of the mechanism of action of hormones that utilize the adenylate cyclase–cyclic AMP and G protein systems. AC, adenylate cyclase; G_s and G_i, G proteins with stimulatory (s) or inhibitory (i) actions; α_s and α_i, activated stimulatory and inhibitory subunits, respectively, of G proteins; H, hormone; I, inhibition; PKA, phosphokinase A; C, catalytic subunit of PKA; R, regulatory subunit of PKA; S, stimulation. For details refer to the text.

$$R_2C_2 + 4 \text{ cyclic AMP} \rightarrow R_2 (\text{cyclic AMP})_4 + 2C$$

Protein kinase: inactive dimer free active catalytic subunits

catalytic subunit: inactive active

The activated catalytic units of the protein kinase phosphorylate serine, and to a lesser extent threonine, residues on the effector proteins. Such a transfer of phosphate groups can change a protein's activity, resulting in a particular biological response. These cyclic AMP-mediated effects include glycogenolysis, lipolysis, an increase in the contractility of the heart, a release of thyroid hormones and corticosteroids, a mobilization of calcium from bone, dispersion of melanin in melanocytes, and an increased reabsorption of water across the renal tubules. The transcription of genes can also be initiated by phosphorylation of transcription factors. Hormones whose effects are mediated by activation of adenylate cyclase include epinephrine (β-adrenoceptors), glucagon, LH, FSH, thyrotropin, corticotropin, MSH, GH-RH, vasopressin (V_2 receptors), secretin, CRH, calcitonin, and parathyroid hormone. The active phosphorylated proteins can subsequently be dephosphorylated by protein phosphatases, and cyclic AMP can be converted to inactive 5'-AMP by phosphodiesterase enzymes.

Hormone responses linked to phospholipase C and phosphatidylinositol

Many receptors that are present in cell membranes can, following binding to their specific hormones, activate adjacent phospholipase C (Putney, 1987; Nishizuka, 1988; Berridge, 1993; Divecha and Irvine, 1995). Such effects

Fig. 4.17. A diagramatic summary of the mechanism of action of hormones that utilize the phospholipase C–phosphatidylinositol system. DAG, diacylglycerol; ER, endoplasmic reticulum; G, G protein; H, hormone; IP_3, inositol-1,4,5-trisphosphate; PIP_2, phosphatidyl-4,5-bisphosphate; PKC, protein kinase C; PLC, phospholipase C; R, receptor. For details see the text.

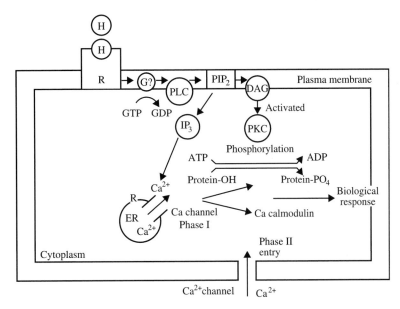

also appear to be transduced by the activities of G proteins. Phospholipase C hydrolyses the membrane lipid PIP_2 to IP_3 and DAG (Fig. 4.17). The response of the cell can then follow two pathways. The IP_3 accumulates in the cytoplasm and activates receptors on the endoplasmic reticulum to open calcium channels and release free calcium ions (Phase I) inside the cell. This calcium, either alone or combined with calmodulin, can activate various enzymes, including protein kinases that contribute to the hormonal response. Following the phase I increase in free (ionized) intracellular calcium ions, the concentrations decline owing to its efflux from the cell. Replenishment then occurs through non-voltage-gated calcium channels in the plasma membrane. The mechanism for this Phase II increase in intracellular calcium is unknown. DAG activates a protein kinase C that by phosphorylating effector proteins, including transcription factors, can also contribute to the hormone's effects. The activation of protein kinase C is facilitated by its migration, under the influence of elevated cytosolic calcium levels, to a site on the plasma membrane where DAG acts. Responses to hormones mediated or modulated in this fashion include hepatic glycogenolysis (α_1-adrenergic responses in some species), contraction of vascular smooth muscle, contraction of the uterus and gall bladder, and the release of growth hormone, thyrotropin, aldosterone, insulin, and LH. Hormones whose effects are mediated by the activation of phospholipase C include epinephrine (α-adrenoceptors), oxytocin, GnRH, TRH, GH-RH, angiotensin II, vasopressin (V_1 receptors), gastrin, CCK, and parathyroid hormone.

The three second messenger systems that are utilized by hormones and which involve adenylate cyclase, phospholipase C and calcium ions frequently all function together in the same cell where they form complex self-regulatory systems (Cooper, Mons, and Karpen, 1995).

The role of calcium and calmodulin

Calcium plays an important role in the mechanism of action of hormones, and it has sometimes, like cyclic AMP, also been called a "second messenger" (Rasmussen and Goodman, 1977; Clapham, 1995). It may be involved in several types of reaction in cells, including changes in the activity of enzymes such as adenylate cyclase, phosphorylase and phosphodiesterase. Calcium also acts as a coupling agent and can link the primary actions of excitants to the ultimate expression of their response. This type of role includes the process of muscular contraction and the secretion of exocrine and endocrine glands. Changes in the levels of soluble "free" intracellular calcium may occur through an activation of calcium ion channels resulting in an increase in the influx of the ion across the plasma membrane into the cell or its mobilization from intracellular stores in the endoplasmic reticulum, mitochondria, or plasma membrane (Bootman and Berridge, 1995). Such an intermediary role of calcium generally appears to result from its interaction with an intracellular binding protein (or calcium "receptor") called *calmodulin* (Cheung, 1980). As a result of this binding, the configuration of this protein changes, so that about 50% of the molecule assumes an α-helical configuration. When in this form it can activate a large number of enzymes, including phosphodiesterase, adenylate cyclase, phospholipase A_2, and calcium-activated ATPase, as well as some protein kinases.

Calmodulin is a protein with a molecular weight of about 16 700, containing 148 amino acid residues. It appears to be present in all animal cells, which is consistent with its ubiquitous function, and it has been identified in species from both the animal and plant kingdoms. Its structure is remarkably conservative and displays few differences in amino acid sequences between such species as cattle and sea anemones. Antibodies to rat testis calmodulin can interact with calmodulin prepared from this coelenterate. "It is likely that calmodulin . . . will provide a link in our understanding of the interactions of calcium and cyclic nucleotides in the control of cellular metabolism" (Means and Dedman, 1980).

Hormonal effects mediated by transcription of nuclear DNA

Steroid hormones (estrogens, progestins, androgens, corticosteroids), $1,25(OH)_2 D_3$ and thyroid hormone all mediate their effects by regulating the initiation of transcription of target genes (Clark and Mani, 1994; Tsai and O'Malley, 1994; Yen and Chin, 1994). These hormones (Fig. 4.18) cross the

Fig. 4.18. A diagramatic summary of the mechanism of action of hormones that utilize the process of genomic transcription. DNA-B, DNA-binding domain; H, hormone, HB, hormone-binding domain; HRE, hormone-response element; IP, inhibitory protein; TA, transcriptional activation domain. For details see the text.

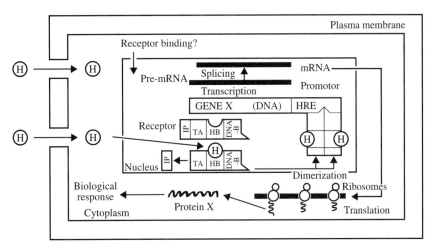

plasma membrane by diffusion and encounter their specific receptors inside the cell. These receptors are sequence-specific DNA-binding transcription factors. This interaction was formerly thought to occur predominantly in the cytoplasm but more precise techniques for the localization of such receptors now indicate that the association occurs mainly in the nuclear compartment (see, for instance, King and Greene, 1984; Welshons, Lieberman, and Gorski, 1984). The glucocorticoid receptor, however, is still thought to bind its ligands in the cytoplasm, from which it is then transported to the nucleus. Some of these receptors (for thyroid hormone and $1,25(OH)_2D_3$) may normally be bound to the genomic DNA, even before they interact with their specific hormones.

The receptors for the steroid hormones, $1,25(OH)_2D_3$ (VDR) and thyroid hormones (TR) are members of a related superfamily (Chapter 3) of large proteins (80 to 100 kDa) containing specific functional zones or domains. There is a ligand- or hormone-binding domain, a DNA-binding domain, and a transactivation-domain (transcription activation). Thyroid hormone receptors also have a potent gene repression function. The DNA-binding domain mediates the interaction of the receptor with its specific binding site on the promoter region of the target gene. The binding sites are specific DNA sequences called the hormone-response elements (HRE). A genetically engineered mutant receptor for human glucocorticoids that contains the expected glucocorticoid DNA-binding domain but a hormone-binding domain coded for estrogens has been cloned. When this hybrid receptor is combined with an estrogen it activates glucocorticoid-responsive genes, not estrogen-responsive ones. The DNA-binding domain on the receptor, therefore, determines which target genes will be affected. In this instance, a glucocorticoid response can be elicited by an estrogen.

It is possible that some receptors may be able to initiate basal genomic responses even in the absence of hormone binding, but their effects may be modified by the presence of the hormone. Following binding to the hormone, the receptor is activated, or transformed, to a state in which an appropriate interaction with its HRE can proceed. This change may involve a covalent modification, such as phosphorylation, a conformational change, or the dissociation of associated proteins. This receptor transformation appears to result from the unmasking of a site that facilitates the formation of receptor dimers which trigger the final response. Some receptors (for estrogens, androgens, progestins, and glucocorticoids, but not for thyroid hormone or $1,25(OH)_2D_3$) may be associated with inhibitory heat-shock proteins (HSP 90, HSP70, HSP56), from which they dissociate following their binding to the hormone. However, although these effects are seen *in vitro* their role *in vivo* is not clear. Thyroid hormone receptors, which function as repressors of transcription in the absence of a ligand, release an associated corepressor protein called SMPT following binding to the hormone (Chen and Evans, 1995). The receptors that interact with the HREs on the gene may be monomers, homodimers (as seen for steroid hormones), or heterodimers (as utilized by thyroid hormone, $1,25(OH)_2D_3$, and retinoic acid receptors). Such combinations appear to increase the diversity and spectrum of possible regulatory responses (Glass,1996).

The precise events that result in the activation of transcription by the target gene are largely unknown but are thought to involve recruitment of RNA polymerase II and general transcription factors (the transcription initiation complex) to the promoter (Beato and Sanchez-Pacheco, 1996). Following processing of the primary transcript, mature mRNAs are exported to the cytoplasm where they are translated to a polypeptide chain on the ribosomes. The proteins formed are often enzymic in their nature and may direct the processes involved in the ultimate response to a hormone. Such an effect may be manifested by changes in intermediary metabolism, growth, and development. Adjustments in mineral metabolism may occur when such proteins furnish ion channels and the enzymes associated with ion "pumps" such as Na–K ATPase and Ca^{2+}-activated ATPase (Ca-ATPase). A thermogenic response to hormones may also be facilitated by the synthesis of Na–K ATPase (Ismail-Beigi, 1993).

Other hormones, including growth hormone, prolactin, insulin and IGF-I, can also influence the synthesis of cell proteins, but their actions on transcription are not as direct as those of the steroidal and thyroid hormones. Detailed information about their mechanisms of action are lacking, but a number of possibilities exist. For instance, cyclic AMP and protein kinases are known to regulate transcription of certain genes. Protein kinase A when activated by cyclic AMP can phosphorylate, and so activate, transcription factors that act at sites on genes. These binding sites on the gene are known as cyclic AMP-responsive elements (CRE) which bind the transcription factor CREB. Phosphorylation of CREB by protein kinase A permits its binding of a coactivator protein (CBP) in which state it affects the chromatin structure of the target gene. Other protein kinases, including protein kinase C, may also promote transcriptional activation of such latent transcription factor proteins. The responses to growth hormone and prolactin involve (Horseman, 1994; Roupas and Herington, 1994; Finidori and Kelly, 1995; Argetsinger and Carter-Su, 1996) the activation of a kinase called JAK 2 (from the *ja*nus *k*inase family, or *j*ust another *k*inase). The JAK family of tyrosine kinases phosphorylates a group of transcription factors with the acronym STATs (*s*ignal *t*ransducers and *a*ctivation of *t*ranscription). Following their phosphorylation, they form homo- or heterodimers that move to the cell nucleus and activate transcription of specific target genes. Several of the "immediate early genes" that are rapidly activated by growth factors code for transcription factors that are referred to as "third messengers" as they can activate or repress other genes that may be involved in a response. Such mechanisms may be utilized by various growth factors and hormones.

Conclusions

In the following chapters, we will be examining the roles of hormones in coordinating different physiological processes in the body. In the present chapter, we have looked at the manner by which the endocrine system itself works. Although information about nonmammals is rather sparse, it appears that the underlying mechanisms of a hormone's synthesis, release, transport, mechanism of action, and its destruction are rather similar in all vertebrates. Even when different hormones are involved, the general underlying processes involved are often similar; but major differences are often apparent between the general types of hormone, especially those made from cholesterol (steroids) and those derived from amino acids. There are a number of interspecific differences in the "life history" of particular hormones in the body and these can be related to the animal's manner of life. The natures of the stimuli that initiate a particular hormone's release are especially variable among the vertebrates and are dictated by the different physiological roles that a hormone may have assumed. In addition, quantitative differences may arise with respect to a hormone's rates of synthesis and destruction, the quantities that are stored in the gland, and the concentrations that appear in the blood. Such differences can arise at distinct stages of the life cycle of an animal, but they are also observed between various species, where they can be related to such characteristics as size, rates of metabolism, and environmental factors like temperature and the availability of different nutrients, salts, and water.

5 Hormones and nutrition

Animals require a continual supply of food in order to sustain life. Such nutrients, in the first instance, are obtained from the external environment. These materials are used as an energy supply, as building blocks for growth and reproduction, and also as a source of certain essential chemicals necessary to the adequate functioning of the metabolic machinery in the body. The processes involved are therefore basic to life and are regulated to a considerable extent by hormones.

The foods that animals obtain from the environments where they live are usually chemically far more complex than can be used by their cells. The original nutrients are transformed in the body into compounds that may sometimes be immediately metabolized by the cells, or they may be converted into substances that can be stored for subsequent transformation into such compounds.

Hormones play an important role in regulating the interconversions of nutrients to metabolic substrates and their stored forms. The endocrine secretions may help to regulate the levels of nutrients by contributing to the control of their absorption from the gut, their levels in the blood, the nature and rate of their storage, their release from tissues, and their assembly into the structural elements of the body.

Animals lead diverse lives in a plethora of environmental conditions. The definitive metabolic processes are basically similar in all animals and lead to the utilization of ATP, for the supply of energy, and the building of cells. Nevertheless, the physiological processes leading to these accomplishments may differ considerably. Such processes are dictated by numerous circumstances and events.

The chemical nature of the foodstuffs that animals obtain from their environments may differ greatly. In their feeding habits, animals may be carnivorous, herbivorous, or omnivorous. Even within these categories considerable differences exist in the types of food animals eat. Some animals may feed principally on invertebrates such as insects, molluscs, and worms that live in terrestrial, fresh water, or marine environments. Other animals feed on vertebrates. Plants from equally diverse situations are also used for food. The possibilities for gastronomic experiments, therefore, appear to be endless, but only a limited number can furnish a particular species with its needs.

Animals have different patterns of feeding. Some eat almost continually, such as cattle and sheep that nibble plants hour after hour. Large predatory carnivores, like lions, snakes, and crocodiles, may only feed intermittently with days or even weeks separating their mealtimes. Circumstances, such as an unexpected drought, may inadvertently result in enforced fasting or even starvation. A dependence on body stores of nutrients for prolonged periods of time may be a fairly predictable part of an animal's life cycle, such as dictated by hibernation during winter, estivation during hot, dry summers, and migrations to more equitable regions for food and in order to breed.

The nutritive requirements of animals may differ considerably. The normal rate of metabolism of different species can differ by more than 100-fold. Warm-blooded homeothermic animals usually have a higher metabolic rate than cold-blooded poikilotherms. Even among homeotherms, the basal metabolic rate differs considerably; for instance, it is about 35 times greater in the shrew than in the elephant. Factors such as size, patterns of activity, and the environmental temperatures experienced contribute to the differences in metabolic requirements of animals. Young, growing animals have special nutrient requirements, whereas breeding and care of the young alter the needs of adults.

Dominating all these differences, and dictating many of them, is the phylogeny of the species. The genetic constitution of a species determines the pattern of its nutrition and the mechanisms involved in the regulation of it. These physiological processes are presumably the result of a prolonged evolution and adaptation to environmental conditions. This is related in the diversity of the endocrine mechanisms that control the metabolism of animals.

Endocrines and digestion

Apart from catching or collecting and then eating food, the first physiological event in the nutritional process is digestion. Food is usually broken down into simpler chemical compounds prior to its absorption from the gut. This process involves the actions of acids, alkalies, and enzymes secreted by glands in the wall of the stomach and intestines as well as the exocrine, or acinar, cells of the pancreas. The orderly flow of these juices is controlled by hormones as well as by nerves (Fig. 5.1). Indeed, the discovery of the role of *secretin* in stimulating the secretion of the exocrine pancreas into the intestine was the first unequivocal demonstration of the role of a hormone in the body, and it was in connection with this discovery that the term "hormone" was initially used. Bayliss and Starling performed the crucial experiment in 1902. A loop of the jejunum of an anaesthetized dog was tied at both ends and was

Fig. 5.1. The role of hormones in controlling gastric acid secretion, pancreatic secretion of salts and enzymes, and the contraction of the gall bladder. *Gastrin*, from the pylorus, initiates secretion of hydrochloric acid by the oxyntic cells in the fundus. The duodenal–jejunal hormones, *secretin* and *cholecystokinin–pancreozymin*, initiate the secretion of, respectively, pancreatic juice and enzymes. Gastric-inhibitory peptide (GIP), from the duodenum–jejunum, inhibits gastric acid secretion. The double open arrows indicate an inhibitory effect; the dashed ones a stimulation.

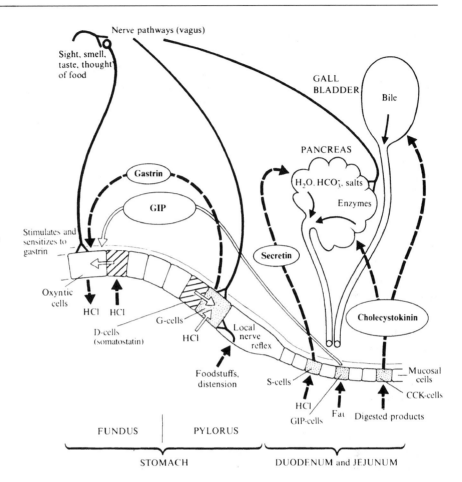

denervated. When acid (0.4% HCl) was introduced into this sac (or the duodenum), secretion of pancreatic juice was stimulated. An extract of the jejunal mucosa was made by rubbing it with sand in the presence of HCl solution. When (after filtering) this was injected into the jugular vein of the dog, pancreatic secretion was also stimulated. As Starling remarked, "then it must be a chemical reflex". This experiment was performed at University College, London, on the afternoon of January 16, 1902 and has been summarized (Sir Charles Martin, see Gregory, 1962): "it was a great afternoon".

Secretin, which is released as a result of the action of acid on the duodenal and jejunal mucosa, stimulates the formation of a voluminous pancreatic juice, rich in bicarbonate and salt but poor in enzymes. The secretion of proteolytic, amylolytic, and lipolytic enzymes (but not water and salt) can be stimulated by the vagus nerve as a result of feeding. Another hormone, however, also assists this process. This is *cholecystokinin–pancreozymin* (it is now usually called cholecystokinin, or CCK), the role of which was not established until 40 years after that of secretin (Harper and Raper, 1943). This endocrine secretion is also formed in the upper parts of the intestine from which it is released in response to the presence of the digestive products. It appears to have a direct effect on the pancreatic acini and an indirect one resulting from stimulation of receptors (CCK-A) on afferent vagal nerve fibers that originate in the gastrointestinal mucosa (Owyang, 1996). The response involves a neural arc to the brain and efferent vagal stimuli to the pancreas.

Gastric secretion is also controlled by nerves and hormones. Stimulation by the vagus initiates the formation of acid by the oxyntic cells and enzymes from the chief cells. The action of the vagus on acid secretion is both direct and mediated by the release of *gastrin* (a hormone formed in the pyloric region of the stomach), which stimulates the oxyntic cells to secrete acid. The vagus also sensitizes the oxyntic cells to the action of gastrin. The complete reflex arc, which is initiated by feeding, initially involves nerve stimulation along cholinergic nerves to the oxyntic cells and the pyloric G-cells, which release gastrin. This hormone then closes the reflex arc and acts on the oxyntic cells. Secretion of gastrin also results from the initiation of a local nerve reflex caused by the presence of food in the stomach. Gastric acid secretion can be inhibited when fat or oils pass into the upper parts of the intestine. This stimulus initiates the release, from intestinal K-cells, of another hormone that can inhibit the secretion of acid from the oxyntic cells. It has been identified as gastric inhibitory peptide (GIP) (Chapter 3).

The presence of fats in the intestine initiates the release of bile from the gall bladder, and this also involves an endocrine reflex through the release of CCK, which contracts the gall bladder and relaxes the sphincter of Oddi (Ivy and Oldberg, 1928). This hormone has several roles in the body and, apart from its actions on the gall bladder and the endocrine and exocrine pancreas, can elicit a sensation of satiety (see below). A variety of intestinal hormones that are released in response to feeding can also promote the release of insulin. These include, glucagon-like peptide I (GLP-I from the L-cells), GIP and, possibly, CCK (Fig. 4.14, p. 203).

The presence of these humoral reflex arcs influencing digestion has been shown in several mammals, but direct evidence as to their presence in other vertebrates is lacking. From the sporadic evidence available, it seems likely that they exist. Indeed, further experiments by Bayliss and Starling in 1903 indicated that this is so with respect to the effect of secretin on the pancreas. They performed experiments on a variety of mammals, including monkey, dog, cat, rabbit, and a bird (a goose "in the process of fattening for Christmas") and confirmed the wider phyletic distribution of this humoral reflex. Secretin-like activity was also shown to be present in the duodenum of humans, ox, sheep, pig, squirrel, pigeon, domestic fowl, tortoise, frog, salmon, dogfish, and skate. This interesting paper by Bayliss and Starling (1903) is entitled *On the Uniformity of the Pancreatic Mechanism in Vertebrata* and must be one of the earliest contributions to comparative endocrinology.

Somatostatin is formed by the D-cells, which are present in various parts of the gastrointestinal tract including the stomach. This peptide (Chapter 3), which is also formed in the hypothalamus and islets of Langerhans, has ubiquitous effects in the body. It can, for instance, block the release of growth hormone, insulin, and glucagon. Somatostatin can also inhibit the release of acid from the oxyntic cells and that of gastrin from the G-cells in the stomach. It appears to be released as a result of low pH in the gastric antrum. The D-cells in the stomach lie in proximity to the G-cells and oxyntic cell so that the secreted peptide may be exerting a local (paracrine) inhibitory effect on secretion of gastrin and acid. Somatostatin has also been identified in the circulation so that it is possible that it also has a more classical type of endocrine role.

The gastrointestinal hormones that contribute to the control of digestion have a wide phyletic distribution but the structures of each type varies. Some phyletic deficiencies, however, exist, such as the apparent lack of gastrin–CCK hormones in some fish. The morphology of the gut of vertebrates displays considerable variation apparently reflecting their diets and phylogeny (Stevens and Hume, 1995). Differences in the endocrine mechanisms controlling digestion would not be unexpected but have not been extensively studied. Such observations usually involve the injection of heterologous hormone preparations and peptides, which sometimes display different responses to those seen in other vertebrates (see Jönsson and Hölmgren, 1989). For instance, gastric acid secretion can be continuously monitored in the cod *Gadus morhua* and high dose of the synthetic peptide pentagastrin (the conserved C-terminal peptide of gastrin) has a weak stimulating effect (Holstein, 1982). However, mammalian gastrin-17 and CCK-8 have inhibitory actions on the acid secretion. This observation may be the result of pharmacological antagonism. The structures of the mammalian hormones may, for instance, differ so much from the cod "gastrin" that while they can combine with the receptors they elicit no effect on acid secretion but do exclude the homologous hormone from its site of action. Alternatively, the teleost hormone could be a natural inhibitor of acid secretion and act as a gastrone. However, such a radical departure from vertebrate conformity seems unlikely.

The transformation of metabolic substrates: the role of hormones

The diversity of intermediary metabolism in vertebrates

Nutrients that are utilized for the production of energy can be classified into three major groups: carbohydrates, fats, and proteins. These materials are also incorporated into the cell structure and so are essential for growth and reproduction.

The nature of the nutrients upon which the animal's metabolism is based depends, in the first instance, on its diet. In carnivores, this consists mainly of protein and fat. The carbohydrates obtained by herbivores may consist of materials such as starches and sugars that can be broken down into simpler sugars by the digestive enzymes. The major organic constituent of most plants is cellulose, which is fermented by microorganisms present in various compartments of the gut (the cecum, the colon, or the rumen) and which produces short-chain fatty acids. These are mainly propionate, butyrate, and acetate. Such fermentation by symbiotic microorganisms is widespread in herbivorous mammals, especially in the sacculated rumen of ruminants (cattle and sheep) as well as the colon of horses and the cecum of lagomorphs, like rabbits, and the sacculated stomach of some marsupials. In nonmammals, the situation is less clear but microorganisms undoubtedly aid digestion in these animals also.

Different species of animals, therefore, show differing dietary dependencies on proteins, fats, and carbohydrates. Proteins can be broken down to their constituent amino acids, fats to fatty acids and glycerol, and carbohydrates to simple sugars, like glucose, or, with the aid of microorganisms, can give rise to fatty acids. The resulting basic subunits can be utilized directly for the production of energy or they may undergo transformations into forms that can be stored and provide a readily accessible reserve. Apart from their reassembly into more complex units, interconversions of one such type of chemical compound into another may also take place in the body. Glucose can thus be readily converted in liver, adipose tissue, and the mammary glands into triglycerides (fats). Some amino acids are transformed by the process of gluconeogenesis, in the liver, to glucose whereas others are changed into fatty acids. The transformation of fatty acids to sugars is not as common, though in ruminants propionate, which is formed in large amounts, is converted to glucose. Glycerol can also be transformed into glucose.

The reserves of protein that are maintained in the body are small and in any case it is not a very suitable substrate for the storage of energy. During starvation, protein may nevertheless make an important contribution to an animal's energy requirements. Substantial amounts of glucose are stored as glycogen in the liver and muscles, but these reserves are inadequate to maintain an animal for prolonged periods of time. Triglycerides provide the most economical and convenient storage form for energy. One gram of fat furnishes 9500 calories whereas the same amount of carbohydrate and protein, respectively, supplies only 4200 and 4300 calories.

Stored fat may be dispersed widely among the tissues in the body but it usually predominates at certain sites. Adipose tissues exist at subcutaneous, mesenteric, perirenal, and periepididymal sites in mammals. The large fat

Fig. 5.2. Effects of various experimental treatments on blood glucose concentrations in amphibians and reptiles. (*a*) Changes in the blood glucose concentrations following pancreatectomy in various reptiles and amphibians. It can be seen that the elevation in glucose occurred relatively promptly in the alligators and toads but was considerably delayed in the lizards and snakes. (*b*) Effects of injected insulin (1 unit/100 g body weight). The amphibians responded more rapidly than the reptiles, but the response was usually more prolonged in the latter. (*c*) Effects of injected glucagon (10 μg/100 g body weight). The reptiles responded more slowly than the amphibians, but the response in the former lasted for a longer time. ⊖—⊖, Alligators; ◇—◇, toads; ●—●, snakes; ○—..—○, lizards, ●·-●, frogs. (From Penhos and Ramey, 1973.)

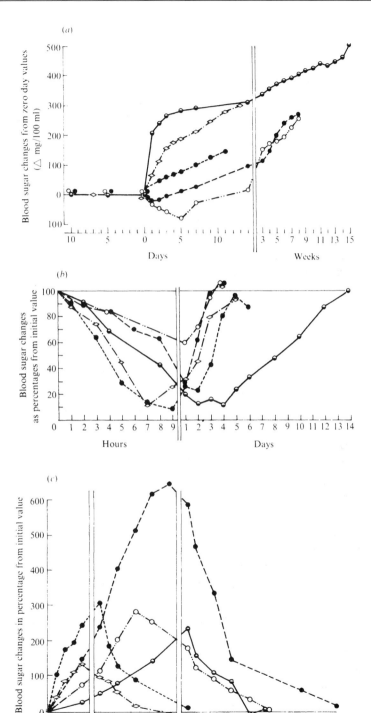

bodies near the gonads in the abdominal cavity of frogs and toads are familiar to student dissectors. The tail of urodeles and lacertilians is also a common site for fat storage. Large quantities of fats may also be stored in the liver of poikilotherms and this is especially important in chondrichthyean fishes, though it is also seen in other vertebrates. In many fishes, fats are stored in close proximity to the muscle fibers that directly utilize their fatty acids.

The transformation, storage, and utilization of fats, proteins, and carbohydrates are regulated to a considerable extent by hormones. The relative differences in the availability and importance of such substrates in different species, not surprisingly, may be reflected in the animal's particular response to hormones. Quantitative, or even qualitative, differences may be observed. Some such variations in the responsiveness of different species of reptile and amphibian to injections of insulin and glucagon, as well as pancreatectomy, are shown in Fig. 5.2. It can be seen that changes in the blood glucose levels following these treatments show considerable interspecific variability in the speed of onset and the magnitude and the duration of the responses.

Hormones that influence intermediary metabolism

When considering hormones that influence the transformation, deposition, mobilization, and utilization of fats, carbohydrates, and proteins in the body, we should be careful to distinguish between those that function physiologically and those actions that probably do not normally occur in the animals (pharmacological effects). For instance, lipolytic activity can be exhibited by at least seven pituitary hormones and also by several secretions from other endocrine glands. It is possible, however, that a pharmacological action of a hormone in one species, or tissue preparation, may reflect a physiological role in some phyletically distant species.

Several hormones contribute to the overall control of the metabolism of carbohydrates, fats, and proteins. Insulin plays a central role, especially in times when food is freely available. It then promotes the storage and subsequent conservation of excess nutrients. Insulin has been called "the hormone of plenty." Glucagon and corticosteroids have important functions during fasting and aid the mobilization of stored nutrients and their conversion to more readily utilizable substrates. They are, therefore, called "hormones of retrieval." Catecholamines, growth hormone, and prolactin have more specialized functions in regulating aspects of intermediary metabolism, such as sudden needs for energy, rapid growth, and the secretion of nutrient milk to feed the young. Thyroid hormones have ubiquitous effects on metabolism and may contribute to the optimal activity of many processes, probably reflecting a modulation of the transcription of various genes. Several hormones usually interact with each other in such metabolic processes; none can function normally in complete isolation from all the others (Fig. 5.3). As so

Fig. 5.3. A summary of the role of hormones in stimulating, or inhibiting, metabolic processes and transformations that control the glucose concentrations in the blood. Solid line, stimulation or increase; dashed line, inhibition or decrease. (Modified from Foà, 1972.)

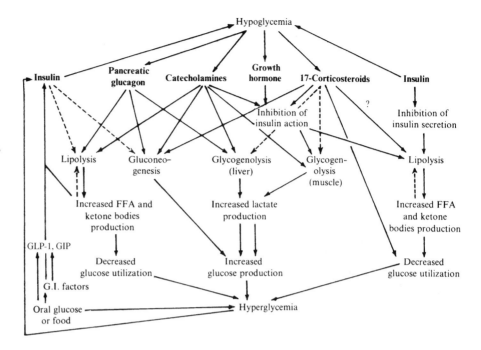

well stated by Tepperman and Tepperman (1970), "It is virtually impossible to separate out one set of signals which control lipogenesis, another which controls gluconeogenesis and a third which controls ketogenesis. All these processes share so much of the metabolic machinery. . . ." With this reservation in mind, we can consider the effects of various hormones on the processes of intermediary metabolism.

Insulin ⎰ Insulin decreases plasma glucose concentrations and, in mammals at least, also reduces free fatty acid levels. In the absence of insulin, muscle wasting occurs because of the excessive mobilization of proteins. These effects are the result of several actions, at various sites, principally the liver, muscle, and adipose tissue. *The processes mediated by insulin* are basically as follows (Fritz, 1972; Larner, 1990; Bressler and Bahl, 1990; Hazelwood, 1993).

1. The rate of uptake of glucose across the cell membranes is increased in skeletal and cardiac muscle, adipose tissue, and mammary gland. Organs like liver, kidney, or brain do not respond in this way to insulin.
2. Glycogen formation from glucose is facilitated. This is partly the result of the more rapid accumulation of glucose, but is also the result of an increased activity of glycogen synthetase in liver, muscle, and possibly even adipose tissue.

3. The mobilization of glycogen is inhibited.
4. Gluconeogenesis from amino acids is inhibited.
5. The accumulation of fatty acids across the cell membranes of adipose cells is increased.
6. Lipolysis of triglycerides to form fatty acids is inhibited.
7. Lipogenesis from glucose is increased in adipose tissue.
8. The uptake of amino acids by muscle and liver cells is stimulated.
9. Insulin inhibits the mobilization of amino acids from protein (related to 4).

The effects of insulin can, therefore, be broadly divided into its actions on the accumulation of nutrients across cell membranes and its facilitation, or inhibition, of metabolic synthesis in cells.

The discovery of the endocrine roles of the pancreatic B-cells and their secreted insulin resulted from histopathological observations in human diabetes mellitus and surgical extirpation of the pancreas in laboratory animals. Specific B-cell cytotoxic drugs, such as alloxan and streptozotocin, provide a more selective, though sometimes controversial, experimental approach in other vertebrates. In 1917, a diabetes-like muscular degeneration that occurs naturally in Japanese carp was found to be associated with the degranulation of the pancreatic B-cells (Nakamura, Yamada, and Yokote, 1971). Morphological aggregation of B-cells into single islet organs (Chapter 2) occurs in a cyclostome, the lamprey *Geotria australis*, and a teleost, the goby *Gillichthys mirabilis*. Surgical removal of these organs results in hyperglycemia in both of these species (Epple *et al.*, 1992; Kelley, 1993). The teleost displayed the other characteristic effects of mammalian diabetes mellitus, such as elevated plasma fatty acid concentrations and an inhibition of growth (Kelley *et al.*, 1993). Injection of bovine insulin corrected the experimental disorders. A likely ancient role for insulin in vertebrates is thus apparent.

Variations in the responsiveness to insulin among vertebrates   The actions of insulin on intermediary metabolism have been principally studied (*in vitro* and *in vivo*) in mammals, usually in laboratory rats. The hypoglycemic action of insulin appears to be widespread in vertebrates (Bentley and Follett, 1965; Falkmer and Patent, 1972; DeRoos and DeRoos, 1979); it occurs among animals ranging from cyclostomes to mammals. Differences in sensitivity nevertheless are apparent, reptiles and especially birds being relatively insensitive to injected mammalian insulin. The pancreas of birds contains little insulin and this is only released "sluggishly" in response to hyperglycemia (Hazelwood, 1973) so that this hormone may not play such an important role in these as in other groups of vertebrates. However, differences in responsiveness to insulin have been observed among birds. Ducks (order Anseriformes) are less sensitive to exogenous insulin than the domestic fowl (order Galliformes) and the latter are less sensitive than mammals (Constans *et al.*, 1991). Some urodeles fail to respond to injected mammalian insulin; yet others do respond (McMillan and Wilkinson, 1972). The crocodilians exhibit a slow response to mammalian insulins; this sluggishness does not appear to reflect the nature of the administered hormone as the responses of alligators were similar whether bovine, avian, or alligator insulins were used (Lance, Elsey, and Coulson, 1993). Among mammals, ruminants are less sensitive than carnivores. Such variations may depend on the type of insulin tested but, more likely, it reflects the metabolic rate and condition of the animals, and the presence of compensatory mechanisms, such as release of glucagon and epinephrine, as well as the normal diet of the animals. Other factors that may contribute to such differences in sensitivity include the numbers of receptors present, their structure, and their affinity for the homologous hormones (see, for instance, Chevalier *et al.*, 1996). The hypoglycemic effect of insulin appears to be a universal one among vertebrates. Differences have been observed, however, in its ability to promote deposition of glycogen in liver or muscle or both. For instance, in lamprey liver, but not muscle, glycogen is increased by insulin injections. Muscle glycogen is increased in the skate (Chondrichthyes) and scorpion fish (Teleostei) (Leibson and Plisetskaya, 1968) where the effects on the liver are less pronounced. The observed lack of glycogen deposition in the liver, however, may reflect, as described in sea bass, an opposing effect of glucagon (Perez *et al.*, 1989). Insulin has unpredictable effects on plasma fatty acids; it decreases them in mammals and teleosts (Lewander *et al.*, 1976; Sheridan, 1994), but may have no effect in birds.

The sensitivity of different species to insulin may be related to their dietary habits. It has been noted that herbivorous mammals withstand an absence of insulin far more readily than carnivorous ones (Gorbman and Bern, 1962; Fritz, 1972). Carnivores only eat periodically so they may have a sudden large intake of nutrients that must be stored for utilization during the fast between meals. Herbivores, by comparison, graze for long periods of the day and are continually absorbing the products, mainly fatty acids, from the large stores of digesting food in their guts. Coordination of the storage and mobilization of nutrients in the body is expected to be more important for periodic eaters like carnivores, than for herbivores (which have aptly been called "nibblers").

Catecholamines   Epinephrine can increase the concentration of both glucose and fatty acids in the plasma. These effects are mediated in the liver and muscles as a result of activation of a phosphorylase enzyme, and in adipose tissue by activation of a lipase. Both effects are mediated by the formation of cyclic AMP. The hyperglycemic effect of epinephrine is seen in species from all the main groups of vertebrates, though the site of its action may differ. For instance, in lampreys, epinephrine mobilizes glycogen in liver but not muscle,

Table 5.1. *Species sensitivity to hormonal stimulation of free fatty acid release from adipose tissue*

	Rat	Mouse	Rabbit	Hamster	Guinea pig	Cat	Dog	Pig	Chicken	Human
Epinephrine or	++	++	0	++	0	+	++	0	0	+
norepinephrine		+			+					
Corticotropin	++	++	++	++	++	+	0	0	+	+
Cortisol	+								0	+
Glucagon	+			+	+				+	
Thyrotropin	+ or		0	0	++	0	+	+		
	++									
MSH (α or β)	0		+	0	+		+	0		
		++								
Vasopressin	0		++	0	++		0	0		

Note: ++, Strong response; +, moderate or weak response; 0, no response.
Source: Shafrir and Wertheimer, 1965.

while in the Chondrichthyes, glycogen from both sites is depleted, as in mammals (Bentley and Follett, 1965; DeRoos and DeRoos, 1972). The hyperlipidemic response to epinephrine has not been studied on such a broad phyletic scale, and even mammalian adipose tissue from all the species studied is not uniformly responsive. Thus, mobilization of fatty acids in response to epinephrine has been shown in the rat, dog, goose, and owl but not in the duck, chicken, rabbit, or pig (Prigge and Grande, 1971; Table 5.1). Epinephrine increases plasma fatty acids in the eel *Anguilla anguilla* (Larsson, 1973), but this is not seen in all fish (Minick and Chavin, 1973) and may reflect an indirect effect by inhibiting release of insulin.

There is some doubt, at least in mammals, as to the efficacy of the circulating concentrations of epinephrine in stimulating glycogenolysis. The concentration usually observed in the plasma appears to be too low to act physiologically on the liver. Adrenergic effects on mobilization of glucose in the liver may, therefore, be mediated by stimulation of the sympathetic nerves. Whether this applies to all tissues, and other species, is unknown.

Glucagon   Glucagon is a "hormone of retrieval" as it mobilizes energy substrates and opposes many of the effects of insulin.

1. Hyperglycemia is promoted.
2. Glycogen is mobilized and converted to glucose in the liver. Its further deposition is inhibited.
3. Amino acids are mobilized and their utilization in gluconeogenesis is enhanced.
4. Lipolysis is promoted and plasma fatty acid concentrations rise.

Glucagon has prominent hyperglycemic effects not only in mammals but also in birds, reptiles, amphibians, and most teleosts (see, for instance, Larsson and Lewander, 1972; Farrar and Frye, 1977; Chan and Woo, 1978). It appears to be relatively ineffective in cyclostomes, chondrichthyeans, and some teleosts (Bentley and Follett, 1965; Patent, 1970). The lipolytic effects of glucagon are prominent in mammals and birds but vary in teleost fish. Such differences may reflect variations between the structures of administered heterologous hormones and the homologous ones, the nutritional condition of the animal, and the season (see, for instance, Plisetskaya *et al.*, 1989; Mommsen and Moon, 1990; Harmon and Sheridan, 1992).

Adrenocorticosteroids   The steroid hormones have profound effects on intermediary metabolism (Vinson, Whitehouse, and Hinson, 1992) (especially in fasting animals), reproductive processes, growth, and lactation. The corticosteroids increase blood glucose concentrations and promote gluconeogenesis and the deposition of glycogen in the liver. They can also promote lipolysis. In excess, corticosteroids promote muscle wasting and a negative nitrogen balance, whereas in young animals they inhibit growth. These two effects reflect their action on protein catabolism. *The actions of corticosteroids on intermediary metabolism* can be summarized:

1. Gluconeogenesis is increased in the liver following mobilization of proteins from skeletal muscle and the deamination of the amino acids that are released. This action is most important, especially during fasting.
2. Glycogen is deposited in the liver because of an increase of the glycogen synthetase reaction.
3. Glycogenolysis is inhibited.
4. Peripheral oxidation and utilization of glucose is inhibited.
5. The conversion of amino acids to proteins and of fatty acids to triglycerides is inhibited.
6. The lipolytic effects of epinephrine, glucagon, and growth hormone are enhanced; corticosteroids appear to have a permissive role in these responses.

Such effects appear to be widespread in the vertebrates, although the information available in nonmammals is sporadic (Chester Jones *et al.*, 1972). Hyperglycemia in response to the injection of corticosteroids has been shown in vertebrates that range phyletically from cyclostomes to mammals. This response is associated with gluconeogenesis and elevation of tissue glycogen levels. The facilitation of gluconeogenesis is associated with increased levels of liver transaminase enzymes of teleost fishes, amphibians, birds, and mammals (though it had once been thought that this increase only occurred in the last two groups of vertebrates) (Janssens, 1967; Freeman and Idler, 1973; Vijayan *et al.*, 1996). An inhibition of growth or loss of body weight has been shown in the domestic fowl and the amphibian *Xenopus laevis*, as well as in two species of teleosts, *Salmo gairdneri* and *Salvelinus fontinalis* (Bellamy and Leonard, 1965; Freeman and Idler, 1973; Janssens, 1967) and in mammals. The actions of corticosteroids on growth and metabolism appear to be basically similar in all vertebrates.

Some differences in the general nature of responses to corticosteroids occur among noneutherian mammals. Three species of kangaroo do not respond to cortisol in the usual manner (McDonald, 1980; Martin and McDonald, 1986). Plasma glucose levels are unchanged by corticosteroids in these macropodid marsupials. Corticosteroids also do not promote gluconeogenesis. However, this is not general to all groups of marsupials; for instance, the brush-tail possum and the sugar glider possum (Bradley and Stoddart, 1990) respond in the more conventional way. The carbohydrate metabolism of a monotreme mammal, the echidna, is also little affected by corticosteroids, but free fatty acid levels in the plasma increase markedly (Sernia and McDonald, 1977).

Growth hormone   Growth hormone is essential for the normal growth and development of the young animal. Its effects have been observed in most vertebrates and are of considerable potential importance for the promotion of growth in farm animals, including fish. Growth hormone can influence many aspects of intermediary metabolism but the effect observed may be influenced by the particular species studied, the stage of its life cycle, and its nutritional state. Interactions frequently occur with other hormones. Not all of the effects of growth hormone are direct ones. Growth hormone has been described as promoting "insulin resistance" and it may have permissive roles in relation to the actions of corticosteroids and catecholamines. "Insulin-like effects" have been described but their physiological significance is doubtful (see Davidson, 1987). Anti-insulin-like actions are, however, observed more often. Growth hormone promotes the formation of IGF-I, which mediates many of its effects, especially on protein metabolism and in bone. A summary of the actions of growth hormone is given below (see Davidson, 1987; Steele and Evock-Clover, 1993).

1. Nitrogen retention is promoted, which is associated with a stimulation of amino acid transport and protein synthesis. These effects are manifested as increased growth of muscle and bone.
2. Decreased lipogenesis occurs and atrophy of fat tissue. Increased lipolysis and fatty acid oxidation may also take place.
3. An increased concentration of blood glucose and deposition of glycogen may be promoted. These actions apparently reflect the glucose-sparing effects of fatty acid utilization. In some species, such as pigs and ruminants, there is an increase in gluconeogenesis, which is also con-

sidered to be an indirect effect, possibly a result of increased transport of amino acids into liver cells.

4. There is a retention of minerals and incorporation of calcium and phosphate into bone.

The overall effect is a carcass that is bigger and leaner, which is a result that is applauded by both farmers and athletes.

Growth hormone has been shown to promote growth in mammals, birds, reptiles and teleost fish (Chapter 3) so that its actions have a wide phyletic distribution among vertebrates. It has also been shown to promote the formation of IGF-I in many such species. The lipolytic actions of growth hormone have been observed in birds and teleost fish (Scanes and Campbell, 1993; Sheridan, 1994) as well as in mammals. The growth-promoting effects of growth hormone have been investigated (McLean and Donaldson, 1993) in fish, especially salmon and trout raised in aquaculture. The homologous salmon and trout hormones, as well as the bovine hormone, have been tested and can increase growth rates by as much as 50%. IGF-I appears to mediate these effects in the fish (see, for instance, Funkenstein *et al.*, 1989; Gray and Kelley, 1991).

Steroid sex hormones   Estrogens and androgens have widespread metabolic effects on the growth and differentiation of tissues, especially the reproductive organs. They may, however, also influence other tissues in the body, principally by promoting the formation of proteins. Androgens when administered have anabolic effects, especially on skeletal muscle (myotrophic action), promoting the formation of proteins. The magnitude of this effect depends on the species, age, and hormonal status of the animal. Estrogens can influence the induction of plasma lipoproteins that sequester fats and cholesterol. The synthesis of high-density lipoproteins (HDL) is promoted while that of low-density lipoproteins (LDL) is depressed. Estrogens also have an anabolic effect in mammals but this is principally confined to the mammary glands and reproductive organs. In oviparous species, estrogens promote the formation of special lipoproteins (vitellogenins) in the liver which are incorporated into the yolk of the egg. Progesterone increases the formation of avidin by the oviduct of the chicken and this is also incorporated into the egg.

Thyroid hormones   Thyroid hormones also contribute to metabolic regulation. Triiodothyronine (T_3) is the principal active hormone. The consumption of oxygen by homeotherms is depressed in the absence of thyroid hormones and they are also necessary for adequate growth and differentiation. The actions of several other hormones are not as pronounced in the absence of the thyroid secretion. Such reduced responses are seen for the catecholamines and corticosteroids. The thyroid gland seems to modulate the

levels and activity of metabolic enzymes in cells; in the instance of epinephrine this action may involve an increase in adenylate cyclase (Krishna, Hynie,
and Brodie, 1968) and β-adrenoceptors (Williams *et al.*, 1977). It can also
promote thermogenesis by contributing to the induction of Na–K Na^+-K^+
activated ATPase (Na–K ATPase) and the mitochondrial uncoupling protein
thermogenin (p. 242). Thyroid hormones, therefore, are often described as
having a "permissive" role in the effects of other hormones and various
enzymes, possibly reflecting a modulation or synergism with the actions of
some transcription factors.

Conclusions   Hormones can be seen to exhibit widespread actions on intermediary metabolism. In some instances, several secretions can exert similar
effects though, in the normal animal, these may not all be physiologically
equivalent. In other cases, hormones may exert opposing effects, either by
acting on different processes or by a more direct inhibition. Hormones can
also directly influence one another's release (Chapter 4) and so mimic or
oppose the actions of other hormones. Intermediary metabolism, although
extremely complex and involving several tissues, many chemical reactions,
and numerous metabolites, is a well-integrated process. This is largely the
result of the actions of hormones at different types of site (some of which are
summarized in Fig. 5.4), both within the same cell and in different kinds of
cell, as well as their ability to act in harmony with each other.

The underlying roles and actions of hormones in regulating intermediary
metabolism appear to be basically similar in all vertebrates. The relative
importance of each action of the hormone may differ, however, depending on
the animal's usual life-style and the particular stage of its life cycle. Events like
fasting (associated with hibernation, estivation, and migration) and reproduction (including the formation of eggs, pregnancy, lactation, and the growth of
the young) are associated with special metabolic needs and hormonally
mediated effects. Some of these will be described in more detail.

Hormones and thermogenesis

The evolution of homeothermy in birds and mammals was the culmination
of many morphological, physiological, and biochemical adaptations. The
maintenance of a constant body temperature by regulating and integrating
the internal metabolic production of heat and its exchanges with the environment created several new roles for hormones and reinforced the importance of
earlier ones (Janský, 1995).

An increased metabolic rate necessitates an additional access to energy
substrates, especially glucose and fatty acids. Their regulation, as already
described, is largely under hormonal control and does not appear to be
basically different from that of cold-blooded vertebrates. The roles of hormones in influencing heat exchanges across the integument will be discussed

in a later section but are not of predominant importance in the maintenance of homeothermy.

Obligatory thermogenesis

The life-sustaining chemical and physical activities of the body inevitably generate excess heat because of their thermodynamic inefficiency. In a resting state this process is called "obligatory thermogenesis" or the "basal metabolic rate." Heat will unavoidably be lost across the integument but such leakage can be limited. Therefore, at an appropriate thermoneutral environmental temperature, an equilibrium may be approached so that the body temperature can be maintained without the need for additional metabolic heat. This obligatory thermogenesis reflects such processes as the contraction of the heart and respiratory muscles, the activities of organs such as the kidneys and liver, the secretory activities of the various glands, and the maintenance of the ion concentration gradients in cells. Physical activities, such as involve skeletal muscles, considerably increase thermogenesis but the conditions will then not be basal. The basal metabolic rate, however, can vary even in the same individual, often reflecting the prevailing environmental conditions but also current physiological activities such as are influenced by diet and preparations for events like reproduction.

Thyroid hormones have a considerable influence on basal metabolic rate. Thyroidectomy can result in a decline of this process by 50%, while an excess of thyroid hormone can result in an increase of a similar magnitude. Such effects are not readily apparent in cold-blooded vertebrates. The precise mechanisms for this effect of thyroid hormone on basal metabolism have not been defined but appear to involve its ubiquitous permissive role in influencing the activities of a variety of metabolically active enzyme reactions. A role for thyroid hormones in mediating the formation of receptor heterodimers that act as "receptor auxillary proteins" is becoming increasingly apparent (DeGroot, 1993; Hawa *et al.*, 1994) and may contribute to such effects. An interesting effect of thyroid hormone on an enzyme system related to thermogenesis concerns its ability to promote the induction of Na–K ATPase (Ismail-Beigi, 1993). This enzyme generates excess energy as a result of its activities in maintaining ion concentration gradients in cells, and in some conditions may contribute to thermogenesis. Thyroid hormones can also increase the number of β-adrenoceptors in cells (Williams *et al.*, 1977) and so it can modulate the actions of catecholamines.

Facultative thermogenesis

Hormones may make a more precise contribution to thermogenesis during the process of the adaptation of an animal to a cold environment. Such stresses can be severe and may be experienced in winter, especially at polar

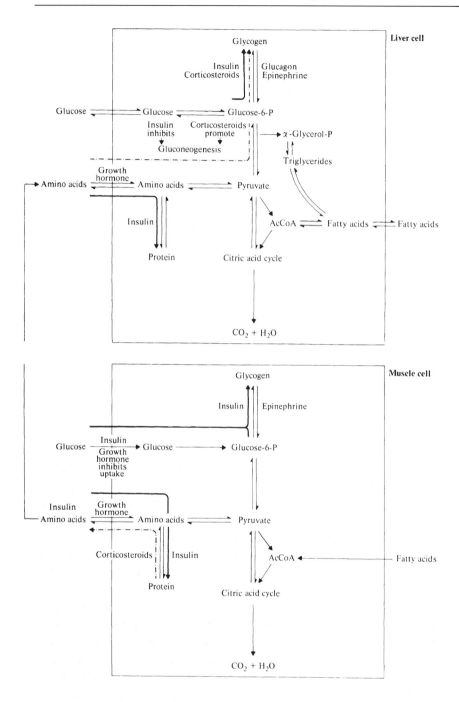

Fig. 5.4. Summary of the effects of hormones on the transfer of metabolites across the cell membranes and their metabolic transformation within liver cells, muscle cells, and fat cells. The principal metabolic transformations that are influenced by insulin are indicated by heavy solid lines and those influenced by corticosteroids, by dashed lines.

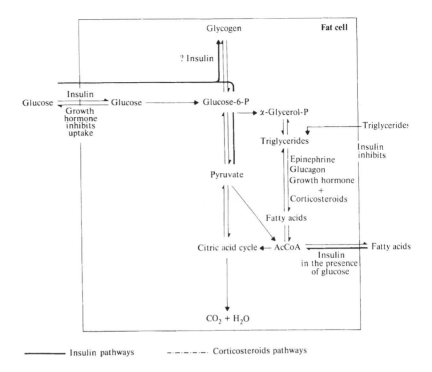

Insulin pathways _.._.._ Corticosteroids pathways

latitudes and at high altitudes. Small mammals and birds, and the newborn, are more prone to heat loss under such conditions than are larger ones. Survival may depend on an ability to generate additional body heat: a process called facultative thermogenesis.

Thyroidectomy of rodents living under cold conditions may severely compromise their ability to survive. Blockade of β-adrenoceptors with antagonist drugs in mice prior to their transfer from an environmental temperature of 25 to 0°C results in the death of the animals in about 3 hours (Estler and Ammon, 1969). Under normal conditions, mice readily withstand such changes in their environmental temperature. The experimental effect reflects the blockade of the action of norepinephrine, which is released from sympathetic nerve terminals.

Large mammals and birds exposed to cold conditions generate additional body heat by rhythmic contractions of skeletal muscle: "shivering thermogenesis" (Horwitz, 1989; Duchamp and Barré, 1993). This process does not directly involve hormones but results from cholinergic nerve stimuli in response to the stimulation of receptors in the skin. Small mammals and the newborn (and even larger mammals when acclimatized to cold conditions) can utilize a novel metabolic mechanism to generate heat under such conditions. This process has been called "nonshivering thermogenesis" (NST) and involves the activity of "brown fat" or brown adipose tissue. This tissue is present in mammals, including marsupials (Loudon, Rothwell and Stock,

1985), but despite some tentative reports is apparently not present in birds (Néchard, 1986) or in monotremes (Griffiths, 1978). However, king penguins and cold-acclimated muscovy ducklings exhibit NST but this occurs in skeletal muscle (Duchamp and Barré, 1993). In ducklings, this response can be stimulated by glucagon (Duchamp *et al.*, 1993).

Brown adipose tissue is named because it has a brownish color as a result of the high concentration of cytochromes in its numerous mitochondria. This tissue is present in the upper back, neck, and axillary regions, as well as perirenally. Brown fat, nevertheless, only makes up 1 to 5% of the body weight (Néchard, 1986). However, when stimulated, it can contribute up to 50% of the total body heat production (Foster, 1986). Small mammals, such as rodents, and especially species that hibernate retain brown fat as adults, but in other species it undergoes a postnatal involution. On acclimatization to cold conditions, however, it may regenerate. Brown fat has a plentiful blood supply and a sympathetic innervation.

The stimulation of heat production by brown fat (Fig. 5.5) occurs as a result of sympathetic nerve stimulation and the release of norepinephrine (Nicholls, Cunningham, and Riall, 1986; Silva, 1993). Under experimental conditions, the injection of norepinephrine and epinephrine are also effective but the physiological role of the latter hormone is uncertain. Injected glucagon is an effective stimulus in rats (Billington *et al.*, 1991). The response to the catecholamines involves the stimulation of β_1-adrenoceptors (and in some species β_3-adrenoceptors) and the activation of adenylate cyclase. The resulting activation of protein kinase A by the cyclic AMP that is formed initiates a lipolysis that supplies the fatty acids used as fuel by the brown fat cells. (It is possible that α_1- and α_2-adrenoceptors are also involved in the response but their precise roles are undefined.) The metabolism of brown fat involves the production of energy that (unlike in other tissues) is not coupled to the synthesis of ATP and hence it results in a greater production of heat per molecule of substrate used. This "uncoupling" results from the presence of a unique protein (molecular mass 32 kDa) that is present in the inner mitochondrial membrane where it exists as a dimer (Ricquier and Bouillard, 1986; Silva, 1993). It apparently can function as a "pore" allowing the entry of protons that bypass the ATP synthetase system, thus allowing uncoupled respiration by the cytochrome enzymes (Fig. 5.5). When bound to a single GDP molecule, the mitochondrial protein dimer is in a closed pore-like configuration; stimulation, possibly mediated by released fatty acids, results in the dissociation of the GDP and the opening of the pore. The protein is called "uncoupling protein" (UCP) or thermogenin and its concentration directly reflects the ability of the brown fat to generate heat.

Thyroidectomized mammals fail to exhibit thermogenesis in brown fat tissue in response to adrenergic stimulation. Administered T_3 can only partly

Fig. 5.5. A diagramatic summary of thermogenesis by brown adipose tissue and the roles of neural stimulation and hormones. AC, adenylate cyclase; CRE, cyclic AMP response element; DAG, diacylglycerol; GlnR, glucagon receptor; InsR, insulin receptor; IP$_3$ inositol trisphosphate; MD, monodeiodinase; PKA, phosphokinase A; PLC, phospholipase C; T$_3$ R, triiodothyronine receptor; UCP, uncoupling protein. (Based on Nicholls, Cunningham, and Rial, 1986; Ricquier and Bouillaud, 1986; Silva, 1993.)

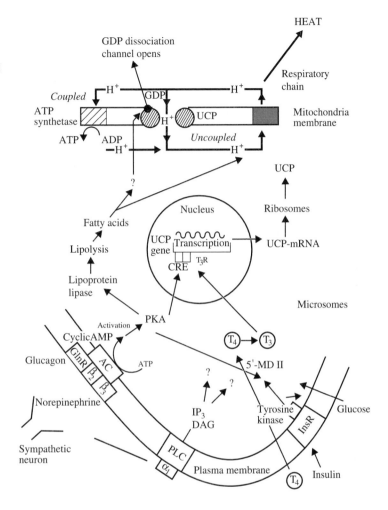

restore the response. Under physiological conditions T$_3$ arises from T$_4$ by the enzyme 5'-monodeiodinase (Chapter 4), which is present in peripheral tissues. Type II 5'-monodeiodinase in brown fat produces high local concentrations of T$_3$ (Bianco and Silva, 1987) following its activation by noradrenergic stimuli, insulin, glucagon, and hypothyroid conditions (Silva and Larson, 1986; Mills *et al.*, 1987). Such a local response can produce enough T$_3$ nearly to saturate adjacent nuclear receptor sites. The uncoupling protein is produced by a UCP gene and transcription is initiated by a synergistic stimulation by the T$_3$ receptor and transcription factors activated by norepinephrine–cyclic AMP (Bianco, Sheng, and Silva, 1988). At least three ancient hormones, T$_3$, insulin, and glucagon, have been utilized in mammals in the processes involved in the action of a novel protein playing a central role in their homeothermy.

Acclimatization to cold

Hormones also contribute to the chronic adaptation to cold conditions (Wunder, 1979; Horwtiz, 1989). Under such conditions, there is an increase in the basal metabolic rate that appears to be related to the activity of the thyroid gland. The capacity for NST may be facilitated by an increase in brown fat and the uncoupling protein. There may be an accompanying increase in 5'-monodeiodinase in the tissue so that the brown fat may then also provide T_3 to the peripheral circulation. Other hormones may contribute to thermogenesis under such conditions, though usually indirectly. Corticosteroids have important effects on intermediary metabolism. Adrenalectomized rats have difficulties in thermoregulating in cold conditions: effects which apparently reflect the deficiency of corticosteroids (Deavers and Musacchia, 1979). The echidna, *Tachyglossus aculeatus*, is a monotreme and, therefore, an extant representative of a "primitive" mammal. It can adapt to cold conditions and may then enter a state of torpor. When it is adrenalectomized and exposed to cold, the echidna can still enter torpor but it subsequently fails to be aroused by warmer conditions and succumbs (Augee and McDonald, 1973). Administered cortisol, however, assures its survival. Glucocorticoids are produced by the echidnas at increased rates at low environmental temperatures and aid the mobilization of fat and may mediate a gluconeogenic response, though the latter effect appears to be relatively small in this species (Sernia and McDonald, 1993). Leptin mRNA levels in white fat of lean Zucker rats decline when they are exposed to 4°C (Bing *et al.*, 1996). This suggested that this hormone may also have a role in the adjustment of metabolism under such conditions.

There are sporadic reports that thyroid hormones may influence oxygen consumption in cold-blooded vertebrates. Such effects of thyroidectomy and administered thyroid hormones have been observed most often in reptiles, especially lacertilians but also a crocodilian and a chelonian (Maher, 1965; Hulbert and Williams, 1988). Such experimentally induced excursions in oxidative metabolism are seen more readily at ambient temperatures of about 30°C rather than 20°C. Administered T_4 at 30°C increased oxygen consumption of a turtle and an alligator by about 25%, and a lizard by 32%. A field study of the iguanid lizard *Dipsosaurus dorsalis* showed seasonal changes in both the thyroid hormone concentration in the plasma and the standard metabolic rate but they did not appear to be related to each other (John-Alder, 1984). Increases in oxygen consumption in response to administered T_4 have been observed in a frog, *Rana pipiens* (McNabb, 1969), and an Indian teleost fish (Peter and Oommen, 1993).

It is notable that epinephrine may, in addition to being thermogenic, help to promote heat loss in some mammals. Sweating during exercise in monkeys is dependent on circulating epinephrine and is considerably reduced following denervation of the adrenal medulla (Robertshaw, Taylor, and Mazzia, 1973). This poses an interesting conflict of interest in the physiological role of epinephrine, its potential thermogenic effect presumably being minimal in circumstances where it also contributes to the dissipation of heat. The role of catecholamines in temperature regulation is even more diverse when one recalls that adrenergic nerves, by a peripheral vasoconstrictor action, can also promote heat conservation. This affords another interesting example of the evolution of a hormone's role.

There is little information about the effects of catecholamines on thermogenesis and oxygen consumption in nonmammals (Harri and Hedenstam, 1972). This effect apparently cannot be demonstrated in birds (pigeon, titmouse, and gull) or in fishes. There seem to be no reports about such an action in reptiles, either. In European frogs, in injections of norepinephrine and epinephrine *Rana temporaria* increased oxygen consumption by 25–35%. The response depended somewhat on the particular frogs used; it was seen in cold- and warm-adapted summer frogs but not warm-adapted winter frogs. The physiological role of catecholamines in thermogenesis of nonmammals remains in doubt.

The storage of nutrients and their utilization during fasting

The diversity of feeding–fasting patterns in vertebrates

As described above, sufficient food may only be available to animals at irregular intervals of time separated by several hours, many months, or even years. In some instances, when climatic conditions are unfavorable, animals may sequester themselves in protected havens where their metabolic activity is minimal. Hibernation during the winter months is well known in many small mammals (especially among the Insectivora, Rodentia, and Chiroptera), which seek refuge in burrows and allow their body temperatures to decline to levels similar to the ambient one. Bears also hibernate but maintain a much higher body temperature. Other animals become inactive during hot, dry periods of drought, which result in a limited food supply and a shortage of water that may produce severe osmotic problems. In the latter instance, called "estivation", the body temperature is also similar to the ambient one, but as this is usually relatively high, metabolism would be expected to be greater than in those animals that are hibernating. Animals may survive for many months or possibly even years under these conditions. A report from Russia (Siberian Correspondent, 1973) described a live newt found in a piece of ice in Siberia that, according to carbon dating, had been entombed for nearly 100 years. African lungfish can survive for 2–3 years in a state of estivation, though more usually it is for 4–6 months between seasonal rains. Such periods of estivation are also common among amphibians that live in hot, dry deserts.

Table 5.2. *Interrelationship of metabolism with the nutritional state of mammals*

Nutritional states	Hormonal states		Liver				Muscle			Adipose		
	Insulin	Glucagon	Glycolysis	Lipo-genesis	Gluconeo-genesis	Keto-genesis	Glucose uptake	Protein synthesis	Proteo-lysis	Glucose uptake	Lipo-genesis	Lipolysis
Carbohydrate-fed	+++	±	++	++	0	0	++	±	0	++	++	0
Protein-fed	++	+++	0	+	++	0	+	0	0	+	0	0
Carbohydrate- and protein-fed	++++	±	+	+++	0	0	++	++	0	+++	0	0
Fasting (low insulin)	+	++	0	0	++	++	0	0	+	0	0	++
Diabetes (absent insulin)	0	++++	0	0	++++	++++	0	0	++	0	0	++++

Notes: + to 0, either concentration of the hormone or rate of function described.
Source: Cahill, Aoki, and Marliss, 1972.

Although hibernation and estivation are associated with minimal activity and metabolic needs, other situations associated with fasting require a high expenditure of energy. Such an occasion is most dramatically seen during seasonal and breeding migrations. Birds may fly many hundreds, or even thousands, of miles from temperate regions at the beginning of winter to warmer tropical climes and then return again in the spring. Fishes, such as lampreys and salmon, when they become mature, migrate from rivers where they grew up into the sea from which they later return in order to breed. Eels make the opposite migration from breeding grounds in the sea to rivers and then later the young return to the sea to breed. Other seagoing creatures, such as turtles and whales, also make long journeys. On many of these occasions, the animals do not feed or do so only infrequently. Reserves of nutrients are amassed in the body in preparation for the migrations during which they are expended. The endocrine glands undoubtedly have a role to play in such storage and the subsequent utilization of nutrients, but the available information is only fragmentary. Further clues can be obtained from the voluminous studies of endocrine function during normal feeding and fasting.

Endocrines and feeding

The release of hormones   Feeding results in changes in the secretion of several hormones. Gastrin, secretin, CCK, GLP-I, and GIP are released from the gut in response to the presence of food and digestive products. The release of insulin is promoted by absorbed glucose and amino acids and this effect is enhanced by GIP, GLP-I, glucagon, and possibly CCK (Fig. 4.14, p. 203). Parasympathetic nerve stimulation, in response to elevated blood glucose levels, also promotes the release of insulin, while sympathetic nerve stimuli and epinephrine inhibit it during hypoglycemia. The release of glucagon is inhibited by the presence of high plasma glucose concentrations, and by insulin and GLP-I. Glucagon is secreted in response to absorbed amino acids and, when appropriate, by both sympathetic and parasympathetic nerve stimuli. The former may be more important during mild hypoglycemia and the latter in severe hypoglycemia.

The endocrine response to feeding differs somewhat with the diet, depending on the relative amounts of protein and carbohydrate present (Table 5.2). A high carbohydrate diet is associated with high insulin and low glucagon levels whereas a predominance of protein elevates the concentrations of both of these hormones. In ruminants, the fatty acids that are absorbed from the rumen may also stimulate the release of insulin (Manns, Boda, and Willes, 1967).

The control of feeding   The control of feeding involves a complex series of physiological and behavioral events that utilize neural and hormonal mechan-

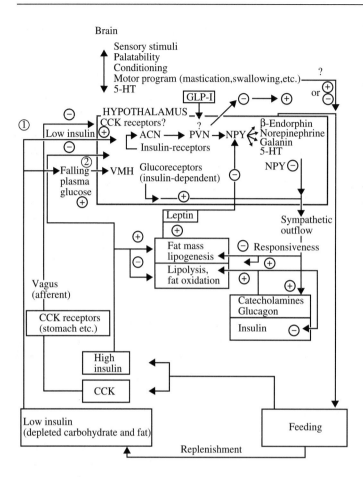

Fig. 5.6. Tentative summary of the current hypotheses regarding the regulation of feeding behavior and its interactions with the liposatic system. 1. The insulin-neuropeptide Y hypothesis. (Berelowitz, Bruno, and White, 1992; Schwartz *et al.*, 1992; Sahu and Kalra, 1993). 2. The glucostat hypothesis (Le Magnen, 1992). These theories principally differ in their proposals regarding the nature of the neurochemical events that take place in the hypothalamus. For further details see the text. ACN, arcuate nucleus; CCK, cholecystokinin; 5-HT, 5-hydroxytryptamine (serotonin); NPY, neuropeptide Y; PVN, paraventricular nucleus; VMH, ventromedial hypothalamus.

isms for coordination. Specific regions of the hypothalamus provide the sites for the receipt, modulation, and transmission of the relevant signals. A tentative summary is given in Fig. 5.6. This description includes two hypotheses, not necessarily exclusive, that principally differ with respect to the details of the hypothalamic mechanisms that may be involved in the control of feeding: the glucostat hypothesis and the insulin–neuropeptide Y hypothesis. The fat and carbohydrate reserves in the body provide the basic signals to eat. The storage of these reserves is reflected by the plasma concentration of glucose and the fat mass, both of which are related to plasma hormone levels. The fat mass is enhanced by the action of insulin, which promotes lipogenesis and inhibits lipolysis. As already described, plasma insulin is depressed by sympathetic nerve stimulation, circulating epinephrine, and glucagon. Under such conditions, there is an increase in lipolysis and caloric fat oxidation. The sensitivity of the fat cells to neural and hormonal stimulation and the release of the hormones that influence these processes are principally controlled by the sympathetic outflow from the autonomic nervous system. The control of the latter, which includes parasympathetic nerves that also influence hormone release, arises from the hypothalamus.

The glucostat hypothesis (Le Magnen, 1992) proposes that lipostasis is regulated by the stimulation of glucoreceptors in the hypothalamus following a falling plasma glucose concentration. The glucoreceptors are thought to be present in the ventromedial hypothalamus and their sensitivity is insulin dependent. Stimulation of the hypothalamic receptors results in an increase in sympathetic outflow with resulting lipolysis and fat oxidation and a decrease in the fat mass. Replenishment of the fat mass involves subsequent feeding behavior and lipogenesis. (In humans, glucoreceptors may be present in the gut (Lavin *et al.*, 1996). It has been suggested that on exposure to elevated intestinal glucose concentrations there is a vagal stimulation and a local secretion of GLP-I and GIP. The resulting release of insulin possibly reduces appetite.)

The insulin-neuropeptide Y hypothesis proposes that low insulin levels in the plasma may enhance feeding (Berelowitz *et al.*, 1992; Schwartz *et al.*, 1992; 1995; Sahu and Kalra, 1993). The insulin upon entering the hypothalamus may interact with receptors in the arcuate nucleus (ACN) which is also a site of synthesis of NPY (Chapter 3). An increase in NPY mRNA associated with feeding, has been observed in the hypothalamus. This peptide is released from the terminals of its nerve axons in the paraventricular nucleus (PVN). When NPY is injected into the hypothalamus, it is a potent orexigenic (appetite-stimulating) substance and is thought to promote feeding. A receptor for NPY (Chapter 3) has been identified in the paraventricular nucleus. It has been characterized as a novel type of receptor called the Y_5-receptor (Gerald *et al.*, 1996). The subsequent response appears to involve other neurotransmitters such as β-endorphin and norepinephrine. NPY can inhibit sympathetic adrenergic outflow from the hypothalamus, an effect that is also predicted to enhance the fat mass. These events constitute part of the so-called "adipostat" mechanism. NPY, however, does not appear to have a unique role in this process and can apparently be replaced or overridden by the actions of other processes. Therefore, mice in which the NPY gene has been disrupted (NPY knockout mice, – / –) are still able to regulate their appetite and maintain a normal body weight (Erickson, Clegg, and Palmiter, 1996). (Such mice are prone to convulsive seizures but survive and breed). Other mechanisms that may contribute to the control of feeding behavior involve the hormones leptin, GLP-I, CRH and CCK, as well as the neurotransmitter 5-HT.

Leptin, which can induce satiety (Chapter 3), is released from fat cells in response to elevated plasma insulin concentrations. It appears to interact with receptors associated with the choroid plexus and hypothalamus and it can inhibit the synthesis and actions of NPY (Stephens *et al.*, 1995; Smith *et al.*, 1996). Nevertheless, it is still able to induce satiety in the NPY knockout mice (Erickson *et al.*, 1996). The response is indeed enhanced in such animals, suggesting that a direct antagonism exists between their effects. Leptin also contributes to a decrease in body fat mass by activating adrenergic sympathetic processes and thus increasing the expenditure of energy (Halaas *et al.*, 1995; Collins *et al.*, 1996).

GLP-I (see Chapter 3), when administered into the cerebroventricular space, produces a cessation of feeding in fasted rats (Turton *et al.*, 1996). This effect can be specifically blocked by one of its receptor antagonists called excendin. The latter also augments the orexigenic effect of NPY. The mRNA for GLP-I receptors has been identified in the hypothalamus, including the paraventricular nucleus, of rats (Shughrue, Lane, and Merchenthaler, 1996). The GLP-I appears to act in the region of the paraventricular nucleus, which is also the site of release of NPY. The response does not involve changes in the synthesis of NPY. CRH is formed in the paraventricular nucleus (Schwartz *et al.*, 1995; Seeley *et al.*, 1996). When it is infused into this site in rats it also produces satiety. The synthesis of CRH in the paraventricular nuclei of rats is promoted by overfeeding and depressed by underfeeding. It appears to inhibit the synthesis of NPY. This action of CRH is apparently unrelated to its effect on the hypothalamus–pituitary–adrenocortical axis. However, elevated concentrations of plasma corticosteroids may inhibit its formation. The precise contributions of GLP-I and CRH to satiety, in relation to that of leptin, have not yet been defined.

The orexigenic effects of low plasma insulin concentrations can be inhibited by CCK. The latter is released from CCK cells and neurons in the gut following feeding. Its effect can be blocked by cutting the vagus (Ritter, Brenner, and Tamura, 1994; Smith and Gibbs, 1994). Peripheral A-type CCK receptors, such as exist in the abdomen and on the vagus, may be mediating its effect via afferent (sensory) vagal nerve fibers directed to the hypothalamus. It is also possible that CCK receptors are present in the hypothalamus (see Moran *et al.*, 1992). Such receptors may be responding to CCK released from adjacent neurons that participate in the response. Experiments with CCK have been mainly performed in mammals, especially rodents. Some species may, however, behave differently as seen with the failure of endogenously released CCK to exhibit an appetite-suppressing effect in the domestic fowl (Choi *et al.*, 1994) and its "weak" effect in humans (Lieverse *et al.*, 1994). Nevertheless, its role in domestic fowl is still contentious (Rodríguez-Sinovas *et al.*, 1997); it has been shown to have a prominent effect in goldfish when injected peripherally or into the brain (Himick and Peter, 1994). In rats, CCK not only suppresses food intake but also drinking (Ebenezer, 1996). This observation has led to the suggestion that the peptide may be having a general aversive effect, not specifically related to appetite. It may, for instance, induce a general feeling of malaise and nausea (see also

Verbalis *et al.*, 1986). The possibility that such nonspecific effects may contribute to the plethora of responses to various putative satiety agents merits further consideration.

Several drugs that influence the metabolism and actions of 5-HT have also been observed to change feeding behavior and weight gain in humans and rodents (Tecott *et al.*, 1995; Spedding *et al.*, 1996). Thus, the 5-HT antagonist clozapine (a sedative drug) can promote weight gain while fluoxetine (Prozac, an antidepressant drug), which enhances serotoninergic neurotransmission, can induce weight loss. Such observations suggested that 5-HT, following its interaction with one of its type of receptor, the 5-HT$_{2c}$ receptor, may also elicit a sensation of satiety. These receptors are present at several sites in the brain including the hypothalamus. Mutant mice that lack the 5-HT$_{2c}$ receptor have an increased appetite and may attain twice their normal body weight. This disorder does not appear to be directly mediated by metabolic responses, such as those involving adipose tissue, but it influences behavioral aspects of feeding. The relationships of this appetite-suppressing response to the stimulation of 5-HT$_{2c}$ receptor responses and the roles of NPY, insulin, leptin, GLP-I, CRH, and CCK are at present unknown.

The frantic feeding activities and accumulations of body fat that occur in birds prior to seasonal migration, and in mammals preparing for hibernation, would appear to involve adjustments of such feeding–lipostatic mechanisms. Genetically obese Zucker "fatty" rats have elevated levels of NPY associated with the arcuate nucleus in the hypothalamus (Schwartz *et al.*, 1992) and possibly provide a model for such situations. A decline in the stimulation or numbers of brain 5-HT$_{2c}$ receptors, as seen in the 5-HT$_{2c}$ receptor-deficient mice, could also be involved. Fat accumulation prior to migration and hibernation may require a change in a "set-point" determining optimal fat mass. Possibly there is an uncoupling of such a relationship to feeding behavior. Subsequently during migration, normal sympathetic outflow from the hypothalamus may mediate the utilization of fat reserves. However, a decline in feeding behavior and appetite (an anorexia) may also occur in such fasting conditions (Mrosovsky and Sherry, 1980). Possibly such an effect is mediated by 5-HT through its brain 5-HT$_{2c}$ receptors.

The disposal of absorbed nutrients

The nutrients that are absorbed from the digestive tract can be disposed of in the body in several ways. They may be used immediately as a source of energy. This process may be relatively direct, such as the oxidation of glucose and fatty acids. Amino acids can be transformed to glucose in the liver. The large amounts of propionate absorbed by ruminants can also be converted to glucose. Certain amino acids can also be changed into fatty acids. These products can be more readily transformed into energy. Gluconeogenesis is stimulated by glucagon and corticosteroids. This process is especially important in ruminants that require glucose to aid utilization of large amounts of fatty acids. As we shall see, gluconeogenesis is also fundamental for homeostasis during fasting in other species. The overall processes of oxidation of nutrients are, at least in mammals, chronically influenced by thyroid hormones but more acutely by the levels of the energy substrates themselves, as well as certain other hormones (see thermogenesis, p.241).

Nutrients that are not required for the immediate production of energy by the animal are stored (usually following their metabolic transformation) and can be subsequently utilized during fasting. Elevated insulin levels play a central role in this process (Fig. 5.7; see also Table 5.2). Insulin facilitates the conversion of glucose to glycogen and triglycerides, of amino acids to protein, and of fatty acids to triglycerides. Simultaneously, insulin inhibits further gluconeogenesis and lipolysis. Other hormones may impinge their influence on these processes but usually in a negative manner. For instance, a decreased secretion of glucagon and epinephrine is favorable to lipogenesis. Growth hormone favors incorporation of amino acids into proteins.

The principal energy store utilized by animals during periods of prolonged fasting is fat, though proteins can also undoubtedly be used. Obese animals fare better and survive longer. The immense fat reserves of migrating birds, fish, and whales are well known. The continual food collecting and feeding activities of mammals prior to winter hibernation are almost legendary. The total fat content of the body is normally controlled within limits that can be considered modest, both physiologically and esthetically.

Migration and fasting

Preparations for migration

Animals periodically show dramatic departures from their normal limits of fat content that may be associated with their potential requirements during a period of fasting. Humpback whales reproduce in warm equatorial waters, after a migration that takes about 5 months. They do not feed during this time but live off their fat stores, which are accumulated as a result of excessive feeding on krill in the Arctic and Antarctic oceans. The maximum accumulations of fat in several species of migratory birds may be equivalent to about 50% of the total body weight. The normal fat levels in such birds are 3–10% of their body weight. The fat is contained in various parts of the body but especially in cutaneous and subcutaneous sites and can be deposited very rapidly in 1 to 2 weeks. The timing of this activity appears to be the result of changes in the length of the day and so is under photoperiodic control. Depending on the season the stimulus may be an increase (as in spring) or a decrease (as in autumn) in the hours of daylight (King and Farner, 1965). Such changes in day length are associated, in birds, with a nocturnal restless-

Fig. 5.7. The role of hormones in intermediary metabolism during *feeding* and *fasting*. Solid line, stimulation or increase; dashed line, inhibition or decrease. (Modified from Foà, 1972.)

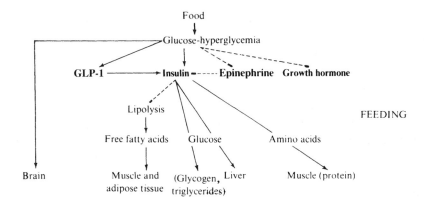

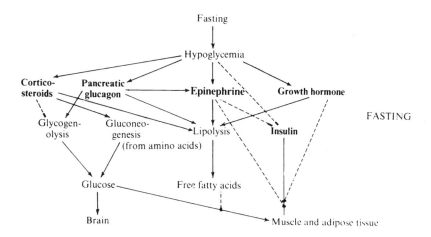

ness and activity (*Zugunruhe*), frantic feeding, development of the gonads, and changes in the pituitary gland. It is, therefore, reasonably suspected that the deposition of fat in these circumstances is associated with endocrine signals, but the evidence is difficult to interpret. The problem is probably largely the result of trying to define the role of single hormones when the interactions of several are involved. The effects of photoperiodic stimulation are (possibly apart from those in the pineal gland) conveyed to the endocrine system through the pituitary gland, which is suspected of playing an early role in the preparations for migration.

In birds, preparations for the vernal breeding migration are initiated by the photostimulation of long days (Rowan, 1925; Chapter 9). Removal of the gonads prior to this time, while the day length is still shortening, results in a suppression of premigration feeding behavior, fattening, and the initiation of migration (Wingfield, Schwabl, and Mattocks, 1990; Rankin, 1991). The injection of testosterone can restore such activities in either sex. In contrast to its action prior to the vernal migration, gonadectomy has no effect on the preparations for the autumnal migration. The activities prior to the vernal migration are associated with increases in secretion by the thyroid gland and they can be inhibited by thyroidectomy (Wingfield *et al.*, 1990; Pant and Chandola-Sarlani, 1993). The ratio of T_3 to T_4 in the plasma rises prior to the vernal but not the autumnal migration (Pathak and Chandola, 1982), suggesting that there may be changes in peripheral 5'-monodeiodinase activity. Other hormones that may contribute to the metabolic changes associated with migration include corticosteroids and prolactin. In some species, such as the white crowned sparrow, *Zonotrichia leucophrys gambelli*, prolactin can promote the deposition of fat (Meier and Farner, 1964). This hormone may also be involved in the preparations of such birds for migration (Stetson and Erickson, 1972). However, such a role for prolactin was not observed in European quail preparing for migration (Boswell *et al.*, 1995). A definitive description of the role of hormones in bird migration is still awaited. Study of the nonbreeding autumnal migrations may be enlightening as it is not complicated by endocrine preparations for breeding.

Knowledge of the endocrine processes controlling deposition of fat is, despite a widespread applicability in animal and human nutrition, not well understood. It seems to involve several hormones but these, and the pattern of their effects, may differ depending on the species, the diet, and the physiological occasion.

Endocrines and fasting

Hormones and the utilization of stored nutrients   During fasting, the animal is dependent on its endogenous stores of nutrients for energy. These must be converted to substrates that can be metabolized, principally glucose, fatty acids, and ketone bodies, the levels of which are increased by a low insulin and a high glucagon level in the plasma (Table 5.2).

During periods of fasting that last only a few hours, or in times of sudden acute need (such as for violent action) mobilization of glycogen stores in the liver and muscles may be sufficient. This transformation is increased by the action of glucagon (on the liver) and, on occasions, also catecholamines (Fig. 5.7).

During longer periods of time, when no food is consumed, the liver and muscle stores of glycogen are usually maintained. Fatty acids are mobilized from triglycerides in adipose tissue and are used to provide energy. This may be a direct process involving β-oxidation or result in the formation of ketone bodies. These substances, acetoacetate, β-hydroxybutyrate, and acetone, are produced mainly in the liver, but in ruminants they are also formed by the rumen epithelium and the mammary glands. Fatty acid mobilization is favored (Table 5.2 and Fig. 5.7) by low insulin levels, resulting from hypo-glycemia as well as elevated glucagon concentrations. Catecholamines, growth hormone, and corticosteroids can also promote lipolysis. Glucose is also necessary for adequate utilization of ketone bodies in the citric acid cycle whereas certain tissues, especially brain, have a specific requirement for glucose. This substrate is obtained during starvation (as a result of gluconeogenesis) from certain amino acids as well as from propionate and glycerol. Gluconeogenesis is stimulated by glucagon and corticosteroids. As a prerequisite, amino acids must be mobilized and proteolysis is favored by low insulin levels and is promoted by corticosteroids.

Although fat is the main source of energy during prolonged fasting in vertebrates, considerable protein catabolism may also occur and, indeed, in the more terminal stages of starvation may be inevitable. This protein utilization may be more important in poikilotherms. During estivation in amphibians and lungfishes, very high concentrations of urea accumulate in the body fluids (Smith, 1930; McClanahan, 1967), indicating substantial protein catabolism (Janssens, 1964). These animals tolerate high urea con-centrations (as much as 600 mM) in their body fluids, levels that would be fatal to a mammal. Aquatic species that form ammonia instead of urea would be more readily able to excrete the toxic by-products of protein catabolism during fasting. In addition, mammals that hibernate and birds that migrate are subjected to circumstances that are often associated with limited supplies of water and may (apart from the associated muscle wasting) find excessive protein catabolism an added disadvantage because of the problem of extra nitrogen excretion.

In mammals and birds the metabolic rate may decrease during prolonged fasting, and this adjustment is accompanied by a decline in the levels of T_3 in the plasma. There is, in addition, a decrease in the number of thyroid hormone receptors in the cell nucleus (Schussler and Orlando, 1978). In rats, the decrease in thyroid hormone is accompanied by a decrease in plasma TSH concentration, suggesting that the hypothalamus–pituitary–thyroid axis is involved in the decline in the metabolic rate (van Haasteren *et al.*, 1996). This response appears to reflect a decrease in the secretion of TRH and, possibly, an inhibitory effect of elevated plasma corticosterone on the synthesis and release of TSH. Other hormones, such as somatostatin and leptin (see below), could also be involved.

Fasting in birds and mammals

Many birds experience a reduced food intake during their reproductive periods. Such fasting may be related to the incubation period for the eggs, the site and accessibility of the nest, the season, and whether one or both parents contribute to incubation "shifts" and the care of the young. Following hatching, the needs of the young may exceed the parents' abilities to tend them and collect food. Gallinaceous birds, sea birds, like albatross, and Canada geese usually lose weight during the incubation period (Lea *et al.*, 1992). Columbid birds (pigeons and doves) initially feed their young with a crop-sac secretion (crop-milk, pigeons-milk) during which time they may lose weight. Penguins undergo prolonged fasts during the subantarctic winter when they incubate their eggs. For the male emperor penguin, this period may extend to about 4 months (Robin *et al.*, 1988), while in the male king penguins it lasts about 50 days over the courting period and the first shift of incubation (Cherel *et al.*, 1988b). After 3 to 4 months of parental care following hatching, king penguin chicks, living in the subantarctic Croizet Archipelago may only be fed irregularly by the parents and endure a period of fasting that lasts for 4 to 6 months through the winter (Le Ninan *et al.*, 1988).

Periods of prolonged fasting in birds and mammals can be divided into three phases (Fig. 5.8).

> *Phase I* is usually only of several days' duration during which time hormone levels adjust to the new conditions. The basal metabolic rate usually declines and lipid mobilization is initiated.
>
> *Phase II* may extend for a prolonged period and is usually a time of relative stability with respect to the basal metabolic rate and lipid mobilization; there is a low rate of protein utilization.
>
> *Phase III* is a period when lipid mobilization declines, reflecting depleted stores and protein is used at an increased rate for gluconeogenesis.

In the fasting male king penguin (Cherel *et al.*, 1988b) plasma T_4 and corticosterone levels decline during Phase I while glucagon concentrations start to rise (Fig. 5.8). In Phase II, T_3 and T_4 decline further, glucagon continues to rise, while the insulin and corticosteroid levels are stabilized. These changes are consistent with maintenance of a low basal metabolic rate, the conservation of body protein, and continued lipid mobilization. Phase III commences about 35 days after the beginning of fasting by the king penguins, when lipid reserves are partially depleted. Protein utilization for gluconeogenesis commences, reflecting climbing glucagon and corticosterone concentrations in the plasma. At this time the ratio of glucagon:insulin climbs to 7.4 compared with about 3 during Phase II.

King penguins also endure a molting fast that lasts for about 40 days, during which time they lose about 60% of their body weight (Cherel, Leloup, and Le Maho, 1988a). The loss of feathers leaves them in a relatively exposed situation when they are unable to enter the water to feed. The hormonal profiles in this fast differ from those in the incubatory fast. Protein utilization for feathering is initially high and the cutaneous exposure limits possible declines in basal metabolic rate. Plasma thyroid hormone concentrations,

Fig. 5.8. The hormonal changes that occur in the plasma during the breeding fast of the king penguin in the Antarctic. As described in the text, this process can be divided into three phases reflecting differences in their intermediary metabolism. (a) glucagon and insulin; (b) corticosterone and aldosterone; (c) T_4 and T_3. (Based on Cherel et al., 1988b.)

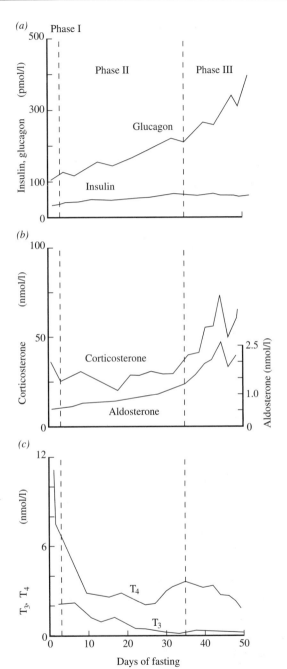

which support these processes, are elevated during Phase I and part of Phase II. Glucagon levels are maintained and mediate the mobilization of lipids in Phase II. Towards the end of this phase and during Phase III there is a second period of protein mobilization and catabolism which, as in the incubatory fast, is accompanied by elevated glucagon and corticosterone concentrations.

As just described, the chicks of king penguins endure an even more remarkable fast than the adults (Le Ninan *et al.*, 1988). During Phase I, protein utilization and basal metabolic rate decline and these changes are accompanied by decreases in plasma corticosterone and T_4. The glucagon concentrations in the plasma are maintained while insulin levels usually decrease. In Phase II, this protein conserving–lipid utilizing regimen is maintained and may continue for 4 to 6 months until regular feeding resumes. Survival of the king penguin chicks depends on prolongation of Phase II, which enhances protein conservation.

In mammals leptin levels in the plasma decline during starvation and this change may contribute to the physiological adjustments that occur under such conditions (Ahima *et al.*, 1996). The genetically obese *ob/ob* mice, which lack an active leptin molecule, exhibit metabolic abnormalities that are reminiscent of starvation, including sterility. Administration of leptin to normal mice during starvation reduces the responses of the neuroendocrine adrenal, thyroid, and gonadal axes to this condition. Leptin can also restore reproductive activity to male and female *ob/ob* mice, apparently by signaling that the fat stores are adequate for the caloric needs of reproduction (Barash *et al.*, 1996; Mounzih, Lu, and Chehab, 1997). It also may contribute to the initiation of puberty in female rats (Cheung *et al.*, 1997). It has been suggested that falling plasma leptin levels may contribute to the normal control of these neuroendocrine responses during starvation. Fertility declines during starvation, including during human anorexias, resulting in speculation that a lack of leptin may be contributing to the failure of ovulation under such conditions. Such an effect could be involved in the decline of fertility that may occur in some very lean female athletes and ballet dancers.

Stored nutrients may be mobilized and oxidized to produce energy at greatly contrasting speeds. A bird during a nonstop migratory flight of up to 2400 km may use up almost its entire fat reserve in 40–60 hours (Odum, 1965). Hibernating and estivating animals, by comparison, have relatively small energy requirements so that the rate of utilization of stored nutrients is much slower than usual. No definitive information is available about the catabolic role of the endocrines in these circumstances.

Migration of fish

The migrations of fish often involves movements to their breeding grounds and those of immature fish to areas where they feed and undergo further maturation. Travel between areas of differing salinities also often occurs. Long distances may be traveled, necessitating intense swimming activity and survival by using stored nutrients. Preparations for reproduction may be occurring concomitantly. Such fish are often considered to be experiencing "stress." It is hardly surprising to observe that many hormonal changes may occur during such piscine migrations.

Changes in the activity of the thyroid gland were probably the first endocrine manifestation of fish migration to be observed. There is, for instance, an increase in the activity of the thyroid gland in migrating Atlantic salmon, *Salmo salar* (Leloup and Fontaine, 1960). Premigratory smoltification in immature salmon is also accompanied by increased thyroid gland activity (Rankin, 1991). During migration, thyroid hormones may facilitate fasting metabolism and aid in the mobilization of stored lipids.

High concentrations of corticosteroids are present in the plasma of migrating and spawning salmon, rainbow trout, eels, and sturgeon (Robertson *et al.*, 1961; Murat, Plisetskaya, and Woo, 1981). Such a release of corticosteroids can occur in response to exercise and stress in salmonids (Fagerlund, 1967; Hill and Fromm, 1968) and could be contributing to the mobilization of lipids, and even to gluconeogenesis. High plasma concentrations of insulin are also maintained during the spawning period of lampreys, sturgeon, and salmon (Murat *et al.*, 1981) even though feeding is not occurring. It may, however, be contributing to the conservation of body proteins. During the final spawning, plasma insulin levels decline dramatically. Exercise and starvation have been shown to result in increased concentrations of growth hormone in the plasma of steelhead trout and rainbow trout (Sumpter *et al.*, 1991). This hormone appears to be promoting lipid mobilization, a role that it could also have during migration. Other hormones may be involved in the osmoregulatory adaptations that may accompany migration in fish. Prolactin and cortisol are released in response to changes in salinity, such as occur when fish move between rivers and the sea (Chapter 8). A plethora of hormones, therefore, may contribute to successful migration in teleost fish.

Lampreys (*Lampetra fluviatilis*) (on histological evidence) have an active thyroid gland at the beginning of their breeding migration from the sea into rivers (Pickering, 1972), although this activity declines as sexual maturity is approached. Sea lampreys (*Petromyzon marinus*) appear to release larger amounts of T_4 during spawning (Hornsey, 1977). Normally, the digestive tract degenerates during the migration but this can be prevented if the lampreys are hypophysectomized, gonadectomized, or injected with estrogens (Larsen, 1969; 1980). It, therefore, appears that gonadal hormones, either by direct action or possibly by a change in a negative-feedback mechanism in the pituitary or hypothalamus, can influence tissue catabolism in migrating lampreys. Unlike in salmon, corticosteroids cannot be detected in the plasma of the migrating sea lamprey *Petromyzon marinus* (Weisbart and Idler, 1970);

as a result, a gluconeogenic role for corticosteroids in migrating cyclostomes is doubted.

Bird migration

We can only speculate about the effects of hormones in the metabolism of birds during migratory flights. Glucagon has a potent effect in increasing free fatty acid levels in the plasma of birds so that this hormone, which is present in substantial quantities in the avian islets of Langerhans, may be important on these occasions. Other hormones that may contribute to lipid mobilization during migratory flights are catecholamines, growth hormone, and corticosteroids (Ramenofsky, 1990). No changes in plasma prolactin levels were observed during the migration of snow geese from Louisiana to Canada (Campbell, Etches, and Leatherland, 1981). The plasma corticosterone levels in migrating garden warblers (*Sylvia borin*) were measured during one of their stop-overs in the Algerian Sahara desert (Schwabl, Bairlein, and Gwinner, 1991). The concentrations of this steroid hormone were low and, in individual birds, negatively correlated with body fat stores. It was suggested that exhaustion of such fat stores may, however, result in elevations of corticosterone levels, as observed in the gluconeogenic Phase III of fasting in king penguins.

Hibernation in mammals

A number of animals, especially mammals, that experience long, cold winters, may undergo periods of reduced activity during which they do not feed regularly. This condition is called hibernation and it may be prolonged for 7 or 8 months. The physiological and behavioral pattern, however, varies in different species. Small animals that hibernate usually reduce their body temperatures to within a few degrees of the ambient one but they briefly "wake up" at regular intervals of about 2–10 days. Their body temperature then returns to normal. Following such arousal some species, such as chipmunks, will then eat food that they have stored during the previous summer and autumn. They then return to the state of hibernation. Other species, such as woodchucks, do not feed during these breaks in their winter "sleep." Bears hibernate in dens for periods of up to 8 months during which time they live off the large fat stores accumulated as a result of long periods of eating in the autumn. In contrast to the small mammals, their body temperatures drop only slightly, about 4 or 5°C, but other bodily functions, such as basal metabolic rate, heart rate, and kidney function, decline dramatically. It would appear that hormones must play a special role in the intermediary metabolism and calorigenesis of such hibernating animals. Information is, however, somewhat limited.

Hibernation and torpor are associated with several hormonal changes related to the mammals' unique nutritional status and, sometimes, concurrent preparations for reproduction. Endocrine events differ during the different stages of hibernation: preparations for the event, its initiation, maintenance, and the arousal to resume a more normal life-style. There are species differences in such hormonal patterns, which do not necessarily reflect a causal role in its initiation of terminal arousal. The hormonal changes often appear to result from endogenous circannual, photoperiodic, and thermoperiodic rhythms. The European edible dormouse, *Glis glis*, like many other mammalian hibernators, commences to accumulate considerable fat stores in July. In their natural environment, this event in the male corresponds to the commencement of a decline of high plasma concentrations of T_4 and testosterone (Jallageas and Assenmacher, 1986). Castration blocks this hibernation cycle so that weight gain occurs less rapidly, while thyroidectomy has no effect. Arctic ground squirrels (*Spermophilus undulatus*) in summer have high plasma insulin levels while glucagon concentrations are low (Feist *et al.*, 1986). Such a regimen is consistent with lipogenesis. The administration of leptin to such ground squirrels reduces their prehibernation eating behavior and the associated increase in weight (Ormseth *et al.*, 1996). The initiation of hibernation in the edible dormouse, garden dormouse, European hamster, and ground squirrels is unaffected by gonadectomy or thyroidectomy (Jallageas and Assenmacher, 1986). A low level of thyroid gland activity is usually maintained during hibernation but may increase towards the time of arousal (see also Hulbert and Hudson, 1976). However, the levels of thyroid hormone in the plasma do not rise in the edible dormouse until after arousal and the rise appears to be temperature dependent. In the woodchuck *Marmota monax*, urinary excretion of catecholamines is at its lowest level just prior to the onset of hibernation (Wenberg and Holland, 1973). Awakening, however, is associated with an increase in their urinary excretion. Castration prolongs the period of hibernation (or "season") in male Turkish hamsters (*Mesocricetus brandti*) and golden-mantled ground squirrels (*Spermophilus lateralis*) (Hall, Bartke, and Goldman, 1982; Dark, Miller and Zucker, 1996). Conversely the administration of androgens inhibits their hibernation (Hall and Goldman, 1980; Lee *et al.*, 1990). The latter treatment had a similar effect on female Turkish hamsters. The effect of estradiol was, however, weak. In small rodents awakening from harsh winter conditions, a recrudescence of the testes appears to presage early spring breeding activity. The secreted androgens may maintain, but not necessarily initiate, a permanent arousal from hibernation. The females usually rest longer than the males and only emerge when the climatic conditions are favorable for the survival of the young. A possible hormonal contribution to their awakening has not been established.

Some courageous observations have been made on hibernating black bears,

Ursus americanus (Nelson, 1980). The nitrogen metabolism of these animals is reduced to very low levels during this winter fast, the animal living principally off its substantial fat reserves. During hibernation, the levels of pituitary TSH and thyroid hormones decline, but these hormones can be stimulated by the injection of TRH. It appears that these animals are in a state of hypothyroidism that is caused by a decrease in the activity of the hypothalamus. As described above, decline in thyroid function has also been observed in hibernating rodents. During their long winter sojourn, bears not only do not feed but they may also give birth and suckle their cubs, a rather remarkable endocrine feat.

Estivation in African lungfish

Estivation may last 2–3 years in African lungfish. During this phase of its life cycle, *Protopterus annectens* only secretes thyroid hormone at about 1/75 of the rate that it does in its normal free-swimming state (Leloup and Fontaine, 1960). The TSH content of the pituitary is similar in both conditions so that the release of TSH may be inhibited by estivation. Godet (1961) has suggested that an inhibition of the adenohypophysis precedes estivation in these lungfish. It was found that when the pituitary was removed after estivation had commenced, there was no effect on the subsequent torpor. An intact pituitary gland was, however, indispensable for the fishes survival on emergence from estivation. Many amphibians also estivate during periods of drought, but their endocrinology does not appear to have been investigated on these occasions.

Hormones and lactation

Lactation, or the secretion of nutrients by the mammary glands in order to feed the young, is an activity confined to the mammals. Somewhat analogous processes, such as the formation of pigeon-milk by the crop-sac of some birds, can occur in other vertebrates. The mammary glands may, however, have evolved from cutaneous tissues. Ancestral homeothermic reptiles possibly developed cutaneous brood patches to facilitate the incubation of their eggs (Mepham and Kuhn, 1994). Oozings from such areas may have constituted prolacteal secretions that helped to nourish the young at a place and in a season when the collection of appropriate food was difficult. It may also have made possible the production of relatively immature young that could be maintained and undergo development on such a diet. Extant oviparous monotremes and marsupials, both of which have well-developed mammary glands, provide tentative analogs for such situations. The preparation of the milk secretion and its delivery to the young is promoted by several hormones that have thus assumed novel roles compared with those of their forebears.

Fig. 5.9. The role of hormones in the growth of the mammary gland and the secretion of milk. In this instance the example is the laboratory rat, but in other species different combinations of hormones (called the "lactogenic complex") may be required (see text). (Based on Lyons, 1958, from Cowie, 1972.)

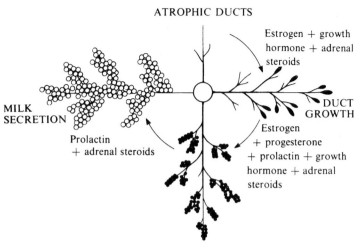

ATROPHIC DUCTS

Estrogen + growth hormone + adrenal steroids

DUCT GROWTH

MILK SECRETION

Prolactin + adrenal steroids

Estrogen + progesterone + prolactin + growth hormone + adrenal steroids

LOBULO-ALVEOLAR GROWTH

Milk is a nutrient solution the composition of which differs depending on the species and the duration of the lactation. (For a complete account see Cowie, Forsyth, and Hart (1980), Mepham and Kuhn (1994), McNeilly, Forsyth and McNeilly (1994).) It contains fats, carbohydrates, and proteins as well as minerals and other essential dietary items. The fat content of milk varies from 0.3% in the rhinoceros to 49% in porpoises; the lactose from 0.3% in whales to 7% in humans and the protein from 1.2% in humans to 13% in whales. The contained energy is equivalent to 500 kcal/kg milk in the rhinoceros to 2773 kcal/kg in the reindeer (Kleiber, 1961). The nutritional requirements of the mother may, therefore, be considerable.

The mammary glands, like adipose tissue, can make triglycerides and also lactose. The proteins present are largely synthesized by the mammary tissue though some, such as serum albumin and immunoglobulins, are transferred directly from the plasma. In order to deliver these nutrients efficiently to the young, the mammary tissue differentiates into a complex system of alveoli and ducts. These processes, morphological differentiation, and the secretion of nutrients are regulated by several hormones so that the mammary glands afford an example of the multiple actions that hormones may have on a tissue.

Figure 5.9 depicts the arrangement of the mammary gland tissues and a summary of the effects of hormones on them. The milk is secreted into the alveoli and passes down the duct prior to release from the teat or nipple. Experiments on animals with an intact pituitary, but with the ovaries removed, indicate that development of the alveoli is influenced largely by estogens and progesterone, whereas the duct system responds principally to estrogens. The neurohypophysial hormone oxytocin is released (Chapter 4) as a result of a suckling stimulus on the nipple. It contracts the myoepithelial cells that surround the alveoli and results in milk letdown (the *galactobolic effect*). Initiation of the secretion of milk (*lactogenesis*) by the alveoli is dependent on prolactin, which is released during parturition and as a result of suckling. Although these hormones are directly essential for the activities mentioned, adequate circulating levels of other hormones, including thyroid hormones, corticosteroids, and growth hormone, are also necessary. The experiments that demonstrate this are complex and involve restoration of the growth and function of the mammary glands in animals from which the pituitary and the ovaries have been removed. No single hormone appears to be effective on one function, but combinations involving ACTH or corticosteroids, prolactin, growth hormone, thyroid hormones, estrogens, and progesterone are necessary. The metabolic effects of the absence of certain of these hormones appear to inhibit or limit the actions of other hormones that act more directly.

Species differences are common. Lactation in many eutherians appears to be initiated following a peripartum decline in progesterone levels, but this hormonal change does not mediate the initiation of milk secretion in marsupials (Renfree, 1994). The maintenance of lactation in most marsupials and eutherians is promoted by prolactin, but in ruminants (cows, ewes, and goats) growth hormone is more important. The galactobolic response to oxytocin and oxytocin-like hormones may also display interspecific differences. Such milk-letdown and nursing takes place about 20 times a day in rats but only once a day in rabbits and on alternate days in tree shrews. In rabbits, there is a sustained release of oxytocin during the nursing period but in most species (rats, ewes, sows, and women, but not cows) it takes place in rapid spurts or pulses. Different species exhibit variations in their preferred hormone concoctions at this time, which is referred to as a lactogenic complex. Marsupials, which produce very immature young, exhibit considerable changes in the composition of their milk as the period of their lactation proceeds. Rather surprisingly, they can maintain two young of very different ages simultaneously on separate teats, which each supply milk of an appropriately different composition. Such asynchronous lactation (Renfree, 1994) occurs in a common systemic endocrine milieu. The response of each mammary gland to the common available hormones then, presumably, are controlled by the tissue itself. This regulation may involve the participation of local tissue hormones and, possibly, differences in the numbers and types of hormone receptor in each part of the mammary gland.

The maintenance of lactation is called *galactopoiesis* and the hormonal requirements of this are often difficult to distinguish from lactogenesis. Species differences again exist. Generally, the following occur: removal of the ovaries once lactation has been established does not influence lactation; indeed, estrogens and estrogens plus progesterone are often used to terminate, or dry up, the secretion of milk; removal of the thyroid, adrenals, parathyroids, or the endocrine pancreas depresses lactation. As described above, prolactin usually has a major role in maintaining lactation, except in ruminants where growth hormone is relatively more important. The administration of thyroid hormone or growth hormone to lactating cows increases the yield of milk. The use of T_4 to increase milk production in dairy cows was initiated in the 1940s. However, its hyperthyroid-like side effects limited its economic use. Bovine growth hormone has been used more recently to increase milk production in cows. It is interesting to recall that growth hormone can promote the peripheral activation of T_3 from T_4. The role of such hormones in maintaining lactation probably reflects their roles in maintaining the metabolic transformations of energy substrates in cells generally. It is not unexpected that such metabolic processes are necessary, directly and indirectly, for the adequate secretion of milk by the mammary tissue.

The mammalian lactational process is a phylogenetically novel one that has arisen in the later stages of vertebrate evolution. Its complexities, not surprisingly, necessitate coordination by the endocrines. The hormones involved probably existed prior to the mammary gland itself, so that an evolution of their role in the body has occurred in this case. Certain processes occur in vertebrates that are analogous to mammalian lactation. The formation of pigeon-milk by the crop-sac of pigeons and doves, with which they feed their young, is an example. This process, which involves the proliferation of the crop-sac epithelium, can be induced by prolactin. In certain teleost fishes, mucous secretion from the skin can be stimulated by prolactin (Bern and Nicoll, 1968). In one such fish (*Symphysodon*), the newly hatched young have been observed to feed off the surface of the parental fish, suggesting another possible "lactational" role for prolactin.

Storage of nutrients in the egg

The early development of the young is supported by nutrients stored in the egg. These materials may be sufficient for the complete embryonic development of the young, as seen in birds and most reptiles, or they may only support a limited differentiation, as seen in most fishes and amphibians, which leads to the emergence of a free-swimming larva. An even more limited embryonic growth may precede the implantation of the ovum in the uterus of viviparous species. Viviparity is most common in mammals but may sometimes occur in fish, especially elasmobranchs, and in reptiles.

Estrogens, progesterone, and gonadotropins may play important roles in the growth and maturation of the egg and these effects are summarized in Fig. 5.10. Progesterone promotes secretions from the oviduct that provide adjunct nutrients and protectants to the eggs. These secretions include the jelly coatings of fish and amphibian eggs and the protein- and mineral-containing fluid that surrounds the yolk in cleidoic eggs of reptiles and birds.

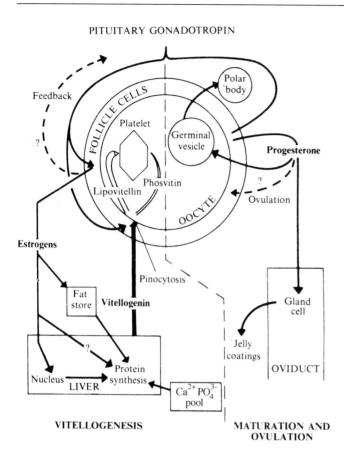

PITUITARY GONADOTROPIN

VITELLOGENESIS

MATURATION AND OVULATION

Fig. 5.10. The role of hormones in the production of eggs in anuran amphibians. The pituitary gonadotropin controls the release of estrogens and progesterone by the ovary. *Left.* The control of the process of *vitellogenesis.* Estrogen, formed by the follicular cells, induces the formation of vitellogenin in the liver, which is incorporated into the yolk of the developing oocytes. The uptake process is stimulated by the gonadotropin. *Right.* The *maturation of the oocyte* and ovulation. These processes are controlled by the pituitary gonadotropin, which mediates the formation of a "maturation agent" that seems to be progesterone. The progesterone also stimulated the oviductal glands to secrete the "jelly" with which the eggs are coated. This summary is principally based on experiments on *Xenopus laevis* and *Bufo bufo.* (From Follett and Redshaw, 1974.)

The yolk of eggs contains most of the nutrients required for the predicted development of the young. In the domestic fowl, about 50% of this comprises fats and proteins. The number of eggs produced successively in one period of time (the clutch) varies considerably in different species, but 20–30 are not uncommon in birds and reptiles. Under domestic conditions, a chicken may produce 200–300 eggs in a year. Even species that produce small alecithal eggs must contribute considerable amounts of nutrients to the eggs, especially as such species often produce many hundreds of such eggs.

The deposition of the yolk proteins is called *vitellogenesis.* Estrogens can promote the appearance in the blood of elevated levels of lipophosphoglycoproteins called *vitellogenins.* Vitellogenesis does not occur in mammals but it has been observed in all other groups of vertebrates. These include cyclostomes, elasmobranchs, teleosts, dipnoans, amphibians, reptiles, and birds. Its presence in lampreys and Pacific hagfish (Pickering, 1976; Yu *et al.*, 1981) suggests that this role for estrogens evolved early. An homologous process of vitellogenesis occurs in invertebrates, including annelids, echinoderms, insects, and nematodes (Byrne, Gruber, and Ab, 1989). In the vertebrates, vitellogenesis occurs in response to the stimulation of estrogen receptors in the liver. Although not usually expressed, vitellogenin genes are also present in male vertebrates, where their transcription can be initiated by the administration of estrogens. The vitellogenic effect of estrogens is very specific and cannot be mimicked by any other type of steroidal hormone. Mammals possess hepatic estrogen receptors that mediate the synthesis of various proteins but not vitellogenins. However, a protein component of mammalian lipid transport proteins, apolipoprotein B-100, has substantial structural homologies to vertebrate and invertebrate vitellogenins (Baker, 1988). In the chicken liver, this protein can even be induced by estrogens. The apolipoprotein B-type proteins have a wide distribution in vertebrates and if they shared a common ancestral gene its duplication and divergence may have occurred even before the origin of the vertebrates. The vitellogenin gene has been identified in several vertebrates and invertebrates. While substantial differences exist they are related and appear to belong to the same ancient multigene family (Byrne *et al.*, 1989).

The general structures of the vitellogenin genes from *Xenopus*, domestic fowl, and killifish are similar (Ho, 1991; Specker and Sullivan, 1994). There are four vitellogenin genes in *Xenopus laevis* (Wahli *et al.*, 1981) each coding for a vitellogenin polypeptide with molecular masses of 200 to 240 kDa (Two such polypeptides make up plasma vitellogenin). Three vitellogenin polypeptides and their genes have been identified in the domestic fowl. They have molecular masses of about 180 kDa (Byrne *et al.*, 1989). Teleost vitellogenins appear to have a somewhat different structure to those of tetrapods. The dimers range from about 300 kDa in tilapia to 540 kDa in salmon (Specker and Sullivan, 1994). Vitellogenins have been identified in reptiles, a turtle, *Chrysemys picta* (Ho, L'Italien and Callard, 1980), and the tuatara, *Sphenodon punctatus punctatus* (Brown *et al.*, 1994).

The effects of estrogens on vitellogenesis in the turtle *Chrysemys picta* can be inhibited by progesterone (Callard, Riley, and Perez, 1990). Progesterone receptors have been identified in the liver of this reptile. It has been suggested that the progesterone receptor and its ligand may interact with a progesterone–responsive element on the vitellogenin gene. Such an effect could interfere with the transcriptional activity of the estrogen-responsive element, thus decreasing the formation of vitellogenin mRNA. Such a mechanism may have contributed to the inhibition of vitellogenesis of ancestral, viviparous, mammals, where placentation provides for the sustenance of the embryo.

The vitellogenins are carried in the plasma to the ovary where they are taken up by the developing ova. Once they are so accumulated they are converted to smaller lipophosphoproteins called *phosvitin* and *lipovitellin*, which are stored to await the needs of the developing embryo. Gonadotropins, apart from stimulating the secretion of estrogens, have a separate effect on vitellogenesis as they have also been shown to increase the uptake of yolk proteins by the developing ova. This effect of these pituitary hormones has been observed in chondrichthyeans, teleosts, and amphibians (Follet and Redshaw, 1974; Craik, 1978; Idler and Ng, 1979; Tyler *et al.*, 1991).

Some birds can deposit steroid hormones, especially testosterone, in the yolk of their eggs prior to ovulation (Schwabl, 1993; 1996). These hormones are present in eggs containing embryos of both sexes and do not influence sexual differentiation. However, the presence of testosterone is associated with an increase in the growth rates of the nestlings, enhanced begging behavior, and aggression towards siblings. In canaries, *Serinus canaria*, the concentration of testosterone in the yolk increases in successively laid eggs. As the most recently laid eggs hatch later, the androgen may contribute to a "catch-up" of growth by the younger chicks. However, in the cattle egret, *Bubulcus ibis*, the situation differs as the first laid egg has the highest concentration of testosterone (Schwabl, Mock, and Gieg, 1997). The domination of the size of the first hatched chicks in this instance is increased, often to the detriment of their younger siblings. This situation appears to enhance "siblicide" which may be an advantage to the species when food is scarce. A reduction in the size of the brood may then promote the survival of at least some chicks in the brood.

Conclusions

The processes of nutrition, whether involving the digestion of food materials, their transformation, storage, and utilization following absorption, or the feeding and growth of the young, are very dependent on the actions of hormones. It is especially notable that several hormones are usually involved in such processes ("multihormonal" effects) probably more often than elsewhere, and these excitants have widespread effects both in respect to the types of tissue that they act upon and the nature of the underlying biochemical events that they influence. Such ubiquitous endocrine effects are well integrated with each other by a series of feedback controls that involve the levels of metabolites and the hormones themselves. The latter indeed often directly influence each others' release. The hormonal pathways that function, or predominate, in the control of nutrition of a particular species seem to depend more on its feeding habits and the stage of its life cycle than on its phylogenetic origins. This probably reflects the common biochemical pathways underlying nutritional processes in all vertebrates. The principal species differences in responsiveness to hormones are quantitative ones, though when novel processes have evolved, such as lactation and homeothermy in mammals, this has been accompanied by adaptations of the roles of hormones to integrate the new systematic features.

6 Hormones and calcium metabolism

Calcium is vital for animal life. In vertebrates, this divalent ion is present at various sites, including the body fluids and structural parts of the cell (especially the mitochondria and endoplasmic reticulum), and in most species it is also a major component of the endoskeleton. The outer shell of the eggs of birds and many reptiles also consists principally of calcium. The physiological role of calcium appears to be the result of a rather unique set of physicochemical properties. In aqueous solution, calcium can exist in a soluble form, which is important for its mobility in the body, and yet, equally essential to its role, many of its salts, including phosphates and carbonates, have a low solubility so that a physicochemical equilibrium may exist between its solid and aqueous forms. The quantity of calcium free in solution is thus restricted in the presence of certain anions. In addition, such relatively insoluble salts in certain of their crystalline forms, principally calcium phosphate and calcium carbonate, can contribute to the mechanical support and stability of biological structures. Calcium also has a ready propensity to associate and combine with proteins. Such combinations are seen in the body fluids, where a considerable portion of the calcium is bound to serum proteins, and in cells, where it contributes to their structural stability by helping maintain essential ionic bridges at vital points in protein molecules; therefore when tissues are placed in calcium-free solutions they tend to disintegrate and the cells swell and fall apart.

Calcium plays an essential role in coordinating many events in the body that may reflect those general properties described above. Calcium stabilizes membranes and this effect can be seen in the hyperactivity of nerve fibers placed in solutions with low calcium concentrations. Such instability and the repetitive electrical depolarization of the nerve cell membrane result in tetanic contractions of the muscles they supply. Muscle contraction requires the presence of calcium; when released from the sarcoplasmic reticulum within the cell, calcium couples the initiating electrical depolarization of the cell membrane to those processes that initiate changes in the contractile proteins. In a comparable manner, intracellular calcium is necessary for the ultimate

Table 6.1. *Calcium and phosphate in serum and in the environment*

Environment and representative species	Total Ca^{2+} (mmole/liter)	Ionic Ca^{2+} (mmole/liter)	Total phosphate (mmole/liter)
Environment			
Pacific ocean	10.0 ± 0.1	10.0	0.00
Brackish water	2–5	–	–
Lake Huron	0.9 ± 0.1	0.9 ± 0.1	0.003
Marine invertebrate Nephrops	11.95	–	–
Cyclostomes			
Hagfish, *Eptatretus stoutii*	5.4 ± 0.1	3.0 ± 0.4	1.5 ± 0.2
Lamprey, *Petromyzon marinus* (from fresh water)	2.60 ± 0.1	1.74 ± 0.2	1.3 ± 0.1
Chimaeroid			
Ratfish, *Hydrolagus colliei*	4.8 ± 0.3	–	2.2 ± 0.4
Elasmobranchs			
Marine shark, *Carcharhinus leucas*	4.50 ± 0.9	3.10 ± 0.4	2.0 ± 0.6
Freshwater shark, *C. leucas nicaraguensis*	3.0 ± 0.2	1.7 ± 0.1	1.6 ± 1.7
Teleost			
Marine, *Paralabrax clathratus*	3.2 ± 0.8	2.0 ± 0.9	2.0 ± 0.2
Freshwater, *Megalops atlanticus*	2.5 ± 0.2	1.8 ± 0.2	1.2 ± 0.4
Mammal			
Human, *Homo sapiens*	2.32 ± 0.08	1.15	–

Source: Copp, 1969.

initiation of many endocrine events such as the release of hormones (Chapter 4) and the responses of their definitive effectors. It has been somewhat belatedly recognized that calcium ions, like hormones, can also act as an extracellular messenger. An interaction with G protein-linked calcium receptors can influence the activities of tissues such as the parathyroid glands, kidneys, and brain (Brown, Vassilev, and Hebert, 1995).

The availability of calcium to animals varies considerably, depending on the environment and their diet. The physiological need for calcium may also vary a great deal. Young, growing animals, especially those that are forming large amounts of bony tissue, have much greater requirements than adults. The latter, however, periodically need more calcium for reproductive processes, as during pregnancy and lactation in mammals, and in birds and reptiles for the production of large cleidoic eggs that are covered with a shell of calcium carbonate. Even fishes and amphibians that do not have cleidoic eggs require substantial amounts of calcium for the production of their eggs. The availability of calcium in the environment varies a great deal. The concentration in sea water is even higher than that in the body fluids of vertebrates but in fresh water, little calcium is usually present (Table 6.1), therefore, vertebrates living in the sea or in fresh water may be expected to experience very different problems with respect to the availability of this ion. Terrestrial animals obtain most of their calcium from their diet, which may include appreciable amounts obtained from certain drinking waters. In times of extra need for calcium, vertebrates, especially birds during egg laying, may consume inorganic calcium-containing minerals, such as the so-called grit fed to domestic fowl. The ultimate acquisition of calcium, whether from the external bathing solutions, drinking water, or food, takes place principally from the gut. Absorption across the intestine is regulated in relation to the animal's needs. Amphibians and fishes possess integuments that may be permeable to calcium. In the former group, such exchanges appear to be quite small, except in tadpoles where a substantial uptake appears to occur across the gills (Baldwin and Bentley, 1980; 1981). In the frog *Rana pipiens* the uptake of calcium across the skin appears to involve an active transport process and displays seasonal variability, being greatest in mid-summer (Stiffler, 1995). The gills of fishes are also permeable to calcium and may be an important site for the accumulation of this mineral from the solutions in which they live.

The calcium that is not absorbed from the intestine, either because of a lack of physiological need for it or because it is present in chemical combination that makes absorption impossible, is excreted in the feces. Additional amounts of calcium are also excreted by the kidney in the urine. Urinary losses of calcium should be viewed not only as an excretory mechanism for ridding the body of an excess of this ion but also as part of an unavoidable loss that results from the formation of urine. The calcium that is not bound to plasma proteins is filtered across the renal glomerulus, but most of this ion is subsequently reabsorbed by the renal tubules. This conservation is less effective in the presence of a high plasma calcium concentration so that extra amounts of this ion are excreted in the urine. Conversely, a hypocalcemia results in an increased renal reabsorption of calcium, which is accompanied by an increased phosphate excretion. The excretion of calcium and phosphate in the urine is thus subject to physiological control.

The bony skeleton possessed by most vertebrates plays a central role in calcium metabolism. Calcium phosphate salts are an integral part of this structure and make the principal contribution to skeletal rigidity. Not all vertebrates, however, have a bony calcareous skeleton; it is absent in the cyclostome and chondrichthyean fishes, which have a more elastic cartilaginous skeleton in which little calcium is deposited. In the bony vertebrates, the skeleton is the predominant quantitative site of calcium in the body though this should not be taken to reflect its relative importance, as its presence is equally vital to the soft tissues. Marine animals have a readily available and inexhaustible supply of calcium in the sea water in which they live. As a result, they do not need large stores of this mineral such as occurs in bones.

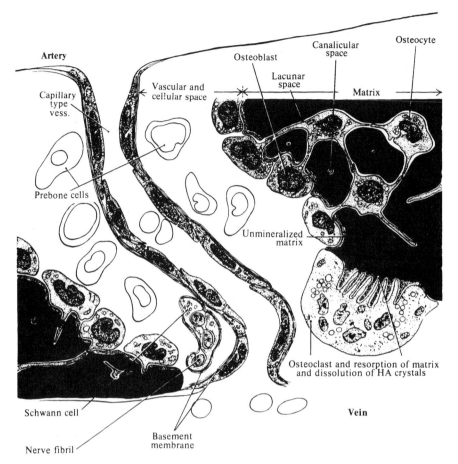

Fig. 6.1. Schematic representation of a physiological unit of bone tissue showing the osteocytes, osteoblasts, and osteoclasts. (From Doty, Robinson, and Schofield, 1976.)

However, species that live in fresh water and on dry land have a more limited access to calcium. The evolution of bone can, therefore, apart from its contribution to animal mechanics, be looked upon as an important potential contribution to the maintenance of body calcium. The calcium in the bones and that in the body fluids and soft tissues exist in equilibrium with each other. Two processes are involved in this: first, there is a relatively static *physiochemical* equilibrium that reflects the solubility (and insolubility) of the calcium salts in the body and which only results in small exchanges of calcium; second, the major exchanges of calcium between the bones and body fluids result from a *dynamic*, physiologically mediated equilibrium that results from the activities of cells in the bone: the *osteoblasts, osteocytes*, and *osteoclasts*.

Bone, despite its mineral-like appearance, is a living tissue (Fig. 6.1). Some fishes, however, possess acellular bone that, once formed, is much less labile than the cellular type of bone characteristic of the tetrapods. This type of bone is present in species of teleosts that are considered to be the most advanced in their evolution; such as tilapia and killifish. Cellular bone is present in less favored teleosts like herring, eels, and salmonids. It has a microscopic honeycomb-like appearance owing to the presence of numerous small chambers, or lacunae, each of which contains an osteocyte bone cell. Numerous fine channels (canaliculi) radiate from the lacunae to the surrounding mineralized tissue, and these provide a pathway through which the bone fluids and the dendritic-like extensions of the osteocytes maintain contact with the tissue. The osteocytes act as sentinels controlling local mineral exchanges. Bone also has an extensive network of blood vessels through which the supply of blood can be regulated according to its metabolic needs. Exchanges of bone minerals take place at any of the bone's free surfaces, the canaliculi, at the inner and outer borders of the limb bones (periosteum and endosteum), the channels through which the blood vessels pass, and special tunnels, called "cutting cones," which are excavated in the tissue. Minerals are principally mobilized from the cortical regions of the shafts of the long bones, but a labile store also occurs in the medullary regions of the bones of birds, where it is an especially important store during the egg-laying cycle.

Bone is formed by the osteoblast bone cells. These cells extrude collagen, which is laid down as a matrix into which calcium phosphate, and some carbonate are subsequently deposited (called "accretion"). It is possible that the osteoblasts also play some direct role in the process of mineral deposition but this is not essential. Minerals are deposited in two phases, initially as an amorphous form of calcium phosphate, which is subsequently changed to a structurally stronger, crystallized form resembling the mineral apatite. The osteoblast, after surrounding itself with such tissue, then is transformed into an osteocyte. The mineralized bone is not necessarily a permanent structure but can be modified and remodeled, and the calcium and phosphate may be returned to the tissue fluids. The process of "resorption" occurs more readily from the amorphous than the crystalized phase but the latter can also be broken down. Resorption of minerals from bone is associated with an increase in the activity of the multinuclear osteoclasts, and the osteoblasts are also involved. The process of resorption involves the secretion of enzymes by the bone cells and these facilitate the dissolution of the minerals. This process can be regulated. In teleosts, calcium can be mobilized by mononuclear resorptive cells, which appear to be more effective in the cellular type of bone.

Physiological control of calcium levels in vertebrates can thus be affected at four major sites: the intestine, by the control of absorption; the kidney, by the regulation of reabsorption from the glomerular filtrate; the gills, across which influx and efflux may be regulated; and the bones, which act as a storage site for calcium phosphate. The very important role of bone in calcium homeostasis cannot be overemphasized, but it should be recalled that this tissue is not present in all vertebrates. The coordination of the exchanges of calcium at these four sites is under the control of hormones. These are parathyroid hormone, from the parathyroids (which are absent in fishes), and calcitonin from the thyroid C-cells and the ultimobranchial bodies. To these hormones can be added another (or others) that are derived from vitamin D_3. These are $1\alpha,25$-dihydroxycholecalciferol ($1,25(OH_2)D_3$) and, possibly $24,25$-dihydroxycholecalciferol ($24,25(OH)_2D_3$). Prolactin from the pituitary gland may play a role in the formation of the former. It is of further interest that removal of the corpuscles of Stannius, which are present in certain bony fishes (Chapter 2), results in an elevation of plasma calcium levels. These tissues secrete a hormone called stanniocalcin (formerly named hypocalcin, Chapter 3) that regulates plasma calcium concentration in a process which includes an ability to control its transport across the gills. Estrogens aid mobilization of medullary bone in birds during egg-laying.

The discovery of parathyroid hormone and calcitonin

The discovery of the roles of the parathyroids, the mammalian thyroid C-cells, and the ultimobranchial bodies in the regulation of calcium metabolism is an interesting endocrine tale. It is not an example of a triumph of "goal-oriented" research but rather a serendipity and the persistent following up of a series of unexpected observations. Species differences in the endocrine tissues have played an important part in establishing the role of these glands.

As we have seen (Chapter 2), two types of endocrine tissue are known to be concerned with regulating calcium metabolism in tetrapod vertebrates. These tissues are present in close morphological association with the thyroid gland. Early efforts to remove thyroid tissue surgically in humans were sometimes seen to result in tetanic muscular contractions such as are associated with low plasma calcium concentrations. Closer examination revealed that in these instances the parathyroid tissues had been removed. Subsequently, other experiments in animals confirmed the importance of the parathyroids in maintaining optimal calcium levels in the blood, and extracts of these glands were shown to exhibit a hypercalcemic action. Such experiments to demonstrate the role of the parathyroids are relatively simple in the rat, where there are two distinct bodies of parathyroid tissue present on the surface of the thyroid gland. In another favored experimental animal, the dog, substantial amounts of parathyroid tissue (as well as C-cells) are present, embedded deeply in the thyroid gland, so that this species is not an ideal one for such experiments and this has, in the past, contributed to some misinterpretations.

As related by Hirsch and Munson (1969), the rat is an ideal species, and favorite subject, for parathyroidectomy. Two methods have been used to remove the rat parathyroids. A simple surgical removal of the two glands can be performed or they can be destroyed with an electrocautery. In the 1950s, the latter procedure was more popular. When, however, a comparison, partly retrospective of blood calcium levels following each type of operation was made, the resulting hypocalcemia was observed to be much greater following electrocautery than following surgical excision. This observation probably did not initially gain the serious attention it deserved. In retrospect it has, especially when it was observed that stimulation of the thyroid gland by the electrocautery at sites removed from the parathyroid tissue also produces a drop in blood calcium concentration. The possibility was then considered that this stimulation resulted in a release of a substance from the thyroid that exhibited a hypocalcemic action. A couple of years prior to the latter observation, D. H. Copp (Copp *et al.*, 1962) proposed the presence of a hypocalcemic hormone in mammals that he called calcitonin, which at the time he considered to be formed by the parathyroid glands. The proposal of the presence of a new hormone was based on experimental observations involving perfusion of the thyroid–parathyroid complex in dogs. Blood containing abnormally high or low concentrations of calcium was perfused through the arteries supplying these tissues and the resulting venous outflow was then passed back into the dog, and the effects on the general, systemic, plasma calcium levels were observed. It was found that, when the perfusing blood has a low calcium concentration, the outflowing blood, when passed into the dog's general circulation, produced a hypercalcemia which could be accounted for by the presence of parathyroid hormone. In the opposite type of experiment, in which the thyroid–parathyroid complex was perfused with blood having a high calcium concentration, the parathyroid hormone level would be expected to decline, as indeed it does. The basic question that was asked was this: is such a decline in parathyroid hormone sufficient to bring about a decline in the calcium concentrations from hypercalcemic to normal calcium levels? In other words, while parathyroid hormone exerts a positive effect in elevating blood calcium concentration, is the mere absence of this hormone all that is needed to adjust the calcium level in a downward direction? Copp found that this was not so. Although the venous perfusate from the thyroid–parathyroid complex that was exposed to high calcium concentrations produced a drop in calcium levels when infused systemically, this hypocalcemia was much greater than could be produced after removing the glandular complex (Fig. 6.2). The implication was drawn that the response is a positive one involving the action of a hormone that has a hypocalcemic action; Copp called this hormone "calcitonin".

It was thought at first that calcitonin came from the parathyroids and, as already commented upon, in view of the intermixture of tissues that occurs in the thyroid region in dogs such an error was not surprising. Hirsch and Munson (1969), on the basis of their experiments on rats, proposed the presence of a hypocalcemic hormone in the thyroid gland itself, which, in order to distinguish it from Copp's hormone, they called thyrocalcitonin. These two hormones are in fact identical and in mammals this hormone originates from the thyroid gland. By the choice of an appropriate species, this time the pig, in which the thyroid contains no parathyroid tissue, appropriate

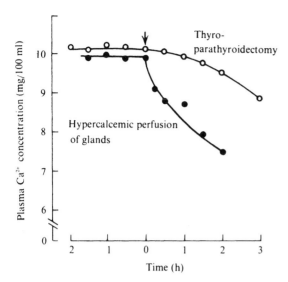

Fig. 6.2. The hypocalcemic response of dogs following perfusion of the thyroid–parathyroid gland complex with hypercalcemic solutions. In intact animals, the decline in plasma calcium concentration (at time zero) occurs promptly, but if the glandular complex is removed, the resulting hypocalcemic response is much slower. This suggests the action of a hypocalcemic hormone in the intact dogs. (From Hirsch and Munson, 1969; from data of Copp *et al.*, 1962.)

perfusion experiments of the thyroid alone, similar to those described, demonstrated the presence of calcitonin in the venous effluent blood.

The question then arose as to the site of origin of calcitonin in the thyroid tissue. Is it also produced by the same tissue that forms thyroxine? The answer is no. Calcitonin is formed by the parafollicular or C-cells, which are present in the mammalian thyroid, and this tissue is quite distinct from that which secretes T_4. The presence of these secretory cells was first described by E. C. Baber in 1876, but their function was unknown. Radioimmunofluorescent studies, using fluorescent antibodies to calcitonin, show quite clearly that the C-cells form this hormone.

The C-cells are present in the mammalian thyroid but not that of non-mammals. Embryologically they are derived from the ultimobranchial bodies that are present in all nonmammals except the cyclostomes. An early clue to the function of this tissue was described by Rasquin and Rosenbloom in 1954, several years before the discovery of calcitonin. It was found that Mexican cave fish (*Astyanax mexicanus*) kept for prolonged periods of time in complete darkness suffered skeletal deformities associated with hyperplasia of the ultimobranchial bodies. Rasquin and Rosenbloom suggested that the ultimobranchials contained a parathyroid-like hormone. Extracts of the ultimobranchial bodies obtained from chickens and, subsequently, from many other species, including even dogfish, showed that they exhibited a hypocalcemic effect when they were injected into rats that was caused by the presence of a calcitonin-like hormone that was not present in the thyroid glands of these species (Copp, Cockcroft, and Keuk, 1967a).

Vitamin D and 1α,25-dihydroxycholecalciferol

Cholecalciferol (Chapter 3 and p. 68) is a vitamin (D$_3$) or prohormone that can become a hormone (DeLuca, 1974; Kodicek, 1974; Norman and Henry, 1974; Gross and Kumar, 1992). This sterol is formed in the skin of mammals from 7-dehydrocholesterol. The reaction is a photochemical one that requires ultraviolet light, normally sunlight. The vitamin D$_3$ that is formed can be absorbed into the blood where it is bound to a plasma protein (Bouillon *et al.*, 1978). It travels to the liver where it may be stored or metabolized to a more active form.

It has been recognized since the 1920s that deficiency of vitamin D can result in a bone disease called rickets. This condition occurs in children who grow up in areas where sunlight and dietary sources of vitamin D may be deficient. The disorder results from inadequate absorption of calcium from the intestine, a process that is normally stimulated by vitamin D$_3$. When vitamin D is administered there usually is a delay of several hours before an increase in the absorption of calcium occurs from the gut (DeLuca, 1971). There are two main reasons for this tardiness. First, the prohormone must be metabolized to a more active hormonal form and, second, the action of the hormone involves genetic transcription in the epithelia cells lining the intestine.

Vitamin D$_3$ undergoes hydroxylation at its 25-position in the liver microsomes and mitochondria to produce 25-hydroxycholecalciferol (25-OHD$_3$). This metabolite is about twice as active as the parent compound and appears in the plasma where it is bound to a protein. It is accumulated by the kidney (Henry *et al.*, 1992) where it is again hydroxylated at the 1α-position to produce 1α,25-dihydroxycholecalciferol (1,25(OH)$_2$D$_3$, calcitriol (see Fig. 3.1). The reaction involves a 1α-hydroxylase that is associated with the mitochondria of the proximal segment of the nephron (Brunette *et all.*, 1978). This enzyme has been identified in the kidneys of most groups of vertebrates, ranging from teleost fishes to mammals (Henry and Norman, 1975). The conversion of 25-OHD$_3$ to 1,25(OH)$_2$D$_3$ is increased by hypocalcemia and hypophosphatemia, but it is decreased by hypercalcemia (Tanaka, Frank, and DeLuca, 1973). The product, 1,25(OH)$_2$D$_3$, exerts a negative (inhibitory) feedback effect on its own further synthesis (Henry *et al.*, 1992). Several hormones, including parathyroid hormone, estrogens, prolactin, and growth hormone, have been shown to promote the activity of the 1α-hydroxylase system but such effects can be variable and some may be species specific. The physiological role of parathyroid hormone in promoting the conversion of 25-OHD$_3$ to 1,25(OH)$_2$D$_3$ in the kidneys of mammals and birds is not in doubt. This response involves the mediation of the adenylate cyclase–cyclic AMP system but whether it initiates an activation or induction of the 1α-hydroxylase has not been established. It is also not clear if the effects of hypocalemia on this process are solely mediated by the released parathyroid hormone. Prolactin and estrogens have been shown to increase the formation of $1,25(OH)_2D_3$ in chicks (Spanos $et\ al.$, 1976a,b; Baksi and Kenny, 1977; Pike $et\ al.$, 1978; Bikle $et\ al.$, 1980). Such actions of prolactin have not yet been unequivocally demonstrated in mammals. As these hormones are involved in reproduction, it is possible that they may have a special role in assuring the additional supplies of calcium that are needed at such times. Growth hormone can increase the absorption of calcium across the intestine, especially in growing animals. In rats, this effect appears to result from an increased formation of $1,25(OH)_2D_3$ by a facilitation of the action of the 1α-hydroxylase (Spencer and Tobiassen, 1981).

A number of other metabolites of vitamin D_3 have been identified, including 24,25-dihydroxycholecalciferol and $1\alpha,24,25$-trihydroxycholecalciferol (Schnoes and DeLuca, 1980). These sterols are formed in the kidney under the influence of a 24-hydroxylase enzyme. They generally have a lower biological activity than $1,25(OH)_2D_3$. The $24,25(OH)_2D_3$ has been shown to influence bone metabolism in humans and it has a longer duration of action than $1,25(OH)_2D_3$ (Kanis $et\ al$., 1978; Ornoy $et\ al$., 1978). Whether such metabolites of vitamin D_3 also have physiological roles is uncertain (Fraser, 1980). They could, for instance, be supplementing the action of $1,25(OH)_2D_3$ or their observed effects may merely be pharmacological curiosities. However, special hormonal functions (such as ones involving distinct morphological sites, durations of effect, and roles in different species) are all possibilities. One cannot avoid being reminded of the somewhat belated discoveries of the important physiological roles of T_3 and aldosterone subsequent to the establishment of the functions of their respective homologs, T_4 and cortisol.

Apart from its actions on calcium metabolism $1,25(OH)_2D_3$ has effects on many tissues, where it can influence cell growth, proliferation, and differentiation (Hewison, 1992; Norman, 1994). Such actions involve the hematopoietic and immune systems, as well as skin and pancreatic B-cells. The responses utilize receptors for $1,25(OH)_2D_3$ that have a ubiquitous presence in numerous tissues, where they may be involved in processes of "gene regulation and genetic circuitry" (Minghetti and Norman, 1988).

Mechanisms and interactions of parathyroid hormone, calcitonin, and vitamin D_3 on calcium metabolism

Regulation of the calcium levels in vertebrates usually depends on the interactions of four effectors: bone, intestine, kidney and, in fish, the gills. These respond to various combinations of parathyroid hormone, calcitonin, and hormonal metabolites of vitamin D; in fishes stanniocalcin is also involved. The most detailed information about the mechanism of action of vitamin D_3

on calcium metabolism is, probably, available on the intestine. The observations mainly encompass experiments on both mammals and birds (Hurwitz, 1989; Nemere and Norman, 1991) and, more recently, on teleost fish (Flik *et al.*, 1993).

Following its two-step activation in the liver and kidney $1,25(OH)_2D_3$ enters its effector intestinal epithelial cells and combines with its nuclear receptors (Chapter 4). Such receptors, which are associated with the chromatin, have been identified in many other types of tissue, especially in mammals and birds but also in teleost fishes (Marcocci *et al.*, 1982). These binding sites, apart from the intestine, are present in kidney, bone, pancreas, the avian shell gland, brain, and teleost gills. Such a distribution suggests that vitamin D_3 has a ubiquitous physiological role not necessarily confined to regulating calcium metabolism. Its known activities with respect to calcium include organs such as intestine, kidney, bone, and gills (see Figs. 6.3 and 6.4). The effect of vitamin D_3 on calcium absorption from the intestine was probably the first of its actions to attract detailed attention. The domestic fowl has considerable need for calcium during the stage of its life cycle that involves egg laying and it provided the first major experimental preparations. Mammals, especially rats, were used a little later.

The precise nature of the mechanisms of calcium transport across the intestinal cell membranes, from the lumen contents to the extracellular fluid, is still somewhat controversial. It is, however, generally agreed that transit of calcium across the mucosal epithelial cells involves a three-step process.

1. Entry into the cells occurs across the apical plasma membrane, down an electrochemical gradient (the concentration of free calcium ions in the cell cytoplasm is only about 10^{-7} M). The precise nature of this process is, however, not clear and could involve a movement of the ions through aqueous pores, an interaction with a "carrier" molecule, or a rearrangement of membrane lipids.

2. Transit across the cytoplasmic compartment to the basolateral plasma membrane poses possible problems. The required maintenance of low concentrations of free calcium suggests that the ion will be sequestered (such as in mitochondria and the endoplasmic reticulum) or that it will be bound to calcium-binding proteins or enveloped by lysozomes. Vitamin D_3-induced calcium-binding proteins (CaBP) are present in the cytoplasm and appear to have a role in this process. Based on cytochemical observations and electron probe analyses, it also appears that lysozomes may play an important role in such transcytoplasmic movements of calcium (see, for instance, Davis and Jones, 1982).

3. The basolateral plasma membrane poses a major energetic barrier for such calcium transport as the calcium ions must move against a substantial electrochemical gradient from the cytoplasm to the extracellular

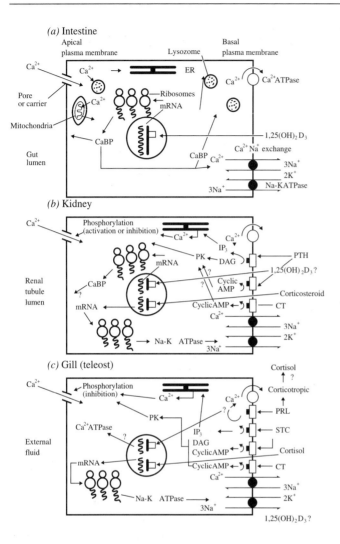

Fig. 6.3. Tentative summaries of the actions of calcemic hormones on the transport of calcium across the epithelial membranes of the intestine (*a*) distal regions of the renal tubule (*b*) and the gills of teleost fish (*c*). These events, are described in more detail in the text. The proposed mechanisms of the effects of the hormones on the gills are the most tentative. CT, calcitonin; DAG, diacylglycerol; CaBP, calcium-binding proteins (calbindins); ER, endoplasmic reticulum; IP$_3$, inositol trisphosphate; PK, phosphokinases A or C, STC, stanniocalcin.

fluid. This process requires a linked expenditure of energy. Two mechanisms are involved; a Ca^{2+}-activated ATPase (Ca-ATPase) "pump" and a sodium–calcium ion exchange process. The latter exchanges three sodium ions from the extracellular fluid for a single calcium ion from the cytoplasm. The energy for this process is ultimately derived from the sodium concentration gradient across the plasma membrane, which is established by the activity of the Na–K ATPase ("sodium pump") mechanism.

Calcium may also move by diffusion across the intestinal epithelium by following the paracellular route between the cells. This ion may occur in either direction and, depending on the concentration gradient for calcium, can even result in a net rate of calcium absorption into the extracellular fluid.

Fig. 6.4. A summary of the effects of calcemic and other hormones on calcium exchanges in bone. For more details see the text. AC, adenylate cyclase; CaBP, calcium-binding protein; CT, calcitonin; DAG, diacylglycerol; IP$_3$ inositol trisphosphate; PLC, phospholipase C; PTH, parathyroid hormone; + facilitation; − inhibition.

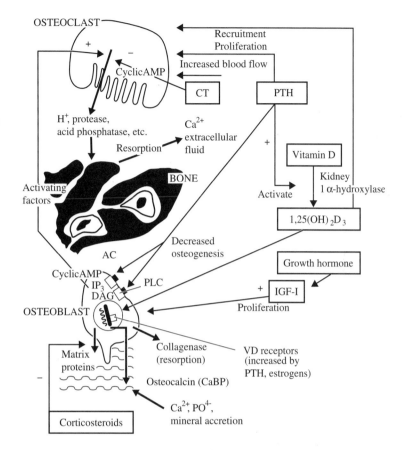

It does not appear to be a process subject to hormonal regulation but it could be influenced by such events as cell damage, and swelling and shrinkage.

The site of action of 1,25(OH)$_2$D$_3$ on active calcium transport across the intestinal epithelial cells may involve any, or even all, of the three main steps described above and involve the apical plasma membrane, the basal plasma membrane, and movement through the interconnecting cytoplasm.

In 1966, Wasserman and Taylor working with chicken intestinal preparations identified a protein that increased in concentration in the presence of vitamin D$_3$. This important observation has provided our main concepts about the mechanism(s) of action of vitamin D$_3$, as this protein is clearly induced in a manner that parallels changes in the calcium transport process. Several such proteins have been identified and they are called CaBPs or calbindin D. The principal avian protein has a molecular mass of 28 kDa and the predominant mammalian one 9 kDa. Such proteins have a widespread distribution in the body, including intestine, kidney, bone, uterus, and avian eggshell gland. By 1992 (Nemere and Norman, 1991) eight avian and 11 mammalian tissues had been identified that contained such CaBPs. Calbindin D$_{28K}$ appears to be main such avian protein and it is also present in the brain of mammals. Calbindin D_{9K} is otherwise the principal one of these proteins present in mammals. Osteocalcin is another vitamin D_3-inducible protein; it is present in the matrix of bone. Calbindin D_{9K} in the rat uterus can be induced by estradiol but not vitamin D_3 (L'Horset *et al.*, 1990). The avian uterine shell gland forms calbindin D_{28K} and this process may also occur independently of vitamin D_3 (Bar *et al.*, 1992; 1996). Calcium and estrogens may contribute to its induction but other factors appear to be involved. (Under *in vitro* conditions $1,25(OH)_2D_3$ has been shown to influence its synthesis and estrogens may act as a coregulator (Corradino, 1993).) Calbindins with molecular masses of 27 to 28 KDa have been identified in various tissues in reptiles, amphibians, and teleost fish (Parmentier *et al.*, 1987). In the fish, the calbindin appeared to be confined to the brain, resulting in the suggestion that it may have originally evolved as a neuronal protein. Calbindin D_{28K} and calbindin D_{9K} have structures that are unrelated and appear to have evolved separately.

Calbindin D has been localized in several regions of the chicken intestinal epithelial cell (Thorens *et al.*, 1982), including the apical plasma membrane, the cytoplasm, and the nucleus. There are several suggestions regarding its possible role in transepithelial calcium transport:

- a facilitation of the diffusion of calcium across the cytoplasm (Bronner, 1996)
- an association with vesicles associated with the entry of calcium ions into the cell across the apical plasma membrane
- a buffering, or protective, effect by maintaining low concentrations of free, ionized calcium in the cytoplasm.
- an activity, as yet undefined, as a metabolic mediator (perhaps analogous to the role of calmodulin) in the regulation of cell metabolism

These observations were admirably reviewed by Nemere and Norman in 1991.

Parathyroid hormone only has an indirect influence on calcium transport across the intestine. It contributes to the activation of vitamin D_3 by promoting the action of 1α-hydroxylase in the kidney. Calcitonin appears to have no role in calcium transport in the intestine. Stanniocalcin has been observed to inhibit calcium absorption from the intestine (*in vitro*) in a teleost fish (p. 299).

Calcium transport across other types of epithelium that contribute to calcium homeostasis appear to conform to the general three-step process described above (Fig. 6.3), but as they often respond to different hormones their control mechanisms differ.

Rapid responses to $1,25(OH)_2D_3$, which occur within a few minutes, have been described in the intestine and other tissues (de Boland and Nemere, 1992). These effects do not appear to involve nuclear receptors. Such actions of the hormone may involve perturbational "membranophilic" effects or an interaction with plasma membrane-associated receptors. Possible transduction mechanisms include G proteins and the activation of protein kinases and calcium channels. The physiological significance of such effects remains to be assessed.

Kidney tubules

In mammals, more than 95% of the calcium in the glomerular filtrate is reabsorbed across the renal tubular epithelium (Brown, 1992; Costanzo and Windhager, 1992). Two-thirds of this conservation occurs across the proximal renal tubule owing to passive diffusional movements of the calcium ions. Another 20% of the calcium is absorbed across the thick ascending loop of Henle. The remaining 10% is subject to hormonal regulation of its reabsorption in the distal regions of the renal tubule. Some such control may also exist in the loop of Henle, but this segment is only present in mammals and birds. Excretion of filtered calcium is relatively greater in reptiles and amphibians, a deficiency possibly reflecting the lack of this renal tubular segment (Laverty and Clark, 1989). Regulation of calcium absorption from the renal tubule mainly involves the action of parathyroid hormone but $1,25(OH)_2D_3$ and calcitonin can also influence this process. The mechanism of calcium transport across the renal tubule is basically similar to that which occurs across the intestine (Fig. 6.3) and appears to involve passive entry of calcium ions across the apical plasma membrane into the cytoplasm followed by their active extrusion across the basolateral plasma membrane. CaBPs have been identified in renal epithelial cells of various species, including amphibians and reptiles (Laverty and Clark, 1989).

The action of parathyroid hormone in stimulating reabsorption of calcium across the distal renal tubule involves an activation of adenylate cyclase and the formation of cyclic AMP. However, a number of observations (see, for instance, Meltzer *et al.*, 1982) indicate that the phospholipase C–phosphatidylinositol pathway is also stimulated by parathyroid hormone. Activation of both protein kinase A and protein kinase C appears to mediate such renal calcium ion absorption (Friedman *et al.*, 1996). This process of calcium reabsorption has been associated with a rise in intracellular calcium levels owing to its influx across the apical plasma membrane (Costanzo and Windhager, 1992) and its extrusion across the basolateral plasma membrane. The entry of the calcium ions into the cell is mediated by the action of the parathyroid hormone, apparently as a result of the activation of a protein kinase, but details of the mechanism are not available. Parathyroid hormone may also be having a more direct cyclic AMP-dependent effect by stimulating the activity of the calcium–sodium exchange mechanism on the basolateral plasma membrane (Bouhtiauy, Lajeunesse, and Brunette, 1991).

Even less is known about the actions of calcitonin and $1,25(OH)_2D_3$ in the kidney. The latter appears to promote calcium reabsorption in the distal renal tubule (Brown, 1992) where the presence of calbindin has been observed. Calcitonin has inconsistent effects on the renal excretion of calcium and in mammals it may have either hypocalciuric or hypercalciuric actions (Costanzo and Windhager, 1992). The latter is observed most often. Calcitonin appears to act on the thick ascending loop of Henle, where it can activate adenylate cyclase. The precise role of this enzyme in the response is, however, not clear. Calcitonin and $1,25(OH)_2D_3$ have prominent effects on bone while the latter also acts on the intestine. Responses at these sites may be predominant and influence losses of calcium to the urine. Differences in such hormone-mediated responses of the kidney appear to occur among the vertebrates.

The mobilization and resorption of calcium from *bone* involves changes in the activity of the osteoblast and osteoclast cells (Herrmann-Erlee and Flik, 1989; Brown, 1992). These processes are influenced by various hormones, and also by local factors, or regulators, that are produced by bone, cartilage, and blood cells. Such regulators include prostaglandins, tumor necrosis factors (TNF), transforming growth factors (TGF), and IGFs. They may exert paracrine and autocrine effects that may be modulated by hormones. The osteoblasts appear to secrete collagenase, which may initiate the absorption of calcium, while the osteoclasts secrete acids, hydrolases, and an acid phosphatase that perpetuate the resorptive process (Fig. 6.4). Parathyroid hormone and $1,25(OH)_2D_3$ facilitate the recruitment and proliferation of the osteoclasts. This action of the parathyroid hormone may be inhibited by estrogens (Kaji *et al.*, 1996). Parathyroid hormone also decreases the osteogenic activities of the osteoblasts while estrogens appear to promote their proliferation (Erikson *et al.*, 1988). The action of parathyroid hormone on bone involves both an activation of adenylate cyclase and phospholipase C and the formation of the corresponding second messengers. Parathyroid hormone promotes the resorptive activities of the osteoclasts but its effect appears to be partly an indirect one through the release of "activators" from the osteoblasts. However, a direct effect of this hormone on the osteoclasts has also been demonstrated (Datta *et al.*, 1996). The $1,25(OH)_2D_3$ interacts with its nuclear receptors in the osteoblasts (Christakos and Norman, 1978) and promotes the formation of bone proteins including osteocalcin and, probably, collagenase. Parathyroid hormone and estrogens enhance the response of the osteoblasts to $1,25(OH)_2D_3$ by increasing the numbers of its receptors (Liel *et al.*, 1992; Krishnan *et al.*, 1995). Calcitonin directly inhibits the calcium resorptive activities of the osteoclasts, an effect which is mediated by cyclic AMP (Wong, Luben, and Cohn, 1977). Apart from increasing bone growth, the IGFs can also promote the resorptive activities of the osteoclasts. Their action is an indirect one that is mediated by the osteoblasts. The latter also respond directly to growth hormone, which promotes their proliferation and differentiation (Slootweg *et al.*, 1996). The process of resorption of calcium from bone, therefore, involves the complex interactions of a variety of hormones and regulators, some of which are summarized in Fig. 6.4.

Cortisol and its more potent synthetic analogs, when administered to humans, may cause a loss of bone mineral. The reasons for this clinical side effect of these steroids appear to be multiple. There may be a loss of the protein matrix of bone, a decreased absorption of calcium from the gut, and an increased excretion of calcium in the urine. In the gut of chicks there is a decrease in the $1,25(OH)_2D_3$-induced CaBP (Feher and Wasserman, 1979) following the administration of cortisol.

Fish gills

The gills are a major permeable interface between the fish and their bathing fresh water or sea water. Apart from respiratory gas exchanges, water and solutes can often quite readily cross this integumental barrier. Such exchanges include those of calcium and can be subject to physiological regulation. Fish living in the sea are exposed to a solution with a relatively high concentration of calcium ions compared with that in their body fluids so that they suffer from the potential risk of a diffusional hypercalcemia. By comparison, most fresh water has a quite low concentration of calcium so that losses from the body fluids may then be expected across the gills. Teleost fish can limit such exchanges of calcium and actively regulate its uptake and excretion by the gills. The site of such calcemic activities is the branchial chloride cell, which is also involved in the regulation of exchanges of monovalent ions (Chapter 8). The mechanism of the transfer of calcium across the gill epithelial cell appears to be similar to that already described for the intestine and kidney (Fig. 6.3) and involves the dual calcium ion extrusion mechanisms situated in the basolateral plasma membrane of the chloride cells. These are the Ca-ATPase and the sodium–calcium ion exchange mechanism. The latter appears to be more important in the teleosts gills when the cytoplasmic concentration of calcium exceeds 10^{-6} M (Flik *et al.*, 1993). The activities of the chloride cells and the properties of the paracellular pathways across the branchial epithelium appear to limit unregulated diffusional influx and efflux of calcium ions. Several hormones appear to be involved in regulating these activities of the chloride cells (Smith, 1956). Stanniocalcin secreted by the corpuscles of Stannius of bony fishes (Chapters 2 and 3) has a hypocalcemic action. Several other hormones are also involved in the regulation of calcium exchanges across the gills of such fish, including calcitonin, cortisol, prolactin, and, possibly $1,25(OH)_2D_3$. There are clearly a plethora of calcemic hormones that may act on the gills of fish. The circumstances when they may act, however, vary and depend on the species, the composition of its environmental solution, the physiological condition of the fish, and the manipulations, such as the dose of an administered hormone, utilized by the observer.

In sea water the hypocalcemic effects of stanniocalcin would appear to be paramount. This hormone appears to act to limit the rate of transfer of calcium across the apical plasma membrane of the chloride cell. Information about its precise mechanism of action is fragmentary, but based on that available (Flik *et al.*, 1993) it appears to involve interactions with receptors for adenylate cyclase and phospholipase C that are situated in the basolateral plasma membrane. A choice of second messengers may be available that includes cyclic AMP, IP_3 and DAG. The final effect, possibly involving protein kinases and phosphorylation of effector proteins, is predicted to be a blockade of pores utilized for the entry of calcium into the cells. Calcitonin, which has a hypocalcemic effect in mammals, is also present in fish (Chapter 3) but its actions on their calcium metabolism are rather unpredictable. However, calcitonin receptors have been identified in the gills of teleosts (Fouchereau-Peron *et al.*, 1981). This hormone may act in a parallel manner to stanniocalcin. Such a duplication of hormonal effects is, however, difficult for some to accept.

Hypophysectomy may result in hypocalcemia in teleosts (p. 300). The hormonal deficiency in such fish can be corrected by the administration of mammalian prolactin (Flik *et al.*, 1986). Homologous teleost prolactins have also been shown to have a hypercalcemic effect in teleosts (Flik, Rentier-Delrue, and Wendelaar Bonga, 1994). This novel action of prolactin has been associated with an enhanced activity of branchial Ca-ATPase and a proliferation of the chloride cells (see, also, Flik *et al.*, 1986; 1993). Cortisol has been shown to have a similar hypercalcemic action in teleosts (Flik and Perry, 1989). It is, therefore, considered possible that prolactin may be acting by eliciting a corticotropic effect (Flik *et al.*, 1994). It is also noteable that cortisol has a well-established ability to induce formation of Na–K ATPase, which in gills contributes to the maintenance of the gradient of sodium concentration that is utilized by the sodium–calcium exchange mechanism. Somatolactin is a novel putative hormone that is secreted by the pars intermedia of the pituitary gland in teleosts (Chapter 3). It has structural similarities to both growth hormone and prolactin and is released in fish placed in solutions with a low concentration of calcium (Kakizawa *et al.*, 1993). Possibly somatolactin has a hypercalcemic action and emulates prolactin.

A calcemic role for vitamin D_3 in fish has been controversial but when it is administered to teleosts it may have a hypercalcemic effect. The hormone $1,25(OH)_2D_3$ has such an action in Atlantic cod and the response may involve the gills (Sundell, Norman and Björnsson, 1993), which are known to possess the appropriate receptors (Sundell *et al.*, 1992).

The gills appear to be responsive to a surfeit of putative calcemic hormones, some of which duplicate the role of others. A sense of endocrine

Fig. 6.5. Summary of the effects of parathyroid hormone, calcitonin and vitamin D_3 on calcium metabolism in mammals. There are three main sites of action: the kidneys, intestine, and bone.

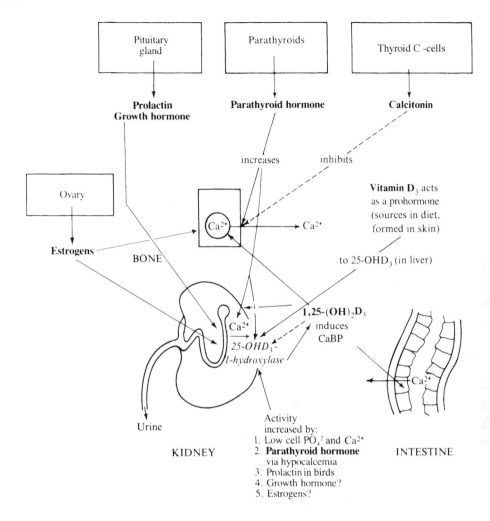

frugality suggests that we must await for some reassessment of the roles of such candidate hormones.

Regulation of calcium metabolism

The hormonal regulation of calcium metabolism as outlined (Fig. 6.5) is based principally on experiments using mammals and, in a few instances, birds. Calcium metabolism in nonmammalian vertebrates, however, may exhibit some interesting differences. Such variations are to be expected on quantitative grounds, as in the hen during its egg laying when relatively vast amounts of calcium are rapidly utilized. Qualitatively predictable differences arise as the parathyroids are absent in fishes, whereas the cyclostomes appear to have neither these glands nor ultimobranchial bodies. The absence of a calcified bony skeleton in the chondrichthyeans and cyclostomes may also be expected to be a matter of some physiological consequence in calcium metabolism.

Phyletic differences in the role of hormones in calcium metabolism

Mammals

Mammalian calcium metabolism has been described in some detail in the preceding section along with some of the anatomical variations in the distribution of the parathyroid tissues. A variety of mammals including human, dog, rat, pig, sheep, and goat have been examined, and the regulation of calcium levels in the blood is related to the activities of the parathyroids and the thyroid C-cells. The former have a vital role to play, as mammals deprived of these tissues suffer tetanic seizures because of a hypocalcemia. The rachitic effects of vitamin D deficiency on bone are also well known though species differences in vitamin D_3 requirements exist. It is, however, not clear how physiologically essential the thyroid C-cells normally are in mammals. During calcium stress, when large increases in blood calcium levels occur, calcitonin undoubtedly facilitates the homeostatic adjustment of the concentration of this ion; however, the role of calcitonin in the regulation of the smaller and more usual changes in calcium concentration is uncertain. Species differences appear to exist.

African mole-rats (Rodentia; family Bathyergidae) lead an entirely subterranean fossorial existence, away from sunlight, consuming a diet of roots, corms, and tubers (see, for instance, Lovegrove and Knight-Eloff, 1988). Such a restricted life-style resulted in speculation regarding their sources of vitamin D_3. Plasma concentrations of this vitamin and $1,25(OH)_2D_3$ were found to be at levels which in other species would clearly indicate rachitic disease. The vitamin D metabolism of two species of these rodents has been studied: the Damara mole-rat (*Cryptomys damarensis*) from the Kalahari desert and the almost hairless naked mole-rat (*Heterocephalus glaber*) from the Horn of Africa. Both species exhibit low plasma levels of vitamin D_3 and its metabolites but they possess the normal complement of vitamin D hydroxylases and the hormone receptors (Buffenstein, Sergeev, and Pettifor, 1993; Sergeev, Buffenstein, and Pettifor, 1993; Pitcher, Sergeev, and Buffenstein, 1994). Supplementation of their diets with vitamin D_3 or exposure to sunlight elevates the plasma concentrations of $1,25(OH)_2D_3$ but makes little or no difference to their bodily calcium metabolism. In their normal state, they exhibit no overt signs of a deficiency (Buffenstein and Yahav, 1991; Buffenstein *et al.*, 1991; Pitcher *et al.*, 1994). These rodents can apparently maintain adequate bodily levels of calcium by utilizing processes that require little or no vitamin D_3. They exist on a high-calcium diet, and in the Damara

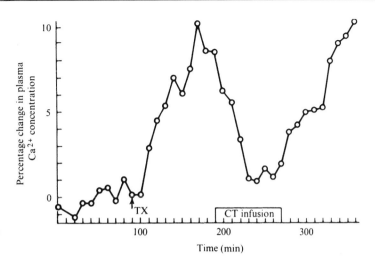

Fig. 6.6. The observed increase in plasma calcium concentration in young pigs, following removal (TX) of the thyroid (and its contained calcitonin-secreting C-cells). Infusion of calcitonin (CT infusion) restored the calcium concentration to normal, but as soon as this ceased the levels climbed again. After about a day, the calcium levels returned to normal, probably as a result of other physiological adjustments, including a decline in secretion of parathyroid hormone. (From Swaminathan *et al.*, 1972.)

mole-rat the intestinal absorption of calcium ions has been found to be a highly efficient process that involves a nonsaturable diffusion mechanism which is independent of vitamin D_3 (Pitcher *et al.*, 1992; 1994). The functioning of other vitamin D_3-dependent sites, especially bone, does not appear to have yet been described.

The question of the normal physiological role of calcitonin in mammals has been examined in young pigs (Swaminathan, Bates and Care, 1972). Removal of the thyroid results in a rapid rise (in 1–2 hours) of the plasma calcium concentration. This hypercalcemia is presumed to be the result of a lack of calcitonin, as it can be corrected by infusing small amounts of this hormone into the animal (Fig. 6.6). It is notable, however, that when these thyroidectomized pigs were allowed to recover, without injections of calcitonin, blood calcium levels returned to normal after 24–48 hours. This recovery is probably the result of an adjustment in the rate of secretion of parathyroid hormone. In adult rats on a normal diet, the evidence to indicate that calcitonin has a physiological role is contradictory (Harper and Toverud, 1973; Kumar and Sturtridge, 1973). Injected antibodies to calcitonin were, however, shown consistently to increase plasma calcium concentration both before and after eating (Roos *et al.*, 1980). The results are consistent with calcitonin having a physiological role in regulating plasma concentration in rats under normal pre- and postprandial conditions.

Stanniocalcin has until recently been considered to be a solely piscine hormone (Chapters 2 and 3). However, it has also been identified in the plasma and several tissues of mammals, including the thick ascending loop of Henle and the distal tubules of the kidney. It may be exerting a local regulatory action on ion transport at these sites (Haddad *et al.*, 1996). Human stanniocalcin has been shown to promote renal phosphate reabsorp-

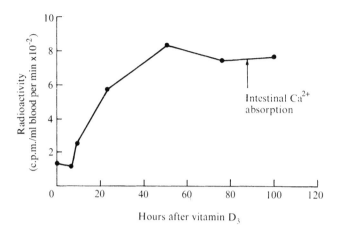

Fig. 6.7. The effect, in young chicks, of an oral dose of vitamin D_3 on calcium absorption from the intestine. The chicks were on a low-calcium diet. (Modified from Harmeyer and DeLuca, 1969.)

tion in rats (Olsen *et al.*, 1996). Such an effect could enhance the deposition of calcium in bone.

Birds

Birds, especially during their egg-laying cycle, have a very high rate of calcium turnover (Hurwitz, 1989). A domestic hen may then utilize an amount of calcium equivalent to 10% of that in its body each day. Most of this calcium is derived directly from the food. When compared on a unit body-weight basis, the domestic hen during egg laying absorbs calcium across its intestine 100 times more rapidly than a human. Vitamin D, as in mammals, increases the rate of calcium absorption from the intestine of the domestic fowl (Fig. 6.7). At other times, such as during the nocturnal fast, calcium is also mobilized from the bones, principally from medullary bone rather than cortical bone. This distinction is of some endocrine importance as estrogens can influence the turnover of calcium in medullary bone whereas parathyroid hormone only acts on cortical bone. Calcium transport by the eggshell gland of quail appears to be stimulated by $1,25(OH)_2D_3$, which is, therefore, a novel effector organ for this hormone (Bar and Norman, 1981).

As described in Chapter 5, estrogens facilitate the formation of vitellogenins by the liver of birds. They appear in the blood, from which they are incorporated into the yolk of the developing eggs. These proteins can bind calcium and hence their presence is associated with elevation of blood calcium concentrations. This interesting hypercalcemic effect of estrogens is prominent in birds as well as other oviparous "bony" vertebrates but is absent in mammals, which may be related to their viviparity and lack of vitellogenins. Such a response is especially appropriate to the needs of vertebrates that produce large megalecithal eggs, which contain a lot of calcium.

Vitamin D metabolism has been extensively studied in domestic chickens. Prolactin and estrogens have hypercalcemic actions in these birds and, as

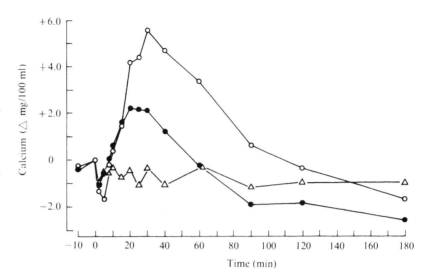

Fig. 6.8. The hypercalcemic effects of parathyroid hormone injections into female domestic fowl. The response was enhanced in those birds receiving a higher supplement of dietary calcium, suggesting that this hormone may be increasing intestinal calcium absorption or the diet decreases basal PHT levels: ○, 31 units PTH/kg, 5.00% dietary calcium; ●, 31 units PTH/kg. 2.26% dietary calcium; △, control. (From Mueller *et al.*, 1973a.)

described earlier in this chapter, these effects may result from an increased formation of $1,25(OH)_2D_3$ in the kidney (Christakos *et al.*, 1981). It has also been observed that ovulation in Japanese quail is associated with an increased synthesis of $1,25(OH)_2D_3$ over the next day (Kenny, 1976). The eggs of hens raised with $1,25(OH)_2D_3$ as their only source of vitamin D fail to hatch (Henry and Norman, 1978). If, however, these birds are also given $24,25(OH)_2D_3$ hatchability is normal, suggesting that this metabolite of vitamin D_3 may have a special role in the nutrition of laying hens.

Both the parathyroids and the ultimobranchial bodies hypertrophy in egg-laying hens, suggesting that they are involved in regulating the calcium turnover in such birds. However, the precise contribution of each of these glands to avian calcium metabolism is still not clear.

Removal of the parathyroids results in hypocalcemia in birds and this is particularly dramatic in young, growing chicks. Injections of parathyroid hormone elevate blood calcium concentrations through an effect on cortical bone. Injected parathyroid hormone has an extremely rapid hypercalcemic action in the laying hen (Fig. 6.8); it acts six to eight times more rapidly than in the dog (Mueller *et al.*, 1973a,b). This rapid initial phase of the response in these birds cannot readily be related to an increased activity of the osteoclasts, suggesting that a dual mechanism of action of parathyroid hormone may exist. Parathyroid hormone preparations when injected have been shown to increase the blood supply to bone and this effect could be important for the rapid and massive mobilization of calcium that occurs during egg laying in birds. Alternatively, an inhibition of the rate of accretion of calcium into bone may be involved (Kenny and Dacke, 1974).

The parathyroid glands have also been shown to influence urinary calcium and phosphate excretion in birds. Parathyroidectomy in starlings results in a

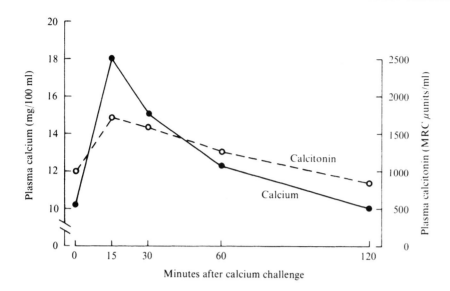

Fig. 6.9. The increase in calcitonin level in the plasma of Japanese quail given intraperitoneal injections of calcium (50 mg/kg). (Modified from Boelkins and Kenny, 1973.)

marked increase in calcium excretion, whereas phosphate excretion declines (Clark and Wideman, 1977). Injected parathyroid hormone rapidly corrects these deficiencies. It appears that in birds the kidney may play an important role in the actions of parathyroid hormone on calcium and phosphate metabolism, but clearly more species need to be studied.

Calcitonin has been identified in the blood of several species of birds including pigeon, goose, duck, domestic fowl, and Japanese quail (Kenny, 1971; Boelkins and Kenny, 1973). The plasma concentrations of calcitonin are much higher in these birds than normally observed in mammals and they can be increased by injecting calcium solutions (Fig. 6.9). In immature Japanese quail, the increase can be as great as 15-fold. It is, therefore, rather surprising that it has not been possible to demonstrate a physiological role of the ultimobranchial bodies in normal calcium metabolism in birds. Injections of calcitonin usually fail to elicit a hypocalcemia, but it is possible that this reflects a high basal rate of calcium turnover and a rapid compensatory release of the endogenous parathyroid hormone. It seems possible that calcitonin could be especially important during egg laying when it could reduce excessive oscillations in blood calcium levels and possibly even influence calcium deposition in the egg. The blood calcitonin levels, however, do not change during the egg-laying cycle of the domestic hens, nor does ultimobranchialectomy significantly influence calcium metabolism in these birds (Speers, Percy, and Brown, 1970). In addition, this operation does not significantly influence the development of the skeleton in growing chickens; therefore, although some people believe that the ultimobranchial bodies may have a role in regulating calcium metabolism in birds this has not been established. Possibly, calcitonin has some other physiological role.

Reptiles

Parathyroidectomy results in hypocalcemia and tetanic muscular contractions in several species of snakes and lizards (Clark, 1972) and a chelonian and crocodilian (Oguro, Tomisawa, and Matuoka, 1974; Oguro and Sasayama, 1976). Such effects are, however, difficult to demonstrate in turtles though a small decline in plasma calcium concentration has been demonstrated following removal of the parathyroids of the Japanese turtle *Geoclemys reevesii* (Oguro and Tomisawa, 1972). The sites of action of parathyroid hormone in reptiles are uncertain. It does not influence urinary calcium excretion in snakes or turtles (Clark and Dantzler, 1972; Laverty and Clark, 1981). Parathyroid hormone has, however, been observed to increase the mobilization of calcium from bone in young turtles (Bélanger, Dimond, and Copp, 1973). A renal effect of parathyroid hormone is also apparently lacking in amphibians (p. 294). It is possible that such an effect of this hormone did not emerge until the evolution of the birds.

Calcitonin is present in the ultimobranchial glands of reptiles and it has been identified in their plasma (Kline and Longmore, 1986). Attempts to demonstrate a mammalian-type hypocalcemia following the injection of this hormone have been largely unsuccessful in a variety of reptiles, including turtles, lizards, and snakes (Copp and Kline, 1989). However, using salmon calcitonin, a hypocalcemic effect has been observed in *Iguana iguana* and in young chuckwallas, *Sauromalus obesus* (Kline, 1981; 1982). Calcitonin failed to influence renal calcium excretion in water snakes (*genus Natrix*) or the lizard *Dipsosaurus dorsalis* (Kiebzak and Minnick, 1982). Reptiles, like birds, exhibit a hypercalcemic response to injected estrogens (Clark, 1967).

Reptiles have common phyletic affinities to mammals as well as to birds so that their endocrine function is of rather special interest. The reptiles share with birds the problems associated with the production of large cleidoic eggs, which in many species are covered with a calcareous shell. It should, therefore, not be surprising to observe similarities in their calcium metabolisms. Reptiles, however, are poikilotherms and this may influence the relative importance of processes that are involved in the metabolic coordination of calcium levels as the speed of the adjustments need not be as great. Reptiles appear to lack significant hormonal regulation of urinary calcium excretion. Little information is available about the roles of the calcemic hormones on reptilian bone or intestine and it is possible that the regulation of body calcium levels resides principally in these tissues.

Amphibians

The amphibians have a special position with regard to our understanding of vertebrate calcium metabolism, as it is in this group that the parathyroid glands first appear on the phyletic scale. The ultimobranchial bodies, however, have persisted from their piscine ancestors. The information that is available is relatively sparse and is not consistent with the special interest that these animals deserve.

Parathyroidetomy has a hypocalcemic effect in a variety of amphibians including frogs and tadpoles, and several urodeles (Cortelyou, 1967; Bentley, 1984). This effect is often quite transient and plasma calcium concentrations return to normal in a few days. In the frog *Rana clamitans* and several newts there is, however, a long-term hypocalcemia that may result in tetany (Oguro and Sasayama, 1978; Wittle and Dent, 1979; Wittle, Augostini, and Chizmar, 1990). The Japanese giant salamander, *Megalobatrachus davidianus* (Oguro, 1973), is, however, unresponsive to parathyroidectomy, and this is reminiscent of early observations in three other species of urodeles in which this operation failed to induce tetany. The giant salamander appears to have parathyroid hormone present in its parathyroids, but it has been suggested that this salamander lacks a target-organ system for parathyroid hormone and this may also be so in some other urodeles. Some urodeles, such as the mudpuppy *Necturus maculosus*, even lack parathyroid glands. Calcium excretion in the urine of amphibians does not appear to be influenced by parathyroid hormone (see, for instance, Cortelyou, 1967; Bentley, 1983; Sasayama and Clark, 1984). Its principal site of action would appear to be bone cells. Removal of the ultimobranchial bodes in *Rana pipiens* results in an initial elevation of the calcium concentration in the plasma that subsequently subsides and is replaced by a hypocalcemia (Robertson, 1969a,b; 1975). Ultimobranchialectomized bullfrog tadpoles display deficiencies in their abilities to maintain plasma calcium concentrations when kept in solutions containing elevated concentrations of calcium (Sasayama and Oguro, 1976). Calcitonin has been identified in the plasma of *Rana pipiens* and the concentration was increased when these frogs were kept in solutions with high calcium levels (Robertson, 1987). Injected calcitonin results in a hypocalcemia in *Rana tigrina* (Srivastav and Rani, 1989) but it had no effect on urinary excretion or plasma concentrations of calcium in the toad *Bufo marinus* (Bentley, 1983). In bullfrog tadpoles, the uptake of calcium from their bathing solutions was reduced by injected salmon calcitonin (Baldwin and Bentley, 1980). This effect appeared to reflect an action on the gills, as has been observed in teleost fish. Parathyroid hormone had no effect on uptake of calcium by these tadpoles but $1,25(OH)_2D_3$ promoted accumulation.

Vitamin D_3 administration has a hypercalcemic effect in *Rana pipiens* (Robertson, 1975) and *Bufo marinus* (Bentley, 1983). Vitamin D_3 can increase the absorption of calcium across the intestine of *Rana pipiens* but as the frogs and toads used above were fasting it seems likely that the response reflects a mobilization of calcium from bone.

Anuran amphibians (*Rana pipiens* has been principally studied) possess "lime sacs" that are novel sites for the storage of calcium in the body (Robertson, 1969a,b). These organs are extensions of the lymph sacs; they extend caudally along the vertebral canal and emerge between the vertebrae. They contain calcium carbonate, instead of calcium phosphate as in bone, and this exhibits a mobility that includes an added storage following the administration of calcium chloride or vitamin D. Such treatment, which produces a hypercalcemia, also results in the hypertrophy of the ultimobranchial bodies (Robertson, 1968). The lime sacs are also present in tadpoles and may constitute a storage site for calcium that can be utilized at metamorphosis. Ultimobranchialectomy limits the tadpoles ability to accumulate calcium in the lime sacs (Robertson, 1971).

The fragmentary information available suggests that amphibians utilize some calcemic hormones. However, like in reptiles, a renal response to parathyroid hormone is in doubt. They possess some novel tissues that may influence their calcium metabolism, such as a permeable skin, lime sacs, and gills in larval forms. As we shall see in the next section, lungfishes, which are phyletically akin to amphibians, do not appear to respond in this manner to any of these hormones.

The fishes

A considerable diversity in the mechanisms for the regulation of calcium metabolism is not unexpected in the fishes. This may result from their great phyletic diversity, their ability to live in aqueous environments containing high (sea water) or low (most fresh water) levels of calcium, the possession of either a bony or a cartilaginous skeleton, and the presence or absence of the ultimobranchial bodies and the corpuscles of Stannius.

Paleontological evidence suggests that the ostracoderms, which were the jawless ancestors of modern fishes, had a bony dermal exoskeleton and, sometimes, also a bony endoskeleton that had the appearance of a tissue that functions as a store for calcium (Copp, 1969; Simmons, 1971). These fishes lived in fresh water, where the calcium concentration was, presumably, low so that calcium storage in the bones may have been physiologically important. When such ancestral fishes returned to the sea, where there was an unlimited quantity of calcium, they lost their bony skeleton, and this situation persists in present-day cyclostomes and chondrichthyeans. We can extend this speculation (and that is all it is!) and consider whether or not such vertebrates possessed an endocrine system for controlling calcium concentration in the body. In extant species of cyclostomes and chondrichthyeans, there is as yet no evidence for such a control mechanism. In the ancestral fresh water ostracoderms such a control could have been more important and may possibly have occurred. It is even possible that the chondrichthyean ultimobranchial bodies represent a survival from those times. The ancestral fresh water ostracoderms need not, however, have possessed such a system for controlling calcium metabolism, as there is, for instance, no evidence for the endocrine control of such processes in some contemporary fresh water fishes like lampreys and the lungfishes (Urist, 1963; 1976; Urist *et al.*, 1972).

Many cyclostome fishes live in the sea but others spend their entire life in fresh water. Lampreys migrate from the sea into fresh water where they survive for several months without feeding and produce large numbers of eggs. Cyclostomes are considered to regulate calcium levels by utilizing a so-called open system whereby calcium transfer takes place across the membranes of the body, such as in the intestine, gills, and kidney. The absence of the endocrine tissues known to influence calcium metabolism in other vertebrates, however, should not be taken to indicate a lack of such control mechanisms. Further investigation of calcium metabolism of these very interesting fish is clearly to be desired.

Despite some earlier reservations about its physiological role, vitamin D_3 is a necessary dietary constituent for teleost fish maintained by aquaculture (Brown and Robinson, 1992). Hypovitaminosis D results in hypocalcemia while hypervitaminosis D is associated with hypercalcemia. Specific nuclear receptors for $1,25(OH)_2D_3$ have been identified in the gills and intestine of both European eels (Marcocci *et al.*, 1982) and Atlantic cod (Sundell *et al.*, 1992). Such receptors were not found in the kidney of the eels or the bone of either species. Administration of vitamin D_3 or $1,25(OH)_2D_3$ has a hypercalcemic effect in fed fresh water American eels and goldfish (Fenwick *et al.*, 1984; Fenwick, 1984). The responses appeared to reflect an increased absorption of calcium from the intestine. Fasted Atlantic cod, which drink sea water, also exhibit a hypercalcemia in response to the administration of $1,25(OH)_2D_3$ but this effect was not as great as observed in the fresh water teleosts (Sundell *et al.*, 1993). It was suggested that this response was mediated by calcium absorption from the intestine and uptake across the gills but was modified by the better developed hypocalcemic mechanisms that are expected to be present in marine, as compared with fresh water, species.

Sporadic, but persistent, observations have uncovered unexpected disparities between the relative activities of the prohormone vitamin D_3 and its various metabolites in teleost fish. Vitamin D_3 has a greater hypercalcemic effect in eels than does $1,25(OH)_2D_3$ (Fenwick *et al.*, 1984). In contrast vitamin D_3 is ineffective in Atlantic cod (Sundell *et al.*, 1993). Unexpectedly, *in vitro* preparations of intestine from these cod failed to respond to $1,25(OH)_2D_3$ but $25(OH)D_3$ was effective (Sundell and Björnsson, 1990). Another metabolite $24,25(OH)_2D_3$ had an inhibitory effect on the intestinal calcium transport (Larsson, Björnsson, and Sundell, 1995). This action was a rapid, apparently nongenomic response. The intricacies of the metabolism and metabolites of vitamin D_3 in teleost fish may yield some interesting endocrine information and, possibly, novel hormones. The administration of vitamin D_3 to elasmobranch fish did not elicit a calcemic response (Urist, 1962). A South American lungfish (genus *Lepidosiren*) also failed to respond to vitamin D_3 or $1,25(OH)_2D_3$ (Urist *et al.*, 1972). This observation is somewhat surprising in view of the more recent information demonstrating hypercalcemic effects of vitamin D_3 in teleost fish and amphibians. Further studies on elasmobranch fish and lungfishes, including the Australian genus *Neoceratodus*, could yield new information about the evolution of the endocrine role of vitamin D_3 and the control of calcium metabolism of vertebrates.

Information about the regulation of calcium metabolism in cartilaginous chondrichthyean fish is scant. Like most nonmammalian vertebrates, they have ultimobranchial bodies that contain calcitonin. The structure of this peptide in these fish is more similar to that of teleosts, such as salmonids, than it is to that in tetrapod vertebrates (Chapter 3). Calcitonin is present in high concentrations in the plasma of several species of sharks (Glowacki *et al.*, 1985). Early observations using tetrapod calcitonin failed to detect any effects of this peptide on the calcium metabolism of cartilaginous fish. However, salmon calcitonin can produce a modest hypercalcemia in leopard sharks (Glowacki *et al.*, 1985). This contrasted with its more conventional hypocalcemic effect in a teleost, the kelp bass. Subsequent studies on other teleosts (the marine grey mullet and rainbow trout adapted to fresh water and to sea water) have, however, suggested that the calcemic response may be dose dependent. Low doses of salmon calcitonin were hypercalcemic in these teleosts, as seen in the leopard shark (Fouchereau-Peron *et al.*, 1987). Higher, less physiological doses had a hypocalcemic effect. Hypercalcemic responses to calcitonin have also been observed in brown trout (Oughterson *et al.*, 1995). High external environmental calcium concentrations were found to reduce the plasma calcitonin levels in these fish, which would not appear to be consistent with a hypocalcemic role. A similar decline in the plasma calcitonin level has been observed in rainbow trout transferred from fresh water to sea water, though this response was preceded by an increase in its concentration (Fouchereau-Peron *et al.*, 1986). Such observations in Japanese eels and young coho salmon, however, failed to show any changes in plasma calcitonin concentrations (Hirano *et al.*, 1981; Björnsson *et al.*, 1989). It has been suggested that the normal physiological response to calcitonin in fish could be a hypercalcemia rather than the hypocalcemia that occurs among tetrapod vertebrates. A final consensus on this interesting aspect of the evolution of the role of calcitonin remains to be concluded. Possibly calcitonin has other physiological functions in fish.

The bony fishes are a very diverse group and information about the regulation of calcium metabolism is almost entirely limited to teleosts. Within this group, few species have been studied but considerable variability seems to exist. One of the reasons for these differences is that although all teleosteans possess a calcified endoskeleton (Simmons, 1971), in some of these fishes, especially marine species, the bone is acellular so that the calcium deposits are relatively immobile. Other teleosts, usually fresh water species, have cellular bone from which calcium can be mobilized more readily. The scales of teleost fish contain substantial amounts of calcium that can be reutilized by the fish. This cutaneous calcium, however, may vary in its magnitude, as in some species (such as eels and catfish) these integumental appendages are quite reduced.

The responses of teleost fish to administered calcitonin are, as described above, variable and, apart from varying with the dose used, may also depend on the environmental and physiological circumstances of the fish, and the species (see, for instance, Fenwick and Lam, 1988; Singh and Srivastav, 1993). A dipnoan, the South American lungfish *Lepidosiren paradoxa*, failed to respond to calcitonin (Pang and Sawyer, 1975). The calcemic responses of fish to this peptide could involve bone, as seen in mammals, and the transport of calcium ions across the gills. As described above, calcitonin receptors have been identified in the gills of teleosts (Fourchereau-Peron *et al.*, 1981). Calcitonin has also been observed to decrease the influx of calcium across the perfused gills of salmon and eels (Milhaud *et al.*, 1977; Milet, Peignoux-Deville, and Martelly, 1979). Measurements of calcium uptake in intact fish are consistent with such an effect and it could be mediating observed hypocalcemic responses. However, other effectors may exist and those that mediate the hypercalcemic responses are unknown.

In rainbow trout, the calcitonin gene-related peptide (CGRP) concentration in the plasma increases markedly when they are transferred from fresh water to sea water (Lamharzi and Fouchereau-Peron, 1996). This secretion, which is more prolonged than that of its shared gene product calcitonin, reaches a peak at the same time as the maximum plasma calcium level. The binding of CGRP to its specific receptors on the plasma membranes of the branchial cells was also greatest at this time. Possibly CGRP contributes to the subsequent decline in plasma calcium concentration in these fish but its precise role is undefined.

The corpuscles of Stannius that are present in most bony fishes (Chapter 2) also may influence calcium metabolism, and an interaction between these tissues and the ultimobranchial bodies seems to occur. M. Fontaine, in 1964, found that removal of the corpuscles of Stannius in the European eel *Anguilla anguilla* results in a marked increase (1.4-fold) in the plasma calcium concentration. This effect of stanniectomy has been confirmed in other teleosts such as the goldfish *Carassius auratus* and the Asiatic and North American eels *A. japonica* and *A. rostrata* (Chan, 1972). These hypercalcemic effects are accompanied by a reduced calcium excretion in the urine, and an increased osteoclastic activity that mobilizes calcium from the bone. The hypercalcemic effect of stanniectomy can be prevented by the transplantation, or the injection of extracts, of the corpuscles of Stannius into the deficient fishes. The hormone (formerly called hypocalcin) has been purified and is called stanniocalcin (Chapter 3). It has been shown to be an effective replacement and can maintain plasma calcium concentrations in stanniectomized European eels (Hanssen *et al.*, 1989). Antiserum to trout stanniocalcin has been raised in rabbits and when injected into trout, American eels, and guppies blocks the effects of the endogenous hormone and so promotes calcium uptake (Fenwick, Flik, and Verbost, 1995).

The activity (from histological observations) of the corpuscles of Stannius of the killifish *Fundulus heteroclitus*, appears to be greater in normal sea water than in an artificial sea water, where the calcium concentration is low (Pang, Pang, and Sawyer, 1973). Stanniectomy in these fish only results in hypercalcemia when they are bathed by solutions with a high calcium concentration; if they are in artificial calcium-poor sea water or calcium-poor fresh water, blood calcium concentrations are unaffected. Hypercalcemia also appears to promote stanniocalcin release in American eels and rainbow trout (Lafeber and Perry, 1988; Fenwick and Brasseur, 1991). Therefore, the corpuscles of Stannius may play a physiological role in teleosts that live in solutions with a high calcium concentration, such as sea water, where their secretions exert a hypocalcemic action.

Stanniocalcin appears to act principally by decreasing the influx of calcium ions across the gills of teleost fish. However, it could be acting at other sites, as suggested by the observation that it can inhibit calcium absorption from the intestine (*in vitro*) of chum salmon and Atlantic cod (Sundell *et al.*, 1992). The branchial effect of stanniocalcin has been observed in various species, including eels and rainbow trout (Milet *et al.*, 1979; So and Fenwick, 1979; Lafeber *et al.*, 1988). Extracts of corpuscles of Stannius of goldfish have also been shown to have a transitory hypocalcemic action in an elasmobranch, the stingray *Dasyatis akajei* (Srivastav *et al.*, 1996). (It will be recalled (Chapter 2) that stanniocalcin-like immunoreactivity has been identified in the plasma of a shark.) Stanniocalcin has also been shown to promote the absorption of phosphate (*in vitro*) from the renal proximal tubule of the winter flounder (Lu, Wagner, and Renfro, 1994). Such an action could facilitate the deposition of plasma calcium ions into bone and so also contribute to the hormone's hypocalcemic effect. Although stanniocalcin has not been found in cyclostomes, it has, unexpectedly, also been shown to inhibit calcium uptake across the gills (*in vitro*) of the hagfish *Eptatretus cirrhatus* (Forster and Fenwick, 1994). In a preparation of the gill pouch, it was found to enhance calcium efflux instead of blocking its influx, as occurs in teleosts.

Both the ultimobranchial bodies and the corpuscles of Stannius appear to be able to limit changes in body calcium by a similar type of action on the gills. Such a duplication of a physiological role is rather unexpected. It should be recalled, however, that the corpuscles of Stannius are not present in all bony fishes but are confined to the Teleostei and Holostei. It has been suggested that the corpuscles of Stannius may be responsible for coarse adjustments in body calcium levels in such fishes whereas calcitonin is involved in finer regulation (Milet *et al.*, 1979).

The contribution of M. Fontaine to the discovery of the role of stanniocal-cin in the regulation of calcium metabolism in teleost fish was preceded in 1956 by his observation that hypophysectomy in European eels resulted in a hypocalcemia (Fontaine, 1956). This discovery has been investigated and confirmed in several other teleosts. When *Fundulus heteroclitus*, kept in artificial sea water with a low-calcium concentration, are hypophysectomized they undergo tetanic muscular contractions associated with a considerable decline in the plasma calcium levels (Pang, 1973). This response is not seen if the fish are kept in ordinary sea water with high concentrations of calcium. The effects in the low-calcium solution can be prevented by injecting the hypophysectomized fish with extracts of the pituitary or by transplanting this gland under the skin. It was suggested that the pituitary hormone was prolactin (Olivereau and Olivereau, 1978; Pang *et al.*, 1978). This hormone was found to be released in fish placed in fresh water that had a low calcium concentration (Wendelaar Bonga, 1978). The release was, however, shown subsequently to be primarily occurring in response to the low osmotic concentration rather than to the calcium present (Wendelaar Bonga and Pang, 1989; Arakawa *et al.*, 1993). Ovine prolactin, when injected, increases the branchial uptake of calcium ions in tilapia in fresh water while decreasing its loss (Flik *et al.*, 1986). This effect on the gills is associated with an increase in cell membrane Ca-ATPase. Homologous tilapia prolactins I and II also display this effect (Flik *et al.*, 1994). It is still uncertain whether the action of prolactin is a direct one. Cortisol can, for instance, also promote branchial calcium uptake in rainbow trout, and it appears to utilize a similar type of mechanism (Flik and Perry, 1989). As described above, somatolactin from the pars intermedia of the pituitary gland could be exerting a prolactin-like effect, but this possibility remains to be investigated.

Although knowledge about the regulation of calcium metabolism in fishes in incomplete, several intrinsically interesting facts are known. We have seen that endocrine control of this process may be related to the corpuscles of Stannius, which are only present among the Osteichthyes. The secretion of the ultimobranchial bodies, calcitonin, may exert a hypocalcemic action, possibly associated with the presence of cellular bone but also involving changes in calcium uptake across the gills. A hypercalcemic effect has, however, also been observed in these fish. Chondrichthyeans possess cal-citonin which, when injected, has a hypercalcemic action. Almost nothing is known about the regulation of calcium metabolism in sharks and rays, while lampreys and hagfishes may be dependent on an "open system" that does not involve the action of hormones. The regulation of calcium metabolism in cyclostomes and chondrichthyeans undoubtedly will provide a very interesting area in which to pursue this subject further.

An extensive comparative account of calcium metabolism in vertebrates has been provided by Dacke (1979) and Pang and Schreibman (1989).

Conclusions

Calcium is essential for the life of vertebrates, but their requirements for this mineral and its availability in the environment vary considerably. This need can be related to vertebrate phylogeny because of the systematic presence or absence of a bony calcareous skeleton and the characteristic life of some groups in the sea, which is rich in calcium. It is, therefore, not surprising to observe that the role of hormones in the regulation of calcium metabolism appears to have changed considerably during the course of evolution. In marine fishes that lack a bony skeleton, hormonal regulation of calcium does not seem to occur. When such an endocrine control system is present, as in marine bony fishes, it is concerned with limiting concentration of calcium and lowering its levels in the blood (hypocalcemic effects). The gills may be an important effector site. This response possibly involves calcitonin (which is present in all vertebrates except cyclostomes) and in teleosts, stanniocalcin, which is secreted from the corpuscles of Stannius. Teleost fishes in fresh water, where the calcium levels are low may utilize a pituitary hormone, prolactin, to help to maintain adequate concentrations of calcium in their body fluids. In mammals and birds, prolactin may also play a role in calcium metabolism by promoting the activation of vitamin D. Tetrapods have acquired a "new" hormone, parathyroid hormone, the role of which is to mediate a hypercalcemia; bone is the major site of its action. A physiological role for calcitonin in calcium metabolism is doubtful in many species. Two other hormones contribute to the regulation of calcium metabolism but apparently in distinct groups of vertebrate; vitamin D has a hormonal function in tetrapods where it facilitates the accumulation of calcium in the body but it is uncertain whether it has such a role in fishes. Estrogens have assumed a "special" endocrine role in many vertebrates, where they assist the deposition of calcium in the developing egg; this effect is absent in mammals.

7 Hormones and the integument

The skin and gills of vertebrates constitute the major external interface between the animal and its environment. This integument is physiologically and anatomically a very important tissue that exhibits considerable diversity reflecting the differences that exist in the physiochemical gradients between the vertebrates and their environments (Bereiter-Hahn, Matoltsy, and Richards, 1986). The integument may, therefore, play a role in the animal's osmoregulation, thermoregulation, and respiration. In addition, the integument provides signs and signals that can promote social and sexual contact and can help the animal to blend in with its surroundings and so protect it from predators, or help it catch its food. Of primary importance is the skin's role as an integumental skeleton by which it contains the animal in a condition that facilitates its locomotion. The relative importance of these various roles of the integument varies in different species and the structure varies accordingly also.

In fishes and larval amphibians, the gills, which function as organs of respiration, make up a large part of the animal's external surface. Exchanges of oxygen and carbon dioxide readily occur across these highly vascularized tissues, which are also the sites of considerable movements of water and salts. Many fishes contain special cells in their gills and skin called "chloride cells" or ionocytes, which are the site for active extrusion of salts. The endocrine control mechanisms influencing the permeability of the gills are described in Chapter 8.

The skin is the major nonbranchial interface between the animal and its environment. In its simplest form, the skin consists of two major layers of tissue: an outer epidermis, which has several strata of cells, and an inner dermis. However, such a simple arrangement does not exist in nature, as various other structures are also included in the skin that modify its properties. These structures include scales, hair, feathers, pigment cells, secretory glands, and certain sense organs. Such accessories contribute to the particular physiological properties exhibited in the integument of each species.

The skin is, therefore, a complex tissue that has different physiological needs that depend on the species, the environment, and the stage of the animal's life cycle. The constitution of the skin is not static but undergoes continual change commensurate with the normal needs of growth and repair. In addition, rapid changes in the physiological properties of the skin also can occur, such as involve an increased blood supply and the secretion of sweat in response to the need to dissipate heat in the mammals, and an increased osmotic permeability to water, as a result of dehydration, in amphibians. Many cold-blooded vertebrates can rapidly alter the distribution of pigment in the skin so that they blend more closely with the shades and hues of their surroundings. Seasonal changes commonly occur in the integument, such as the changes in pigmentation that may be associated with breeding and alteration of the color, length, and density of fur and feathers in summer and winter. Such changes in the fur and feathers may alter the insulative properties of the integument and contribute to the animal's camouflage.

While the skin has a considerable innate ability to regulate its functions, it is also dependent on the nervous and endocrine systems with whose aid it can coordinate its activities with the rest of the body. The skin has a plentiful nerve supply that mediates its sensory functions and regulates its blood supply. The secretions of cutaneous glands are also predominantly under neural control, though circulating catecholamines exert some effects on them. Rapid changes in the distribution of pigment may be controlled by nerves, but hormones are also very important. The endocrines help in the maintenance of the nutritional and anatomical integrity of the skin as well as such processes as molting, pigmentation, and the function of certain cutaneous glands. Hormones that influence cutaneous function include several from the pituitary such as prolactin, MSH, vasotocin, ACTH, LH, and TSH, and also T_4, the catecholamines, corticosteroids, gonadal steroids, MCH, and melatonin. Some of the actions of these hormones are confined to relatively few species while the effects of a hormone on the skin may be quite different in one species compared with another. The variation that is observed in the cutaneous effects of particular hormones suggests that considerable evolution has occurred in their special roles. It is also often difficult to decide whether the actions following an excess or deficiency of a hormone are the result of its direct action on a specific cutaneous effector or are merely the result of a more diffuse, indirect effect such as may result from general changes in the animal's physiological and nutritional status.

Hormones and molting

The epidermis is regularly renewed as its outer layers drop off and are replaced by new cells that are formed from the underlying epithelium. This may be a more or less continuous process, as is common in mammals, or it may take place suddenly at regular intervals varying from a few days, as in many amphibians, to several months in certain lizards and snakes. The hair of mammals and feathers of birds are also subject to such periodic renewal, and this may also occur at precise times of the year such as at the onset of winter or spring or just before, or after, the breeding season (pre- and postnuptial molts). In reptiles the shedding of the epidermis is often called *sloughing*, and the shedding of the pelage in mammals, the plumage of birds, and the epidermis in amphibians is called *molting*. The molting cycle is considered to involve three processes: tissue proliferation, differentiation, and shedding, or ecdysis.

It is generally considered that the regular cyclical molting that occurs in fish, reptiles, and amphibians reflects an autonomous rhythm in the skin upon which the actions of hormones can impinge in a permissive manner (Ling, 1972). The seasonal molts that occur commonly in birds and mammals are closely allied to the external stimuli, principally the photoperiod, but they are also modified by the external temperature and the nutritional condition of the animal. The pituitary, gonads, thyroid gland and pineal are the principal endocrines that influence molting in vertebrates.

Removal of the pituitary usually prevents, or considerably prolongs, the length of the reptilian and amphibian molting cycles and blocks the seasonal molts observed in many birds and mammals. This effect of hypophysectomy is the result of the absence of several hormones, a lack of TSH, with its tropic effect on the thyroid, prolactin, and corticotropin. In lacertilian reptiles, urodele amphibians, birds, and mammals the thyroid hormones accelerate the molting process; thyroidectomy has an inhibitory effect. It is interesting that in ophidian reptiles (snakes) removal of the thyroid results in a decrease in the length of the sloughing cycle that is in direct contrast to what is observed in their lacertilian relatives (Maderson, Chiu, and Phillips, 1970; Chiu and Lynn, 1972).

Based on studies on the tokay lizard, *Gekko gecko* (Chiu *et al.*, 1986), it appears that the frequency of shedding in reptiles may be an endogenous rhythm that is generally influenced by metabolic factors and external temperature. Hormones may only contribute indirectly by the fine tuning of the animal's metabolism. Differences in the effects of hormones on molting occur within the Amphibia, for whereas this process if facilitated by the thyroid gland in *Ambystoma mexicanum* and *Notophthalmus viridescens* (Urodela) this is not so in *Bufo bufo* (Anura) (Jørgenson, Larsen, and Rosenkilde, 1965; Hoffman and Dent, 1977). In *Bufo bufo*, however, corticotropin and the corticosteroids (corticosterone is most active) are necessary for successful molting (Jørgenson, 1988). To further complicate any attempts to define a phyletic uniformity, it has been found that corticotropin completely *inhibits* sloughing in the lizard *Gekko gecko* (Chiu and Phillips, 1971a).

Molting in birds is usually associated with their reproduction and occurs following the hatching and rearing of the young when the plasma levels of gonadotropins and steroid sex hormones are low. This event is associated with increased plasma concentrations of thyroid hormones (Smith, 1982; Groscolas and Leloup, 1986; Hoshino *et al.*, 1988). The administration of T_4 can promote the molt while thyroidectomy can delay it. Such observations have been made on a variety of domesticated and wild birds. Whether or not thyroid hormones can actually initiate the molt is controversial, but increases in their plasma concentrations usually accompany it and may precede it. Thyroid hormones can facilitate the growth of new feathers and may also contribute to the associated mobilization of body proteins (Groscolas and Leloup, 1986; Cherel *et al.*, 1988b). Some birds may experience cold conditions during the molt, which necessitates an additional need for thyroid hormones for thermoregulation. King penguins in Antarctic regions undergo a postnuptial molt during which time they fast as they cannot go to sea to feed (Cherel *et al.*, 1988a). During the initial period of this molting, their plasma thyroid hormone levels, especially T_4, are considerably elevated. This hormonal change appears to be related to the molting and, possibly thermoregulation, as when these penguins undergo a fast associated with breeding and incubation of the egg the thyroid hormone concentrations in the plasma decline. The respective roles of T_4 and T_3 in molting are not clear, but the T_4 levels are usually increased. It is probable, however that the T_4 is converted to T_3 at local sites, a process that could be enhanced during the molt. In domestic hens, which are not usually allowed to incubate their eggs, a "forced molt," to restore laying, may be induced by limiting their food supply and changing from a lighting schedule of long to short days (Etches, 1996). An early hormonal change in the forced molt is a rapid rise in plasma corticosteroid concentration, which is followed by a decrease in gonadotropin and estrogen levels. The precise role of the corticosteroids in this process has not been described. Thyroid hormone concentrations rise at this time.

Prolactin has diverse actions in vertebrates and it is especially notable that many of its effects are on the integument (Dent, 1975) or derivatives of it, most notably the mammary glands (which are merely modified sweat glands). It has been shown that the injection of prolactin decreases the length of the sloughing cycle in the lizard *Anolis carolinensis* and *Gekko gecko* (Maderson and Licht, 1967; Chiu and Phillips, 1971b). Prolactin has been found to facilitate the growth and increase the appetite of lizards so that its effect could reflect their nutritional condition. In this respect, it should also be remembered that hypophysectomized lizards usually do not eat and are not in perfect health. Prolactin injections have also been shown to accelerate molting in a urodele amphibian, the red eft, *Notophthalmus viridescens* (Chadwick and Jackson, 1948), though such an action has not been demonstrated in other amphibians. This effect is related to an increase in the mitotic activity of the epidermis (Hoffman and Dent, 1977). It is interesting that, in this newt, prolactin promotes the transition (or metamorphosis) from the terrestrial form into the aquatic breeding stage when it returns to water ("water-drive effect" of prolactin), and this is associated with cutaneous changes. Whether the effect on the skin is a primary one is uncertain, as the prolactin could be stimulating some noncutaneous process concerned more generally with metabolism and metamorphic change.

The seasonal changes that occur in the pelage of mammals and the plumage of birds appear to be principally under photoperiodic control. Changes in the hours of light are transmitted via the eyes, pineal, and hypothalamus to the pituitary. Such photoperiodic changes also control the gonadal cycles so that it may be difficult to separate the two events. However, it has been shown that the testicular cycle and the winter molt in Djungarian hamsters are triggered by different critical day lengths (Duncan et al., 1985). European badgers, red foxes, and mink molt at the end of their reproductive periods. In the males, this corresponds to a decline in the plasma testosterone levels. Implantation of this hormone at this time delays the onset of the molt for several weeks (Maurel and Coutant, 1986). Like in birds, plasma T_4 levels are also elevated at the time of the molt. In the badger, thyroidectomy abolished the molt but it could be restored by the administration of T_4 (Maurel, Coutant, and Boissin, 1987). The molt in badgers is promoted by long spring days and is associated with a rise in the concentrations of prolactin in the plasma (Maurel, Coutant, and Boissin, 1989). Prolactin may trigger the molt in such species as mink in spring, and the Djungarian hamster in autumn (p. 329). Implants of melatonin, which mimic the hormone's release during short days, delays the molt in badgers and suppresses the plasma concentrations of prolactin. In the badgers, it was concluded that while prolactin favored the molt, its effect may be synergistic with that of other hormones, and it did not necessarily trigger the response. It appears that an optimal balance of testosterone, T_4 and prolactin, as well as melatonin, may promote the molt in some mammals but the particular relationship of such hormones in the process may differ between species.

The precise manner in which hormones influence molting is not understood. Cyclical changes in molting in poikilotherms are usually characterized by brief periods of cellular activity and rapid cell division interspersed by periods when little activity occurs, which are referred to as the "resting phases". It is thought that T_4 and prolactin shorten the resting phase (except in snakes!) during which time the skin's activity is normally reduced by lower levels of these hormones. In newts prolactin has been shown to promote active cell division in the epithelium and this is reminiscent of its action on the crop-sac epithelium in pigeons. In toads, the absence of the pituitary does not prevent the formation of new layers of epidermis (or sloughs) but prevents the shedding, or casting off, of these cells. This shedding is promoted in toads by corticotropin and corticosteroids. It is interesting that some frogs that estivate during periods of drought are protected from excessive dehydration by a cocoon composed of accumulated layers of epithelial cells (Lee and

Mercer, 1967), and it seems possible that this may reflect a decline in their pituitary and adrenal function.

Hormones and skin glands

Vertebrates possess several types of secretory gland in their skin, which serve a variety of functions (Quay, 1972). These glands can be classified into two major groups: the mucous glands and the proteinaceous glands.

The proteinaceous-type glands have undergone considerable evolutionary modification and include (to name only a few) a variety of venom glands in fishes and amphibians, the uropygial (or preening) gland that is present in many birds, and the sebaceous glands, sweat glands, and mammary glands of mammals. The sweat glands play an important role in temperature regulation in mammals. The sebaceous glands are associated with hair follicles and the fatty sebum that they secrete serves to protect the hair from wetting. The sebaceous glands and the sweat glands sometimes secrete special odoriferous substances that may play an important role in territorial behavior (by defining territorial limits) and also act as sexual attractants. Such scent glands may become enlarged and congregate in distinct areas of the body. Examples of this include the "side glands" on the heads of shrews, the "anal glands" and submandibular "chin glands" in rabbits, the "ventral gland" in gerbils, and the forehead skin of roebucks. In sheep, the sebaceous glands secrete a pheromone that accumulates in the wool of the rams (Knight, Tervit, and Lynch, 1983). When wool from Dorset rams is placed over the muzzles of the ewes their breeding cycle is accelerated and they usually ovulate within a few days (Chapter 9). Male goats also produce such pheromones and these attactants have been identified as a series of fatty acids (including 4-ethyl-octanoic, 4-ethyl-decanoic, and 4-ethyl-dodecanoic acids). When applied to the muzzles of ewes these fatty acids had the same effect as the rams' wool. Steroidal pheromones from boars' urine were ineffective in the ewes.

The maturation and function of the sweat glands and sebaceous glands in mammals are influenced by hormones (Strauss and Ebling, 1970). The odoriferous scent glands are commonly observed to be larger in the male than the female whereas maturation of sebaceous and sweat glands occurs during puberty in humans. It therefore, seems likely that they are influenced by sex hormones, and androgenic steroids are undoubtedly involved. Natural and experimental differences in the development of the submandibular chin glands in rabbits are shown in Fig. 7.1. These glands are much larger in sexually mature than immature rabbits, and they decrease in size when the testes are removed. Injections of testosterone promote the development of the chin glands while estrogens inhibit it.

Castration results in a decrease in the development of sebaceous glands and a reduced secretion of sebum in the skin of mammals (Ebling, 1974).

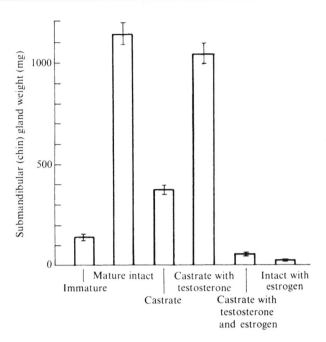

Fig. 7.1. The effects of gonadal hormones on the weight of the submandibular (chin) glands in male rabbits. It can be seen that these glands exhibit considerable increases in weight after sexual maturity, an effect that is prevented by castration. Testosterone injections overcome this effect of castration while injecting estrogens brings about an involution of the glands. (From Strauss and Ebling, 1970.)

Hypophysectomy has a similar effect but the response does not merely reflect an absence of gonadotropins, which can stimulate the secretion of androgens (Shuster and Thody, 1974). The administration of testosterone to hypophysectomized rats that have also been castrated does not completely restore the activity of the sebaceous glands. Removal of the posterior lobe of the pituitary gland has a similar effect to complete hypophysectomy. Neither oxytocin nor ADH stimulates the sebaceous glands in such animals, but it was found that α-MSH completely restored their activity. It was suggested that α-MSH may be acting as a sebotropic hormone. Androgens and α-MSH appear to act together to enhance each other's action (synergism) on the sebaceous glands (Ebling *et al.*, 1975; Thody *et al.*, 1976). The androgens appear to increase their growth and size whereas α-MSH promotes the secretion of lipids.

In primates, especially humans, the watery secretion of the eccrine sweat glands plays an important role in providing water than can be evaporated from the body surface and so aid in the dissipation of heat (Chapter 8). Sweat contains dissolved salts, including sodium and potassium, the concentration of which is influenced by aldosterone; the sodium/potassium ratio declines under its influence. In addition, although the secretion of sweat is primarily under neural control in primates it is also increased, during exercise, by circulating epinephrine.

Mucous glands are present in the integument of fishes and amphibians. Their role is contentious but seems to be related to an aquatic life, where it has been suggested that the mucous secretion may have a protective action on the skin as well as serving certain special functions such as the formation of a cocoon in estivating African lungfish and providing food for the young of a cichlid teleost. In the latter fish, this mucous secretion is promoted by the injection of prolactin (Egami and Ishii, 1962). Injected T_4 increases the number of mucous cells in the skin of the guppy, *Poecilia reticulata* (Schwerdtfeger, 1979). It has also been suggested that prolactin may limit the permeability of teleost fish and urodele amphibians to water and sodium (Chapter 8) by an action on the mucous glands, but the evidence for this is equivocal.

Knowledge about the role of hormones in regulating growth, development, and secretion of the integumental glands is incomplete. Clearly, however, some such glands do respond to hormones though the primary importance of neural secretory stimuli should not be forgotten. The functions of such glands vary considerably in different vertebrates, and it is interesting that many have attained a responsiveness to certain hormones. The nature of these effects appears to be related to the other basic functions of the hormones in the body; thus, gonadal steroids influence skin glands that are involved in sexual activities, and aldosterone acts in a manner commensurate with its role as a hormone that conserves sodium in the body.

Information about hormone effects on skin glands in nonmammals is sparse. Epinephrine stimulates chloride secretion from the skin glands of European frogs and this may reflect a direct action of such hormones in the body or a mimicking of a stimulation of the sympathetic nerves. Epinephrine, as well as vasotocin, has also been shown to stimulate the secretion of a sticky, milky-white material from the proteinaceous skin glands of the African clawed toad, *Xenopus laevis* (Ireland, 1973). The effect of the catecholamine, but not vasotocin, can be prevented by an α-adrenergic-blocking drug.

Hormones and pigmentation

The integument of most vertebrates contains pigment that makes a major contribution to what is often a very colorful appearance. Pigment may be present within the epidermis or dermis itself or color the integumental appendages, such as scales, hair, and feathers. Apart from contributing to our esthetic delight in contemplating nature, it seems likely that an animal's coloration may be useful to its physiology (Hadley, 1972). Appropriate pigmentations may contribute to the animal's camouflage, protect the internal organs from solar radiation, and promote the absorption or reflection of heat and light, so aiding in photoreception and contribute to the synthesis of vitamin D in the skin. Integumental colors also provide signs that are important for appropriate dimorphic sexual behavior and reproduction.

Pigments of different colors are usually present in the skin in cells called chromatophores. These cells commonly contain a black or brown pigment called melanin and are called melanocytes or, if the intracellular distribution of pigment can be changed, melanophores. The yellow and red pigments (xanthines and carotenes) that also occur in the skin of vertebrates are contained in, respectively, xanthophores and erythrophores. Some chromatophores also contain pteridine platelets that reflect light, giving an iridescent appearance and are thus called iridophores. The complex and beautiful colors of many vertebrates are the result of blending the colors reflected by the various chromatophores.

Many vertebrates can alter their coloration in response to environmental and behavioral needs. Such changes may take place in a relatively slow manner, as when the total amount of pigment in the epidermis, or its appendages, changes. The result is a relatively static coloration that is attained over a period of days or weeks. This process is called *Morphological color change* (which may involve melanocytes or melanophores) and is seen when we tan in the sun or when an animal changes the color of its pelage or plumage with the onset of summer or winter or in preparation for the breeding season. In addition, many cold-blooded vertebrates can rapidly change their color, a process that only takes a few minutes or at the most several hours. This relatively rapid response is called *physiological color change*. Both morphological and physiological color changes are influenced by the actions of hormones, especially the pituitary MSH.

The melanophores are cells with long dendritic-like extensions that radiate from a central core. In shape they resemble nerve cells, from which they are derived. The melanin is contained within cellular organelles called melanosomes. Darkening and lightening of the skin, as occur in physiological color change, reflect a migration of the melanosomes in dermal melanophores so that they are widely distributed in the cell (dark color, the melanin is said to be dispersed) or they aggregate in small globs in the center of the cell (light color, the melanophore is said to have a punctate appearance) (Fig. 7.2). The grosser effects of these changes in the color of frog skin are shown in Fig. 7.3. The dispersal of the melanin in the melanophores may depend on a microtubular system in the cell, as certain drugs (for example cytochalasin B) that break such tubules also prevent dispersion of the pigment. The other chromatophores have a rather similar structure to the melanophores but contain different pigments. Iridophores that respond to MSH do so in the opposite manner to that of the melanophores; the platelets of reflecting materials aggregate so that the cell has a punctate appearance. Not all chromatophores exhibit a physiological color change response to MSH, and indeed this is not usually seen in the xanthophores and erythrophores (Bagnara, 1969; Taylor and Bagnara, 1972). In epidermal melanocytes (unlike the dermal melanophores), the pigment is relatively fixed in its position so that differences that occur usually reflect the total quantities of pigment that are present (morphological color change).

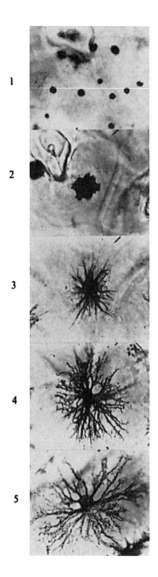

Fig. 7.2. The microscopic appearance of the dermal melanophores of the dogfish *Scyliorhinus canicula*. When the fish is maximally dark, as in 5, the melanosomes are dispersed throughout the cell, which can then be seen in outline: when the fish is pale they are aggregated in the central region (as in 1) so that definition of the cell outline is obscured. The numbers 1 to 5 correspond to the "melanophore index". (From Wilson and Dodd, 1973a.)

The different types of chromatophore present in colorful animals, including many frogs and lizards, are arranged in layers, and changes in the distribution of pigment within these zones alter the transmission of light and the color that is perceived. The innermost layer consists of melanophores and these are overlain by the reflecting iridophores. The xanthophores form a layer closer to the surface of the skin. Changes in the density of the melanophores, which absorb light, and the iridophores, which can reflect it, thus can alter the color of a frog from a light to a dark color and influence the

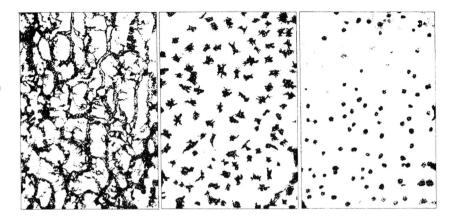

Fihg. 7.3. The gross appearance of amphibian melanophores (under a low-power microscope) when the animal is (left to right): dark, intermediate, and pale in color. (From *The Pigmentary Effector System* by L. Hogben (1924). With permission of the publishers Oliver & Boyd, Edinburgh.)

display of the colorful pigments in the superficially placed xanthophores. The melanophores, iridophores, and xanthophores together make up what is called a "dermal chromatophore unit."

The mechanism of release of MSH

As we shall see in the succeeding sections, the release of MSH from the pituitary melanotrope cells plays a most important role in both physiological and morphological color change in vertebrates. A distinction between the release and the particular roles of α-MSH and β-MSH has not generally been made but α-MSH appears to be predominant. The N-acetylated forms (Chapter 3) of this peptide are the most active. The principal stimulus for release, especially in cold-blooded species, is the receipt of light, usually by the lateral eyes, but the pineal may also function as a photoreceptor in some species. In addition to directly influencing MSH release in acute situations, light may also contribute to a cyclical, photoperiodic release of this hormone in some mammals that seasonally change the color of their pelage. The intimate mechanisms that control the release of MSH from the pars intermedia are only partly understood and appear to be quite complex (Howe, 1973; Kastin, Schally, and Kostrzewa, 1980; Hadley, 1980; Vaudry *et al.*, 1994). When the pars intermedia is transplanted ectopically to another part of the body away from the hypothalamus, or if its connections to this part of the brain are severed, MSH is secreted in an apparently uncontrolled manner (Iturriza, 1969; Penny, Tilders, and Thody, 1979). The regulation of hormone release, therefore, appears to be mainly under an inhibitory control originating in the hypothalamus.

The pars intermedia has a nerve supply that comes from the base of the brain and contains both aminergic and peptidergic neurons. Cholinergic nerves have also been identified. The vascular supply usually shows considerable species variability but it is quite poor, though portal vessels coming from the hypothalamus and neural lobe have been described. The inhibitory controlling stimuli appear principally to involve the aminergic nerves that originate in the hypothalamus. Their effects may, however, be modulated by the peptidergic neurons, which may decrease or enhance the release of MSH.

The neural stimuli that are responsible for regulating the secretion of MSH from the pars intermedia may be mediated by a variety of neurotransmitters. Dopamine, which has an inhibitory role and hyperpolarizes the melanotrope, appears to be predominant but the α_2-adrenergic actions of norepinephrine may also contribute to the response (Wilson and Dodd, 1973a; Bower, Hadley, and Hruby, 1974; Olivereau, 1978; Vaudry et al., 1994). Other substances that may influence the release of MSH include GABA and adenosine, which are inhibitory, and acetylcholine, CRH, mesotocin, atrial natriuretic peptide and TRH, which are stimulatory. Species differences in such responses exist and a clear delineation of pharmacological from physiologically relevant responses has not always been possible. TRH can promote the release of MSH in anuran amphibians and teleost fish, but not in urodele amphibians, rats, or the lizard Lacerta vivipara (Dauphin-Villemant, Tonon and Vaudry, 1992). The action of TRH in frogs (Tonon et al., 1983) results in cell depolarization and a stimulation of phospholipase C and adenylate cyclase. Norepinephrine can also promote the release of MSH by interacting with β-adrenoceptors and activating the adenylate cyclase–cyclic AMP system. Over a period of more than 20 years a number of candidate peptide MSH-releasing and MSH-release-inhibiting factors have been described. However, a definitive one that mediates a physiological response has not yet been agreed upon. There are currently some interesting possibilities that may even conclude the search. The frog brain contains a member of the pancreatic polypeptide family, neuropeptide Y (NPY, see Chapter 3). It only differs from human NPY by one substitution in its 36 amino acid residue chain (Chartrel et al., 1991). This peptide has been identified in neurons associated with the pars intermedia in frogs and it can inhibit the release of MSH from the melanotropes (Valentijn et al., 1994). It has been called melanostatin and acts by inducing a hyperpolarization of the hormone-secreting cells. Another member of the pancreatic polypeptide family has been identified in the skin of frogs. It is related to peptide tyrosine tyrosine (PYY) and is called skin PYY or SPYY (Mor et al., 1994). It can also inhibit the release of MSH. Frog skin also contains TRH (Chapter 3) and it has been suggested that SPYY and TRH in association with skin melanophores may be part of a regulatory control system between the pars intermedia and the skin of frogs. Such an interesting possibility remains to be explored and could, if confirmed, even establish the frog skin as an endocrine organ.

Studies on the mechanism of release of MSH were initially performed on amphibians but have since been extended to mammals, reptiles, and fishes. The observations have generally indicated that a similar system for the control of secretion of this hormone exists throughout the vertebrates. Therefore, the evidence for an inhibitory role of dopamine and α-adrenergic stimulation appears to be consistent in species as phyletically separate as dogfish and rats. Pharmacological observations on a teleost fish, eels, and a reptile (the lizard *Anolis carolinensis*) suggest that nerves that secrete 5-HT may stimulate the secretion of MSH (Olivereau, 1978; Levitin, 1980). Whether this mechanism is unique to these groups of vertebrates is unknown.

Physiological color change

Physiological color changes occur in many cold-blooded vertebrates, from the cyclostomes to reptiles. These changes in the distribution of the pigment in the skin occur in response to a variety of conditions and stimuli. Many vertebrates exhibit a diurnal rhythm in the degree of aggregation and dispersion of melanin in the melanophores. They turn pale at night and dark during the day. This change may reflect the perception of light and be mediated by receptors in the eyes and the pineal, and sometimes can result from a direct stimulation of the melanophores by light. In other instances, such as in the lizard *Anolis carolinensis*, a diurnal rhythm can even be seen when the animals are kept in complete darkness. As it is not seen in these lizards after they are hypophysectomized, it probably reflects an inherent diurnal rhythm in the activity of the pituitary gland. Superimposed on such rhythmical changes in skin color are direct, and adaptive, responses to external stimuli. These stimuli include the perception of certain light patterns caused by the color and shade of the substrate on which the animal is placed (*background response*) and, to a lesser extent, the external temperature and "excitement". The last two effects, which have been observed more commonly in lizards, may override the background response.

The first recorded observations of physiological color change are more than 2000 years old, but our understanding of the mechanism involved is quite recent. An appreciation of the role of hormones in these responses principally resulted from the pioneering studies of L. Hogben and F. R. Winton in the 1920s. An excellent account of the work of the Hogben school in England and that of many others, including G. H. Parker in the USA, has been given by Waring (1963). Waring joined the Hogben school in the 1930s, and his account of the processes involved in regulating color change is an ideal example of the stringent analytical approach and the application of formal logic that we should all aspire to in scientific investigation. Although physiological color change does not occur in mammals or birds, the elucidation of its mechanism has contributed a great deal to our understanding of the role of hormones in physiological coordination. The following account is largely based on Waring's, but one should also consult the book by Bagnara and Hadley (1972).

Types of melanophore response

The nonvisual response

a. Coordinated response.   This type of response may be abolished by denervation of the skin or the removal of the pituitary or the adrenals.

b. Uncoordinated response.   This occurs when the melanophore (or possibly a skin receptor close to it) directly responds to a stimulus. This type of response can be seen rather clearly in the horned toad, *Phrynosoma blainvilli.* If these lizards are blinded, hypophysectomized, and the pineal eye is covered, and they are then placed in a black box with no light, they become a pale color. When, however, a think beam of light is focused on a piece of denervated skin in these lizards this darkens in comparison to the rest of the integument. A localized response to temperature can also be demonstrated in *Phrynosoma*, for when an area of the skin of a maximally dark lizard is exposed to water at 37 °C it pales in that region. Similarly, maximally pale skin will darken locally at a temperature of 1 °C. Chameleons also exhibit dramatic localized changes in skin color; the skin of blinded animals turns dark in light but if a certain area is shaded by an object, a lighter colored "print" or outline of this object can be seen. Such responses do not involve hormones or the ordinary nerve supply and appear to reflect a direct response of the melanophore; however, a local nerve reflex initiated from a nearby cutaneous receptor or a release of a local hormone could be involved.

The visual response   This response is the result of the reception of light by the lateral eyes or the pineal in certain species, including some larval amphibians and cyclostomes, and possibly even some lizards. The responses may be a generalized lightening (in the dark) or darkening (in the light) of the skin or be influenced by the color of the background: the background response. When the animal is on a white substrate with overhead illumination, it may turn a pale color and if on a black background (also with overhead illumination) it may turn a dark color. These changes are called the white (or tertiary) and black (or secondary) ocular-background responses. The different effects of light in these two sets of circumstances appear to result from the stimulation of different parts of the retina; thus, the eye of a frog in a black tank of water (Fig. 7.4) receives light only on the more basal parts of the retina, the "B" (for black) area, but in a white tank, where the light is reflected into the eye from all sides, the entire retina, including a "W" (for white) area, is stimulated. Such special receptor areas for light in the retina have also been found in teleost fishes and lizards.

Quantification of the melanophore response   Early observations of vertebrate color change have been described in general subjective terms such as

Fig. 7.4. The manner by which the reception of light initiates the dispersion of melanin in the melanophores of a frog (in this instance *Xenopus*) sitting in a tank of water with a black background and overhead illumination. As the tank has black sides all light will enter the water from above and enter the lens at an angle that is the critical angle for air and water (49°). (From Waring, 1963.) Using these data, as well as the dimensions of the eye and the refractive index of the lens (*Xenopus* has its eyes on the top if its head), Hogben provided a diagram (*b, i, ii*) that shows the area of the retina receiving such light rays. All these conditions result in a darkening of the skin, this has been called the B-area (for black) as opposed to the W-area (for white), which initiates skin lightening in frogs on a white background when light reaches wider areas of the retina. (From Hogben, 1942. Reproduced by permission of the Royal Society.)

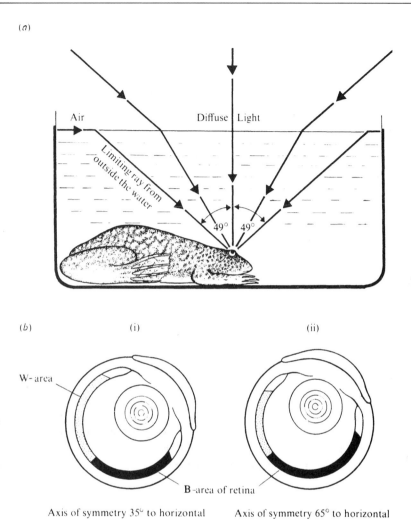

Axis of symmetry 35° to horizontal   Axis of symmetry 65° to horizontal

"pale" and "a tint rather dark than pale" that lack adequate precision for a proper scientific analysis and make comparisons of results from different laboratories almost impossible. Hogben introduced a more stringent quantitative description called the melanophore index (or MI) (Figs, 7.2 and 7.5), which has a gradation of 1, for maximally pale, with the melanin fully aggregated, to 5 when the animal is dark and the melanin is fully dispersed. In lizards, this can be translated, as in chameleons, to 1 = yellow, 3 = medium-green, and 5 = black. This simple standard of measurement allowed considerable advances to be made in the analysis of the mechanism of color change. Today, electrophotoreceptive devices are also used to quantify the melanophore responses.

	Appearance of melanophores	Melanophore index (MI)
Intact anurans		
In complete darkness		2.5 to 3.0
Light overhead:		
White background		1.5
Dark background		4.5 to 5.0
"Complete" hypophysectomy		
(light or dark background)		1.0
Denervation of pars intermedia		
(light or dark background)		4.5
Eyeless (light or dark background)		2.5 to 3.0

Fig. 7.5. The melanophore responses, mediated through the eyes, of anurans in relation to the receipt of light when on a white or a dark background. The effects of surgical changes of the pituitary of these responses have been summarized. The melanophore index (MI) in relation to the degree of dispersion, or aggregation, of melanin in the melanophores is given in the lower section. For a description and explanation of these responses the text should be consulted. (Based on Bradshaw and Waring, 1969.)

Chart of melanophore index for amphibians:

1　2　3　4　5

Color change in amphibians

The earliest observations on the role of hormones in vertebrate color change were made on European frogs, *Rana temporaria*, and subsequently the African clawed toad, *Xenopus laevis*. With overhead illumination (Fig. 7.5) on a white background, these amphibians are pale (MI about 1.5) and on a black background they are dark (MI about 4.5). When placed in complete darkness, or if they are blinded, they have an intermediate shade (MI = 2.5). In *Xenopus*, this change in melanin distribution in the melanophores is seen as a white or a black coloration, but in frogs that have overlying layers of yellow-green pigment this appears as a pale green or yellow to a black color. When amphibians are "completely" hypophysectomized so that no pituitary tissue remains (such remnants commonly *do* remain, as in *Xenopus*), the animals become maximally pale, the MI is 1 and they cannot respond to changes in the background color. If the pars distalis is removed carefully so that the pars intermedia and pars tuberalis remain intact, the background responses are retained. This operation is relatively simple to perform in *Rana spp.* but it is more difficult in *Xenopus spp.*, where the pars tuberalis is usually removed together with the pars distalis. This results in an inability of *Xenopus* to display a background response and it becomes permanently dark (MI = 5). The pars intermedia has a nerve supply coming down from the hypothalamus and when this is cut the anurans also have an MI of 5. Removal of the pars

Fig. 7.6. A composite diagram tentatively summarizing the humoral and neural control of colour change in vertebrates. MCH, melanin-concentrating hormone. For details see the text.

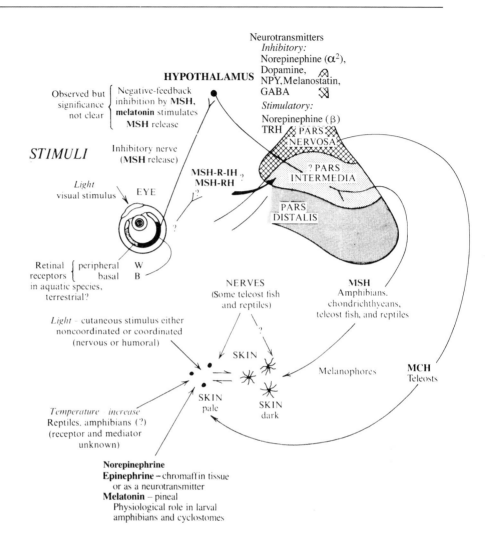

tuberalis appears to be associated with an interference of the hypothalamic connections to the pars intermedia and, as these are of an inhibitory nature, a sustained release of its secretion, MSH, occurs. MSH when released from the pars intermedia is carried in the blood to the melanophores where it promotes a dispersion of melanin so that the animal darkens.

The sequence of events resulting in the black background response is summarized in the following section (Fig. 7.6). Light from an overhead source, in frogs on a dark background, falls on the B-area of the retina where it stimulates receptors that transmit messages along the optic nerve. These messages, traveling along pathways that are as yet unknown, inhibit the normal inhibitory effects of the nerves supplying the pars intermedia and this results in a release of MSH. When the frogs are on a white background, light also falls on the retina but on the receptors in the W-area and this reduces the release of MSH. Normally, anurans kept in the dark, as well as blinded animals, have an MI of about 3, which appears to reflect a sustained, but submaximal, release of MSH. Stimulation of the "W" retinal receptors in some way inhibits the release even further, possibly by increasing the inhibitory nerve impulses to the pars intermedia.

Pallor of the skin is, therefore, usually thought to result from a decline in the levels of MSH in the blood. In the amphibia, in contrast to some teleosts and reptiles, nerves are not involved in the aggregation of melanin in the melanophores. The injection of epinephrine into *Rana spp.* produces a skin lightening and an aggregation of melanin in the melanophores, which is an α-adrenergic response. The normal physiological importance of this effect is unknown.

When dried bovine pineal glands are fed to anuran tadpoles, the melanin in the melanophores on the body, but not the tail, aggregates and the animal's body pales (Bagnara and Hadley, 1970). Under these conditions, the tadpole's internal organs can be seen clearly. This effect is the result of the action of melatonin formed in the pineal gland. Normally, tadpoles, such as those of *Xenopus laevis*, pale at night and darken during the day, and this can be prevented if the pineal, but not the lateral eyes, is removed. Formation and release of melatonin occur in darkness and this appears to mediate the diurnal rhythm of color change in these tadpoles. The effect of melatonin on the melanophores is a direct one and is not mediated through the pituitary as has sometimes been suggested, as the effect of melatonin is not prevented by hypophysectomy. This physiological effect of melatonin is confined to tadpoles and does not contribute to skin lightening in adult amphibians.

Color change in the Chondrichthyes

Many sharks and rays exhibit dramatic changes in color depending on the shade of the background; with overhead illumination they become dark on a black background and pale on a white one. Two dogfish (*Scyliorhinus canicula*) in their dark and pale phases are shown in Fig. 7.7. Waring, in 1936, found that when he transplanted a dogfish pituitary into another dogfish that was pale in color, it turned dark owing to a dispersion of melanin in the melanophores (Waring, 1936; 1938). The release of the MSH is caused by the absence of the hypothalamic neural inhibitory control mechanism present in these fish (Wilson and Dodd, 1973a). An analysis of the color change in these fish, *Squalus, Scyliorhinus, and Raja*, shows that they exhibit white- and black-background responses that are mediated humorally by MSH, just as in amphibians. The intermediate lobe of the pituitary of dogfish contains three forms of MSH (Chapter 3): α, β and γ. Acetylated α-MSH is its most potent form and administration of antibodies to this hormone to grey-adapted *Scyliorhinus canicula* promotes skin palor (Sumpter *et al.*, 1984). Antibodies

Fig. 7.7. Two dogfish, *Scyliorhinus canicula*, in their dark and pale color phases (×.33). (J. F. Wilson, personal communication.)

to γ-MSH were without effect. (Antibodies to β-MSH were not available). The active form of the hormone is probably the acetylated form of α-MSH. Direct neural control of the melanophores does not appear to occur in the Chondrichthyes.

The background response is not seen in blinded dogfish, though the fish show a slight paling in darkness that suggests the presence of a nonvisual response (Wilson and Dodd, 1973a). When kept in total darkness, the pallor exhibited by these dogfish is not seen if the pineal is removed and they become darker. The pineal may, therefore, contribute to nonvisual color change as observed in tadpoles and cyclostomes (see below).

Color change in teleosts

Although the chondrichthyean fishes and amphibians that have been examined all have humoral control of their color change, the teleosts, which lie phyletically between these two groups, may also possess a neural coordinating mechanism. The teleosts, as has become apparent in the comparison of their other biological systems, exhibit considerable interspecific differences in the control of color change that presumably reflect the systematic diversity within this large group of fish.

Stimulation of nerves controlling melanophores usually results in an aggregation of the melanin and a paling of the skin color in teleosts. A dispersion of melanin, in response to neural stimuli, however, may also occur in some species but the evidence for this is equivocal. The neural responses usually appear to be adrenergic ones associated with sympathetic nerves. The melanin-aggregating effects of such stimuli are mediated by α_2-adrenoceptors but a β-adrenergic melanin-dispersing action has not been excluded.

Like amphibians and elasmobranchs, many, but not all, teleost fish utilize MSH as a hormone to disperse melanin in the melanophores and so promote skin darkening. This hormone has been identified in the plasma of such fish. In European eels and rainbow trout, MSH is secreted at an increased rate when they are kept on a dark background (Baker, Wilson, and Bowley, 1984). Measurable changes of MSH in the plasma of these fish, however, took about 2 days to occur and neural mechanisms may predominate in the interim. All species of teleosts do not respond in this way. The pituitary melanotropes of flounder, mollys, and blennys are unresponsive to changes in background color and they do not appear to utilize MSH for physiological color change.

For many years there was speculation about the possible presence of a pituitary hormone that might oppose the effects of MSH and promote skin palor. Such a hormone has been identified in teleost fish (Baker and Ball, 1975; Rance and Baker, 1979; Baker, 1994). It is secreted from the neurohypophysis where it originates in neurons coming from the hypothalamus. This hormone was called melanin-concentrating hormone (MCH) and it has subsequently been found to have a wide phyletic distribution and is even present in mammals (Baker, 1994; and Chapter 3). The ability of MCH to promote aggregation of melanin in melanophores is, however, confined to the teleosts. In trout it is secreted at increased rates in fish kept on a white background (Kishida, Baker, and Eberle, 1989). Apart from directly antagonizing the effect of MSH on the melanophore it may also block the release of this hormone.

One can readily foresee the prospective biological advantages of an ability to change color rapidly as this may help protect the animal from a predator or assist it to catch food. Rapid color change may be especially important in animals that live in places where the background colors are variegated and across which the animals constantly travel in their search for food and sexual companionship. Many teleost fishes that roam about gaily colored reefs may find such rapid color change an especial advantage.

In the Teleostei, color change can be mediated by three types of mechanism: (1) a humoral one, (2) a neural one, or (3) a combination of both neural and humoral processes. It should be emphasized that the effects of hormones on physiological color change are relatively slow compared to those responses mediated by neural mechanisms.

Anguilla (the eel)   Eels exhibit black- and white-background responses but only change their color slowly, like amphibians. Hypophysectomy abolishes the full expression of these responses and as the pituitary contains a material that, when injected, disperses melanin in the melanophores, the response is considered to be predominantly a humorally mediated one.

Following hypophysectomy, the eel is not maximally pale but has a melanophore index of 1–2 and this has contributed to speculation that a MCH is also present in the pituitary, as, if only MSH were involved, one would expect a MI of about 1. It has been suggested that, alternatively, melanophore-dispersing nerve fibers are present. As described above, it is now known that MCH is present in the pituitary gland of eels (Gilham and Baker, 1984). This hormone is released in response to light in fish kept in a white background.

In contrast to amphibians, hypophysectomized eels continue to exhibit a small background response; the MI is 3.5 on a black background and 1.8 on a white one. This response is abolished by the severance of cutaneous nerves that are known to innervate the melanophores; therefore, although the predominant mechanism mediating color change in eels is a dual hormonal one there is an underlying neural control that only becomes apparent after the pituitary is removed. The melanin-aggregating effect of injected epinephrine in eels is probably an α_2-adrenergic response that is mimicking the action of sympathetic nerve stimulation

Fundulus heteroclitus (the killifish)   Killifish exhibit the usual black- and white-background responses but these are *not* abolished following hypophysectomy. *Complete* darkening, or dispersal of melanin, does not occur following this operation, indicating that pituitary MSH may be necessary for the full expression of the black-background response.

The overriding control is nevertheless a neural one. Injections of MSH into pale fish does not disperse melanin and electrical stimulation of cutaneous nerves in dark fish evokes pallor. The injection of MSH into pale fish that have had parts of their skin denervated results in a melanin dispersion of these localized areas. Extracts of *Fundulus* pituitaries can evoke such dispersion of melanin.

In *Fundulus sp.*, color change occurs in response to neural stimuli to the melanophores; there are melanin-aggregating nerve fibers, and possibly even "dispersing fibers". Underlying this mechanism, but generally overridden by it, is an ability to respond to MSH, and this hormone is necessary for a maximal darkening of the fish.

Phoxinus phoxinus (the European minnow)   In minnows, there is little evidence for a role of endogenous hormones in the dispersion of melanin. Black- and white-background responses are not prevented by hypophysectomy, though it has been observed that such fish cannot sustain a black coloration as readily as intact fish. Denervation of the skin abolishes the background responses and the melanin fully disperses. Nerve stimulation evokes an aggregation of melanin and aggregating nerve fibers undoubtedly exist. There is also some evidence that suggests the presence of melanin-dispersing fibers.

The injection of extracts of the pituitaries from *Phoxinus* does not disperse melanin in either the intact or denervated skin of these fish. MSH from anurans will, however, darken the denervated skin of *Phoxinus*. There is no evidence for a melanin-dispersing hormone in *Phoxinus* and indeed pituitary extracts have an opposite, aggregating effect, apparently reflecting the presence of a MCH. Color change in teleosts may be influenced by three types of mechanism.

1. A predominantly humoral one that overrides a neural mechanism, as in *Anguilla*. It may involve MSH and MCH.
2. Predominantly neural coordination that overrides a humoral process but which is still important, such as in *Fundulus*.
3. A neural coordinating mechanism with no evidence for an effect of endogenous MSH, as in *Phoxinus*. The presence of an MCH has, however, not been excluded.

Color change in reptiles

The reptiles have either humoral or neural mechanisms coordinating their color change. Most observations, certainly the most detailed ones, have been made on lacertilians, which often display very dramatic changes in color, as epitomized by the chameleons.

Snakes and crocodilians also possess chromatophores, and a chelonian, *Chelodina longicolis*, has also been shown to exhibit background responses that are mediated by MSH (Woolley, 1957).

The lizard *Anolis carolinensis* exhibits both visual and nonvisual background responses that are abolished following hypophysectomy. The nonvisual response, which can be overridden by the visual one, may be the result of photostimulation of the pineal.

Excitement, such as results from electrical stimulation of the mouth or cloaca, results in a mottling of the skin color patterns in *A. carolinensis* because of a dispersion of melanin in some melanophores and an aggregation in others. This effect can be mimicked by the injection of epinephrine, which may normally mediate the response. The darkening appears to result from β-adrenergic effects and the lightening from α-adrenergic ones. The melanophores are not innervated, and cutting the general nerve supply to the skin does not influence color change.

Chameleons, *Chamaeleo pumila* and *Lophosaura pumila*, exhibit rapid and dramatic changes in color that are either visual responses or mediated by receptors that are apparently present in the skin. The observations that have been made on these responses suggest a neural control of the melanophores; nerve stimulation results in an aggregation of melanin. It is also possible that melanin-dispersing nerve fibers are present. It, therefore, seems that the control of color change in chameleons is a neurally coordinated process though, as no experiments were done on hypophysectomized animals, a subsidiary role of MSH could not be completely excluded. More recent observations (Okelo, 1986) on the East African grass chameleon, *Chameleo gracilis*, suggest that in this species a dual neural and hormonal control system is present. Hypophysectomized chameleons are usually dark green (MI 3 to 4, see Waring, 1963) but never turn completely dark, except when injected with MSH. A yellow or light green color (MI 1 to 2) can be evoked by nerve stimulation or the injection of epinephrine. The color-change control mechanism appears to have similarities to that of the teleost *Fundulus*, but injected MSH can cause skin darkening in the intact lizards. The dark green color of the hypophysectomized lizards could indicate the absence of another pituitary hormonal factor with a melanin-concentrating action. Recorded changes in color in these lizards were completed in less than 30 minutes, a speed which is consistent with a dominant neural control process.

Color change in cyclostomes

The control of color change in the lampreys (Petromyzontoidea) and hagfishes (Myxinoidea) is intrinsically very interesting because of their lowly phyletic position on the vertebrate scale.

A background-color response, pale on a white substrate, dark on a black one, has been described in the hagfish, *Myxine glutinosa*, but the coordinating mechanism for this change is unknown.

Lampreys appear to lack a background response but exhibit a diurnal rhythm in color (dark in the day and pale at night) that in some species is mediated by the pineal gland. Young, in 1935, found that removal of the pituitary abolishes this rhythmical color change in adults and ammocoete larvae of *Lampetra planeri*, and the lamprey then becomes permanently pale in color. In the ammocoetes, pinealectomy also abolished this diurnal rhythm but, in contrast to hypophysectomy, the animals were permanently dark. These observations have been extended (Eddy and Strahan, 1968) to two species of Australian lampreys. These antipodean cyclostomes also exhibit a diurnal rhythm in color that stops following hypophysectomy. In larval *Geotria australis*, pinealectomy also abolishes the rhythm, but in metamorphosing larval *Mordacia mordax*, the lateral eyes must be removed to see this effect.

As described above, the pineal is the site (especially in the dark) of formation of melatonin, which in anuran tadpoles is a very potent stimulant of melanin aggregation and so pales the skin. It has been found that the injection of melatonin into larval *Geotria* also results in skin pallor, though this effect is absent in *Mordacia*. In addition, if the pineal is transplanted under the skin of *Geotria* a local paling is observed. The pineal may, therefore, be involved in regulating the rhythmical changes in color seen in the ammocoetes of *Geotria*, both by the production and release of melatonin and by acting as a photoreceptor organ. Following pinealectomy, the ammocoetes are permanently dark, which may reflect either the lack of an inhibitory effect of melatonin on the release of MSH or, more likely, the absence of a direct antagonism to the action of MSH on the melanophores. It also seems likely, in retrospect, that the same mechanism(s) regulates color change in the ammocoetes of *Lampetra planeri*.

The involvement of melatonin in color change in some larval cyclostomes, some amphibians, and a chondrichthyean is most interesting. Melatonin does not seem to have this role in many vertebrates, but nevertheless it is present in representatives of all the vertebrate groups. The propensity of the pineal to respond to diurnal changes in light makes it a potentially valuable gland for mediating endocrine rhythms dictated by changes in the seasons. As we shall see in Chapter 9, the pineal may in this manner contribute to the control of reproductive cycles in many mammals.

Evolution of color change mechanisms in vertebrates

From the preceding observations on the mechanisms controlling color change, we may make some guesses about the evolution of this process in vertebrates. Information of a comparative nature about extant species from diverse phyletic groups can be interpreted in a manner that may help us to reconstruct the past. Waring has summarized the available information about vertebrate color change in Fig. 7.8, but states that he has been sternly warned "about pressing this kind of thing too far".

The original underlying mechanism coordinating color change in verte-

Fig. 7.8. The geological age of the different vertebrate groups in relation to the types of mechanism (humoral, neural, or both) they utilize to coordinate their melanophore background color responses. The solid bars represent the possession of predominant humoral mechanisms; the broken bars, mixed humoral and neural mechanisms; the small squares, predominantly neural control. To this diagram could be added the Chelonia (*Chelodina oblonga*) and the Ophidia (*Crotalus*), both of which have a humoral control mechanism and are identifiable from the Triassic Period and the chameleon *Chameleo gracilis*, where there is evidence for a dual neural and humoral control mechanism. It can be seen that humoral mechanisms appear to be the oldest, suggesting that neural control is a later acquisition. (From Waring, 1942.)

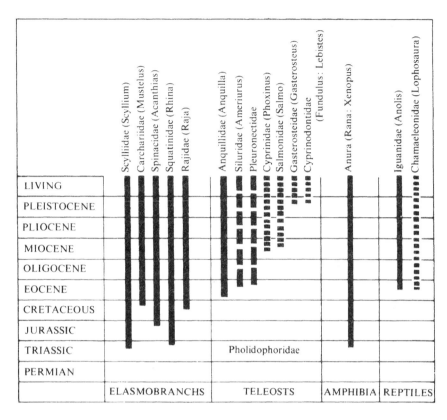

brates would appear to have been a humoral one as this has been observed exclusively in the Chondrichthyes (elasmobranchs) and has persisted in the Anura and Chelonia, all of which can be traced back to the Triassic Period. Superimposed, and probably subsequent to this, has been the evolution of a neural control of the melanophores that is seen in some teleosts and reptiles. In the teleosts, there is evidence of transitional changes as some species appear to utilize both neural and humoral mechanisms. The eels (Anguillidae) are specially interesting in this respect as they are normally completely dependent on humoral control, but there is also evidence for the presence of a neural mechanism that normally does not appear to contribute to color change. We can only speculate as to whether the eel represents a stage in the evolution toward a neural control of color change or is an evolutionary regression from this development. At the bottom of the vertebrate phyletic tree are the cyclostomes, which possess a MSH that helps mediate the diurnal rhythm in color change, but there is little evidence to suggest that these lowly fishes have an adaptive background color response. We can speculate even less as to whether or not the role of melatonin in controlling color changes in larval cyclostomes reflects a primeval effect of this hormone, though it might!

Morphological color change

Several hormones influence morphological changes in the color of the integument. They include MSH and sex hormones; prolactin and melatonin may also be involved. Such humoral effects on pigmentation are less well characterized than those of MSH on physiological color changes. Considerable interspecific differences in humoral effects on morphological pigmentation occur, which make it impossible to give any generalized definition of their respective roles.

Melanin is formed from tyrosine by a complex chain of reactions that initially involves a copper-containing enzyme called tyrosinase. This enzyme may be in a soluble form in the cytoplasm of the melanocyte but when active it is attached to the melanosomes. A genetic absence of tyrosinase results in albinism, though other factors, including ultraviolet light and several drugs, can also contribute quantitatively to changes in pigmentation. Increased levels of integumental melanin are associated with increased tyrosinase activity, which can be influenced by hormones.

The morphological pigmentation of the skin, as well as fur and feathers, is in the first instance the result of the formation of melanin in the epidermal melanocytes, where it is attached to the melanosomes. In the skin, each melanocyte is associated with several keratinocytes to which the melanosomes, with their attached pigment, can be transferred. This functional association is called the "epidermal–melanin unit" (Quevedo, 1972). In birds and mammals, the melanin is passed from the melanocytes associated with the feather tracts of hair follicles to the developing feathers or fur.

Dramatic changes in pigmentation associated with endocrine function have been noted in mammals that show seasonal differences in coat color, in birds that display sexual dimorphism in plumage color, and in mammals suffering from endocrine imbalances. Many monkeys display prominent changes in color of parts of their skin associated with the sexual cycle; these areas are called the "sex skin." At ovulation, the buttocks enlarge and become red in color because of an engorgement of blood in the large venous sinuses. There is also an accumulation of mucopolysaccharides in the skin. This development of the sex skin is under the control of estrogens. These are only a few examples of the pigmentary change that may occur in vertebrates and which are influenced by the action of hormones.

Hormones and seasonal changes in fur color

Short-tailed weasels (*Mustela erminea*) change their coat color from brown to white with the onset of winter (Rust and Meyer, 1968). This change is a photoperiodic response resulting from changes in the length of the daylight hours and can be prevented by hypophysectomy. The latter operation results in a permanent white coat, but the growth of new brown fur can be promoted when the weasels are injected with MSH. In addition, if the pituitary is transplanted to the kidney, where MSH release is increased owing to a lack of hypothalamic inhibition, brown fur also grows on previously white animals.

If melatonin in a "slow-release vehicle" of beeswax is implanted under the skin, weasels undergoing a normal spring molt from a white to brown pelage regrow white, instead of brown, fur (Rust and Meyer, 1969). It is possible that this response reflects a decline in MSH secretion. Melatonin has been shown to directly inhibit melanin synthesis (see later) which provides another possible explanation. Observations on Djungarian hamsters (*Phodopus sungorus*) suggest that another type of mechanism may exist in this species. These Siberian hamsters undergo similar seasonal color transformations and molts to the weasels. They change from a golden brown fur in summer to a white coat at the beginning of winter and then regain their colored fur the following spring (Duncan and Goldman, 1984). These changes have been related to the daylength and the activity of the pineal gland (Badura and Goldman, 1992). Injected prolactin, but not MSH, promotes the growth of the golden brown summer coat in normally white animals. Suppression of endogenous prolactin secretion by injecting a dopamine-like drug (bromoergocryptine) prevented the spring molt to the colored summer coat. Pinealectomy, which blocks the autumnal decline in prolactin secretion, prevented the molt to the winter coat. The physiological determination of fur color in short-tailed weasels and Djungarian hamsters, therefore, appears to differ. While melatonin plays a central role in the seasonal changes in both species, the acquisition of a colored summer coat appears to be related to the secretion of MSH in the weasels but prolactin in the Djungarian hamsters. As described below, the synthesis of the yellow-brown pigment phaeomelanin, in contrast to the brown-black eumelanin, does not appear to be controlled by MSH. Therefore, the difference in the basic type of pigmentation of the two species may be contributing to the observed difference in hormonal control. Possibly phaeomelanin synthesis is controlled by prolactin. A direct effect of melatonin in inhibiting melanin synthesis at the beginning of winter may also be occurring in the hamsters (Weatherhead and Logan, 1981).

Hormones and morphological color change in cold-blooded vertebrates

Cold-blooded vertebrates that undergo physiological color changes have also been shown to increase the levels of melanin in their skin in response to a continual environmental "black-background" stimulation of MSH secretion. Such a change has been observed in amphibians, teleosts, and chondrichthyeans. In the goldfish, *Carassius auratus*, corticotropin increases the cutaneous levels of melanin and the activity of tyrosinase (Chavin, Kim and Tchen, 1963), an effect that cannot be mimicked by mammalian MSH. In contrast, MSH, but not corticotropin, increased cutaneous melanin synthesis in the killifish, *Fundulus heteroclitus* (Pickford and Kosto, 1957) and one must, therefore, be careful not to draw any general systematic conclusions about the role of such hormones in melanin synthesis in the Teleostei. As the hormone preparations used in the fishes were of mammalian origin, the differences in the responses could reflect the degree of similarity of these exogenous hormones to the particular endogenous MSH present in each species of fish.

The effects of changes in the level of endogenous MSH on melanin levels in the melanophores have been observed in the dogfish *Scyliorhinus canicula* (Wilson and Dodd, 1973b). Removal of the neurointermediate lobe of the pituitary in this chondrichthyean resulted in an almost complete loss of melanin from the skin. In the converse experiment, when increased circulating levels of MSH were promoted by severing the inhibitory hypothalamic connections to the intermediate lobe, there was an increased concentration of melanin in the skin.

Cold-blooded vertebrates may also exhibit morphological changes in skin color and pattern that are of a rather colorful nature. A teleost, the stoplight parrot fish, *Sparisma viride*, can change its sex from female to male and it then assumes a brightly colored integument. This changeover is associated with a decline in plasma estrogen concentrations and a sharp rise in the levels of the potent teleost androgen 11-ketotestosterone (Cardwell and Liley, 1991). Injection of the androgen into the female fish induces the sex change and the accompanying color change. Some female lizards develop colored, orange and orange-red, spots on various parts of their bodies during the period of the development of their eggs in the body; these spots are, therefore, called "pregnancy spots" (Cooper and Ferguson, 1972; Ferguson and Chen, 1973; Medica, Turner, and Smith, 1973). Such changes in color have a hormonal basis, which has been examined in the collared lizard, *Crotaphytus collaris* and the leopard lizard, *C. wislizenii*. The injection of progesterone induces such pigmentation in ovariectomized lizards, and estrogen increases the response though it is ineffective alone. The natural levels of these hormones change during the growth of the eggs, and their circulating concentrations have been measured in such lizards and can be correlated with the development of the pregnancy spots. The injection of FSH also induces the formation of such pigmented areas in leopard lizards. It would appear that under natural conditions a release of gonadotropin stimulates the development of the ovarian follicle together with a release of gonadal steroids (Chapter 9) and these directly mediate the response. The role of pregnancy spots in these lizards is uncertain, but it has been suggested that they may deter the males from inappropriate amorous advances.

Hormones and sexual dimorphism in avian plumage color

The mechanisms by which hormones mediate the seasonal sexual dimorphism in the color of plumage in weaver birds, *Steganura paradisaea*, and the nonseasonal differences in domestic brown leghorn fowl have been studied by Hall (1969).

The male weaver bird grows prominent black feathers just before (prenuptial) the breeding season. When areas of white feathers are plucked from these birds, the injection of pituitary LH results in the appearance of melanin granules in the feather tracts of these areas and a related growth of black feathers. The formation of melanin is associated with an increased activity of tyrosinase in the feather tracts. The effect of LH is direct and is not mediated through any action on the gonads as it is still seen in castrated birds.

The male house finch, *Carpodacus mexicanus*, has red or orange feathers on its crown, throat, and belly, which is in contrast to the female in which this plumage is brown. When the colored feathers are plucked from castrated males, the new, regrown feathers are of the female, brown type, which contrasts with the renewal of the colored plumes in intact birds (Tewary and Farner, 1973). It thus appears, that as in the male weaver bird, the more gaily colored plumage of the male is determined by the presence of male sex hormones. It seems, however, that these hormones are from the gonads of the finches, though an indirect action that could involve pituitary is also possible.

The male domestic brown leghorn fowl has black feathers on its neck and breast; those in the female are pinkish brown. In this instance, the coloration of the male plumage is not hormone dependent but that of the female results from the action of estrogens. The injection of estradiol increases melanin formation and the activity of tyrosinase. The estrogens act directly on the feather tracts as their action is a local one at the site of the injection (Greenwood and Blyth, 1935).

Mechanisms of hormone-mediated changes in integumental melanin distribution

Physiological color change is an alteration in the dispersion of melanin in the melanophores, such as that mediated by MSH and epinephrine, and this is related to the level of cellular cyclic AMP. The levels of this nucleotide are changed in response to the presence of MSH, which increases it, and sympathetic nerve stimulation, which usually decreases it. The MSH activates adenylate cyclase (Novales, 1972) while neural α_2-adrenergic stimuli reduce the enzyme's activity (Anderrsson, Karlsson, and Grundström, 1984; Morishita *et al.*, 1993). An activation of melanophore adenylate cyclase and a dispersion of melanin have also been described in response to β-adrenergic stimuli in the spadefoot toad, *Scaphiopus couchi*, the lizard *Anolis carolinensis*

(Abe *et al.*, 1969; Goldman and Hadley, 1969) and, possibly, the toad *Xenopus laevis* (van Zoest *et al.*, 1989). The adrenergic responses may occur in response to epinephrine in the plasma. The mechanism of the melanin-aggregating effect of MCH in teleosts does not appear to have been described but its synergistic effect with norepinephrine (Green and Baker, 1989) suggests that they may have different, but interacting mechanisms. The response of the teleost melanophore may also involve a rise in intracellular free calcium ions in response to IP_3 formed by an activation of the phospholipase C–phosphatidylinositol system (Fuji, Wakatabi, and Oshima, 1991). Such proposed changes in calcium have not, however, been detected in all teleosts (Sammak *et al.*, 1992). The melanin-containing melanosomes can move centripetally (aggregation) or centrifugally (dispersion) along the melanophore microtubular system. A response to cyclic AMP is mediated by the regulation of cell proteins; kinesin producing a dispersion and dynein an aggregation of melanin (Sammak *et al.*, 1992).

Comparison of the effects (*in vitro*) of α-MSH on the melanophore of a reptile, the lizard *Anolis carolinensis*, an amphibian, the frog *Rana pipiens*, and a teleost, the Brazilian fresh water eel *Synbranchus marmoratus*, suggests that there are intergroup differences between the receptors mediating the responses to the hormone (Castrucci *et al.*, 1989). The α-MSH molecule includes a heptapeptide segment, α-MSH-(4–10) Met–Glu–His–Phe–Arg–Trp–Gly) which, with a minor exception, appears to be completely conserved among the vertebrates (Chapter 3). The acetylated analog of this peptide, Ac-α-MSH-(4–10)NH$_2$, was active in all three species tested, though it was somewhat less potent than the intact hormone. The minimally active sequence in the reptile and amphibian was Ac-α-MSH-(6–9)NH$_2$ while in the teleost it was a larger peptide, Ac-α-MSH-(5–10)NH$_2$. The tetrapods appeared to have similar recognition sites on their receptors but that of the teleost was different. Accoutrements in excess of the "minimal sequence" may influence other aspects of the hormone's action, such as its accessibility and ease of binding to its receptor sites, and its survival in the plasma under *in vivo* conditions.

The paling response of the melanophores of larval amphibians and lampreys to melatonin appears to be mediated by a specific G protein-coupled receptor that can inhibit adenylate cyclase. Such receptors were first cloned from dermal melanophores of the toad *Xenopus laevis* (Ebisawa *et al.*, 1994; Reppert and Weaver, 1995). These melatonin receptors belong to a distinct family, the homologs of which have been identified at various tissue sites in several mammals, chickens, and zebra fish. They may initiate several responses. Such receptors have, for instance, been found in the pars tuberalis and suprachiasmatic nucleus of sheep, where they may mediate circadian effects on reproductive rhythms (Reppert, Weaver, and Ebisawa, 1994; see also Chapter 9).

Morphological color change involves an increased synthesis of melanin in the cutaneous melanocytes under the influence of tyrosinase. Two types of melanin may be formed: brown-black eumelanin and yellow-red-brown phaeomelanin. In a strain of mice in which the juveniles have a dark brown pelage, MSH promotes the formation of eumelanin but not the phaeomelanin, which occurs in the adults (Burchill, Thody, and Ito, 1986; Burchill and Thody, 1986). The effect of MSH depends on an activation of adenylate cyclase. The response involves G protein-coupled receptors that have been cloned from mouse and human melanoma cells (Mountjoy *et al.*, 1992). Defective MSH receptors are present in a laboratory strain of mice which then, even in the presence of MSH, only form a yellow pigment (Robbins *et al.*, 1993). The nature of the change in tyrosine activity has been controversial as it could involve an activation of extant enzyme or the *de novo* synthesis of additional enzyme. In mouse melanoma cells, the latter occurs (Fuller and Viskochil, 1979; Halaban *et al.*, 1984). Regulation of a post-tyrosinase step in melanin synthesis, however, may still take place. On the basis of *in vitro* observations on mice melanocytes, it has been suggested that melatonin may inhibit melanin formation in this way (Weatherhead and Logan, 1981).

In birds (Hall, 1969), the action of LH in increasing cutaneous tyrosinase activity does not appear to involve the synthesis of a new enzyme, as its effect is not inhibited by puromycin. The effect of estradiol on pigmentation in the brown leghorn fowl, however, *is* inhibited by puromycin so that the action of this hormone may involve the induction of tyrosinase. The many effects of hormones on integumental pigmentation, therefore, may be reflected in a diversity in the mechanisms by which they exert their effects. At the present time, however, it would appear that the activity of tyrosinase is central to their morphological actions

Has there been an evolution of the roles of MSH and MCH?

Although MSH has well-established effects in mediating color change in cold-blooded vertebrates, its normal role in birds and mammals in uncertain. MSH undoubtedly, in some circumstances and in certain species, stimulates the synthesis of melanin in the skin of mammals, but whether this is its only physiological role is not clear. It is possible that MSH may exhibit such a function, especially in species like the weasel that seasonally change the color of their coat. Such an effect is, however, limited to a relatively few species, and yet mammals, as well as birds, apparently possess MSH and sometimes in several molecular forms (Chapter 3). Does it then have other physiological effects in these animals? The numerous possible physiological roles of MSH have been reviewed by Lincoln and Baker (1995). Despite a widespread search and considerable informed speculation, unequivocal answers have not yet emerged.

The phyletic persistence of MSH-like activity in extracts of the pituitary gland of birds and mammals has formerly suggested that such a molecule probably endows the animal with some selective advantage and could reflect its role as a hormone. However, it is now known that α-, β- and γ-MSHs are integral amino acid sequences of a single large precursor protein, pro-opiomelanocortin (POMC), which is the product of a single gene (Chapter 3). It would, therefore, seem that the evolutionary survival of these small peptides need not necessarily provide an important individual selective advantage. They are linked to a complex of other hormones in some of which they are even a constituent part. The MSHs nevertheless may have other physiological roles apart from those concerned with color change and pigmentations. Alpha-MSH has been identified at sites in the brain (Oliver and Porter, 1978), leading to speculation that it may have neural functions. Its propensity to promote melanin synthesis by activating or inducing tyrosinase, which is an enzyme also involved in catecholamine synthesis, has suggested the interesting hypothesis that it may be involved in brain metabolism and function (Shuster *et al.*, 1973). MSH injected into the brain of rats has been shown to result in a decrease in the sensation of pain (Walker, Akil and Watson, 1980). The melanotrope cells of the mammalian pars intermedia are known to release their analgesic endorphin-related peptides in response to stress or the injection of epinephrine (Berkenbosch, Tilders, and Vermes, 1983). A co-secretion of such β-endorphin and α-MSH has been observed from the pars intermedia in Soay sheep (Lincoln and Baker, 1995). A seasonal pattern of such secretion apparently related to daylength occurs, with the highest rates in summer and the lowest ones in winter. These changes may be related to other cyclic endocrine events, such as changes in the release of prolactin, FSH and corticosteroids, as well as fat storage and the activities of skin glands. Teleost fish kept on a black background have an elevated plasma level of MSH and an enhanced ability to release cortisol in response to stress (Gilham and Baker, 1985). It was suggested that MSH may be facilitating the release of CRH or corticotropin. MSH, therefore, may modulate the activities of neurons, especially those with neurosecretory roles. Should it be classified as a neuropeptide?

The only established role of MCH is related to color change in teleost fish but it has a widespread distribution in neural tissues of many vertebrates and its structure has been highly conserved. It is clear that it must have other functions and it has been suggested that these may be concerned with neurotransmission and neuromodulation (see Baker, 1994). Secretion of MCH (*in vitro*) occurs from the hypothalami of rats and this process can be enhanced by cyclic AMP and corticosteroids (Parkes and Vale, 1992), responses that could reflect such neural roles. Other suggestions regarding the function of MCH include an influence on gustatory and auditory sensations, and a release of hormones involved in osmoregulation, lactation, and adaptation to stress. It is considered likely that MSH and MCH have other physiological functions apart from those involved in controlling an animal's hue, and they need not necessarily be endocrine ones.

Conclusions

The integument is a very complex tissue that may be involved in several physiological phenomena, including osmoregulation, color change, temperature regulation, and reproduction. There are many characteristic processes involved in such mechanisms that have an exclusive systematic distribution (for example, sweat glands, mammary glands, and branchial chloride-secreting cells) so that when hormones are involved, as they often are, their effects follow phyletic suit. Such responses must have also arisen at distinct time during vertebrate evolution. It is interesting to observe that there is often a definite relationship between the nature of the particular hormone and the general physiological process involved: sex hormones influence sexual processes in the skin, as well as elsewhere, and adrenocorticosteroids regulate electrolyte movements in sweat glands and across the amphibian skin, as well as in the kidney (Chapter 8). Some hormones have a special propensity to mediate processes in the integument. MSH influences pigmentation in nearly all groups of vertebrates commencing phyletically with the cyclostomes, and, on occasions, it may even promote formation of melanin in mammalian skin. Its action, however, appears to be rather "conservative" as it has no other established effect on any other types of process in the body nor for that matter in the integument either. There are, however, some theories as to other possible roles. In contrast, prolactin is "versatile", as apart from effects at nonectodermal sites, it influences many integumental processes including molting cycles, the secretions from the mucous and mammary glands, proliferation of the crop-sac and development of the brood-patch in birds, and the control of water and salt movements across the gills of fishes. Some of these effects will be described in the next chapters.

8 Hormones and osmoregulation

About 70% of the body weight of animals is water, in which is dissolved a variety of solutes, the presence of many of which is vital for life. Within the body, the solutions inside the cells differ from those that bathe the outside, and the composition of each of these solutions must be maintained so as to provide an environment with an electrolyte content and osmotic concentration suitable for life. These intra- and extracellular fluids provide the framework in which life exists.

The physiochemical properties of the body fluids in animals usually differ greatly from those of their external environment. Animals continually suffer exposure to the whims of the exoteric conditions and this will tend to change the composition of their body fluids. In addition, although the intra- and extracellular fluids have identical osmotic concentrations, there are qualitative differences in the solutes they contain, and equilibration, through diffusion, will tend to occur. Such animals, however, maintain the gradients between their body fluids and the environment, an equilibrium that is maintained as a result of a complex pattern of physiological events. These processes involve the cells, and special tissues and organs that are concerned with osmoregulation. The integration of the functions of these homeostatic tissues relies largely on hormones. The nervous system makes little direct contribution to such regulatory processes, though at the cellular level itself considerable autoregulation, independent of hormones, exists. Hormones do ultimately influence some cellular processes, of course, but they generally appear to do this in effector tissues like the kidney, gills, and gut, which are especially concerned with the overall osmoregulation of the animal. For a more complete account of the role of hormones in osmoregulation the book by Bentley (1971) and the more recent compendium collected by Pang and Schreibman (1987) could be consulted. Animals occupy diverse osmotic environments; the major ones are the sea, fresh water, such as rivers and lakes, and dry land. Differences exist between the availability of water and salts within these environments and this is particularly apparent to animals that lead a terrestrial life. Water may be relatively freely available to some terrestrial species that live in areas where rainfall is high, and lakes, ponds, and rivers exist in close proximity to where they live. Other animals, however, live in dry, desert regions where water may only be available sporadically and in limited quantities. Salts are freely available to marine animals, but in fresh water the supplies are more restricted, and some terrestrial animals may occupy regions where a low salt content in the soil may be reflected in a salt-deficient diet. It is, therefore, not surprising that the processes for controlling the water and solute content of the body, called osmoregulation, can differ considerably between species that habitually occupy diverse ecological situations. These differences may be manifested as a tolerance to osmotic changes but are principally seen as differences in the functions of the tissues and organs concerned in maintaining the composition of the body fluids. Not unexpectedly, the evolution of the tissues and organs concerned with osmoregulation has been accompanied by changes in the role of the endocrine glands that help integrate their functioning.

Homeostatic events that contribute to osmoregulation may involve changes in either the rates of loss of water and solutes or in the processes of their accumulation

Osmoregulation in terrestrial environments

Terrestrial vertebrates may lose water as a result of evaporation from the skin and respiratory tract (Fig. 8.1). This loss is greatest in hot, dry air in which little water vapor is present. In homeotherms, evaporation from the respiratory tract and the skin may be increased. In the latter this may be owing to the activity of the sweat glands, as a result of the need to dissipate heat. While evaporation will be the predominant route for water loss in animals living in hot, dry conditions, additional quantities pass out of the kidneys in the urine, and out of the gut in the feces. Reproductive processes, such as egg laying and lactation,will also result in increased water losses. Water is gained in the food from which substantial quantities may be absorbed across the intestines, and as a result of drinking. In addition, amphibians, like frogs and toads, can absorb water across their permeable skin, from damp surfaces and pools of fresh water. Any excessive water that may be gained in such ways is excreted by the kidneys.

Salt losses in land-living vertebrates occur in the urine and feces and, in mammals, in the sweat gland secretions. Some birds and reptiles possess special glands in their heads called cephalic salt glands, which have an ability to secrete concentrated solutions of salts. Some regulation of the losses from the sweat glands and gut occurs, and the secretion of the kidneys undergoes a rigorous process of conservation so that salts that are deficient in the body may be conserved. Additional conservation of urinary salts can occur from urine during its storage in the urinary bladder of amphibians and some reptiles. Birds and many reptiles lack a urinary bladder, but in such animals the urine passes into the cloaca and up into the large intestine, where some of

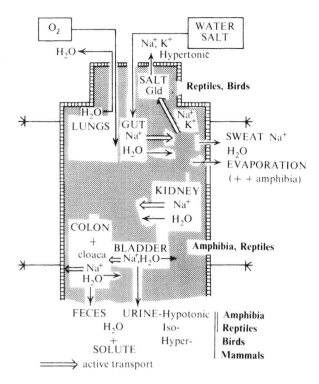

Fig. 8.1. A diagrammatic summary of the pathways of water and salt losses and gains in vertebrates living on dry land. (From Bentley, 1972.)

its contained salts and fluid can be transported back into the blood. Salts are mainly gained in the diet of terrestrial animals, and the drinking of brackish water may also result in salt accumulation. The latter process, however, occurs only rarely in nature. Excesses of accumulated salts can be excreted by the kidneys and in many birds and reptiles by the cephalic salt glands.

Osmoregulation in fresh water

Many species of vertebrates live in fresh water, a solution that is hypoosmotic to the body fluids and which only contains small amounts of dissolved solutes. Water will therefore tend to be gained by osmosis across the integument of such animals. Amphibians have a relatively permeable skin and can take up large amounts of water in this way. Fishes respire with the aid of gills, which, apart from allowing the exchange of oxygen and carbon dioxide, are also permeable to water and so are an additional route for the accumulation of water in the body (Fig. 8.2). The skin of reptiles and mammals that frequent fresh water is usually quite impermeable so that little water is accumulated through this channel and, as they breathe with the aid of lungs, they do not suffer the osmotic problems associated with the presence of gills.

Vertebrates living in fresh water may be prone to a greater salt loss than their terrestrial or marine relatives. When the integument is permeable, such

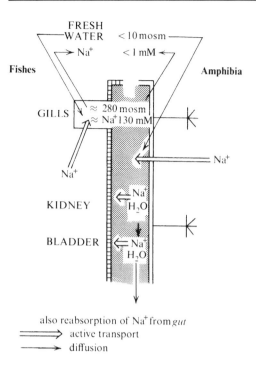

Fig. 8.2. A diagrammatic summary of the pathways for water and salt losses and gains of fishes and amphibians that live in fresh water. (From Bentley, 1972.)

as the skin in the amphibians and the gills of fish, salt loss may be expected to occur as a result of diffusion. Such potential losses, however, are rather limited and are much smaller than may be expected on a simple physiochemical basis. In addition, salt losses may continually occur in the urine because of the necessity for the excretion of water that has accumulated by osmosis.

The solute excretion in the urine is reduced by its reabsorption from the renal tubules. In aquatic amphibians and reptiles (such as turtles) and some fishes, sodium is also reabsorbed from the urine that is stored in the urinary bladder. In species lacking a urinary bladder, such as crocodiles and birds, solutes may be reabsorbed in the posterior regions of the gut following the passage of urine into the cloaca.

Salts are principally obtained from the food of aquatic vertebrates. However, additional gains of sodium chloride may be made as a result of its active transport, against electrochemical gradients, across the skin of amphibians and the gills of fish. It has also been suggested that some turtles may actively accumulate sodium across their pharyngeal and cloacal membranes during their irrigation by the external fresh water solution. Primary active salt transport across the gills and rectal gland appears to involve chloride ions with sodium ions following along the created electrochemical gradient.

Fig. 8.3. A diagrammatic summary of the pathways for water and salt losses and gains in osteichthyean and chondrichthyean fishes that live in the sea. (From Bentley, 1972.)

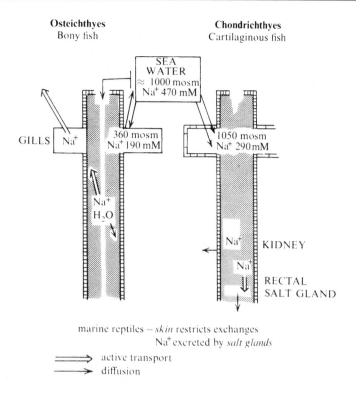

Osmoregulation in the sea

Most species of fishes as well as a number of reptiles, mammals and at least one frog (the crab-eating frog, *Rana cancrivora*) live in the sea. This solution is strongly hyperosmotic to the body fluids of most vertebrates. The exceptions are the hagfishes (Agnatha), the sharks and rays (Chondrichthyes), and the coelacanth, which are isoosmotic, or slightly hyperosmotic, to sea water. The crab-eating frog is also slightly hyperosmotic to the sea water in which it lives.

Most of the bony fishes (Osteichthyes) tend to lose water by osmosis, especially across their gill membranes. The Chondrichthyes and myxinoid Agnatha, by comparison, may gain small amounts of fluid in this way. The sodium chloride concentration in the sea water is much higher than that in the blood of osteichthyean or even chondrichthyean fishes so that an accumulation of salt will tend to occur (Fig. 8.3).

The mechanisms utilized to maintain osmotic balance in sea water are varied (Kirschner, 1980). Marine teleost fish drink sea water and much of the salt that is present is absorbed across the gut wall; water follows this solute by osmosis. The salt is excreted by special cells called chloride-secreting cells, or simply chloride cells, present in the gills and skin. Divalent ions are excreted in the feces or urine. The gain of salts by marine chondrichthyeans is small compared with that of teleosts, and this is excreted in the urine and as a

Table 8.1. *Target organs for osmoregulatory-type responses to hormones in vertebrates*

The responses in *italics* appear to be physiological ones while others are either only pharmacological or the evidence for their normal, *in vivo* role is as yet equivocal. The responses are not necessarily present in all members of the orders of vertebrates that are indicated

Target organ	Phyletic distribution of target organ	Stimulatory hormone	Nature of response	Phyletic distribution of responsiveness
Kidney	All vertebrates	Aldosterone	Decreased Na+ excretion (tubular)	Mammals, birds, reptiles
		Angiotensin	Diuresis or antidiuresis (GFR)	Various vertebrates
		Natriuretic peptides	Natriuresis, diuresis (GFR, tubule)	Various vertebrates
		Vasotocin	Decreased GFR and increased renal tubular water reabsorption: decreased urine flow (antidiuresis)	Amphibians, reptiles, and birds
			Decreased GFR: antidiuresis	Teleost (eels)
			Increased GFR: diuresis	Some teleosts and lungfishes
		Vasopressin	*Increased renal tubular water reabsorption: antidiuresis*	Mammals
Urinary bladder	Most vertebrates (except birds, some reptiles, and many fishes)	Vasotocin	*Increased water and Na+ reabsorption*	Amphibians (mostly anurans)
		Aldosterone	Increased Na+ reabsorption	Amphibians, reptiles, and mammals
		Prolactin	Reduced water permeability, increased Na+ reabsorption	Teleost (starry flounder)
		Urotensin II	Increased Na+ reabsorption	Teleost (goby)
Gills	Fishes	Cortisol	*Increased outward Na+ secretion*	Marine teleosts
			Increased inward Na+ absorption	Fresh water teleosts
		Growth hormone	*Proliferation of chloride cells*	Teleosts
		Natriuretic peptides	Increased Na+ efflux	Teleosts (SW)
		Epinephrine	Decreased α or increased β secretion by chloride secreting cells	Teleosts

Organ	Group	Hormone	Action	Species
		Prolactin	*Decreased Na⁺ diffusion outward and water accumulation (inward)*	Euryhaline teleosts (FW)
		Urotensin I	Decreased Na⁺ extrusion	Euryhaline teleosts (SW)
		Urotensin II	Increased Cl secretion (*in vitro*)	Teleost (SW) chloride-cells
			Decreased Cl secretion (*in vitro*)	Teleost chloride cells
		Vasotocin	Increased inward or outward movements of Na⁺	Fresh water (inward) or marine (outward) teleosts
Skin	All vertebrates	Vasotocin	*Increased water and Na⁺ absorption*	Some amphibians
		Aldosterone	*Increased Na⁺ absorption*	
		Prolactin	Decreased permeability to water and Na⁺	Urodele amphibians (?)
Sweat glands	Mammals	Epinephrine	Increased Cl secretion	Teleost (killifish)
		Aldosterone	*Reduced Na⁺ in sweat*	
Salt glands	Some birds and reptiles (nasal and orbital glands) and chondrichthyeans (rectal glands)	Epinephrine	*Increased secretion during exercise*	
		Aldosterone	Reduces Na⁺ secretion	Lizard
		1α-Hyroxycorticosterone	Increases Na⁺–K ATPase	Chondrichthyeans
		Vasoactive intestinal peptide	Increases secretion (*in vitro*)	Chondrichthyeans
		Natriuretic peptides	Increase secretion	Chondrichthyeans and birds
Salivary glands	Mammals	Aldosterone	*Decreases Na⁺ and increases K⁺ loss*	
		Epinephrine	*Dries up secretion*	
Intestine	All vertebrates	Cortisol	Increased NaCl absorption	Teleosts
		Aldosterone	*Increased NaCl absorption in colon*	Mammals, birds, amphibians

GFR, glomerular filtration rate; SW, sea water; FW, fresh water.

concentrated solution from a tissue unique among the vertebrates, the rectal salt gland. This salt-secreting organ is situated in the nether regions of the gut. Marine reptiles appear to have a relatively impermeable integument that restricts the gain of salt, but excesses, such as may be gained in the food, can be excreted by the cephalic salt glands that are often modified tear, or orbital, glands.

It can be seen that the osmotic problems of vertebrates differ considerably and depend on the environment where they live as well as the anatomical and physiological wherewithal that is conferred by their phylogeny. The maintenance of osmotic homeostasis is dependent on a variety of tissues and glands, some of which, like the kidney, are present in all the major phyletic groups of vertebrates, while others, like cephalic salt glands and the rectal gland, have a more restricted distribution. The activities of many of the organs and tissues involved in osmoregulation are controlled by hormones. A summary of these is given in Table 8.1.

Osmoregulation in vertebrates is dependent on the active participation of such tissues as the kidneys, gills, skin, urinary bladder, gut, and certain salt-secreting glands. These tissues contribute to the excretion and conservation of water and salts, and their roles and physiological significance are not the same in all species.

The hormones that influence osmoregulation most directly are the neurohypophysial hormones, principally vasotocin and ADH (vasopressin), the adrenocorticosteroids, the catecholamines, atrial natriuretic peptides (ANP), and prolactin. Corticotropin and angiotensin are indirectly involved because of their roles in controlling the release of adrenocorticosteroids. Angiotensin may also have a more direct effect on some membranes while the urophysis, which has a putative endocrine function, may also be involved.

Active transport and secretion of ions, especially sodium, potassium, bicarbonate, and chloride, across or from the epithelial membranes that make up the tissues that effect osmoregulation are basic to their physiological function. Such membranes are osmotically permeable to water, which can pass across them with an ease that may vary, depending on the membrane and the physiological conditions. The adequate functioning of these osmoregulatory tissues is ultimately dependent on their blood supply. Hormones may, therefore, influence the activity of osmoregulatory tissues by actions at several sites.

● Hormones may alter the processes of active sodium and chloride transport and the secretion of hydrogen ions, bicarbonate, and potassium. Cortisol and aldosterone can alter sodium and potassium movements across many epithelial membranes. Catecholamines can increase transport and secretion of chloride ions in several tissues and also can inhibit the effects of ADH. All these hormones act directly on the cells

involved. Vasotocin, guanylins, angiotensin, and ANPs may also directly influence ion transport across epithelial membranes but the physiological significance of such effects is often not clear.

- Osmotic and diffusional movements of water and sodium across epithelial membranes can be changed by hormones. Vasotocin and ADH
may increase the permeability to water of the renal tubule (in some
species), as well as that of amphibian skin and urinary bladder. Prolactin
can reduce the permeability of the gills of certain teleost fishes to
sodium, thus limiting diffusional losses of this ion in fresh water.
- The catecholamines, adrenomedullin, angiotensin, ANPs, and neurohypophysial peptides can alter the diameter of blood vessels and so may
influence the functioning of osmoregulatory tissues by virtue of their
vasoactive actions. The urine flow, especially in nonmammals, may be
influenced by changes in the rate of filtration of plasma across the
glomerulus and this process can be altered by these hormones. Alterations in secretion and absorption of ions across the gills of fishes may
also be changed by such hormonally mediated variations in the regional
blood flow.
- The functional differentiation of chloride-secreting cells in the teleost
gills may be promoted by growth hormone and cortisol. Such an effect
on cells at other osmoregulatory sites is possible

While this chapter is principally concerned with hormones, the role of
nerves in osmoregulation should also be mentioned. Neural integration is not
common, though it does occur. The cephalic salt-secreting glands in birds
and reptiles are stimulated to secrete as a result of the stimulation of autonomic cholinergic nerves (Peaker and Linzell, 1975). The sweat glands of
mammals secrete in response to the need to dissipate heat, and this usually
occurs following stimulation of autonomic cholinergic or adrenergic nerves.
However, sweat gland secretion during exercise may depend on circulating
epinephrine. The vasoconstrictor tone of blood vessels is primarily dependent
on the activity of adrenergic nerves that can thus, indirectly, alter the
functioning of tissues. This effect is sometimes observed in the kidney of
animals but is probably not a usual physiological mechanism.

The role of hormones in osmoregulation

Mammals

These vertebrates have a complement of hormones, similar to other vertebrates, that can influence osmoregulation, though the roles of such secretions
and their precise structures may differ somewhat from those in nonmammals.

Vasopressin   Vasopressin is unique to the mammals. It reduces urinary water losses (antidiuresis) as a result of an increased osmotic reabsorption of water from the kidney tubules. Vasopressin's phyletic forebear, vasotocin, as we have seen, has a slightly different chemical structure that confers on it a pronounced ability to contract smooth muscle, such as that in the oviduct and uterus, and also, when injected, to promote contractions of the myoepithelial cells in the mammary glands. Injected vasotocin also has an antidiuretic action in mammals and it could, therefore, conceivably function in such a physiological role. Vasopressin, however, lacks the prominent effects that vasotocin has on nonvascular smooth muscle contraction and so exerts a more specific action in the body. Its evolutionary perpetuation in mammals is, therefore, not surprising. Arginine-vasopressin (AVP, ADH) is the most common antidiuretic hormone among mammals (Chapter 3). However, lysine-vasopressin (LVP) can serve this function in some Suiformes (including the domestic pig) and macropodid marsupials (kangaroos). The latter also may possess another homolog, phenypressin, (Phe^2-AVP), which can also function as an antidiuretic hormone. Both pigs and marsupials have well-developed antidiuretic mechanisms and differences in the structures of such hormones do not result in major declines in their antidiuretic activity, even when they are administered to heterologous species. (LVP has been extensively used in replacement therapy for AVP in humans.) Such structural changes possibly could be compensated for by adjustments in their receptor sensitivity or a more generous release into the circulation. Vasopressin does not appear to have any other physiological role, on other organs, in mammals except, possibly, to help promote the release of corticotropin. It can, however, exert other effects, such as increasing the blood pressure, contracting the uterus, raising blood sugar levels, and may possibly facilitate memory and learning when it is administered.

Adrenocorticosteroids   Adrenocorticosteroids play an important role in controlling sodium and potassium metabolism in mammals. The absence of the adrenal cortex in mammals quickly results in death, resulting mainly from losses of sodium and an accumulation of potassium. Aldosterone is the most effective of the adrenocortical hormones that exhibit actions on sodium and potassium metabolism in mammals, though the others, especially corticosterone, can also exert such effects. Sodium excretion from the kidney, sweat glands, and salivary glands is reduced while potassium loss is increased; there is a drop in the ratio of sodium/potassium in the secreted fluids. Aldosterone can promote sodium reabsorption from the large intestine and the mammary gland ducts (Yagil, Etzion, and Berlyne, 1973; Dolman and Edmonds, 1975). The reabsorption of sodium from the rabbit urinary bladder is also promoted by aldosterone (Lewis and Diamond, 1976). The osmoregulatory effects of the corticosteroids in mammals are, therefore, all directed to the same general purpose (i.e. sodium conservation and potassium excretion) and seem to involve at least five different target issues. Such mineralocorticoid effects are mediated in their target tissues by type 1 corticosteroid receptors, (type II are involved in glucocorticoid responses). Following the isolation and cloning of these receptors it was found that under *in vitro* conditions type I receptors not only bound aldosterone but also the glucocorticoids cortisol and corticosterone. The question then arose as to how aldosterone exerts its selective mineralocorticoid effects *in vivo*. Differential binding to plasma proteins may contribute (aldosterone is poorly bound) but this difference only provides a partial explanation. A tissue enzyme, 11-hydroxysteroid dehydrogenase (11-HSD) is present in target tissues, such as the kidney (Funder, 1993). This enzyme can readily inactivate cortisol and corticosterone by converting them to 11-ketosteroids. This reaction, however, is prevented in the aldosterone molecule by the presence of its aldehyde group at the nearby C-18 position. (Such an effect has also been found in the gut of the domestic fowl (Grubb and Bentley, 1992).)

Catecholamines   Epinephrine has a less prominent role in osmoregulation. It can stimulate the secretion of sweat glands in some species. In addition, epinephrine can antagonize the release of ADH, and its effects on the kidney. Such inhibition is an α-adrenergic action that can be demonstrated in experimental animals, though its possible physiological importance is not clear. It can also promote the release of renin (β-adrenergic effect).

Renin–angiotensin system   Angiotensin, apart from initiating the release of aldosterone, has been shown to promote sodium reabsorption from the kidney tubule and the rat colon *in vitro* (Davies, Munday, and Parsons, 1970; Munday, Parsons, and Poat, 1971). However, the normal physiological significance of this effect is in doubt (Dolman and Edmonds, 1975). Another interesting effect of angiotensin is its ability, when injected, to promote drinking (Fitzsimons, 1972; 1979). Drinking is elicited by the sensation of thirst that arises in the brain in a number of circumstances, including a reduction in the volume of the extracellular fluids. This latter response is reduced if the kidneys are removed, suggesting that the renin–angiotensin system may be involved. Indeed, the injection of small amounts of angiotensin II into the region of the "thirst center" in the anterior diencephalon of the brain promotes drinking. These effects are mediated by the hormone's action on the subfornical organ and organum vasculosum in the brain. Receptor sites for angiotensin II have been directly observed in the latter, using a fluorescent analog of the peptide (Landas *et al.*, 1980). The drinking of salt solutions (sodium-appetite) can also be promoted by the administration of angiotensin II into the brain or plasma of several mammals including rats, mice, sheep, cattle, and rabbits (Bryant *et al.*, 1980; Weisinger *et al.*, 1987; Thunhorst and Fitts, 1994; Denton, McKinley, and Weisinger, 1996;

Mosimann *et al.*, 1996). The sodium-appetite occurs normally in response to sodium depletion and this effect can be reduced by the administration of antagonists of the action and inhibitors of the synthesis of angiotensin II. The site of action of angiotensin II appears to be hypothalamus in a region that is distinct from the thirst center (Weisinger *et al.*, 1993). However, the precise mechanism of action and physiological significance of this interesting effect of angiotensin is not yet clear. This "Na-appetite" can also be induced in sheep by the injection of ACTH, an effect that depends on the release of corticosteroids (Weisinger *et al.*, 1980). It seems possible that these two hormones may normally interact to adjust the animal's taste for salt.

Atrial natriuretic peptide   ANP is a recently discovered acquisition to the mammalian osmoregulatory armamentarium (Chapter 3). (It is also present in nonmammals.) This hormone is released from the atria of the heart in response to distension resulting from expansion of the intravascular space (Brenner *et al.*, 1990; Atlas and Maack, 1992). It has a vasodilator effect and decreases the cardiac output. There is also a prominent diuresis and natriuresis resulting from several actions on the kidney.

- The glomerular filtration rate (GFR) is increased through a dilatation of the afferent glomerular arteriole and constriction of the efferent arteriole. Other effects, however, contribute to the responses, as emphasized by the observation that an aglomerular teleost, the toadfish *Opsanus tau*, also exhibits a natriuresis in response to administered ANP (Lee and Malvin 1987).
- There is a decrease in the secretion of aldosterone owing to a direct inhibition of its synthesis as well as a decline in the release of renin.
- An inhibition of the reabsorption of sodium ions from the distal renal tubule occurs, apparently through blockade of sodium channels on the apical plasma membrane.

The natriuretic effect of ANP is probably only utilized when there is an excessive accumulation of salt, such as from some diets, the drinking of salt solutions or, most often, in some diseases of the heart, liver, and kidneys. A role for the natriuretic peptides in normal salt regulation in mammals is, however, usually considered to be unlikely. Transgenic mouse models in which the ANP gene is expressed in the liver, resulting in a lifelong elevation of ANP in the plasma, appears to confer no special deficiencies on their ability to regulate body sodium levels (Sonnenberg, 1994). These mice exhibited a relative hypotension and it was suggested that the principal role of ANP in mammals may be a cardiovascular one.

Guanylin, uroguanylin and adrenomedullin   Two other types of peptide with putative hormonal roles may influence the regulation of body fluid volume and electrolyte levels in mammals (Chapter 3). *Guanylin* and *uroguanylin* can promote chloride secretion by activating chloride channels in the intestine and renal tubules of rats and humans (Forte and Hamra, 1996). *Adrenomedullin* like natriuretic peptides, produces a vasodilatation and natriuresis (Schell *et al.*, 1996). It has been shown to decrease sodium-appetite and the secretion of aldosterone and corticotropin, and to increase renal blood flow. These actions could all contribute to the regulation of body fluid volume and composition. Such peptides have been identified in the blood but their hormonal roles remain to be defined. The precise control of water and electrolyte balance in mammals may be a more complex multihormonal process than hitherto envisaged.

Birds

Birds possess the same types of hormone that contribute to the osmoregulation of mammals. However, differences in their precise chemical structures exist (Chapter 3) and they may be utilized in a somewhat different manner. A basic summary of avian osmoregulation has been provided by Skadhauge (1981) and more recent information has been collected by Hughes and Chadwick (1989).

There are several salient differences in the osmoregulation of birds compared with mammals.

1. The presence of nasal salt glands in many species. These organs may concentrate salts in higher concentrations than those seen in the urine.
2. A hypertonic urine can be formed but the concentrations are usually not as great as seen in many mammals. The hypertonicity is achieved with the aid of a similar renal tubular mechanism as seen in mammals, and involves the loop of Henle. This segment is not, however, present in all the nephrons in avian kidneys (Dantzler, 1992). Nephrons in which it is present are called "mammal-type" nephrons while the others are called "reptile-type" nephrons. In mammals the GFR is relatively stable but in birds it can be quite variable, usually reflecting intermittent functioning of the reptile-type nephrons.
3. Birds lack a urinary bladder. The urine passes directly into a cloaca and can move in a retrograde direction into the posterior regions of the gut, where its composition can be modified in the coprodeum, colon and rectal ceca. Schmidt-Nielsen and his collaborators (Schmidt-Nielsen *et al.*, 1963) proposed that such a reabsorption of salts and water from the posterior regions of the gut may allow birds with nasal salt glands to gain osmotically "free" water owing to subsequent secretion of the salts by the nasal salt glands. The concentrations of solutes that are attainable by the salt glands are much greater than in the kidneys.

The actions of hormones are intimately involved in these novel osmoregulatory mechanisms in birds.

Secretion by the nasal salt glands is, however, principally controlled by a neural cholinergic mechanism that responds to changes in plasma salt concentration. An adequate blood supply to the gland is necessary and local vasoconstriction, such as can be elicited by epinephrine, can block the response. Administered angiotensin II can also inhibit the normal response of the salt glands in ducks and kelp gulls, probably by acting in this manner (Wilson, Van Pham, and Tan-Wilson, 1985; Gray and Erasmus, 1989a). The physiological significance of such vascular effects is not clear. It was formerly thought that corticosteroid hormones exerted a direct effect on nasal salt gland secretion in birds, but this interesting hypothesis is no longer tenable (Butler, Siwanowicz, and Puskas, 1989). Previous observations on the effects of experimental adrenocortical insufficiency appear to reflect changes in the blood supply and substrates supplied to the salt glands. However, as will be described below, the integrated activities of the nasal salt glands with salt absorption in the nether regions of the gut do involve hormones.

Vasotocin   Urine flow and concentration in birds is principally regulated by arginine-vasotocin (AVT) which is a homolog of the mammalian hormone arginine-vasopressin (Chapter 3). It is released in response to an increase in the osmotic concentration of the plasma in the domestic fowl (Koike *et al.,* 1977) and various other birds including kelp gulls (Gray and Erasmus, 1989b) and an Australian parrot, the galah *Cacatua rosei capilla* (Roberts, 1991). Vasotocin increases water reabsorption from the renal tubule of the domestic fowl, as occurs in mammals; at slightly higher, but still physiological, concentrations it decreases the GFR (Skadhauge, 1969; Stallone and Braun, 1985). In the domestic fowl, the tubular effect appears to be the most important one in normally limiting urine flow. The glomerular response reflects a closure of the reptile-type glomeruli (glomerular intermittency), apparently as a result of a constriction of the afferent glomerular arteriole. Such an effect has been described in other birds, including Gambel's quail (Dantzler, 1992) and kelp gulls (Gray and Erasmus, 1988). The kelp gulls, however, respond differently to vasotocin than do the domestic fowl. The antidiuresis resulting from low graded doses of vasotocin in kelp gulls results from parallel changes in both water reabsorption from the renal tubule and decreases in the GFR. The kelp gulls live in marine coastal areas of South Africa and possess nasal salt glands. Their osmotic challenges would appear to be more rigorous than normally experienced by the domestic fowl, which, apparently, originated in tropical jungles. The greater emphasis on action of vasotocin on the GFR in the gulls may reflect an adaptation to their different osmotic circumstances and is more like that seen in reptiles.

Atrial natriuretic peptide   A diuresis may be seen in birds following the infusion of isosmotic solutions that expand the extracellular space. Such a response appears to reflect a release of ANP. This response to ANP was originally described in mammals but has since been observed in many other vertebrates. In birds, such a diuresis and natriuresis has been described in ducks and the domestic fowl (Gray, 1993; 1994). A vascular effect resulting in an increased GFR accompanied by a decrease in the reabsorption of sodium from the renal tubule occurs. The latter effect involves an inhibition of aldosterone synthesis as seen in mammals (Rosenberg, Pines and Hurwitz, 1988; Gray, Schütz, and Gerstberger, 1991). A paradoxical action of ANP has, however, been observed in the adrenocortical tissue of turkeys where an increase of aldosterone synthesis occurs (Kocsis, McIlroy, and Carsia, 1995). An infusion of ANP can also increase the secretion from the nasal salt glands of ducks (Schütz and Gerstberger, 1990). These effects of ANP probably only occur rarely and in response to unusual situations but they could be important when some sea birds drink sea water.

Corticosteroids   The principal corticosteroids in birds are corticosterone and aldosterone. Adrenalectomy in ducks results in an excessive loss of salt in the urine (Thomas and Phillips, 1975), which reflects a lack of corticosteroid hormones (Chapter 3). Such effects have also been observed in the domestic fowl and pigeons and appear partly to reflect a reduced absorption of sodium from the renal tubule. (As described below, a further deficiency may occur owing to a decrease in the reabsorption of sodium from urine that passes into the posterior regions of the gut.) Secretion of corticosteroids, especially aldosterone, occurs in response to a low level of sodium in the diet (Rice *et al.*, 1985; Arnason *et al.*, 1986). Both corticotropin and the renin–angiotensin system (Wilson, 1989) appear to be involved in this response, but the precise contributions of each remain to be delineated.

Conservation of sodium chloride and water by birds results from their reabsorption from the contents of the gut as they pass from the duodenum to the colon (Grubb, Driscoll, and Bentley, 1987; Bentley and Grubb, 1989). Reabsorption of urinary salts and water also occurs from the coprodeum, colon, and rectal ceca following the reflux of urine up into the lumen of the gut (Skadhauge, 1989). This process is promoted by the action of aldosterone, which increases the reabsorption of sodium. In the domestic fowl, this effect of aldosterone appears to extend for a considerable distance along the gut, as it is also seen in the ileum (Grubb and Bentley, 1987). Although a diet that has a low content of sodium increases the release of aldosterone in the domestic fowl, it has little effect on the secretion of corticosterone. However, corticosterone is normally present in the plasma in very much higher concentrations than aldosterone and it can also promote (*in vitro*) the absorption of sodium ions from the colon and rectal ceca (Grubb and Bentley, 1989; 1992). Its potency and concentrations in the plasma suggest that it may also contribute to the conservation of sodium in the gut.

Prolactin   Increases in the osmotic concentration of the plasma of birds have been consistently found to result in a release of prolactin (Harvey *et al.*, 1989). This secretion may be mediated by the prolactin-releasing effects of endogenous TRH. The prolactin release is independent of that of vasotocin. There has been considerable speculation regarding the possible role of such released prolactin in the osmoregulation of birds, especially as this hormone has been shown to be involved in such processes in some teleost fish (p. 367). Effects on water and salt exchanges in the nasal salt glands, kidneys, and intestine have been considered possible. However, such reports have been inconsistent. Prolactin can promote feeding behavior, which could be contributing to some of the observed effects. In addition, some heterologous preparations of prolactin that have been used experimentally were commonly contaminated with other hormones such as vasopressin (Harvey *et al.*, 1989). Prolactin's possible role in avian osmoregulation is currently "on hold." Its release in response to osmotic stimuli may sometimes constitute a "stress" effect and the hormone may then have other roles.

Renin–angiotensin system   Birds, like most vertebrates, drink in order to maintain their osmotic balance. Angiotensin II has been shown to promote the sensation of thirst and drinking in a number of birds including the white-crowned sparrow, Japanese quail, and several species of Australian parrots (Wada, Kobayashi, and Farner, 1975; Takei, 1977; Kobayashi, 1981). Three species of the Australian parrots belonging to the genus *Barnadius* showed differences in their sensitivity to this hormone. The twenty-eight parrot lives in quite wet areas while the Port Lincoln parrot lives in more arid zones, but it is considered to have evolved from a species from more temperate regions. The mallee ringneck parrot, by comparison, has lived in very dry areas for a long time. The former two species were very sensitive to injected angiotensin II while the latter was quite insensitive. These observations suggested that birds originating from dry areas have a poor thirst and drinking response to angiotensin II compared with those from wet regions.

Reptiles

Reptiles are poikilothermic, which profoundly influences their osmoregulation, and they represent a substantial metabolic departure from the birds and mammals. These animals live in diverse osmotic habitats including the sea, fresh water, and terrestrial situations ranging from deserts to tropical rain forests. Their territorial conquests have been facilitated by the possession, unlike their amphibian forebears, of a relatively impermeable integument. In many reptiles, urinary water loss is also limited by excreting waste nitrogen as uric acid, which has a low solubility in the urine. Reptiles, however, do not form a hypertonic urine but some possess cephalic salt glands, which can secrete such concentrated solutions. Their hormonal armamentarium is similar to that of other nonmammalian tetrapods. The roles of hormones in the osmoregulation of reptiles has been reviewed by Bradshaw (1986; 1997).

Vasotocin   Vasotocin release into the plasma occurs in response to dehydration and has been described in an Australian lizard, *Varanus gouldii*, and a Brazilian snake, *Bothrops jaracaca* (Bradshaw and Rice, 1981; Rice, 1982; Silveira *et al.*, 1992). Neurohypophysial peptides were first observed to exhibit antidiuretic effects in reptiles over 60 years ago. Responsive species include the alligator *Alligator mississippiensis*, the turtle *Chrysemys picta*, the water snake *Natrix sipedon*, and the lizard *Varanus gouldii* (see, for instance, Bradshaw and Rice, 1981; Yokota, 1990). This antidiuresis reflects both a decline in the GFR, owing to closure of glomeruli, and apparently an increase in water reabsorption from the renal tubule. The glomerular effect of vasotocin reflects its action in constricting the afferent glomerular arteriole and the tubular effect is, probably, an action on membrane water permeability. The latter has been suggested as vasotocin increases renal free water clearance, but a direct effect on the tubules has not been demonstrated (Bradshaw and Rice, 1981). A vasotocin-sensitive adenylate cyclase mediates an increase in the permeability of epithelial membranes to water (Chapter 4). Such an enzyme has been identified in the renal tubular collecting ducts of the Australian arid-zone lizard *Ctenophorus* (*Amphibolurus*) *ornatus* (Bradshaw and Bradshaw, 1996). This observation suggests that vasotocin is acting in a manner that is similar to that of vasopressin in mammals.

Corticosteroids   In reptiles, as also seen in mammals, the plasma levels of aldosterone have been observed to be inversely related to those of sodium (Bradshaw and Rice, 1981; see Bradshaw, 1997). In addition, high plasma potassium concentrations enhance the secretion of aldosterone while high sodium levels increase that of corticosterone.

A renin–angiotensin system is present in reptiles and appears to contribute to the regulation of aldosterone secretion. It has, for instance, been shown that angiotensin II promotes aldosterone synthesis in the Nile crocodile *Crocodylus niloticus* (Balment and Loveridge, 1989) and the Australian lizard *Tiliqua rugosa* (Bradshaw, 1978). Plasma corticosterone concentrations were also increased in the lizard but in the crocodile the effects on synthesis of this steroid were quite variable. Aldosterone can increase the reabsorption of sodium from the renal tubule, as observed in the water snake *Natrix cyclopion* (Elizondo and LeBrie, 1969) and the lizard *Varanus gouldii* (Rice, Bradshaw, and Prendergast, 1982).

A postrenal modification of the urine, including a reabsorption of urinary sodium, may occur in some reptiles. Such salt conservation has been observed during the storage of urine in the cloaca of the South American crocodilian

Caiman sclerops (Bentley and Schmidt-Nielsen, 1965) and in some lizards (Bradshaw, 1986; 1997). However, unlike in birds, a role for corticosteroids in regulating this process has not become generally apparent. Nevertheless, aldosterone has been shown to increase the transport of sodium ions across the colon (*in vivo* and *in vitro*) of the Canary Island lizard *Gallotia galloti* (Diaz and Lorenzo, 1991; 1992). Chelonian reptiles and some lizards can store water in a capacious urinary bladder. Aldosterone increases sodium transport across bladders (*in vitro*) in the tortoise *Testudo graeca* (Bentley, 1962) and the turtle *Pseudemys scripta* (Le Fevre, 1973).

Many reptiles can utilize cephalic salt-secreting glands in order to excrete salts as hypertonic solutions. These glands in reptiles have various embryonic origins. In chelonians they are modified tear glands, in some lizards they are nasal glands, and in crocodiles they originate in the tongue. It seems that, like in birds, their secretory activities may involve neural stimuli but there is little information about this possibility. Numerous experiments have been performed on reptile salt glands that suggest a hormonal influence (Bradshaw, 1997). However, the observed effects mainly indicated that corticosteroids influence the composition (sodium/potassium ratio) of the secreted fluid rather than its volume. Aldosterone appears to influence secretion in herbivorous species, which principally utilize such glands for potassium excretion while conserving sodium. Such a response is of a classical mineralocorticoid type in which corticosteroids decrease the ratio of sodium/potassium, as seen in mammalian sweat glands and salivary glands. It is also possible that in some species corticosterone may enhance the secretion of a sodium-rich solution when this ion is in excess (Bradshaw, 1997).

Vasoactive intestinal peptide   Secretory activity of the lingual salt gland of the salt water crocodile *Crocodylus porosus* can be promoted by the administration of VIP (Franklin, Holmgren, and Taylor, 1996). (This effect is reminiscent of the response of the elasmobranch rectal salt gland to this peptide). Nerves supplying the lingual gland contain VIP immunoreactivity. This peptide, therefore, could be functioning as a neuropeptide, either acting directly on the secretory cells of the gland or increasing its local blood flow.

Renin–angiotensin system   Reptiles drink, a talent that does not occur in their amphibian forebears. As in mammals and birds, the administration of angiotensin II can promote drinking in a variety of saurian reptiles (Fitzsimons and Kaufman, 1977; Kobayashi *et al.*, 1979) and in the Nile crocodile, *Crocodylus niloticus* (Balment and Loveridge, 1989). Possibly, the reptiles were the first tetrapods to evolve a special thirst mechanism as an aid to terrestrial life, and to utilize angiotensin in this process.

The amphibians

Osmotically, the amphibians are a very interesting group as they bridge the gap between the fishes and the tetrapods. Phyletically, they represent the first terrestrial vertebrates yet they are still largely dependent on the ready availability of fresh water and have aquatic larvae. Amphibians have a permeable integument and, therefore, are prone to water loss by evaporation. They cannot form a hypertonic urine. To add to their osmoregulatory woes frogs and toads (as well as newts and salamanders) do not normally drink (Bentley and Yorio, 1979), and such behavior cannot be elicited by injecting them with angiotensin II (Kobayashi *et al.*, 1979). The Amphibia, therefore, are a group of considerable interest both with respect to osmotic regulation and the endocrine mechanisms they utilize for this process and a terrestrial life.

Vasotocin   Vasotocin has an antidiuretic effect in most, but not all, amphibians. Mesotocin, the other amphibian neurohypophysial hormone, when injected, can even have a diuretic effect (Pang and Sawyer, 1978; Galli-Gallardo, Pang, and Oguro, 1979). As in the birds and reptiles, vasotocin initiates both the reabsorption of water from the renal tubules and a decrease in the GFR (Sawyer, 1972a). The tubular response may, however, be lacking in the urodele amphibians (Brown and Brown, 1980) such as the mudpuppy *Necturus maculosus* (Pang, Furspan, and Sawyer, 1983). An antagonist ("KBIV 24") of the vascular (V_1 receptor)-mediated glomerular effects of vasotocin can completely block its antidiuretic effect in the mudpuppy but produced only a partial inhibition in bullfrogs. The difference reflects the renal tubular response to vasotocin in the bullfrog. The generation of cyclic AMP in response to neurohypophysial peptide hormones is an epithelial membrane type response involving the V_2 receptors. Such an effect on cyclic AMP was observed (*in vitro*) in the bullfrog but not in the mudpuppy kidney.

The magnitude of the antidiuretic effect of the neurohypophysial hormones also varies in different species. Vasotocin, for instance, does not reduce urine flow in the African clawed toad, *Xenopus laevis*. This toad is aquatic, and consequently such a response to released vasotocin would be physiologically inappropriate and could even lead to death resulting from hyperhydration. Aquatic urodele amphibians, like the mudpuppy, *Necturus maculosus*, the mud eel, *Siren lacertina*, and the congo eel, *Amphiuma means*, exhibit antidiuretic effects after injections of vasotocin, but the responses are small and the amounts of hormone required are large so that the physiological significance of this effect is doubtful (Bentley, 1973). Young tadpoles do not exhibit water retention (reflecting an antidiuresis) in response to injected vasotocin, but as bullfrog (*Rana catesbeiana*) tadpoles approach metamorphosis such a response becomes increasingly apparent, though it does not reach its full expression until after metamorphosis has occurred (Alvarado and Johnson, 1966).

Vasotocin exerts several other interesting effects with respect to the osmoregulation of amphibians. The skin of amphibians is permeable to water, which moves across it by osmosis. In anurans (frogs and toads), water crosses the skin much more readily when the tissue is stimulated by vasotocin. This hormone appears to make the integument less waterproof, an action that is also seen on the renal tubule. This direct cutaneous effect of vasotocin is equivocal in urodeles (newts and salamanders) and is not seen in some anurans, including tadpoles and *Xenopus laevis*. The ability of vasotocin to increase the skin's osmotic permeability appears to be greater in species that normally occupy dry, rather than relatively wet or damp, habitats. Different parts of the skin of each animal do not necessarily respond in a quantitatively uniform manner to vasotocin. Therefore, skin from the ventral surface of several species of tree frogs (Hylidae) shows a marked increase in permeability to water, but that from the dorsal surface is unresponsive to the hormone (Yorio and Bentley, 1977).

As amphibians do not drink, this increased rate of water accumulation across the skin may aid rehydration in some species, such as those from desert areas, where water is only available sporadically.

The crab-eating frog, *Rana cancrivora*, lives in the sea water in coastal mangrove swamps in southeast Asia. As described above, this interesting amphibian maintains its blood plasma at a hypertonic concentration with respect to the external solution by retaining additional salts and urea. The skin of these amphibians is not responsive to vasotocin, but this hormone increases the permeability of the urinary bladder to urea (as well as water) and so, by permitting its reabsorption from the urine, apparently contributes to the conservation of this solute in the body (Dicker and Elliot, 1973). A cutaneous response would be a disadvantage to such animals as it would increase the rate of water loss if they entered hyperosmotic solutions.

Amphibians usually possess a large urinary bladder in which they can store water equivalent to as much as 50% of their body weight. This water can be reabsorbed in times of need and so constitutes a store that may be very useful to some species. Vasotocin, or dehydration, which releases this hormone into the blood, increases the rate of water reabsorption across the urinary bladder of many amphibians.

The tetrapod urinary bladder has no true embryological analog in the fishes and appears to have first evolved in the Amphibia. It is interesting that some species of this group utilize it for water storage, and possibly urea conservation, with which use it has also acquired a responsiveness to vasotocin. A comparable effect on the bladder has not been demonstrated in any other vertebrate group so that this endocrine adaptation is unique.

When amphibians, in water, are injected with neurohypophysial hormones, they gain weight owing to a water retention. This action is known as the "Brunn effect," or "water balance effect," and is caused by stimulation by vasotocin at the three distinct sites described above: the kidney, the skin, and the urinary bladder. A single hormone, therefore, has multiple effects, all of which are directed to the same physiological purpose, namely the conservation of water.

Catecholamines   The catecholamines have ubiquitous effects on tissue that may involve their blood vessels and permeability. Injected norepinephrine can exert an antidiuretic effect in bullfrogs as a result of a decreased GFR (Pang, Uchiyama, and Sawyer, 1982). An α-adrenergic blocking drug (phenoxybenzamine, which opposes adrenergic vasoconstriction) when injected into the portal blood supply of one kidney resulted in a unilateral diuresis in that kidney. These effects of the catecholamines and the antagonist may reflect a role for the adrenergic nerves supplying the kidney. Epinephrine has also been shown to promote cutaneous uptake of water by the toad *Bufo melanostictus* (Elliot, 1968). The actions of catecholamines can be quite complex and have also been found (*in vitro*) to block the effects of neurohypophysial hormones on water transfer across the urinary bladder of the toad *Bufo marinus* (Handler, Bensinger and Orloff, 1968). This action could involve an α_2-adrenergic response that inhibits the formation of the second messenger cyclic AMP. The physiological significance of such observations in amphibians in unknown, but catecholamines could be contributing to their osmoregulation and may modulate the actions of vasotocin.

Corticosteroids   The sodium metabolism of amphibians can also be regulated by hormones, which can act at several different sites in the body. Vasotocin, *in vitro*, stimulates sodium transport across the skin, from the external media to the blood, and its reabsorption from the urinary bladder. These effects, though prominent, do not seem to persist for a prolonged time and are difficult to reconcile with normal physiological regulation of sodium in the intact animal. Aldosterone has more persistent effects in increasing such sodium transport across the skin, the urinary bladder, and the colon in frogs and toads (Crabbé and de Weer, 1964; Cofré and Crabbé, 1965). It seems likely that aldosterone also acts on the urinary bladder of urodeles, but the evidence for this is equivocal. As this corticosteroid is released in response to sodium depletion in amphibians and is effective at low concentrations, it seems likely that it normally adjusts sodium transport at these sites. As we have seen, aldosterone stimulates sodium reabsorption from the kidney tubules in mammals, birds, and possibly reptiles, but despite frequent attempts to demonstrate it, this action appears to be lacking in amphibians. This is an interesting endocrine situation, as aldosterone makes its initial phyletic appearance in the lungfishes and the amphibians, yet the important renal action of this hormone does not appear to have evolved until much later.

Renin-angiotensin system   The renin–angiotensin system is present in the Amphibia and, as in other tetrapods, contributes to the release of aldosterone (Nishimura, 1987). The renin concentration in the kidney of salt-depleted frogs increases (Capelli, Wesson, and Aponte, 1970). When renin from bullfrog kidney is injected back into these frogs, the aldosterone concentration in the plasma rises (Johnston *et al.*, 1967). This effect is also seen in the edible frog *Rana esculenta* (Dupont *et al.*, 1976). However, corticotropin also promotes the secretion of aldosterone, so that a dual control for this hormone may exist. In amphibians, the secretion of corticosterone appears to be increased by the renin–angiotensin system (Nishimura, 1987), suggesting that it has a general corticosteroidogenic action.

Prolactin   Some urodele amphibians, such as the North American eastern spotted newt, *Notophthalmus viridescens*, and the Japanese newt, *Cynops pyrrhogaster*, undergo a "second metamorphosis" in spring and migrate to ponds and streams in order to breed. This behavior can be promoted by the injection of prolactin and it is called the "newt water-drive effect" (Brown and Brown, 1987). The preference for an aquatic habitat and the associated courtship can be blocked by the administration of an antiserum to prolactin (Toyoda *et al.*, 1996). The response to prolactin is associated with a thickening of the skin and a decrease in the cutaneous permeability to sodium and water (Brown and Brown, 1973). Such a change appears to be an appropriate one for an amphibian about to undertake life in fresh water. It is reminiscent of the action of prolactin on the gills of some teleost fish (p. 367). An analysis of the electrical behavior of the skin of the Japanese newt to the administration of prolactin indicates that there is a blockade of sodium channels in the skin (Takada and Shomazaki, 1988). Other aquatic urodeles exhibit similar responses to prolactin (Brown and Brown, 1987). The neotenic mudpuppy *Necturus maculosus* loses sodium and suffers a decline in its plasma concentration following hypophysectomy (Pang and Sawyer, 1974). The administration of prolactin can prevent this loss of sodium.

The fishes

The fishes are phyletically very diverse and contain several distinct groups that osmoregulate differently. These include the Osteichthyes (bony fishes), the Chondrichthyes (elasmobranchs, cartilaginous fishes), and the Agnatha (cyclostomes or jawless fishes, lampreys, and hagfishes). Most of the available information about the role of hormones in osmoregulation applies to a single order of the Osteichthyes, the Teleostei. Some fish are only able to survive in restricted salinities in fresh water rivers and lakes or in the sea. Others may tolerate a greater range in salinity in estuaries or they may migrate between rivers and the sea in order to breed and further exploit available food supplies. Such euryhaline species utilize a number of osmoregulatory strategies some of which have been summarized earlier in this chapter. Central to the regulation of their water and salt content are the activities of:

- the kidneys, which can excrete osmotically gained water as a hypotonic urine (fish cannot form a hypertonic urine)

- the gills and opercular membranes, across which osmotic exchanges of water and a diffusion of salts occurs. These tissues are the site of salt-secreting chloride cells that can transport ions against their diffusional gradients. Fish in sea water can, therefore, drink and excrete the excess salt through their gills, thus gaining osmotically free water.

These processes may be modulated and even controlled by hormones.

When euryhaline teleosts adapted to sea water or fresh water are transferred between these solutions there are, as a result of osmosis and diffusion of salts, transitory increases or decreases in the solute concentrations of their body fluids. Following this change, which may last for a day or two, a restoration of optimal solute concentrations begins that is usually achieved during the next few days. These adjustments in osmotic concentration by the fish mainly reflect the excretion of water by the kidneys when they are in fresh water, appropriate changes in the permeability of the gills to water and solutes in fresh water or sea water, and an excretion of salt by the branchial chloride cells in sea water. The enzyme Na–K ATPase is essential for the functioning of the chloride cells and its levels in the gills rise during this period of adjustment in sea water and decline during adaptation in fresh water.

The chloride cells proliferate and differentiate in fish placed in sea water. The mature cells are rich in mitochondria and have a complex microtubular system. They have interconnections with neighbouring cells and form salt-secreting multicellular complexes (Foskett et al., 1983; Karnaky, 1986). The basolateral membranes of the chloride cells have high concentrations of Na–K ATPase. Chloride cells were discovered by Ancel Keys (Keys and Bateman, 1932) in the gills of the European eel Anguilla anguilla. They were named to describe the secretion of chloride ions that was observed to occur from the perfused gills when they were exposed to sea water. Subsequently there has been uncertainty as to whether chloride ions are the primarily transported ion or whether it is sodium ions. (Whichever the situation, the other ion would be expected to follow to maintain electrical neutrality). Studies, in vitro, on the activities of chloride cells in teleost opercular membranes (Zadunaisky, 1984; Karnaky, 1986) indicate that it is the chloride ion that is primarily transported. The sodium ions follow through paracellular pathways as a result of the transepithelial electrical potential difference established as a result of the

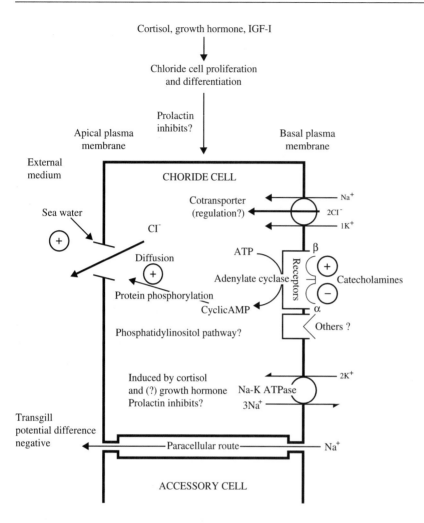

Fig. 8.4. A summary of the possible roles of hormones on chloride ion transport across the gills of teleost fish. For more detailed explanations see the text.

active chloride transport. A tentative model based on the experimental behavior of the chloride cell is shown in Fig. 8.4. Chloride ions are thought to enter the chloride cells across their basolateral plasma membranes from the extracellular fluid. This process appears to be an electrically neutral one and involves a cotransporter in this membrane in which $2Cl^-/Na^+/1K^+$ are associated. The continued activity of this mechanism is dependent on the maintenance of the sodium and potassium ion content of the intracellular fluid of the chloride cells. These ion concentrations are controlled by the activity of the "sodium pump" (Na–K ATPase), which extrudes internal sodium ions in exchange for external potassium ions across the basolateral membrane. The inhibition of this mechanism with drugs such as ouabain blocks the active chloride transport by the cell. The exit of chloride ions from the chloride cell into the external bathing solution is thought to occur as a result of diffusion down an electrochemical gradient across the apical plasma membrane. This step can be influenced by catecholamines and cyclic AMP and appears to involve a specific site on the membrane. It could, for instance, be a "pore" or "channel" provided by a protein in the membrane. Some drugs and chemicals can influence the activity of the chloride cotransporter in the basolateral plasma membrane but a possible hormonal action at this site is at present uncertain.

Water movements across the integument of fish principally involves the gills. Osmotic and diffusional forces determine such exchanges (Isaia, 1984). The available surface area and distance traversed may contribute to interspecific differences in the permeability of the gills to water and changes that occur in different environmental circumstances. Vascular changes may influence the recruitment of gill lamellae and so change the total available surface area (Fig. 8.5). Water must move along nonlipid pathways such as may exist between cells, and within the mainly lipid plasma membranes of the branchial respiratory and nonrespiratory surfaces. Hormones can influence the movements of water across the gills but their precise mechanisms and sites of action are generally unknown except when vascular changes are involved. As a result of the action of vasopressin or vasotocin, specific proteinaceous aqueous channels can be inserted into the plasma membranes of renal tubular cells of many tetrapods and amphibian urinary bladders. Such water channels, called aquaporins, have been identified in many vertebrate tissues (Verkman *et al.*, 1996) but apparently not yet in fish gills. An initiation of a "disorder" in lipid organization in the plasma membrane could also provide a mechanism facilitating the permeability of the gills to water, but the details of such a process are obscure.

The adaptation of teleosts to changes in the salinity of their environmental solutions involves the actions of several hormones that principally act on the gills. Interspecific differences, however, exist that may reflect the fishes normal rate of turnover of water and solutes and their particular branchial morphology and physiology.

Neurohypophysial hormones   Arginine-vasotocin is present in representatives of all the main groups of fishes including the cyclostomes (Chapter 3). Except for the latter (lampreys and hagfishes) it is accompanied by other related peptides: mesotocin in lungfishes, isotocin in teleosts, and a variety of oxytocin-like peptides in chondrichthyeans. Most information available about the actions of these peptides concerns vasotocin. Sensitive and selective radioimmunoassay procedures have demonstrated that vasotocin is present in the plasma of a variety of teleosts (flounder, cod, saith, and carp), elasmobranchs (dogfish and rays), and a cyclostome, the Japanese lamprey (Warne *et al.*, 1994; Uchiyama *et al.*, 1994). Previous methods had usually failed to provide unequivocal demonstrations of such a release of vasotocin into the plasma and had, therefore, limited speculation regarding its possible physio-

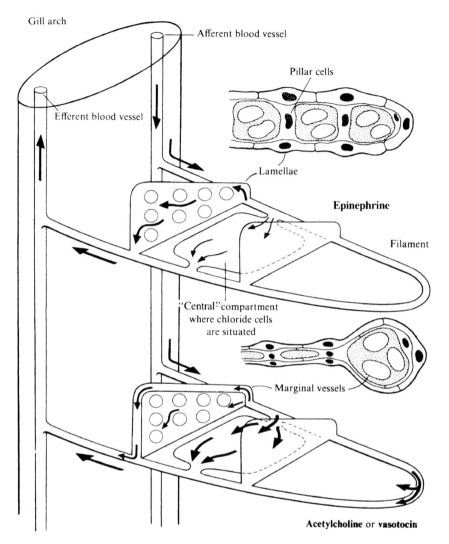

Gill arch

Afferent blood vessel

Efferent blood vessel

Pillar cells

Lamellae

Epinephrine

Filament

"Central" compartment
where chloride cells
are situated

Marginal vessels

Acetylcholine or **vasotocin**

Fig. 8.5 A hypothetical model illustrating how hormones may alter the distribution of the blood flow in the gills of teleost fish and so alter their respiratory and osmoregulatory functions. Epinephrine increases the blood flow to the *lamellae* of the gills, which have a predominant respiratory function, by causing a relaxation of the *pillar cells* (PC). Acetylcholine or neurohypophysial peptides (including vasotocin and isotocin) constrict the lamellae (possibly by contracting the pillar cells) and divert blood to the central compartment. The chloride-secreting cells are situated in the interlamellar region of the central compartment so that their function is facilitated by the presence of the neurohypophysial hormones. (From Rankin and Maetz, 1971; see also Bennett and Rankin, 1986.)

logical roles in the fishes. Such speculation, however, remains, but it is now more informed.

Clues about the physiological role of such peptides in fish largely depend on observing the effects of their administration under different experimental conditions. A consistent effect of vasotocin in fish, and all other vertebrates, is an ability to influence the contractility of vascular smooth muscle, which usually results in an increase in the blood pressure (Pang *et al.*, 1983). This action has been observed in several fishes including European eels (Chan and Chester Jones, 1969; Rankin and Maetz, 1971), African and Australian lungfish (Sawyer, 1970), and Japanese lampreys (Uchiyama and Murakami, 1994). It has been suggested that this response in fish represents the primeval action of this peptide (Pang *et al.*, 1983). Such vascular responses may

Fig. 8.6. A schematic diagram showing the phyletic distribution of the responses of the kidney to neuro-hypophysial hormones in vertebrates. Small amounts of vasotocin may also have an antidiuretic effect as seen in the European eel (see text). (From Sawyer, 1972b.)

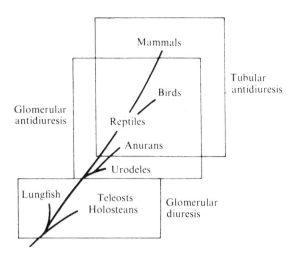

influence osmoregulation in fish by increasing the systemic blood pressure and thus the GFR, altering hydrodynamic pressure locally in the glomeruli by altering the contractility of its arterioles, and by changing the flow and distribution of blood to the gills.

The earliest observed renal response to vasotocin in fish was a diuresis in goldfish (Maetz *et al.*, 1964), European eels (Chester Jones, Chan, and Rankin, 1969), South American lungfish (Sawyer, Uchiyami, and Pang, 1982), and European lampreys (Logan, Moriarty, and Rankin, 1980). These effects were the result of an increase in the GFR, probably caused by rises in the systemic blood pressure (Pang *et al.*, 1983). There is no evidence that vasotocin influences water reabsorption across the renal tubules of fish as it does in tetrapod vertebrates. In many of the experiments on fish, large doses of vasotocin were administered, but when smaller amounts were given to European eels in fresh water (but not sea water) an antidiuresis was observed (Babiker and Rankin, 1973). It was suggested that this response may be caused by changes in glomerular recruitment. Whether such an effect on the kidney reflects a physiological role for vasotocin is still in doubt (Sawyer, 1987), especially as no differences in its plasma concentrations have been observed in teleosts or Japanese lampreys maintained in solutions of differing salt concentration (Warne *et al.*, 1994; Uchiyama *et al.*, 1994). Nevertheless, there have been persistent descriptions of changes in the storage of vasotocin in neurohypophyses of fish in response to placing them in solutions of different osmotic concentration. Transfer from fresh water to sea water has been associated with declines in such activity in flounder, rainbow trout, medaka, and Japanese lampreys (Hyodo and Urano, 1991; Haruta, Yamashita, and Kawashima, 1991; Perrott, Carrick, and Balment, 1991; Uchiyama *et al.*, 1994).

The responses of fish to injected vasotocin provide us with a possible scenario regarding the evolution of the osmoregulatory roles of the neurohypophysial hormones (Fig. 8.6). The ancestral response may have been a vascular one mediated by V_1 type receptors (Pang *et al.*, 1983). An increase in the sensitivity of such receptors associated with the afferent glomerular arteriole may have allowed a primitive type of antidiuresis. Such a response is seen in some extant fish given low doses of vasotocin and it is also the only antidiuretic mechanism present in some amphibians, such as mudpuppies. This primeval response persisted in some tetrapods but was subsequently supplemented by a V_2 type receptor response on the permeability of the renal tubules.

Neurohypophysial peptides, when injected, have also been shown to increase the turnover of sodium chloride in some teleosts, such as eels, when they are transferred from fresh water to sea water. Such hormones may also increase active ion uptake in some fresh water teleosts. It is possible that these hormones have a direct effect on the permeability of the gill epithelium that would be analogous to some of their actions in tetrapods. Specific binding sites for vasotocin have been identified in the gills of European eels (Guibbolini *et al.*, 1988). These sites could be receptors mediating a direct effect of the peptide hormone. Alternatively, it has been shown (Maetz and Rankin, 1969) that regional changes in the branchial blood flow occur that could mediate alterations in ion transfer; thus epinephrine, which increases the blood supply to the respiratory area of the gills, produces a decline in salt excretion from the chloride cells. This change may be the result of a shunting of blood away from the central part of the gill filaments that contain the chloride cells. Neurohypophysial peptides (and acetylcholine) have the opposite constrictor action on the blood supply in the gills (Fig. 8.5) and so could facilitate the functioning of the chloride cells. We can, therefore, conjecture that the vasoactive effects of the neurohypophysial peptides on the gills may also be phyletically older than their direct actions on the permeability of epithelial membranes (Maetz and Rankin, 1969).

Catecholamines   Epinephrine and norepinephrine are present in the plasma of teleost fish and the latter is also involved in adrenergic nerve transmission. The vascular actions of these catecholamines may influence osmoregulation by changing the blood supply to the gills and the kidneys. Their role in the latter in fish is obscure but in the gills they may not only change the blood flow but also initiate a recruitment of lamellae and hence change the functional branchial surface area (Fig. 8.5). They may also have more direct actions (α- and β-adrenergic) on gill epithelia, including the activities of the salt-secreting chloride cells (Fig. 8.4). Catecholamines may either increase or decrease chloride secretion from these cells. Probably the first observation of such an effect was that of Keyes and Bateman (1932) on the perfused gills of eels. When their external surface was bathed in sea water the epinephrine inhibited chloride secretion (α-adrenergic?). The administration of catecholamines to teleost fish undoubtedly influences transbranchial movements of chloride and sodium ions, and also that of water (Pic, Mayer-Gostan, and Maetz, 1973; Mayer-Gostan, Wendelaar Bonga, and Balm, 1987).

In sea water, injected epinephrine reduces the active extrusion of sodium and chloride from the gills. It increases branchial permeability to water in either fresh water or sea water. The effects on ion movements can be prevented by α-adrenergic blocking drugs and that on water by β-adrenergic inhibitors. Beta-adrenoceptors mediate the increases of blood flow to the central lamellar regions of the gills (see the last section in this chapter) and this effect could be contributing to the osmotic change. A more direct effect on the permeability of the gill epithelium to water is, however, considered to be a more likely mechanism (Isaia, Maetz, and Haywood, 1978; Bennett and Rankin, 1987). The ionic responses to epinephrine also appear to be caused by a direct effect on the tissue resulting from an inhibition of the activity of the chloride-secreting cells (see, for instance, Shuttleworth, 1978; Zadunaisky, 1984).

The skin on the head of a number of teleosts has, like the gills, been shown to contain chloride cells. These salt-secreting cells may be present on the inner surface of the operculum and the sides of the jaws. Pieces of such skin can be removed and studied *in vitro*, when a transepithelial chloride transport can be observed. This active transport process can be increased or decreased by catecholamines, the particular response depending on the hormone or drug used and its concentration. Increased chloride transport follows β-adrenergic-type stimulation while a decrease results from α-adrenergic effects (Zadunaisky and Degnan, 1980; Marshall and Bern, 1980). These effects are clearly direct ones on the tissue and do not involve changes in its blood supply.

The overall effects of injected epinephrine on fluid balance in teleosts are hypernatremia and hyperosmolarity in sea water and an accumulation of water in fresh water. Stress, such as that associated with laboratory handling or forced swimming, results in elevated levels of catecholamines in the blood of teleosts and also produces disturbances in fluid balance. It, therefore, seems likely that catecholamines may influence osmoregulation in unusual circumstances, but it is not clear whether or not they have such a role in more equitable conditions. However, it has been suggested (Foskett, 1987) that they may mediate a "rapid hormonal control" mechanism and initiate the decline in chloride secretion (α-adrenergic) that occurs soon after euryhaline teleosts move from the sea into fresh water.

Thyroid hormones   Thyroid hormones were probably the first such excitants to be implicated in developmental and migratory events in fish that ultimately result in changes of their osmoregulatory systems (Fontaine, 1954; Grau, 1987). Increases in the activity of the thyroid gland, measured either histologically or by measurements of hormone levels in the plasma have been found to correspond to various stages in the development of salmonids and their transformation from the fresh water parr to the smolt that will migrate to the sea. Regular patterns of changes in the plasma concentrations of T_4 and T_3 have been measured during such smoltification in Atlantic salmon and coho salmon (Prunet *et al.*, 1989; Young *et al.*, 1989). Such excursions on the concentrations of thyroid hormones accompany those of growth hormone, prolactin, and cortisol. A direct relationship of thyroid hormones to the development of osmoregulatory abilities is, however, not readily apparent. Even synergistic interactions with other hormones known to be involved in such adaptations, such as cortisol, appear to be lacking (Madsen, 1990a), though the evidence is still equivocal (Dangé, 1986).

Adenohypophysial hormones   Hypophysectomy in teleosts has a variety of effects on their osmoregulation depending on such factors as the species, the stage of its life cycle, and whether it is adapted to fresh water or sea water. At least three hormones secreted by the adenohypophysis may make important contributions to the osmoregulation of fish, especially teleosts. They are growth hormone, prolactin, and corticotropin. Their actions may involve other peripheral hormones and growth factors. The ultimate sites of their actions are mainly the gills. Interactions between such hormones may contribute to the final response, which may be an adaptation to life in fresh water or to sea water.

Growth hormone.   In 1956 D. C. Smith described how small doses of mammalian growth hormone considerably enhanced the abilities of brown trout, *Salmo trutta*, to adapt to their transfer from fresh water to sea water. This prescient observation prompted the subsequent comment that "the action of this hormone may be the result of an increased rate of development, possibly an accelerated maturation, of the branchial chloride secreting cells" (Bentley, 1971). Sporadic observations at that time confirmed the ability of growth hormone to enhance the adaptation of other salmonids to sea water (see, for instance, Komourdjian, Saunders, and Fenwick, 1976; Clarke, Farmer, and Hartwell, 1977; Bolton *et al.*, 1987). The definite observations were, however, not made until much later.

A release of growth hormone occurs in response to the transfer of rainbow trout and coho salmon from fresh water to sea water (Sweeting and McKeown, 1987; Sakamoto, Ogasawara, and Hirano, 1990). Elevation of plasma growth hormone levels have also been observed in Atlantic salmon and coho salmon at the stage of their transformation from the fresh water parr to migratory smolt (Prunet *et al.*, 1989; Young *et al.*, 1989). In addition, tilapia adapted to sea water display an enhanced activity of adenohypophysial somatotrope cells (Borski *et al.*, 1994). Administration of growth hormone to brown trout following transfer from fresh water to sea water enhanced their survival and this response was found to be associated with an increase in the number of the chloride cells and a rise in branchial Na–K ATPase levels (Madsen, 1990b). Growth hormone was found to have a similar action in Atlantic salmon (Prunet *et al.*, 1994). The gill Na–K ATPase increased and the chloride cells proliferated and displayed extensive morphological differentiation. These changes were comparable to those observed in salmon parr prior to their transformation to the smolt form (Prunet *et al.*, 1989). Adaptation to sea water and branchial Na–K ATPase were also enhanced by the administration of growth hormone in tilapia, which is a nonsalmonid teleost (Sakamoto *et al.*, 1997).

In mammals, many of the actions of growth hormone are mediated by IGF-I (Chapter 3). In rainbow trout and Atlantic salmon, the injection of bovine IGF-I enhances the ability of the fish to osmoregulate when they are transferred from fresh water to brackish water (McCormick *et al.*, 1991; McCormick, 1996). However, in the salmon the effect was not as great as that of growth hormone. mRNA for IGF-I was found to increase in the gills of rainbow trout in brackish water (Sakamoto and Hirano, 1993).

As will be described below, the proliferation of the chloride cells and the rise in concentration of branchial Na–K ATPase elicited by growth hormone can also be promoted by cortisol, which in some respects appears to duplicate the actions of growth hormone.

Prolactin. Pickford and Phillips in 1959, found that when killifish, *Fundulus heteroclitus*, were hypophysectomized they were able to survive in salt water but they soon died when placed in fresh water. If, however, they were injected with mammalian prolactin, their survival in fresh water was considerably prolonged. No other hormones were found to have this effect. Many species of fishes die in fresh water following hypophysectomy, though others, like the goldfish, eel, and trout, can survive for considerable periods of time. The importance of the pituitary for survival in fresh water varies considerably among the teleost fishes; thus 18 species of the order Antheriniformes were found to be unable to survive hypophysectomy if kept in fresh water though many of the order Ostariophysi survive (Schreibman and Kallman, 1969).

Death in fresh water following hypophysectomy was found in *Poecilia latipinna* to be accompanied by a considerable loss of sodium that was prevented by the injection of prolactin (Ball and Ensor, 1965; 1969). This effect has been confirmed in other species of fish and is principally caused by an excessive loss of sodium across the gills (Maetz, Mayer, and Chartier-Baraduc, 1967; Ensor and Ball, 1972; Lam, 1972). The lack of such sensitivity in some fish, like eels and goldfish, to hypophysectomy seems to be the result of the more restricted permeability of their bodies to sodium, but even in these species prolactin can be shown to decrease branchial sodium loss (efflux).

The synthesis and release of prolactin is influenced by the salinity of the external solution bathing teleosts. These processes may be reflected by changes in the storage of prolactin in the adenohypophysis. When the molly, *Poecilia latipinna*, is transferred from brackish water to fresh water these stores of prolactin are rapidly depleted (Ball and Ingleton, 1973). After a period of adaptation lasting for about 8 days, prolactin levels rise markedly, apparently reflecting an increased synthesis. Subsequent to these important early observations on the role of prolactin in adaptation of teleosts to fresh water, the hormone levels have been measured in the plasma using radioimmunoassays. In coho salmon smolts the release of prolactin is increased in fresh water and decreased in sea water or brackish water (Young *et al.*, 1989). This effect has also been observed in two species of tilapia, *Oreochromis mossambicus* and *O. niloticus* (Auperin *et al.*, 1994a; Yada, Hirano, and Grau, 1994). The morphological transformation of the parr of Atlantic salmon and coho salmon to smolts is associated with a decline in plasma prolactin concentrations (Prunet *et al.*, 1989). Tilapia, like some other teleosts, possesses two prolactins (Chapter 3): "large" prolactin with 188 amino residues ($tPRL_{188}$), and "small" prolactin with 177 residues ($tPRL_{177}$). Both hormones are released in response to transfer of the fish from brackish water to fresh water, but $tPRL_{188}$ shows the greater change. It also has a more pronounced effect on sodium retention by the fish. Prolactin receptors have been identified in the gills of *O. niloticus* (Auperin *et al.*, 1994b) but they do not distinguish between the two hormones. The numbers of these receptors, somewhat unexpectedly, increases when the fish are transferred from fresh water to brackish water (Auperin *et al.*, 1995). Such an upregulation of receptors exposed to decreased concentrations of their ligands (prolactin in this instance) is well known to pharmacologists. It was suggested that in tilapia living in environments where the salinity may fluctuate such an increase in receptors may constitute a precautionary preadaptation to sudden dilutions of their bathing media.

Circulating levels of prolactin are low in fish adapted to sea water. The importance of this is emphasized by the observation that when sea water-adapted *Oreachromis mossambicus* are injected with this hormone their plasma sodium concentration increases (Dharmamba *et al.*, 1973). This treatment, if continued, would probably kill these fishes. The accumulation of sodium is the result of a reduced rate of sodium chloride secretion from the branchial chloride-secreting cells, possibly as a result of an inhibition of Na–K ATPase. The reduced activity of the chloride cells helps effect the adaptation of these fish to fresh water, but in sea water such an action would be disastrous.

Prolactin may also influence the permeability to water of the gills, renal tubule, intestine, and urinary bladders of teleosts (see, for instance, Utida *et al.*, 1972; Ogawa, Yagasaki, and Yamazaki, 1973; Morley, Chadwick, and El Tounsy, 1981). A decrease in the osmotic permeability of such membranes has been observed under *in vitro* conditions. However, inconsistencies between *in vivo* and *in vitro* observations, and the possibility of contamination of some of the early heterologous prolactin preparations used, has prompted caution regarding the possible physiological significance of such observations (Rankin and Bolis, 1984).

Prolactin appears to have an important role in the migration of sticklebacks, *Gasterosteus aculeatus*, between the sea and rivers (see Lam, 1972). These fish normally spend the autumn and winter in the sea but in the spring migrate into rivers to breed. Sticklebacks caught in winter, in the sea, soon die if they are placed in fresh water, but if they are first injected with prolactin they survive much longer. This increased survival ability is also seen in fish in the autumn if they are kept under conditions of long daylength. It seems that, owing to photoperiodic stimulation (the lengthening of the day), there is an increase in the activity of the prolactin cells, which prepares the fish for its future migration into fresh water, during which the hormone is released into the blood.

The precise nature of the effects of prolactin on the permeability of the gills of teleosts is unknown. The recent identification of its receptors in gills is a promising step in elucidating such mechanisms. An inhibition by this hormone of proliferation of chloride cells and a promotion of their dedifferentiation has been suggested (Foskett *et al.*, 1983; Foskett, 1987). Electrical measurements across the gills indicate that there is a decreased diffusion of sodium ions along paracellular pathways. A suggested inhibition of gill Na–K ATPase could be contributing to an inhibition of chloride transport and cell swelling, which could also influence the integrity of paracellular paths. The hypercalcemic effect of prolactin in teleosts (Chapter 6) could also be influencing changes in the permeability of the cell membranes of branchial cells.

There have been many suggestions and indications that the effects of prolactin on the permeability of epithelial membranes to ions and water may not be confined to amphibians and teleosts (Ensor, 1978). Sporadic observations in birds and mammals suggest analogous effects on the kidney, gut, and salt glands may be occurring. Confirmation of these intriguing possibilities remains to be accomplished.

Adrenocorticosteroids   The principal corticosteroids in lungfish are corticosterone and aldosterone (Chapter 3), which is consistent with their relationships to amphibians. In teleost fish cortisol is predominant and it appears to fulfill the role of both a mineralocorticoid and glucocorticoid. Elasmobranchs have a unique corticosteroid, 1α-hydroxycorticosterone. The synthesis and secretion of corticosteroids in fish appears to be largely under the influence of corticotropin (Balment, Hazon, and Perrott, 1987). However, other hormones may also contribute to such regulation. Despite earlier negative observations it now seems that corticosteroidogenesis in the Australian lungfish, *Neoceratodus forsteri*, can be promoted by angiotensin II (Joss *et al.*, 1994). The synthesis of both aldosterone and corticosterone were increased though there was a difference in the sensitivity of each process to different analogs of angiotensin II. (It will be recalled that in mammals angiotensin II only stimulates synthesis of aldosterone, but such a specificity seems to be lacking in many nonmammals). The administration of angiotensin II has also been shown to increase the synthesis of cortisol in European eels (Henderson *et al.*, 1976) and flounder (Perrott and Balment, 1990). Similarly angiotensin II promotes increases in the plasma concentrations of 1α-hydroxycorticosterone in two species of dogfish (Hazon and Henderson, 1985; Balment *et al.*, 1987). Angiotensin II, therefore, has a widespread effect on corticosteroidogenesis among the fishes. The urophysial peptide urotensin I, which has structural homologies to CRH, promotes the synthesis of cortisol in flounder and rainbow trout (Arnold-Reed and Balment, 1989; 1994). Such an effect may contribute to the observed increase in plasma cortisol that follows their transfer from fresh water to sea water. Such responses of fish to exogenous angiotensin II and urotensin I suggest possible physiological roles for these peptides.

The possible role of cortisol in teleost osmoregulation has been studied by observing the effects of interrenalectomy in eels. (This procedure is only possible in these fish). In addition the patterns of release of this corticosteroid have been followed during the adjustments of such fish when transferred between sea water and fresh water.

Interrenalectomy in fresh water European eels results in a loss of sodium ions that can be corrected by the injection of small amounts of cortisol (Chan *et al.*, 1967; Henderson and Chester Jones, 1967). In the sea water-adapted eels, this surgery results in an excessive accumulation of sodium (Mayer *et al.*, 1967). The depression of the salt secretion by the branchial chloride cells, which is the cause of the gain of salt, can be alleviated by the administration of cortisol. This corticosteroid, therefore, appears to contribute to either the accumulation or the secretion of sodium by these fish, depending on the particular osmotic environment it is occupying.

Transfer of North American eels from fresh water to sea water results in an initial rise in plasma cortisol concentration and an increase in the levels of Na–K ATPase in the gills (Forrest *et al.*, 1973a,b). The time courses of these changes is shown in Fig. 8.7 and suggests that they are causally related. The administration of cortisol to eels has also been shown to induce branchial Na–K ATPase, which is associated with its chloride cells (Epstein, Cynamon, and McKay, 1971; Kamiya, 1972). A proliferation and differentiation of

Fig. 8.7. A diagram showing the relationships of the changes in plasma sodium and cortisol concentrations (*b*), the efflux of sodium, and the branchial Na-K ATPase in North American eels (*Anguilla rostrata*) following transfer from fresh water to sea water (*a*). It can be seen that the cortisol concentrations in the plasma initially rise and this is followed by increased Na–K ATPase levels in the gills, which parallels added sodium efflux and the resulting decline in the plasma sodium concentration. (From Forrest *et al.*, 1973a,b.)

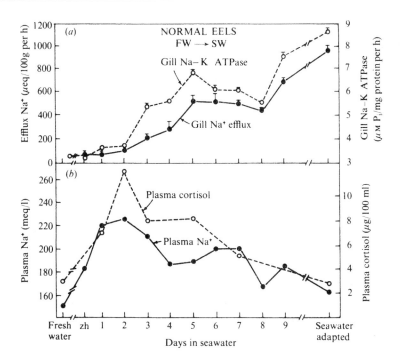

chloride cells also occurs in response to injected cortisol (Thompson and Sargent, 1977; Foskett *et al.*, 1983). It has, therefore, been suggested that the primary effect of cortisol on teleostean osmoregulation may be to promote such proliferation and differentiation of branchial chloride cells (Foskett *et al.*, 1983; Foskett, 1987). In mammals, corticosteroids have a well-defined ability to suppress proliferation and growth of many types of cell and tissue so that such an effect on the gills of teleosts is somewhat unexpected. It could, therefore, represent a novel type of action for this hormone. The newly formed chloride cells do not secrete spontaneously but require the additional stimulus of exposure to sea water. Other hormones may be involved in this stage of adaptation and β-adrenergic actions of catecholamines are one possibility. However, in killifish, the acute, immediate secretion of chloride that occurs in response to exposure to sea water appears to reflect a direct effect of plasma hypertonicity on the volume and activity of the chloride cells (Zadunaisky, 1996).

Not all teleosts respond to cortisol in this manner. The smolts of Atlantic salmon display a drop in plasma cortisol concentrations following transfer to sea water (Langehorne and Simpson, 1986). There was also no relationship between observed increases in gill Na–K ATPase and cortisol levels. In this species, the development by the smolt of an ability to osmoregulate in sea water appears to be more closely associated with the secretion of growth hormone (Prunet *et al.*, 1989). However, administered cortisol has been shown to have a synergistic effect with growth hormone in enhancing the salinity tolerance and levels of branchial Na–K ATPase in these fish (McCormick, 1996). In coho salmon, by comparison, the plasma levels of both cortisol and growth hormone may be associated with the development of such an osmoregulatory ability (Young *et al.*, 1989). The process of the adaptation of fish to life in solutions of different salinity is most likely a multihormonal process. Administration of both cortisol and growth hormone to brown trout enhanced their salinity tolerance in such a manner as to indicate that they may have a synergistic interaction (Madsen, 1990b). Interspecific differences in the relative contributions of hormones to osmoregulation undoubtedly exist.

Corticosteroids may also influence the salt metabolism of teleosts by actions at extrabranchial sites. Hypophysectomy of killifish results not only in a decline of Na–K ATPase in the gills, but also that in the intestine and kidneys: deficiencies that can be corrected for by the administration of cortisol (Pickford *et al.*, 1970). However, there appears to be no evidence that corticosteroids can influence renal tubular salt reabsorption in fish (Nishimura and Imai, 1982). Information on the effects of cortisol on ion transport across the intestine of teleosts is limited (Collie and Hirano, 1987). It may promote salt absorption across the intestine of Japanese eels. In goldfish, the presence of cortisol favored sodium absorption from the intestine but a direct effect was not considered likely (Ellory, Lahlou, and Smith, 1972).

There is little information about the role of corticosteroids in osmoregulation in nonteleost fish. Injected corticosteroids appear to have no effect on renal salt excretion in lampreys, *Lampetra fluviatilis*, in fresh water though there was a decrease in the loss of sodium across the gills (Bentley and Follett, 1963). In the dogfish, *Scyliorhinus canicula*, plasma concentrations of 1α-hydroxycorticosterone rise when they are transferred from sea water to more dilute solutions or if they are placed on a low protein diet (Armour, O'Toole, and Hazon, 1993). The dietary change reduces plasma urea levels, which normally contribute to their osmotic solute balance in sea water. It was suggested that the corticosteroid may contribute to compensatory salt retention under such conditions, by actions on the gills, kidneys, and rectal gland. Adrenalectomy results in a decline in the salt-secreting activity of the rectal gland of the skate *Raja ocellata* (Holt and Idler, 1975; Idler and Kane, 1976). However, more direct control of the functioning of this gland may involve the actions of VIP (Stoff *et al.*, 1979) or, as described below, an ANP.

The renin–angiotensin system   The tissue morphology and chemical structures of the hormonal components of the renin–angiotensin system (RAS) has been described in Chapters 2 and 3. Its presence has been established in bony fishes, especially teleosts, and more recently in elasmobranchs, but it appears to be absent in cyclostomes.

The release of renin from the kidneys of fish can occur, like in tetrapods, in response to declines in blood pressure (Nishimura, 1987; Hazon *et al.*, 1989). In mammals the release of renin is also promoted by hyponatremia but this response appears to be lacking in fish. The transfer of North American and European eels from sea water to fresh water results in a decline in plasma sodium concentrations and renin activity (Henderson *et al.*, 1976; Nishimura, Sawyer, and Nigrelli, 1976). However, the smolts of Atlantic salmon and rainbow trout exhibit an increased plasma renin activity when they are transferred from fresh water to sea water (Smith *et al.*, 1991). The release of renin in these fish appeared to be the result of dehydration. Eels experience a decline in blood pressure on transfer to sea water and this response could also be contributing to the release of the renin (Henderson, *et al.*, 1976; Nishimura *et al.*, 1976).

Administered angiotensin has a vasoconstrictor action and increases blood pressure in teleosts, lungfish, elasmobranchs, and the holostean *Amia calva* (Henderson *et al.*, 1976; Nishimura and Sawyer, 1976; Sawyer *et al.*, 1976; Hazon *et al.*, 1989; Butler, Oudit, and Cadinouche, 1995). Such effects may influence kidney function. Large doses of angiotensin, sufficient to increase blood pressure, evoked a diuresis and natriuresis in North American eels adapted to either fresh water or sea water, and in the Australian lungfish *Neoceratodus forsteri* (Nishimura and Sawyer, 1976; Sawyer *et al.*, 1976). (It is noteable, however, that a similar vascular effect and renal response is seen in the aglomerular goosefish *Lophius americanus* (Churchill *et al.*, 1979).) In rainbow trout, angiotensin may contribute to the decline in urine volume, GFR, and glomerular intermittency that is observed to occur on their transfer to sea water (Brown *et al.*, 1980; Kenyon *et al.*, 1985). Injection angiotensin or renin can also promote the release of cortisol in teleosts and elasmobranchs (Henderson *et al.*, 1976; Balment *et al.*, 1987; Perrott and Balment, 1990) and aldosterone and corticosterone in the Australian lungfish (Joss *et al.*, 1994).

Angiotensin II can promote drinking in mammals, birds, and reptiles, but not in amphibians (Nishimura, 1987). This effect has, however, been observed in many teleosts and some elasmobranchs, but not in cyclostomes. Teleosts that respond in this way to the administration of angiotensin II are usually species that may experience increases in the salinity of the water in which they live. The response may, therefore, represent an emergency one for promoting rehydration in such fish (Kobayashi *et al.*, 1983). The effect of angiotensin on drinking has been carefully examined in the flounder *Platichthys flesus*, which is euryhaline and migrates between fresh water and the sea (Carrick and Balment, 1983). Flounder placed in sea water display increased drinking and this behavior can be imitated by the injection of angiotensin I. Both of these effects, but not that of angiotensin II, can be inhibited by a drug (captopril) that blocks the conversion of angiotensin I to its active form,

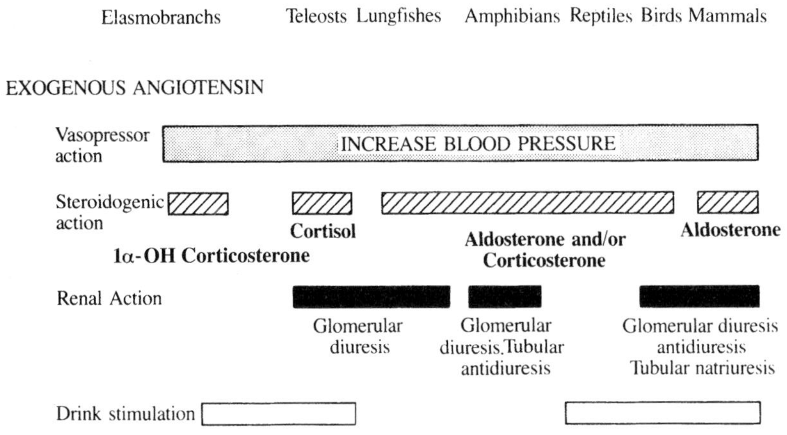

Fig. 8.8. Summary of the phyletic distribution of the effects of injected angiotensin II in vertebrates. (Adapted from Nishimura, 1978; see also Nishimura, 1987.)

angiotensin II. An elasmobranch, the dogfish *Scyliorhinus canicula*, normally drinks little sea water but it can be induced to do so by the injection of angiotensin I or II, or a drug (papaverine) that decreases the blood pressure (Hazon *et al.*, 1989). The effects of angiotensin I and papaverine were blocked by captopril. It seems likely that the observed effects of angiotensin on drinking in fish reflects a physiological role. The renin–angiotensin system has persisted in vertebrates from the elasmobranch fishes to the mammals and it has retained many of its primeval roles but variations do exist (Fig. 8.8).

The natriuretic peptides   These hormones were originally identified in the heart muscle and brains of tetrapod vertebrates and they are also present in fish. They have been identified in cyclostomes, elasmobranchs, and teleosts (Chapter 3). They include peptides of the type A (ANP) and type C (CNP) groups. The CNPs are prevalent among the elasmobranchs and could represent the ancestral forms. Another type of natriuretic peptide has been identified in the ventricles of Japanese eels and has been called ventricular natriuretic peptide (VNP)

As the natriuretic peptides have a special propensity to facilitate the excretion of sodium in tetrapods there has been considerable interest regarding such a possible role in marine fish (Evans, 1990; Takei and Balment, 1993). In eels they are not only released into the plasma in response to hypervolemia, as occurs in mammals, but hyperosmotic stimuli are also effective (Kaiya and Takei, 1996a). Natriuretic peptides have been identified in the plasma of cyclostomes, elasmobranchs, and teleosts where their concentrations are usually higher when the fish are in sea water compared with more dilute solutions (see, for instance, Westenfelder *et al.*, 1988; Evans, Chipouras, and Payne, 1989; Freeman and Bernard, 1990; Smith *et al.*, 1991). Nevertheless, such differences are not invariably observed: North American and Japanese eels display no differences in plasma natriuretic hormone levels when adapted to either fresh water or sea water (Epstein *et al.*, 1989; Kauya and Takei, 1996b). Possibly these fish utilize other mechanisms for osmoregulation more fruitfully.

The actions of the natriuretic peptides in fish generally reflect those that have been observed in mammals. There are, however, some differences. Cardiovascular effects include a relaxation of vascular smooth muscle and a decline in blood pressure (see, for instance, Evans *et al.*, 1989; Evans, 1991; Olson and Duff, 1992). Their administration also results in a diuresis and increased sodium excretion even (as described earlier) in the aglomerular toadfish *Opsanus tau* (Lee and Malvin, 1987). In teleosts, there has been special interest in a possible salt-secreting effect on the gills. Receptors for natriuretic peptides have been identified at such sites, including the gills of hagfish (Toop, Donald, and Evans, 1995a). They have also been found in the glomeruli, renal tubules, and aorta of such cyclostomes (Kloas *et al.*, 1988; Toop, Donald, and Evans, 1995b). In hagfish and eels (Balment and Lahlou, 1987; Broadhead *et al.*, 1992) activation of these receptors results in the formation of the second messenger cyclic GMP. (It will be recalled that in mammals the ANP receptors are the sites of guanylate cyclase). In the hagfish, the receptors interact with both ANP and CNP and appear to be of the GC-A type that is found in mammals (Chapter 3). It would appear that not only the peptides but also the specialized receptors for the natriuretic peptides have an ancient lineage.

The branchial receptors for the natriuretic peptides in teleosts are associated with the chloride cells and the local arterioles. An action at both sites could influence salt secretion. ANP (*in vitro*) stimulates a chloride-dependent electrical current (the short-circuit current) across the chloride cell-rich opercular membranes of the killifish (Scheide and Zadunaisky, 1988). The administration of ANP has also been shown to stimulate the efflux of sodium ions from several teleosts, flounder, dab, and plaice, a response that appears to involve the gills (Arnold-Reed, Hazon, and Balment, 1991). It is possible that cortisol is contributing to this response as ANP has been shown to stimulate its synthesis in flounder (Arnold-Reed and Balment, 1991). This effect on steroidogenesis is of special interest as it contrasts with its inhibitory action in tetrapods. The physiological result is, however, the same: salt excretion.

The natriuretic peptides, including an homologous one from eel atria, have also been shown to inhibit the absorption of sodium chloride from the intestine (*in vitro*) of the winter flounder (O'Grady *et al.*, 1985) and the Japanese eel (Ando, Kondo, and Takei, 1992). This effect, which could be mimicked by a cyclic GMP analog 8-bromo-cyclic GMP, may modulate absorption of salt following the drinking of sea water by marine teleosts. In elasmobranch fishes, accumulated excess sodium chloride can be secreted by the rectal salt gland. Such secretion occurs in response to an increased volume of the body fluids and is not dependent on the intervention of external neural stimuli. It may be a hormonal response. Under *in vitro* perfusion conditions, VIP can, by utilizing a cyclic AMP-dependent mechanism, stimulate secretion by the rectal gland (Greger, Gögelein, and Schlatter, 1988). However, this peptide is ineffective *in vivo* and appropriate changes in its plasma levels have not been observed (Solomon *et al.*, 1985). Administered natriuretic peptides, including homologous CNP, have been shown (both *in vitro* and *in vivo*) to stimulate salt secretion by the salt gland of the spiny dogfish *Squalus acanthias* (Solomon *et al.*, 1992). VIP is associated with an internal neuronal network in the rectal gland and its release can be promoted by natriuretic peptides (Silva, Solomon, and Epstein, 1996).

The urophysial peptides The urophysis is an aggregation of neurosecretory cells in the caudal region of the spinal cord of many teleosts (Chapter 2). It contains a number of biologically active substances including two peptides called urotensin I and II. Urotensin I has a structure which is similar to CRH, while urotensin II shares some chemical features with somatostatin (Chapter 3). In 1956, M. Enami observed histological changes in the urophysis of the loach *Misgurnus anguillicaudatus* exposed to increased salinities (see Fridberg and Bern, 1968). It was, therefore, suggested that the urophysis may be contributing to osmoregulation in the fish (Larson and Bern, 1987). A voluminous literature on this subject has since been accumulated but unequivocal physiological conclusions have, so far, not been made.

Many studies have confirmed the changes in histological appearance of the urophysis and the stores of its peptides in fish exposed to alterations in osmotic concentration. Some (see, for instance, Owada *et al.*, 1985; Arnold-Reed *et al.*, 1991) have shown increases in its putative secretory constituents following transfer from fresh water to sea water, while others (Takasugi and Bern, 1962; Larson and Madani, 1991) have observed similar changes when fish are moved in the opposite direction, from sea water to fresh water. The difference in the responses may reflect interspecific variation or simply a response to a change in the environment of the fish. It is difficult to predict what the alterations in the storage products in the urophysis indicate with respect to their possible secretion. For instance, an increased accumulation could reflect a decreased secretion or an increased synthesis. We must await direct measurements of the levels of the urophysial products in the plasma to resolve this problem.

The venous blood vessels from the urophysis open into the renal portal system and also supply the urinary bladder and parts of the intestine. A released hormone would, therefore, be in an eminently favorable position to exert actions on these osmoregulatory organs. Pharmacological observations, often in mammals, show that urotensin I can have a vasodilatatory effect and can lower the blood pressure, while urotensin II can increase blood pressure in fish (Chapter 3). Such effects could be mediating osmoregulatory responses in fish. More direct actions on salt-transporting epithelia have been observed. Chloride transport across the skin (*in vitro*) of the marine teleost *Gillichthys mirabilis* is increased by urotensin I and decreased by urotensin II (Marshall and Bern, 1981). Sodium transport across the urinary bladder and intestine (*in vitro*) of this fish is increased by urotensin II (Loretz and Bern, 1981; Loretz, Howard, and Seigel, 1985).

Urotensin I can promote the release of corticotropin from the adenohypophysis (*in vitro*) of the sucker *Catostomus commersoni* (Fryer, Lederis, and Rivier, 1983). The effect may reflect its chemical similarity to CRH. Administration of urotensin I to flounder promotes a release of cortisol, a response which is reminiscent of the effects of exposure to sea water (Arnold-Reed and Balment, 1989). Under *in vitro* conditions, urotensin I was observed to have a direct effect in promoting cortisol synthesis by trout interrenal tissues (Arnold-Reed and Balment, 1994). Corticotropin was also effective and the two peptides may be exerting a synergistic action. Urotensin II, which has somatostatin-like effects, may also influence pituitary function. It has, for instance, been found to inhibit the release of prolactin from the pituitary (*in vitro*) of tilapia (Grau, Nishioka, and Bern, 1982). The possible implications of this interesting observation in the role of prolactin on teleost osmoregulation does not appear to have been further explored. Measurements of urotensin concentrations in the plasma are needed to establish the putative osmoregulatory role of the teleost urophysis.

Conclusion   Our knowledge of the nature of the processes controlling osmoregulation in fishes is incomplete, and much of the information about the role of hormones in this process is still speculative. It is therefore still an exciting field of study for the comparative endocrinologist.

Conclusions

The regulation of water and salt content of vertebrates is primarily mediated by the kidney, but several other glands and tissues are also involved. Many of these "accessory" osmoregulatory organs have a distinct systematic distribution, for instance, gills in fishes, cephalic salt glands in birds and reptiles, and sweat glands in mammals. The osmoregulatory functions of most such organs are controlled by hormones that each tend to contribute to this process in a specific manner. Thus, the neurohypophysial hormones increase osmotic permeability whether it be in the tetrapod kidney or the amphibian skin and urinary bladder. Similarly, adrenocorticosteroids help regulate sodium metabolism by increasing transmural sodium transport in the renal tubule of mammals, the skin and urinary bladder of amphibians, the sweat, salivary, and mammary glands of mammals, and the gills of many fishes. Natriuretic peptides can promote sodium excretion by actions on the kidneys of mammals, and in fishes they can have a similar effect, not only on the kidneys but also on the gills of teleosts and the rectal gland of elasmobranchs. The renin–angiotensin system has a role in controlling steroidogenesis in most vertebrates and it can also promote drinking, but not in amphibians or all the fishes. Thus, the same type of hormone is often concerned (or is "utilized") with coordinating the same general physiological process, though in somewhat different ways in various groups of vertebrates. These are, however, exceptions, and systematically unique features, like the role of the gills in osmoregulation of fishes, are accompanied by what may be a unique type of action of a particular hormone, in this instance the ability of prolactin to control the permeability of the branchial (and other) epithelial membranes in teleosts. In mammals, this hormone has quite different roles in processes that include lactation and the functioning of corpora lutea in the ovary. It is suspected that it may also be capable of influencing osmoregulation in some tetrapods. The physiological significance of the effects in the latter is still in doubt and could represent a "vestigial" endocrine response.

9 Hormones and reproduction

The reproductive process is not essential for the life of the individual, though it may make it more interesting, but it is necessary for the perpetuation of the species. In many so-called lower forms of life, reproduction may be an asexual process. A notable disadvantage of this type of reproduction is a diminution in the chances of genetic variability, and the transmission of such inherited changes to other individuals, so that evolutionary adaptation is hampered. Reproduction is a complex process and this is especially true in species that occupy environments where the conditions are variable and large physico-chemical changes occur. The young, developing animal is not usually as adaptable as the adult to such changes in the environment, and so must either be protected from these deviations by the parents or be produced on occasions that are most suitable to its more limited physiological capabilities. In vertebrates both conditions usually prevail; the embryo may develop to a quite advanced stage before becoming independent of the parent and it is usually produced during a season when such conditions as the temperature and food and water supply are favorable.

Reproduction in vertebrates, therefore, involves considerable physiological coordination. The sexual process that requires the union of the sperm and ova necessitates complex physiological, social, and morphological arrangements to ensure that these gametes each ripen at a similar time, and that the two sexes then meet and effect their union. The growth and differentiation of the fertilized egg often involve complex parental care, which may occur *in utero*, within the parent itself, or in an egg that is specially produced to meet the potential needs of the embryo. Care of the young often continues for a period of time following such initial development in the egg or *in utero*. The foregoing events may not be possible, or successfully accomplished, except during certain seasons of the year when the conditions are favorable.

In vertebrates, the coordination of all the processes outlined above involves hormones and the degree of complexity in their actions directly reflects the intricacies of the reproductive processes in a particular species. The endocrine control of reproduction in humans is therefore more involved than in a jawless fish, like the lamprey. The basic pattern, however, is remarkably uniform and involves the endocrine secretions of the pituitary and the gonads. The influences of the hypothalamus on the pituitary gonadotropin release are vital in most groups of vertebrates, but this control may be lacking in the cyclostomes and chondrichthyeans. The pituitary gonadotropic hormones in vertebrates are chemically analogous but have undergone evolutionary changes in their structures. The gonadal steroids, by comparison, are almost identical throughout the vertebrate series. One of the most notable endocrine differences among the vertebrates is the ability of the placenta of eutherian mammals to produce hormones that are similar to those secreted by the gonads and the pituitary. Otherwise, the evolutionary changes are largely a matter of detail. These adaptations include modifications of the gonadal ducts, such as may assist in the processes of fertilization, the production of different types of egg, and the internal incubation of the embryo. A plethora of secondary sex characters, involving such morphological factors as size, color, and scent glands have appeared in the different vertebrate groups and these help to ensure that the sexes meet and mate in the breeding season. Also involved in the mating procedure are a multitude of different patterns of prenuptial behavior. The precise manner by which the time of breeding is controlled also varies in different vertebrates and may involve differences in the length of the gonadal cycle and adaptations to the receipt of different environmental signals, such as light, which is predominant in birds, and temperature, which seems to be more important in reptiles. Variation in the functioning of the hypothalamus, pituitary, and gonads may reflect such differences in the manner of timing of breeding.

There are several major differences in the patterns of reproduction in vertebrates that have considerable effects on the endocrine control of reproduction. These concern the manner by which fertilization is accomplished and the site where the embryo differentiates and grows.

Life in an aqueous medium, such as the sea, lakes, and rivers, provides a relatively stable physicochemical environment for ova, sperm, and the fertilized eggs. Many fishes ensure their reproduction by producing vast numbers of ova, often thousands at a time, which are extruded, before fertilization, into the external solution. The reproductive activities of the male are coordinated with this oviposition so that vast numbers of sperm are released among the eggs and *external fertilization* occurs. This process, which also takes place in most amphibians, has the advantage of simplicity but is only possible in aqueous situations. A general physiological corollary of external fertilization is that usually the eggs are small and contain relatively few nutrients for the support of the young. The animal thus, nutritionally, can afford to produce the vast numbers of ova necessary to assure fertilization on a scale that is adequate for the survival of a sufficient number of the young. Primarily terrestrial species, like the reptiles, birds, and mammals, must of necessity resort to *internal fertilization*. This process requires the production of fewer eggs but more intimate contact and collaboration between the sexes. Internal fertilization also occurs among fishes including some bony fishes, all of the chondrichthyeans, and some amphibians.

There are also considerable differences, again with important endocrine repercussions, between the eggs of different vertebrates and the processes that assist in their successful transposition into viable young. These differences are partly related to the nature of the fertilization process (because of the number of eggs that must be produced) and to life in a terrestrial environment. Eggs produced by most fishes and amphibians are highly prone to evaporation and so cannot readily survive on dry land. Some frogs deposit such eggs in damp burrows, but this is unusual. Birds, most reptiles, and prototherian mammals, such as the platypus, produce eggs that are covered with a protective shell which limits evaporation, and they contain large amounts of nutrients that are sufficient to sustain the young until it reaches a stage of development when it can fend for itself. Similar eggs with a horny shell and large amounts of nutritive yolk are also produced by the chondrichthyean fishes and hagfishes. The production of eggs from which the young develop in the external environment is a process called *oviparity* and the eggs are termed alecithal and megalecithal eggs, depending on the amounts of yolk nutrients that they contain. In many species, including some chondrichthyean and teleost fishes (where the ovum is often fertilized within the follicle in which it develops), amphibians, and reptiles, the eggs may be retained for prolonged periods in the oviduct, during which time the young develop in a relatively protected and secluded situation. This is called *ovoviviparity*. A more intimate contact between the eggs and the wall of the oviduct or uterus may occur whereby the developing young can exchange respiratory gases and even gain fluids and nutrients. This condition is called *viviparity* and as nutrients may be gained from the parent, the eggs usually contain far less yolk. Viviparity has evolved many times in nature and is present among chondrichthyeans, teleosts (where the young are usually contained within a hollow ovary), amphibians, reptiles, and mammals. There are many endocrine variations that result from these different ways of providing for the development of the young including hormonal influences on the maturation and formation of the different types of egg, the morphological development and physiological behavior of the oviducts, and the triggering, at the appropriate time, of the expulsion of the egg (oviposition) or young (parturition).

The eggs and young of many species receive little or no parental care once they are separated from the mother. In some species, however, they are cared for and this may be necessary for their survival. Some teleost fishes are known to deposit their eggs in specially prepared nests that they protect and over which they may circulate water. Others such as the teleost *Oreochromis mossambica* keep a brood of hatched young in the fastnesses of a large mouth that the young can emerge from or retreat to. Several species of frogs (*Pipa pipa, Gastrotheca marsupiata*) and the seahorse (*Hippocampus*) keep young in a pouch, or marsupium, on their backs while other frogs incubate them in modified vocal sacs. Some snakes and most birds personally incubate their eggs, and the care with which birds feed and protect the newly hatched young is well known. Birds usually collect food, which they present to their young and often this is predigested. Pigeons and doves produce a special pasty secretion that contains a high concentration of fat and protein from their crop-sac, the so-called pigeon-milk. The formation of such a special milk secretion with which to feed the young is a characteristic systematic feature of mammals. Such processes, whether it is "tender loving parental care", "broodiness" in birds, or lactation in mammals, are all largely controlled by hormones.

The reproductive apparatus of vertebrates

The gonads of vertebrates have a dual function, as they not only produce the germ cells but also some of the hormones that control the reproductive process; thus, the testis in addition to being the site of formation and maturation of the sperm also produces androgens, principally testosterone and androstenedione in the interstitial tissue (or Leydig cells) and, probably, also in the Sertoli cells. The ovary contains vast numbers of germ cells (primordial follicles) some of which, following a period of growth and maturation, ripen into ova. The follicles in which this latter process occurs are also the site of formation of estrogens. Following extrusion of the mature ovum (ovulation) from its follicle, the tissue "heals" and this may involve an invasion and proliferation of lutein cells so that a corpus luteum is formed. In many species, this structure is the site of formation of progesterone, which can also sometimes be produced by the interstitial tissue of the ovary. It should be remembered, however, that such sites are not exclusive ones for the formation of such steroid hormones. The corpus luteum may, for instance, also form estrogens and the follicles may form progesterone. A more detailed description of the structure of the gonads and the hormones that they produce is given in Chapters 2 and 3.

Associated with the gonads are the duct systems through which the germ cells are delivered to the outside of the animal. Discrete gonadal ducts are absent in the cyclostomes where the eggs and sperm are shed directly into the body cavity from which they escape to the exterior through pores that are formed in the region of the cloaca.

In the teleost fishes, the ovarian ducts represent extensions of the gonadal tissue, but in other vertebrates they are modified Müllerian ducts that differentiate in the embryo to form the oviducts or uterus. The oviducts and uterus are surrounded by a sheath of smooth muscle fibers that, by rhythmical contractions, can propel objects, like eggs, toward the exterior. The musculature may be relatively weak, as in the amphibians, or, as in mammals, be capable of very strong contractions and make up a major portion of the uterine wall. In mammals, this muscle layer is called the *myometrium*. The contractility of such muscle can be influenced by hormones, especially those from the neurohypophysis, which stimulate their contraction. Underlying the muscles of the gonoducts is an inner lining of cells that in mammals is referred to as the *endometrium*. This inner layer of tissue contains numerous glandular cells and may be modified in various ways so that it contributes to the well-being of the egg and, if internal fertilization occurs, the survival of the sperm and its union with the ovum. In oviparous species, segmental differences in function may occur along the length of the oviduct, such as are associated with the formation of albumin and the secretion of a hard outer shell. In amphibians, the jelly-like secretion so characteristic of clumps of frogs' spawn is secreted by glandular cells in the oviduct. In ovoviviparous and viviparous vertebrates, the lining of the oviduct is modified so as to furnish an appropriate environment for maintaining the retained egg or to allow for the implantation and development of the blastocyst and the formation of a placenta. The activity of the surrounding musculature is reduced on such occasions. The female gonoducts (oviduct or uterus and vagina) therefore undergo considerable structural change during the reproductive cycle, which is mediated by the action of estrogens and progesterone.

The sperm is conveyed from the testis along the vas deferens, which is also a tube surrounded by an outer layer of smooth muscle, during which time they may be mixed with secretions from certain accessory sex glands that include the prostate or prostate-like glands. The maintenance of the structure and function, and cyclical changes of the male gonoducts, their associated accessory sex glands as well as the external genitalia, such as the penis, are mainly the result of the action of testosterone. In the absence of this hormone, structural and physiological degeneration of such tissue takes place.

Secondary sex characters in vertebrates

The secondary sex characters are so named because they are not primarily involved in the formation and delivery of the sperm or ova. They nevertheless may play an important part in the prenuptial and nuptial events and contribute to the behavioral and functional synchronization necessary for the fertilization of the ripened ova, the mechanical success of copulation, and the survival of the young. The secondary sex characters differ in each sex and contribute to the dimorphism of the male and female. The differences in appearance between the sexes are basically controlled genetically, and the expression of them may be influenced by the actions of sex hormones. Broad differences, such as those of size, are usually independent of the continuous action of sex hormones, while in other instances only the initial differentiation of a sexual character during early life may depend on hormones.

Hormones are, however, not necessarily continuously needed to maintain such organs after their differentiation; for example, the changes in the larynx of boys at puberty, which results in a deeper voice, require the presence of testosterone though subsequent castration does not result in a return to the prepuberal soprano condition. In other instances, however, a continuous supply of hormones may be required to maintain a secondary sex character, as seen in the instance of the penis in man and the breasts in women. Some secondary sex characters may undergo periods of development and involution that correspond to the changes in the sexual cycle and the differences in the rates of hormone production that occur during these periods. The seasonal development and subsequent shedding of the antlers of deer are well-known examples of this but there are numerous others. Dodd (1960) and Parkes and Marshall (1960) give an excellent account of these structures in cold-blooded vertebrates and birds.

In *cyclostome fish*, the endocrine control of secondary sex characters has been studied in lampreys (Larsen, 1965; 1969; 1973). One cannot distinguish, from their external appearance, between the male and female lamprey during the early part of their autumnal breeding migration into the rivers. With the approach of spring and the onset of breeding, an anal fin appears in the female while the dorsal fins of the male heighten. Such morphological changes do not occur if the animals are hypophysectomized or gonadectomized. There is also a swelling of the cloacal region in both sexes just before breeding and the urinogenital papilla grows larger in the male. These latter characters do not, however, appear to depend on the presence of steroid sex hormones.

The *chondrichthyean fishes* also display few dimorphic secondary sex characters. In some species, the most notable difference is the presence of a pair of copulatory organs called claspers in the external cloacal region of the male. These rod-like organs develop at puberty and their size can be increased by the administration of testosterone, but this effect is not large.

The *osteichthyean fishes* show a considerable range of dimorphic sexual differences. In some species it is difficult to detect such variations, but in others there may be prominent differences in size. The male is sometimes much smaller than the female; variations in color may occur and there may be differences in the size of the fins. Such diversity may only arise, or be accentuated, during the breeding season. The dorsal fins of the bowfin, *Amia calva*, become a brilliant green prior to breeding while the belly turns a pale green. In the female the latter is white. The anal fins sometimes become enlarged in fishes such as *Gambusia affinis* and *Xiphophorus*, where it is called the "sword" and functions as an organ for copulatory intromission, called a gonopodium. The injection of testosterone into the female fish may result in the development of secondary sex characters just like those in the male.

During the breeding season, the male South American lungfish, *Lepido-*

siren paradoxa, develops long finger-like outgrowths from the fore- and hindlimbs. These organs are bright red in color owing to a rich blood supply, and it has been suggested that they may function as respiratory organs. The eggs of these lungfish are laid in burrows where the oxygen tension of the water is low. It has been surmised that the male, who guards this nest and fans the eggs, may secrete oxygen from these organs into the oxygen-poor water of the burrow. The growth of these so-called limb gills can be induced by the injection of testosterone (Urist, 1973), which confirms their nature as that of a secondary sex organ.

Amphibians often display prominent sexual differences during the breeding season. The bright orange coloration and the dorsal crest of the male crested newt, *Triturus cristatus*, can be induced in nonbreeding animals following the injection of testosterone. The development of the nuptial thumbpads and the vocal call-producing muscles in frogs is under hormonal control. Testosterone not only stimulates the development of these tissues but, in *Bufo fowleri*, can also promote the development of the vocal sacs in immature males and prompt them to give their characteristic mating calls or croaks.

It is usually rather difficult to distinguish between the sexes in *reptiles*. The males of many lizards, however, possess appendages about their heads and throats that have a fan-like appearance and can be erected for display. Dorsal crests are also present in some lizards. The tails of male turtles often grow longer than those of the females and aid in copulation. Some reptiles, especially snakes, possess erectile penises that differentiate from their cloacal tissues. The males in many species of snakes and lizards also possess a special secretory segment in the distal part of the renal tubule, called the "sexual segment" and the development of this is under androgenic control.

The colorful dimorphic differences in the plumage of *birds* are well known and generally (though not always) appear to be under the control of estrogens in the female; thus, the bright coloration of the male in domestic fowl, quail, and pheasants is not dependent on androgenic hormones, but the duller, more conservative plumage of the female is under estrogenic control. The injections of estrogen into male birds can stimulate the formation of female plumage. The color of the beaks of birds also is influenced by sex hormones during the breeding season. The beak of the male house sparrow is normally black in the breeding season and this coloration can be induced at other times of the year, or in castrates, by administering androgens and FSH or LH and FSH (Lofts, Murton, and Thearle, 1973). The domestic fowl possesses a red fleshy structure, called the comb, on its head. This is much more developed in roosters than in hens but it regresses following castration (caponization). The comb of the domestic fowl is extremely sensitive to the presence of androgens, which induce its hypertrophy. This response in capons has been widely used to identify androgenic materials and indeed provided the first unequivocal evidence for the presence of androgenic hormones in the mammalian testis.

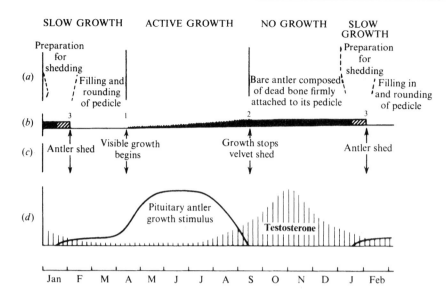

Fig. 9.1. The seasonal cycle in the growth of the antlers of the male Virginia deer (*Odocoileus v. borealis*) in relation to testosterone. This is controlled by light mediating the (hypothetical) release of pituitary hormone(s). (a) The three major phases: growth is slow, fast, and then ceases. (b) The changes in the physical size of the antlers during the periods in (a). (c) The principal events in the cycle. (d) The increases in the pituitary growth stimulus that rises in the spring and subsides in the autumn when it is inhibited by the rising levels of testosterone. The precise nature of the pituitary stimulus is not clear but growth hormone levels peak in early spring and plasma prolactin is high throughout the summer (Bubenick, 1986). (Modified slightly from Amoroso and Marshall, 1960, taken from Waldo and Wislocki, 1951.)

It is unnecessary to recall the secondary sex characters in humans. In other *mammals* differences in size and coloration commonly occur. The red "sex skin" of the buttocks of some female monkeys during the sex cycle has been referred to in Chapter 7. Scent glands, which are under androgenic control, are also common in many mammals. The wild boar and male members of the cat family secrete odoriferous materials (pheromones) into the urine from special cells in the kidney and these glands are controlled by testosterone. In sheep sudiferous glands in the skin of the ram secrete a pheromone that can initiate the ovarian cycle in formerly quiescent ewes (Chapter 7 and p. 402). The antlers of deer start to grow in the spring, apparently under the influence of pituitary hormones. In the autumn, when these animals breed, the antlers lose their covering of "velvet," stop growing, and come under the control of the rising testosterone levels in the blood. When the concentrations of this hormone subsequently decline, the antlers are shed, usually at about the end of January (Fig. 9.1).

Finally, it should be stressed that the behavior of vertebrates during the breeding season is also a secondary sex character (a most important one) that is influenced by hormones. *Sexual behavior* (Herbert, 1972) leads to the cohabitation of the male and female animal and the union of their genes, as provided in the sperm and ova. Subsequently, it may be utilized to foster the well-being and growth of the young. Apart from the use of attractants, such as the color of the integument, special appendages such as crests and fins, inviting smells and scents, and ethereal call sounds, animals also adopt active patterns of behavior to find and attract the opposite sex. Such solicitations become more frequent, impetuous, and insistent at the onset of breeding seasons when ripening of the ova is imminent. Complicated courtship displays may be provided that facilitate pairing, copulation, and preparations for the future young. Hormones play an important role in such processes and influence appropriate neural events that originate in the brain. Castration and ovariectomy usually lead to a decline in such sexual behavior. Caponized roosters do not crow and strut, spayed cats do not call and will actively reject the male, castrated bulls are not aggressive and ignore the solicitations of friendly cows. This sad picture is the result of hormonal imbalance: a lack of estrogens, androgens, and progestins, and possibly a disordered secretion of pituitary gonadotropins and prolactin. The particular hormones involved vary in different species. The injections of estrogens can restore the sexual behavior of many ovariectomized members of the Carnivora, but in the Rodentia progesterone is also needed. In primates, androgens are needed in the female and they play a major role in women. These steroids may be obtained in adequate amounts from the adrenal cortex. In the male, however, more androgens are needed and must normally be supplied by the testes. However, if sexually mature animals are castrated the masculine pattern of behavior, and even an ability to copulate, may persist for some time. Such activities can usually be fully restored by the administration of androgens. Chemical compounds that have an antiandrogenic activity, such as cyproterone, have even been utilized in attempts to reduce the sex drive in humans suffering from satyriasis.

Apart from courtship and copulation, other behavioral activities that are a preliminary to breeding may be influenced by hormones. In birds this often involves the building and preparation of a nest. Nest building can be promoted in canaries that have been ovariectomized by injecting them with estrogens, though they must also be sensitized by exposure to long-day photoperiods (Hinde, Steel, and Follett, 1974). The nature of the photoperiodically induced "factor" is unknown, but the injection of LH into these birds could not mimic it. Normal nesting behavior can be induced in budgerigars if they are injected with prolactin in addition to estrogen (Hutchison, 1975). The special role that prolactin has in the incubation of eggs by birds will be described below. Ringdoves have an elaborate courtship prior to egg laying. Ovariectomy abolishes this behavior, though it can be partially restored by modest doses of estrogens (progesterone is also needed for nest-building behavior). Nest building and brooding of the eggs is not confined to terrestrial vertebrates but also occurs in teleost fishes. In male fish, this activity is often part of their courtship behavior and subsequently provides the focus for the brooding and parental care of the fry. Garibaldi fish, *Hypsyops rubicundus*, which live in rocky coastal areas of Southern California, build nests of red algae during courtship. The female is enticed into this bower during spawning. The male subsequently guards the eggs and fans them with his pectoral fins during the brooding phase. Plasma testosterone and 17-ketotestosterone, especially the latter, show characteristic changes during breeding and decline following the initial courtship to much lower levels during brooding (Sikkel, 1993). Male sticklebacks, *Gasterosteus aculeatus*, build even more elaborate nests during their courtship phase by utilizing a "glue" that is secreted by testosterone-dependent cells in the kidneys (Brown, 1985). After hatching, the young are tended by the male. Castration of these fish abolishes courtship and nest building but these activities can be restored by the administration of testosterone. The egg-fanning activities can be promoted by low doses of prolactin. The aggressiveness of the male declines during the brooding phase, possibly reflecting a decline in androgens. Cannibalism of the fry is inhibited during the parental care phase and it has been suggested that the young may provide inhibitory stimuli. It is possible that such behavior is elicited by a hormonal–pheromonal mechanism.

Periodicity of the breeding season: rhythms in sexual activity

Most vertebrates only breed periodically but, nevertheless, at fairly precise times of the year (see, for instance, Turek and van Cauter, 1994). In temperate zones, this more usually occurs in the spring but in some species, like deer, sheep, goats, badgers, and grey seals, it occurs in the autumn. In equatorial regions where the climate and food supply are relatively similar throughout the year, breeding may often take place at any time. Similarly, some domesticated species, like the laboratory rat, the domestic fowl, and humans may breed throughout the year, a situation that appears to reflect continuously favorable circumstances and, possibly, and endogenous circannual reproductive rhythm.

An ability to reproduce during predictable seasons of the year clearly may be of considerable advantage as the young can then be produced at a time when such factors as the environmental temperature and the food and water supply are adequate. The chances for the survival of the young will thus be enhanced.

How is such precise timing possible? In temperate zones, the environmental conditions that prevail in a certain season are usually fairly predictable; thus, the animals can be expected to take their "cues" and make their reproductive preparations on the basis of the solar calendar. Changes in the length of the day are a direct reflection of these events so that the length of the periods of light and darkness may furnish an excellent calendar to work by. Indeed, such photoperiodic stimulation is basic for the control of the reproductive cycle in many vertebrates. The first clear indication that light influences vertebrate gonadal function was made in 1925 by a Canadian zoologist called William Rowan. He found that the gonads of the junco finch, which normally enlarge when the days grow longer in the spring, could be stimu-

Fig. 9.2. Photoperiodically stimulated growth of the testes of the Japanese quail. For the first 40 days the birds were subjected to long, daily photoperiods of 20 hours of light (L) and 4 hours of dark (D)(20L/4D) and after this to a short daylength of 6 hours of light and 18 hours of darkness. The diameter of the seminiferous tubules (for a given testis weight) is given on the scale on the right. At the top of the diagram (Roman numerals) the changes in the development of the sperm are given: I, spermatogonia only; II, spermatogonia dividing, a few spermatocytes; III, numerous spermatocytes; IV, spermatocytes and spermatids; and V/VI, spermatids and mature sperm. (Modified from Lofts, Follett, and Murton, 1970.)

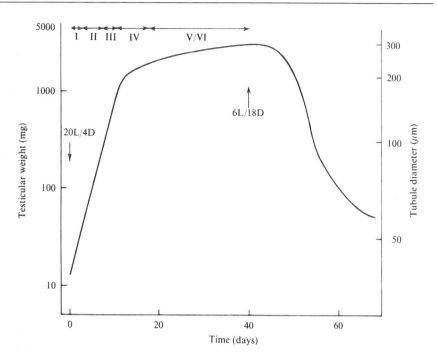

lated to grow even in winter when the birds were subjected to artificially prolonged periods of light. Other factors, however, can also impinge on the onset of the reproductive cycle and even override it. These include temperature, the nutritional condition of the animal, and the related availability of supplies of food and water. There is also evidence for the presence of an internal, inherent rhythm in the sexual activity of some species. It is often difficult to disentangle these various factors and to decide which is predominant.

The effects of light on reproduction have been studied in many species of birds and mammals but few cold-blooded vertebrates. Preparations for spring breeding often commence about the end of December when the length of the daylight hours starts to increase. As shown by Rowan, these conditions can be copied in the laboratory and dramatic increases in the activity of the gonads can then be shown to occur; in the Japanese quail subjected to long-day photoperiods of 20 hours of light and 4 hours of darkness, the weight of the testes increases from 8 mg to 3000 mg in about 3 weeks (Fig. 9.2). The subsequent substitution of a short-day photoperiod of 6 hours of light and 18 hours of darkness results in a decline in the weight of the testes. The gain in testicular weight results mainly from an increase in the length and diameter of the seminiferous tubules, though increases in the activity of the Leydig and Sertoli cells also take place. Comparable increases in development also occur in the ovaries of birds.

The retina of the eye is usually the primary receptor for the photoperiodic detection of light. The information is then transmitted to the brain along pathways that may include the pineal, as seen in many mammals, and ultimately the hypothalamus. Blindness, therefore, usually blocks such responses. However, birds may be unaffected by this condition and still respond to the photoperiod. Apparently light is transmitted through translucent areas of the skull to, as yet, unidentified photoreceptive areas associated with the brain.

The reproductive cycles of all birds or mammals do not necessarily respond to light. Such photostimulatory effects are absent in rabbits, guinea pigs, ground squirrels, and guinea fowl. These differences in response to external stimuli may reflect the effects of domestication or, possibly, in the case of guinea pigs and guinea fowl, their origin from equatorial regions where animals do not experience large changes in day length. Tropical deer that normally breed all year round also persist in this habit after many years in Europe even though they experience cold winter conditions. Deer from equatorial regions that normally have a seasonal cycle also persist in their pattern of reproduction when moved to Europe.

Amoroso and Marshall (1960) have classified animals into those that have a "long-day" and "short-day" breeding season. Long-day animals, which breed in spring, include most birds, as well as horses, donkeys, ferrets, cats, and raccoons. Goats, deer, and sheep are short-day species, which breed in the autumn when the day-length is declining.

The effects of light in stimulating development of the gonads and the timing of reproduction are not seen in the absence of the adenohypophysis or when the hypothalamic connections to the median eminence are cut. Differences in the length of the daily photoperiods of light and darkness control the release of GnRH from the median eminence, which in turn initiates the release of gonadotropins from the pituitary. The gonadotropins, FSH, and LH (and sometimes also prolactin), exert their various effects on the development of the germ cells and the formation and release of the gonadal steroid hormones.

As mentioned in the foregoing, the reproductive rhythms of all animals are not responsive to light. The environmental temperature, for instance, may also play an important role. While birds and mammals often will not breed in extremely hot or cold conditions, thermal changes are usually not of great importance in determining breeding in such homeotherms. In poikilotherms, however, such effects may be more significant. Spallanzani (1784) considered that reproduction in reptiles and amphibians may be related to the environmental temperature, and this still seems to be correct, though light may also contribute. Licht (1972; 1984) has carefully analyzed the role of temperature in controlling reproduction in reptiles and considers that it supplies the most important stimulus. Such stimuli could be acting at several sites.

Fig. 9.3. The effects of the environmental temperature on the responses of the ovaries and oviducts of lizards (*Xantusia vigilis*) to injections (on alternate days) of ovine FSH. It can be seen that at 14 °C there was little change in the weight of the tissues, there was a rather small effect at 20 °C, while at 30 °C the growth of the ovaries and oviduct was marked. (From Licht, 1972.)

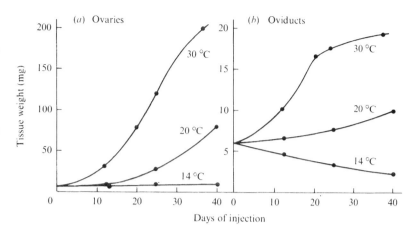

- A direct action on the brain could influence the release of hormones from the median eminence.
- Temperature could be exerting a direct action on gonadotropin formation and release in the pituitary itself. Such an effect has been demonstrated *in vitro* in the carp *Cyprinus carpio* (Lin, Lin, and Peter, 1996). At an incubation temperature of 20° to 25 °C, compared with 10 °C, the basal rate of secretion of GTH-II is increased and there is an enhanced response to GnRH.
- When the temperature is increased it has been shown that the responsiveness of the target tissues to gonadotropins increases. This is shown in Fig. 9.3, where the responses of the ovaries and oviducts of the lizard *Xantusia vigilis* can be seen to increase considerably at higher temperatures.
- It is possible that changes in body temperature may indirectly alter the levels of hormones by changing the rate of their inactivation.

Temperature has also been shown to influence the reproductive cycle in fishes. In poikilotherms, the effect of temperature may be of a "permissive" nature, for in the presence of a low body temperature metabolism is depressed and could be at such a low level that an action of light, or other stimulating factors, may be ineffective.

The availability of food and water can have dramatic effects on the breeding cycle. Many birds that live in the dry desert areas of Africa and Australia (so-called xerophilous species) rapidly come into breeding condition following unpredictable seasonal rains. Breeding in most amphibians, even those from temperate regions, is also finally determined by rain and the availability of water in their breeding ponds. The African toad *Xenopus laevis* is thought to be unresponsive to light; reproduction is determined by an optimal nutritional condition and the availability of water. Domestic animals such as sheep are often fed a special protein-rich diet to bring them into breeding condition. It should be remembered, however, that not all species breed when they are in best physiological condition, as breeding may occur shortly after prolonged periods of hibernation or estivation, as seen commonly in amphibians and lungfishes, or at the end of a prolonged fast that follows a migration, such as in salmon and lampreys. At present it is not known how such a nutritional state and the availability of food and water influence breeding. The breeding cycle of vertebrates is also influenced by their behavior and the visual, olfactory, tactile, and auditory stimuli that they can provide and evoke. Animals kept in captivity in, apparently, fine physical fettle often will not breed and may be missing such stimuli. Interactions between individuals have been especially noted in birds, where territorial aggression and courtship endeavors can further promote the hormonal changes that are necessary for reproduction (Wingfield, 1994). Self-stimulation can even occur, as seen in female ring doves. When these birds take part in their "bow–coo" courtship routine they respond to their own evoked sounds by further ovarian development. Social influences can be very important (Kelley, 1978) and it has been seen that reproduction is promoted in colonies of seabirds when the numbers grow past a critical level. The mechanism for such effects is unknown but would appear to be mediated by the central nervous system and the hypothalamus.

The synthesis of melatonin by the pineal gland exhibits a circadian rhythm with synthesis increasing during the hours of darkness and declining in daylight (Chapters 3 and 4). The retina acts as the receptor of such photoperiodic information. The duration of the increased secretion of melatonin reflects the number of hours of darkness and so it can convey a message regarding the length of the day, and the time of the year. The pineal gland can, therefore, function as a transducer for photoperiodic information, which it can convey to various rhythmically controlled processes in the body, including the neuroendocrine reproductive system. Early studies, especially those on the Syrian hamster (see below), suggested that the melatonin was exerting a general anti-gonadotropic effect. However, exceptions were subsequently identified in which pro-gonadotropic actions were observed. It has since become apparent that the particular response to the circadian release of melatonin depends on such factors as the species (not all even exhibit such nocturnal rhythms of secretion), whether it is a short- or long-day breeder, the time of the year, the current phase in the development of the reproductive tract, and (if melatonin is administered) the time of the day and the manner of this procedure. Differences may be observed if melatonin is given as a short-acting or a long-acting preparation, or in a pulsatile manner. The basic role of melatonin is the conveyance of information regarding the photoperiod and its entrainment by such processes as the reproductive cycle. It can, therefore, have anti- or pro-gonadotropic effects depending on the factors described above (Reiter, 1991b).

Fig. 9.4. The role of the pineal gland in the annual reproductive cycle of the male Syrian hamster. Breeding is normally initiated by the onset of days with long periods of light. Under these conditions, the testes enlarge and reach their maximum size in midsummer and then, as the days get shorter, they regress and are smallest in midwinter. This regression is prevented if the hamster's pineal gland is removed. (From Reiter, 1980.)

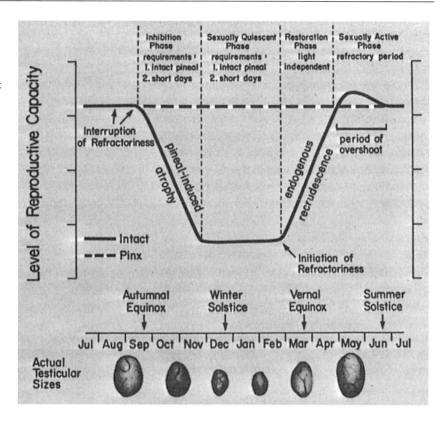

The historic prototype describing the action of the pineal and melatonin on seasonal reproduction concerns a small burrowing rodent, the Syrian, or golden, hamster, *Mesocricetus auratus*. These animals range from Rumania, through the Middle East, to Iran. Two pairs were taken to Great Britain from Syria in 1930. They were found to be friendly pets, adapted well, and bred during their captivity. A stock was retransferred to the USA in 1938 where their descendants yielded their pineal secrets. When these hamsters were adapted to long-day photoperiods and were then placed in darkness their testes decreased in weight, from 3000mg to 500 mg and this regression could be prevented by pinealectomy. Daily injection of melatonin could also interrupt the development of the gonads of these hamsters kept on long-day photoperiods, in a manner that mimics the effects of keeping them on short-days (Tamarkin *et al.*, 1976). The role of the pineal gland in seasonal changes in testicular development of these hamsters is illustrated in Fig. 9.4.

A role of the pineal and melatonin in the seasonal reproduction of vertebrates has been observed most often among mammals. Such species include several rodents, ruminants, such as sheep and deer, and some marsupials. Although many birds exhibit prominent photoperiodic control of their reproduction, a consistent role of the pineal in this process has not become apparent (Follett, 1984). Cold-blooded vertebrates often utilize environ-

mental temperature to signal appropriate reproductive activity and effects of
the pineal in these species have not usually been observed.

The precise site and mechanism of action of melatonin on reproductive
processes in the hypothalamo–pituitary axis is uncertain. It is generally
considered that a modulation of the release of hormones or neurotransmitters
is involved. The most likely site of action is generally considered to be the
hypothalamus, but receptors for melatonin have also been identified in the
pituitary, including the pars tuberalis, higher centers in the brain, and even in
the gonads (Ayre, Yuan and Pang, 1992; Kennaway and Rowe, 1995;
Maywood and Hastings, 1995). Melatonin receptors are found in the
mediobasal hypothalamus (MBH) and lesions to this area in male Syrian
hamsters block the inhibitory effects of short-day photoperiods and the
administration of melatonin on testicular growth (Maywood and Hastings,
1995). The effects of lesioning of the mediobasal hypothalamus is specific and
has no influence on the rhythmical release of prolactin. It is possible that the
melatonin is acting to regulate GnRH release. It will be recalled that the
arcuate nucleus, which is associated with the GnRH pulse generator, is also
present in the mediobasal hypothalamus.

The nature of the stimuli that control reproduction is complex and we do
not yet fully understand how they exert their effects but some excellent
accounts of the events and mechanisms involved are available (see, for
instance, Wingfield and Kenagy, 1991; Turek and van Cauter, 1994). The
endocrines, in close association with the brain, principally mediate the
response of the reproductive system to such stimuli. The eminent British
physiologist F. H. A. Marshall was the first to emphasize the importance of
such an interrelationship in controlling breeding. Some years ago, he summarized the situation (Marshall, 1956) as follows: "that [the] generative
activity in animals occurs only as a result of definite stimuli, which are partly
external and partly internal, while the precise nature of the necessary stimuli
varies considerably in different kinds of animals according to the species, and
still more according to the group to which the species belong."

Maturation of the gametes–the gonadal cycles

As we have seen, animals come into breeding condition at different times of
the year depending on the stimuli they receive and react to, both from the
external and their internal environments. If these "cues" are sufficiently
appropriate and are processed correctly, then breeding will be attempted.
This process involves a complex series of changes in the body that are, to a
considerable extent, mediated by altering the concentrations of hormones in
the blood. The sperm and the ova then mature, or ripen, in preparation for
their eventual union. As these preparations are proceeding, the changing level
of hormones contribute to the other physiological changes that are necessary

to assure the fertilization of the ovum and, if this process is successful, the continued development of the egg and the embryo.

Such cycles in gonadal activity are relatively simple in the male when compared with those in the female. Sperm that can fertilize the ovum may be continually available for a period of many weeks, or even, as in humans and feral pigeons, at all times of the year. The female, however, only produces ova available for fertilization periodically and, if not fertilized, ova usually survive for less than a day. Such a periodic production of ova is an important event as it may not then occur again for many months. To mark this somewhat unique occurrence and make it clear to the male that he is at last acceptable, the female may send out various external signals and even actively seek male company. These "signs" include "calling," as in the cat, the production of a scent, as in the urine of the bitch, and the adoption of certain inviting sexual postures.

In mammals this period of sexual receptivity by the female is commonly called "heat," or, by physiologists, *estrus*. The preparatory period that precedes this is proestrus, but if the animal is in a quiescent state, when no ova are being produced that are available for fertilization, it is called anestrus. The period during which the ova are being specially prepared for fertilization is called the *estrous cycle*, which varies from 4 days in the laboratory rat to 27 days in kangaroos, and (in its equivalent form, the human menstrual cycle) 28 days in women and rhesus monkeys.

Many animals only experience a single estrous cycle in a year (called monoestrous) while others may have several such waves of ova production (polyestrous) spread out over several months of the breeding season or even for the entire year. Whether or not a single estrous cycle will be succeeded by another often depends on whether fertilization has occurred. If not, then there may be (though not always) another chance for successful reproduction within the overall range of the general breeding season.

Testicular cycles in vertebrates

Certain male domestic animals exhibit continual spermatogenesis and sexual readiness throughout the year, though this is not usual except in vertebrates from tropical regions and humans. Seasonal breeding in a species is accompanied by a periodic maturation of the sperm (as well as the ova) along with such accessory and secondary sexual characters that facilitate its delivery on an appropriate occasion. Sperm may be available at all times during the reproductive season or mature in a single or several succeeding waves. The cystic type of spermatogenesis, where large numbers of sperm develop in unison inside envelopes that eject their contents into the seminiferous tubules, is most usual in amphibians and fishes and is especially suited to those species where massive numbers of sperm are suddenly required for external fertiliz-

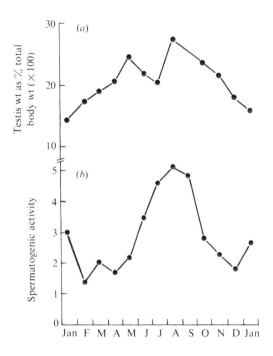

Fig. 9.5. Seasonal changes in the testicular weight (*a*) and spermatogenetic activity (*b*) in the European frog *Rana esculenta*. The decline in testicular weight that commences in May reflects spermiation during the breeding season. This sperm is that formed during the previous summer. Subsequent to this, spermatogenesis proceeds during the succeeding summer months but declines with the onset of winter hibernation. (From Lofts, 1964.)

ation. In reptiles, birds, and mammals, sperm mature from cells in the lining of the seminiferous tubules and this may be a more or less continuous process, though it may also occur in waves. This acystic spermatogenesis is thought to be more suited to internal fertilization, which may be attempted several times during a breeding season.

The maturation of the sperm may proceed in several different ways that are dictated by whether the species is a seasonal breeder and whether it is poikilothermic or homeothermic.

Postnuptial spermatogenesis is the more usual situation in seasonal poikilothermic breeders. This is illustrated by the frog *Rana esculenta*. Spermiation normally occurs in March in Northern Europe, and this is associated with a decline in testicular weight. Soon after this, however, the weight of the testis again increases and spermatogenesis proceeds throughout the summer but is halted during hibernation in winter (Fig. 9.5) though it gradually increases again in the spring. The major spermatogenetic events, therefore, occur in the summer preceding the breeding season and following the nuptial pairing in the spring of that year. Such a pattern of testicular activity is seen in many fishes and reptiles, though considerable variations can occur. In some instances, sperm may mature fully prior to the winter hibernation; in other species, spermatogenesis may be halted at some intermediate stage of development and go on later, in the spring, or again sometimes it merely slows down in winter and proceeds more slowly.

In homeotherms, *prenuptial spermatogenesis* is usual. Testicular activity

Fig. 9.6 A diagrammatic representation of the various patterns in seasonal development of the testes of lizards. I through IV represent variations in the relative testicular activity of lacertilians that live in temperate regions: V, some tropical *Anolis* lizards. "Emerge" and "retreat" indicate the times that temperate species leave and enter winter hibernation. The vertical lines indicate the period when most temperate lacertilians show no spermatogenetic activity. It can be seen that prenuptial spermatogenesis occurs in Types I and II (in spring), but in Types III and IV spermatogenetic activity commences in the autumn. (From Licht, 1972.)

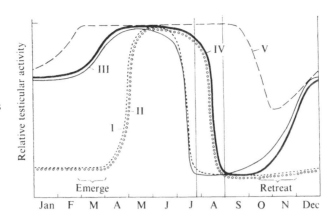

following the breeding season, during the winter months, may be slight but there is a rapid increase in activity when the nuptials become imminent. This pattern is usual in mammals and birds that breed periodically, though some species (such as bats) may store mature sperm in the epididymis for several months, during a period of hibernation for instance. Most birds exhibit characteristic "refractory" periods following the breeding season, when the testes fail to respond to photoperiodic stimuli or administered gonadotropins. The reptiles show a considerable diversity in testicular cycles. Chelonians usually exhibit amphibian-like postnuptial spermatogenesis, but the Lacertilia have several different testicular cycles (Fig. 9.6) and a prenuptial spermatogenesis is common.

The cyclical pattern of spermatogenesis described above is also termed *discontinuous spermatogenesis*, in contrast to *continuous spermatogenesis*. The latter, apart from being present in some domestic temperate species, is common in animals that live in tropical areas where climatic conditions are relatively favorable at all times of the year (Basu, 1969; Lofts, 1984). The frog *Rana esculenta* indeed has a continuous spermatogenetic cycle in warm Mediterranean regions whereas it has a discontinuous cycle in the more northern parts of Europe. This frog has therefore been classified as a *potentially continuous breeder*, or a continuo–discontinuous type. The environmental temperature determines which pattern will persist in such species. It should be noted, however, that it is not possible to alter a discontinuous spermatogenetic cycle to a continuous one simply by raising the temperature; it will not, for instance, change in *Rana temporaria*, where the tissues appear to undergo an inherent rhythm in their ability to respond to hormonal stimuli.

The testicular cycle (like the ovarian cycle) is controlled by the adenohypophysis. Removal of the pituitary abolishes such cyclical activity and results in a regression of the testis that involves the germ cells in the seminiferous tubules as well as the endocrine interstitial tissue (Leydig cells). In mammals, this is thought to involve the action of FSH on the seminiferous tubules and LH on the interstitial tissue. Differences, however, occur: in reptiles an FSH-like gonadotropin appears to have a predominant ("nondiscriminating") effect on testicular function while in amphibians an LH has a comparable role (Licht, 1979). In teleost fish, the "maturational" gonadotropin (GTH II) is involved in the process of sperm maturation. While LH stimulates the production and release of testosterone, the mode of action of FSH is less certain. It is usually necessary for the full maturation of the sperm but this may not always be a direct effect. Hypophysectomized mammals, birds, and fishes, in which the testes atrophy, can produce sperm following the administration of testosterone. However, testosterone apparently cannot restore spermatogenetic activity in amphibians, or at least in *Rana pipiens* (Basu, 1969; Lofts, 1968). In primates, FSH has a separate role in spermatogenesis but there is a synergism between its action and those of androgens (McLachlan *et al.*, 1996).

Testosterone or related steroids are, nevertheless, usually necessary for the maturation of the sperm, but it is not yet clear whether FSH also acts by stimulating the production of an androgen. Unfortunately, the spermatogenetic effects of the administration of testosterone are often variable. It seems likely that FSH may act on the Sertoli cells to produce androgens, which in turn mediate the maturation of the sperm (see Lofts, 1968) but it may also have an independent effect, at least in mammals (McLachlan *et al.*, 1996).

Spermatogenesis is a prolonged and complex process that requires pre- and postnatal maturation of the gonocytes, mitotic divisions of the spermatogonia, and meiotic reduction divisions to form the spermatocytes, spermatids, and the final (spermiogenesis) production of spermatozoa. Androgens, and possibly FSH, are required for certain of these steps to proceed in a normal manner, but there is considerable interspecies variation as to the stages of sperm maturation at which these hormones act. The endocrinology of gametogenesis is an important subject about which we know little. Dodd (1960; 1972b) has summarized what is known about this process in vertebrates. In the rat, testosterone may be necessary for early pre- and postnatal development of the gonocytes, and it also promotes the meiotic division of the spermatocytes later on. FSH is required for the maturation of the spermatids. This pattern is, however, not the same even among the mammals, and the information that is available is rather meager. In lampreys (Cyclostomata), hypophysectomy has little effect on the final stages of the maturation of the sperm. When this operation is performed in late winter or spring, spermiation still occurs, but if hypophysectomy is carried out earlier, in October for instance, there is a considerable delay and spermiation may not take place at all (Larsen, 1973; 1978).

A similar slowing action on the development of the testis has been observed following hypophysectomy in the Pacific hagfish, *Eptatretus burgeri* (Patzner and Ichikawa, 1977). In chondrichthyean fishes, hypophysectomy also inhibits the earlier stages (meiotic divisions of the spermatogonia) of sperm maturation. Selective removal of the ventral lobe of the pituitary of the dogfish results in a slow degeneration of the testis (Dobson and Dodd, 1977a,b). Extracts of the ventral lobe of these fish, but not mammalian gonadotropins, maintained the activity of the gonads. Seasonal changes in the development of the testes of teleost fish are also inhibited following hypophysectomy (Peter and Crim, 1979). This effect appears to reflect the action of the maturational GTH II, which can stimulate spermatogenesis in hypophysectomized flounder (Ng and Idler, 1980; Ng, Idler and Burton, 1980). Teleosts provide excellent models, *in vivo* and *in vitro*, for studying the role of hormones in spermatogenesis (Billard, LeGac, and Loir, 1990; Miura, Kobayashi, and Nagahama, 1994). Species that have been studied include goldfish, Atlantic salmon, and Japanese eels. The process appears to be initiated by GTH II, which promotes the synthesis of androgens by the Leydig cells. The potent teleost steroid 11-ketotestosterone is often involved and may stimulate the Sertoli cells, resulting in the final maturation and spermiation. However, other mediators of the final events may be formed. Thus GTH II can also promote the formation of a steroid, $17\alpha,20\beta$-dihydroxy-4-pregnen-3-one ($17,20\beta$-P) by the testes, which can promote spermiation (Stacey *et al.*, 1994). This maturation-inducing steroid is also produced in the ovary (see below) and it may act as a pheromone in some fish. It is released across the gills or, in its sulfated form, in the urine (Vermeirssen and Scott, 1996). This steroid can stimulate olfactory receptors in male goldfish and salmon where it triggers the release of GTH and a further synthesis of $17,20\beta$-P by the testes (Sorenson *et al.*, 1995; Waring, Moore, and Scott, 1996). The promotion of appropriate sexual behavior and spermiation follows. The testes of the Japanese frog *Rana nigromaculata* synthesize a similar steroid, $17,20\alpha$-dihydroxy-pregen-3-one, which induces spermiation in this amphibian (Kobayashi *et al.*, 1993).

Until recently, little direct information was available about the circulating levels of testosterone and gonadotropins in seasonally breeding amphibians. Changes in the concentrations of such hormones were inferred from histological examination of secondary sexual characters. The Leydig cells and their analogs, the boundary cells, show a seasonal pattern in their histological appearance (Lofts, 1968). In the periods that precede breeding, these cells enlarge and accumulate lipids and cholesterol. These cells become depleted of these materials at the height of the breeding season, and this is thought to reflect the secretion of androgens, which utilize lipids as their substrates. These histological changes are associated with breeding behavior and the development of secondary sex characters and can be imitated by the injection of gonadotropins. Such changes have been observed in fishes, amphibians, reptiles, birds, and mammals. An example of this can be seen in Fig. 9.7 where the height of the thumbpad epithelia in the frog *Rana esculenta* can be seen to

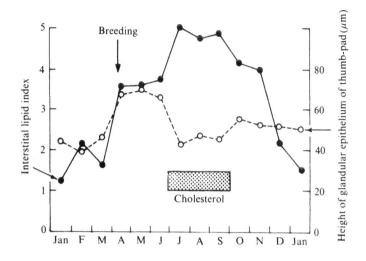

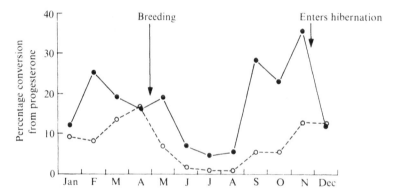

Fig. 9.7 Seasonal changes in the amount of lipid observed, histologically, in the interstitial cells of the testis of the frog *Rana esculenta*. This reaches a maximum in midsummer, after breeding has occurred, when the cellular cholesterol levels are also greatest. These changes are thought to reflect a decline in the synthesis of androgens (which utilize the lipids as substrates for their formation). This change is consistent with the decline in the development of the thumbpad, which is under androgenic control. (From Lofts, 1964.)

Fig. 9.8 Seasonal changes in the ability of the testis of a snake, the cobra, to convert (*in vitro*) progesterone (which acts as a substrate) to androgens. Testosterone production declines following breeding in May but rises again in late summer and autumn, during postnuptial spermatogenesis, only to decrease once more as the snakes go into winter hibernation: ●, testosterone; ○, androstenedione. (From Lofts, 1969.)

decline in June when the lipid and cholesterol content of the interstitial tissue is greatest. A seasonal pattern in the ability of the testis of the cobra to convert progesterone to androgens, *in vitro*, is shown in Fig. 9.8. This androgen synthesis reaches an initial maximum during breeding in May but drops subsequently as the testes atrophy. When spermatogenesis is again initiated in the autumn, the rate of progesterone to androgen conversion increases but declines again with the onset of winter hibernation. In opportunistic aseasonal breeding frogs that live in the tropics the basal plasma concentrations of androgens appear to be maintained at lower levels than those seen during breeding in seasonal species (Emerson and Hess, 1996). Such a strategy may be energetically economical. The androgens in such opportunistic breeders may have a "permissive" role in maintaining spermatogenesis and reproductive behavior and only become elevated acutely in response to proximate behavioral stimuli.

Dogfish, *Scyliorhinus canicula*, have a plasma testosterone concentration of 2ng/ml in February and this rises to 6 ng/ml in August (Dodd, 1975). It was surprising to observe, however, that removal of the various pituitary lobes, including the ventral one, did not change the plasma level of this androgen. It is possible that in these fish the testis enjoys a relative degree of independence from the pituitary.

In the rough-skinned newt, *Taricha granulosa*, plasma androgens rise markedly in March, when breeding occurs, and then decline but rise again in August (Specker and Moore, 1980). The low levels correspond to the period of spermatogenesis and the second increase to the beginning of spermiation. The levels then decline over the winter period. The Australian lizard *Trachydosaurus rugosa* (Bourne and Seamark, 1973) breeds in spring when the weight of its testes is about 1300 mg compared with 180 mg in summer. During the breeding season, the androgen concentration in its plasma is 33 ng/ml, but it is only 10 ng/ml at other times of the year. Similar seasonal changes in plasma androgen levels have been observed in other lizards and also in snakes and turtles (Callard *et al.*, 1976; Bona-Gallo *et al.*, 1980; Courty and Dufaure, 1980). A 10-fold increase in the plasma testosterone concentrations has been observed in starlings (*Sturnus vulgaris*) during the breeding season (Temple, 1974). When Japanese quail are put on a 20-hour light/4-hour dark photoperiod, as occurs at the height of summer in northern latitudes, plasma testosterone levels rise after only 4 days (Follett and Maung, 1978). A periodic decrease in the hormone-secretory interstitial (as well as the spermatogenetic) tissue occurs in all nonmammals that breed periodically. In mammals, however, there is usually permanent hormone-secretory tissue in the testes though they may also undergo seasonal changes in size.

The Sertoli tissue, which has been identified in all groups of vertebrates, has for a long time excited speculation as to its function. The histological appearance of this tissue shows changes in parallel to those of the Leydig cells (Lofts, 1968) and spermatogenesis. An accumulation of lipids, in the Sertoli cells, follows spermiation, but these materials are depleted when spermatogenesis is occurring. In animals that normally breed continually, like laboratory rats, hypophysectomy results in an accumulation of lipids in the Sertoli cells, and this is thought to reflect a lack of their stimulation by FSH. It is now widely accepted that the Sertoli cells produce steroids, inhibin, and, possibly, other "inducing factors" that may influence spermatogenesis and spermiation.

Ovarian cycles in vertebrates

The maturation of the ovum in the ovary and its extrusion (ovulation) and passage into the oviduct or uterus involve the coordinated activity of GnRH, FSH, LH, sometimes prolactin, and the secretion of estrogens, progesterone, and, possibly, even small amounts of androgens from the ovary. The ovarian cycle results from increases and declines in the circulating concentrations of these hormones, and this is largely the result of their interactions in stimulating or inhibiting each other's release, through a negative and positive feedback to the hypothalamus and pituitary.

These hormonal rhythms have been closely analyzed in mammals, and even these results are usually confined to more domesticated species.

Placental mammals   Three general patterns have been identified in the ovarian cycle of placental, or eutherian, mammals and these have been described (1) in the sheep, pigs, and cattle, (2) in the laboratory rat, and (3) in higher primates.

Ovarian cycle in sheep   Sheep are cosmopolitan mammals of a size and docile disposition that has resulted in their frequent use for studying the endocrinology of reproduction. They are also valuable economically. Those breeds living in temperate regions and at higher latitudes usually commence breeding in the autumn so that the young are born in spring when the availability of grass promotes their survival. Sheep have, therefore, been classified as short-day breeders as the proximate cue to commence their reproductive cycles (they are polyestrus) is a decrease in the length of the daily photoperiod. However, the bulk of the world's sheep do not live in such equitable geographic regions so that other cues, especially the available food supply, may influence their breeding (Martin *et al.*, 1994). Australian merino sheep can, if conditions are favorable, reproduce at any time of the year and are called "opportunistic" breeders. The introduction of an attractive novel ram to the flock can also promote reproductive cycles in ewes that were previously in anestrus. This "ram effect" reflects the emanation of pheromones from the ram's cutaneous sudoriforous glands (Goodman, 1994; Martin *et al.*, 1994). The reproductive cycles of sheep, including the testicular one, can, therefore, be influenced by photoperiodic, nutritional, and social cues. There appears to be an underlying endogenous (circannual?) rhythm that can entrain such stimuli. Control ultimately resides in the brain, which can influence the activity of the GnRH pulse generator in the hypothalamus (Chapter 4). The pulse generator can, however, be modulated by the receipt of stimuli resulting from the photoperiod, nutrition, and social contact. It also incorporates negative- and positive-feedback mechanisms that respond to peripheral hormones. The testicular cycle of the ram appears to be controlled in a similar way and can be influenced by pheromones from the ewe ("the ewe effect").

Photoperiodic information is received by the eyes and is transmitted to the pineal, which releases melatonin in amounts reflecting the duration of the period of darkness. Nutritional information may be provided by the concentrations of fatty acids in the plasma (Martin *et al.*, 1994; Martin and Walkden-Brown, 1995), while the rams' pheromones stimulate olfactory receptors.

Fig. 9.9. The estrous cycle of the ewe. Estrus (E) occurs at time *zero* and is followed by ovulation. The concentrations of plasma progesterone (from the jugular), estrogens (ovarian vein), and LH (from the jugular or ovarian vein) are shown in relation to these events. It can be seen that on day 12 there is a decline in progesterone that is accompanied by a climb in estrogen concentration that initiates a "surge" in release of LH, resulting in ovulation. Dashed line, progesterone; dotted line, total estrogens; solid line, LH (jugular or ovarian vein). (From Hansel and Echternkamp, 1972.)

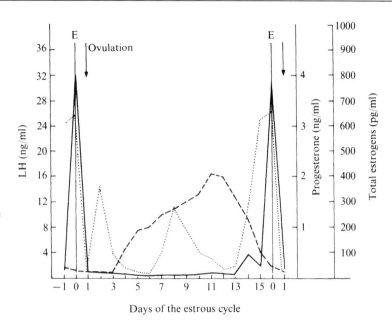

Days of the estrous cycle

The estrous cycle of the ewe is initiated by an increase in the pulse frequency of GnRH release, which is reflected by a release of LH. (FSH is also secreted but appears to be less responsive to GnRH). The cycle lasts for 17 days and the hormonal changes that occur are summarized in Fig. 9.9. The onset of estrous is taken as time zero in the ovarian cycle and this lasts for about 24 hours, ovulation occurring toward the end of this time. Following ovulation, the blood supply to be ruptured Graafian follicle increases and the granulosa cells luteinize to form the corpus luteum. This structure reaches a maximum size on about day 8. Luteinization of the follicle is initiated by the action of LH, and the secretion of progesterone is also stimulated by this hormone. LH is luteotropic, an effect that is seen in most mammals. Progesterone secretion from the corpus luteum rises until about day 11 of the cycle and then on day 12 undergoes a precipitous decline.

Accompanying these events are the development of the Graafian follicles and the maturation of the ova. This process proceeds under the influence of FSH, which parallels LH, and the estrogens secreted by the follicular cells, which are also stimulated by LH. Thus LH appears to have a general steroidogenic effect on the ovarian tissues. During the preovulation phase of the cycle, estrogen levels are moderate but as can be seen in Fig. 9.9 they may display some periodic changes. The LH level is low but sufficient to maintain the secretion of steroid hormones. The estrogens and progesterone that are produced act on the accessory sex organs, especially the uterus and vagina, to get them into condition for the prospective fertilization, the implantation of the egg, and pregnancy. The release of LH is kept low in the preovulatory follicular period as a result of a negative-feedback inhibition of its release that is exerted by progesterone and estrogens on the activity of the GnRH pulse generator and the pituitary gland. Progesterone and estrogens appear to have a synergistic effect in blocking GnRH release. The secretion of FSH is also influenced by GnRH. In sheep, estrogens exert a negative feedback inhibition on the release of this gonadotropin (Price, 1991). Inhibin, which is formed by the ovarian follicle, also inhibits release of FSH, and its concentration in the plasma rises during the follicular phase of the estrous cycle.

Between days 12 and 16 of the cycle, dramatic changes take place in the hormonal concentrations that result in ovulation. There is a rapid decline in progesterone that reflects a breakdown of the corpus luteum. This change decreases the block on GnRH release, and LH levels, therefore, rise. A massive increase in plasma estrogen concentration results. This estrogen, in its turn, exerts a positive-feedback effect on the release of both GnRH and LH so that a "surge" in the plasma concentration of the latter occurs. The LH induces ovulation.

After ovulation, LH stimulates the follicle granulosa cells to luteinize and if fertilization does not occur the cycle will then recommence. In the event of fertilization and an ensuing pregnancy, the corpus luteum, as we shall see below, will persist for a much longer time and contribute to the events of the gestational period.

The corpus luteum can, therefore, be seen to play a commanding role in the estrous cycle. The reason for the decline in the activity of the corpus luteum during the latter part of the estrous cycle has only recently been elucidated. It has been known for many years that when the uterus of guinea pigs is removed the corpus luteum persists for a much longer period of time. This effect can also be seen in the ewe, as well as the cow and sow, but not in women, the rhesus monkey, badgers, or marsupials (Anderson, 1973). The nonpregnant uterus, in some species, appears to produce a substance that has been called a *luteolysin*, which causes the corpus luteum to atrophy. This substance is prostaglandin $F_{2\alpha}$ ($PGF_{2\alpha}$). It is formed in the uterine endometrium and can travel to the ovary by entering the uterine vein and passing locally into the ovario-uterine artery. The mechanism controlling the release of $PGF_{2\alpha}$ is still controversial; estrogens may be involved but rising plasma progesterone levels are now considered to be a more likely stimulus in the ewe (see Goodman, 1994). Ovarian oxytocin may also contribute as it can be released at this time and can promote the synthesis of $PGF_{2\alpha}$ by the endometrium. This $PGF_{2\alpha}$ may have a positive-feedback effect on the ovary and further increase the release of oxytocin. The orderly demise of the corpus luteum is thus assured and it will be noted that it appears to control its own fate, and, therefore, subsequent events in the cycle. As a result it has been dubbed the "pelvic clock."

Seasonal reproduction of sheep in autumn is alternated by a period of anestrus in summer, which is triggered by the long days following the winter solstice. This effect, like that for their estrous cycles, is transduced by melatonin from the pineal gland. Manipulation of plasma melatonin concentrations by pinealectomy followed by a timed administration of the hormone can mimic the effects of either long or short days. Estrous cycles or seasonal anestrus can be initiated in this manner. Melatonin appears to act on the hypothalamus in sheep to modulate the activity of the GnRH pulse generator (Goodman, 1994). As observed in birds entering their seasonal period of refractiveness, thyroidectomy in sheep prior to the onset of seasonal anestrus results in a failure to enter this later phase so that estrous cycles continue unabated (Dahl *et al.*, 1994; Karsch and Dahl, 1994; Karsch *et al.*, 1995). This effect has also been observed on the reproductive activities of the ram (Parkinson and Follett, 1995). The decline in the pulsatile release of GnRH that normally occurs at the onset of seasonal anestrus does not occur following the thyroidectomy. However, the rhythms of release of melatonin and prolactin are unaffected by this operation. The GnRH pulse generator, therefore, appears to become uncoupled from photoperiodic stimuli. The nature of the role of the thyroid is uncertain but it appears to be a "permissive" effect and could involve the induction of proteins concerned with morphological changes seen in neurons in the hypothalamus and which mediate and sustain seasonal anestrus (Karsch and Dahl, 1994).

If sheep are exposed to a day length maintained at that which occurs at the winter solstice, breeding does not occur in the following autumn. The receipt of normal light during the lengthening late winter and spring days appears to be necessary for the timing and ultimate emergence of the breeding season triggered by the shortening days of the following autumn (Malpaux *et al.*, 1989). The underlying endogenous reproductive rhythm may be synchronized during the previous spring to the projected breeding events. The short days of autumn sustain the new breeding season, which is initiated by an increase in the frequency of the GnRH pulse generator. This response results from a relief of a strong negative-feedback effect of estrogens, which is predominant during the summer anestrus (Fig. 9.10). During the breeding season this negative-feedback effect of estrogens is low and progesterone is the major hormone regulating the activity of the GnRH pulse generator.

Ovarian cycle in the laboratory rat   The laboratory rat, despite some unique specializations, has probably provided most of our basic knowledge of the mammalian ovarian cycle. The rat is an albino mutant of the brown rat, *Rattus norvegicus*, which before its various migrations, usually as a stowaway, lived in Japan and east Asia. The genus *Rattus* is a large one containing 137 species, mostly from tropical and subtropical regions. Brown rats live in towns and agrarian areas where their survival is enhanced by their commen-

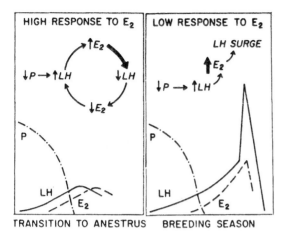

Fig. 9.10. The possible hormonal basis for the initiation of the breeding season in ewes. This event occurs in response to short daylengths in the autumn and winter; the sheep are in anestrus during the summer. It has been proposed that the seasonal onset of the estrus cycles results from a change in the sensitivity of the brain to the inhibitory effects of estrogens (E$_2$) on the release of LH. In the summer, the sensitivity of the negative-feedback system in the hypothalamus to estrogens is high so that LH and resulting estrogen (E$_2$) levels are low (left panel). With the initiation of the breeding season, however (right panel), the sensitivity of the inhibitory mechanism to estrogen declines so that LH is released more readily: estrogen levels then climb and can then initiate the surge in LH release that precedes ovulation. (From Legan, Karsch, and Foster, 1977.)

sality with humans, and a remarkably effective process of reproduction. They are spontaneous ovulators, polyestrus, and under favorable conditions will breed throughout the year. In the wild, they tend to favor spring and midsummer. Their estrous cycle lasts for about 4 days and gestation is 20 to 22 days. They bear 8 to 10 young and lactation lasts about 28 days during which time they can again become pregnant. These rats may have six to eight litters a year. As they attain early sexual maturity, about 38 days of age in the female, they can be considered to be very effective breeding machines.

The major events of the laboratory rat's estrous cycle (Rowlands and Weir, 1984; Freeman, 1994; McNeilly, Forsyth and McNeilly, 1994) follows a diurnal rhythm and can be precisely predicted to the hour of the day (Fig. 9.11). If the rats are kept in continuous light, the cycle is disrupted and they fail to breed. The estrous cycle can be divided into four periods during which times specific hormonal changes occur. Estrus, variously considered as Day 0 or Day 1, lasts for 25 to 27 hours. Ovulation and mating occur at this time. Plasma concentrations of hormones (estradiol-17β, progesterone, LH, FSH, and prolactin) subside during estrus from their high levels in the preceding proestrus, and luteinization of the empty follicles is initiated. The succeeding diestrus lasts for 55 to 57 hours and is usually divided into diestrus-I and diestrus-II. Plasma progesterone and estradiol levels rise through diestrus-I and diestrus-II. If mating has not occurred, the corpora lutea start to regress in late diestrus-I and progesterone concentrations fall in diestrus-II. This process reflects the influence of uterine luteolysins. The corpus luteum of the rat in the absence of mating has a remarkably brief life of only 2 to 3 days. If mating does not occur, a very short cycle, therefore, results, which can immediately be repeated, hopefully, to achieve a fruitful outcome.

If mating occurs the lives of the corpora lutea are prolonged. They are "rescued" as a result of the unique luteotropic effects of prolactin. Mating, or even the mechanical stimulation of the cervix, initiates a rhythmical pattern of

Fig. 9.11. The principal events in the rat estrous cycle in relation to the time of the day. This cycle is precisely timed on the basis of a diurnal rhythm. Ovulation can be seen to occur shortly after midnight. Other events including mating behavior, the LH "surge", and the development of the uterus are depicted. (From Armstrong and Kennedy, 1972.)

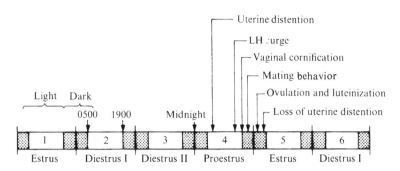

release of prolactin that reaches a peak twice each day between 5 and 7 p.m. and 3 and 7 a.m. In the absence of conception, this pattern persists for the 12 to 14 days of "pseudopregnancy," and during pregnancy for 20 to 22 days. (After 14 days of gestation, the prolactin is supplemented by a placental luteotropin).

In the absence of pregnancy and pseudopregnancy, diestrus-II is followed by proestrus. This period is characterized by a peak in the plasma level of estradiol and a second peak of progesterone (predominantly 20α-hydroxy-progesterone). The estradiol exerts a positive-feedback effect in the hypothalamus to initiate a neurogenic stimulus that releases GnRH, which promotes the LH surge. The sensitivity of the pituitary gonadotrope cells to GnRH may also be increased by the estradiol. This LH surge, which is maximal between 5 and 7 p.m., induces ovulation in the succeeding estrus. The effects of the estradiol may be modulated by progesterone. Almost simultaneous peaks of FSH and prolactin also occur in response to the estradiol. The release of the latter may involve a decline in the release of dopamine and/or a rise in TRH in the hypothalamus.

Parturition in rats is followed in 24 to 36 hours by a postpartum estrus. If fertilization then occurs the blastocysts undergo a period of quiescence lasting for 5 to 7 days prior to their implantation. The next litter are, therefore, born about 28 days after the birth of the previous one, at a time after weaning has occurred. If the postpartum estrus has not been fruitful, but the litter survives, then the cycle remains in abeyance for up to 20 days. This inhibition is caused by the release of prolactin during suckling and the maintenance of the corpora lutea of lactation. Follicular development is inhibited. This effect of prolactin appears to be caused by an inhibitory effect on the release of GnRH (McNeilly et al., 1994). If, however, the young are lost and postpartum conception has not occurred, normal estrous cycles are usually resumed after 5 to 6 days. In rats, the release of GnRH and the gonadotropins during the estrous cycle are controlled by a negative-feedback inhibition by estrogens acting on the hypothalamus and, probably, the pituitary. The effects of the estrogens are enhanced by progesterone. The release of FSH is also inhibited by inhibin secreted by the ovarian follicles. Its plasma levels are inversely related to those of FSH (Freeman, 1994).

The rat has evolved a remarkably efficient physiological system to assure its reproductive success. It utilizes unique mechanisms including an evanescent corpus luteum and a luteotropic role for prolactin that is initiated by mating.

The ovarian cycle of higher primates: the rhesus monkey   Humans and a number of other primates have a similar ovarian cycle, lasting for about 28 to 33 days, which includes a unique process of shedding remnants of old endometrial tissues. This type of ovarian cycle is called a menstrual cycle (*mensis*, month). Apart from humans (Hominidae), it is also exhibited by the Old World monkeys (family Cercopethecidae, which includes macaque and rhesus monkeys) and great apes (family Pongidae: the gorillas and chimpanzees). The New World monkeys (family Cebidae: marmosets, squirrel monkeys, capuchins, spider monkeys, etc.) do not exhibit this type of ovarian cycle.

Nonhuman higher primates usually live at lower latitudes and in areas where seasonal changes in food supply are not common. Gorillas and chimpanzees, like humans, therefore, do not exhibit seasonal reproduction. However, other species, including the macaque and rhesus monkeys, often have irregular menstrual cycles in summer and tend to favor the short days of autumn for breeding (Spies and Chappel, 1984). Possibly they are responding to an underlying circannual rhythm. Melatonin release from the pineal does not appear to contribute directly to their reproductive cycles. Seasonality in some primates is signaled in the female by the development of brightly colored patches on the buttocks and by sexual advances.

The ovarian cycle in the gorilla is 31 days, in the chimpanzee 30 days, and in most Old World monkeys about 28 days. (In humans it is also 28 days). The most studied nonhuman primate is the rhesus monkey, *Macaca mulatta*. It comes from India and although not usually very friendly it adapts well to captivity. Studies on its reproductive physiology have not only included the interactions between the gonads and reproductive tract but also detailed studies of its hypothalamic and pituitary control mechanisms. The hypothalamic "pulse generator" that drives the release of GnRH and initiates the cycle was originally discovered and exhaustively studied by E. Knobil and his collaborators in the rhesus monkey. This species probably provides the most appropriate experimental prototype for describing the primate menstrual cycle, though some may prefer to consider the human condition. The following account of the ovarian cycle in the rhesus monkey (Fig. 9.12) is based on the reviews provided by Spies and Chappel (1984) and Hotchiss and Knobil (1994).

The beginning of menstruation is often taken as Day 1 of the ovarian cycle. (It is also described in the context of ovulation being Day 0 with positive and negative numbering on the days following or preceding this event.) Ovulation occurs near the middle of the 28 day cycle of the rhesus monkey and humans. In the rhesus monkey, there is a small rise in plasma FSH at the commence-

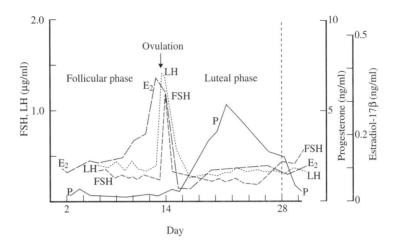

Fig. 9.12. A summary of hormonal changes in the plasma of the female rhesus monkey (*Macaca mulatta*) during its reproductive cycle. E_2, estradiol-17β; P, progesterone. (Based on Spies and Chappel, 1984.)

ment of the cycle (Day 1), probably heralding the commencement of the growth of the Graafian follicle and the once hourly (circhoral) oscillations of the hypothalamic pulse generator and release of GnRH. In the rhesus monkey, and humans, the pulse generator appears mainly to play a permissive role, maintaining the output of the pituitary gonadotropins. It is, however, subject to modulation from higher centers in the brain. The positive- and negative-feedback effects of peripheral estrogens on gonadotropin release are mainly mediated by their direct actions on the pituitary gland. A small decline in release of FSH may occur during the follicular phase of the cycle while sufficient LH levels are maintained for the developing follicle and its synthesis of estradiol. The estradiol promotes the growth of the uterine myometrium and endometrium. This follicular phase of the cycle (lasting for about 14 days) enters its final stages about 48 hours prior to ovulation. There is then a large increase in release of estradiol, which initiates a surge of FSH and LH. The latter induces ovulation about 12 hours later. In the rhesus monkey, like humans but in contrast to sheep, the rate of oscillation of the pulse generator is unchanged prior to ovulation but its amplitude increases. The rising tide of estradiol on passing a critical threshold acts on the pituitary gonadotrope cells to increase their responsiveness to GnRH, possibly involving changes in the sensitivity of its receptors. Sexual receptivity and advances to the males are greatest during this periovulatory period, reflecting the actions of estradiol and also, possibly, the presence of adrenocortical androgens.

The next stage of this primate ovarian cycle is called the luteal phase and it also lasts for about 14 days. Under the influence of LH a luteinization of the collapsed follicle occurs with the resulting formation of the corpus luteum. This tissue secretes progesterone, and also some estrogens. Plasma progesterone levels increase dramatically, promoting a differentiation of endometrial glands and an initiation of their secretory phase. Follicular development is suppressed and sexual receptivity diminishes.

The oscillation rate of the pulse generator declines during the luteal phase, a change that may reflect a negative-feedback effect of progesterone in the brain. After about 8 days of this phase, if conception has not occurred, the plasma progesterone levels start to decrease, reflecting the demise of the corpus luteum. The proximate cause of this event is not clear in the rhesus monkey. Uterine prostaglandins do not appear to be involved. Estrogens can promote luteolysis in the rhesus monkey, a process that could involve the induction of ovarian prostaglandins, but this possibility is controversial. Possibly the life expectancy of the corpus luteum is preprogramed. However, if fertilization occurs, its life is extended into pregnancy as a result of its "rescue" by chorionic gonadotropin. In the absence of pregnancy, the ovarian cycle of the rhesus monkey terminates after 14 days of the luteal phase and a new cycle may commence.

The timing mechanism during the menstrual cycle appears to reside in the ovary. Transplantation of the ovaries of rhesus monkeys to castrated male monkeys, which also normally exhibit a circhoral release of GnRH, results in such ovaries exhibiting their usual cycle (Norman and Spies, 1986). It has been suggested that the ovary may contain a "circamensual clock" containing two units that reflect the predetermined lives of the Graafian follicle and the corpus luteum.

During pregnancy in the rhesus monkey, the activity of the corpus luteum declines and the placenta becomes the principal source of progesterone. However, late in pregnancy it commences to re-establish itself. Following parturition there is no postpartum ovulation, as occurs in many nonprimates. If lactation and suckling occur, the life of the corpus luteum is prolonged and may contribute to the period of ovarian quiescence seen at this time. The principal site of the "block" of further cycles appears to involve a decreased release of GnRH. It has been suggested that prolactin that is released in response to suckling may be mediating this effect, but the evidence is equivocal (McNeilly *et al.*, 1994).

The menstrual cycles of rhesus monkeys and humans are remarkably similar and cross-extrapolations are often made. In some respects, this cycle in the higher primates is simpler than those in other mammals, partly reflecting the lack of an input from photoperiodic and diurnal signals. The hypothalamic pulse generator also appears to be more autonomous with respect to its sensitivity to peripheral stimuli.

Ovulation   The mechanisms of initiation of ovulation may differ among the mammals though they all appear to involve a surge in the release of LH. In the examples described, ovulation takes place in response to an internal programing that controls hormone release so that ovulation is then said to be *spontaneous*. In rats and sheep, the release of LH in this type of ovulation results from a rise in plasma estrogen concentrations, which increases both the release of GnRH and the sensitivity of the pituitary gonadotrope cells to its action (Speight *et al.*, 1980; Kesner, Convey, and Anderson, 1981). In the rhesus monkey, estradiol can stimulate gonadotropin release in the absence of GnRH, provided that the pituitary gland has been suitably primed by the latter peptide hormone. Estradiol may therefore have a direct role in initiating the release of LH at the time of ovulation.

In other mammals, ovulation can be *induced* as a result of copulation and sexual excitement. This latter type of ovulation is known to occur in such species as the rabbit, cat, ferret, mink, and raccoon, and it is suspected that it may also sometimes occur even in women. In such species, estrogen is released from the developing ovarian follicles, which are under the influence of FSH and LH (see Schwartz, 1973). This estrogen indicates when the follicles are ripe and results in mating behavior. This behavior is in contrast to spontaneous ovulators in which progesterone is also necessary. If copulation takes place, this initiates a surge of LH release as a result of neural stimulation of the hypothalamus and pituitary, and ovulation occurs. This event is accompanied by a rise in progesterone levels and takes place several hours after coitus when the sperm are ensconced in the oviduct. In nonmammals, the proximate hormonal cause of LH release may be progesterone rather than estrogens. The mere presence of the male or even some substitute may be all that is necessary to initiate ovulation. Apart from gallinacious birds like the domestic fowl (as well as domestic geese and ducks), most birds do not produce eggs in the absence of the male. It has, however, been reported that some birds, such as pet parrots, will lay eggs if suitably stroked and tickled. Copulation may, therefore, not always be necessary, and courting behavior and sexual display may be effective stimulants of ovulation.

The ovarian follicles are enveloped by an outer membrane that includes the theca externa and which contains structural components such as collagen. It is supplied by a microcirculatory system. When ovulation, which has been triggered by LH, approaches, the follicle swells, reflecting a decline in the tensile strength of the theca externa and an increase in its blood supply (Tsafriri, Reich, and Abisogun, 1994). Rupture of the follicle involves proteolysis, principally caused by the fibrinolytic action of plasmin and the effect of collagenase. LH stimulates a plasmin-activating system in the granulosa cells of the follicle, and also a local release of collagenase. Such processes appear to involve synthesis of prostaglandins.

Following parturition, several species of eutherians, including the rabbit, ferret, mink, and raccoon, come into a postpartum heat when they copulate, and this, as indicated, results in ovulation. Copulation is not always needed to precipitate ovulation in these circumstances for, as we shall see in the next section, postpartum ovulation is common in marsupials where it is a spontaneous event and occurs at a time that merely reflects an extension of the normal estrous cycle.

Delayed implantation　Pregnancy usually persists for a precise and predictable period of time. Some interesting and, at first, mystifying exceptions have, however, been encountered. Animals that conceive in the autumn and deliver their young in spring can, on some occasions, such as when the length of the daylight period is artificially increased, produce their young much earlier. There have been other instances described, especially in kangaroos, where a female has been taken into captivity and without any contact with a male has, many months later, given birth to a young one. Faced with the necessity for an explanation, some people were even forced to consider the possibility of virgin birth! The cause is, nevertheless, quite a reasonable one. In a number of mammals, especially the mustelid Carnivora (such as weasels, skunks, badgers, and sable), fur seals, roe deer, and macropodid marsupials (kangaroos), development of the fertilized egg can sometimes cease when a blastocyst, containing about 100 cells, has been formed. This blastocyst lies dormant for a time that may extend for several months, but it can be subsequently stimulated to continue development. The delay is called an *obligatory* one when it is determined by external conditions, such as light, as seen in badgers, pine marten, weasels, roe deer and some marsupials. In mustelids, the activity of the corpus luteum is lost prior to implantation; increases in progesterone levels then appear to trigger further development (Wade-Smith *et al.*, 1980). In other species, such as the mouse, rat, and some podid marsupials it is *facultative* and controlled by more physiological events. As will be described in more detail in the later discussion on marsupials, this inhibition results from the effects of suckling and lactation. In rats and mice, development appears to be initiated by estrogens but the marsupials probably utilize progesterone for this purpose (Heap, Flint, and Gadsby, 1979).

Pregnancy　An excellent account of the role of hormones in this process is given by Heap (1972) and Heap, Perry, and Challis (1973). When the fertilized egg is retained in the oviduct or uterus and the subsequent development of the young occurs at this site, pregnancy is said to be occurring. This term is usually assumed to include the viviparous condition but may also encompass ovoviviparity. The internal incubation of the young is also called *gestation.* The condition of pregnancy appears to have reached its highest state of organization in placental mammals, though little information is available about this process in nonmammals. Pregnancy is not a uniquely mammalian phenomenon as it occurs in some chondrichthyeans, teleosts, reptiles, and amphibians, though not in birds. Gestation may occur for quite long periods of time in placental mammals but this is not unique as it may extend for 2 years in some viviparous sharks and is of 1 to 4 years' duration in the ovoviviparous urodele *Salamandra atra.*

As we have seen, the hormonal preparation of the mammalian uterus for the reception, fertilization, and implantation of the egg is initially stimulated by estrogens and progesterone, the latter usually having the subsequent dominant action, though both steroids act in simultaneous collaboration. Subsequently during pregnancy, these favorable uterine conditions need to be maintained and even modified from time to time as the fetus grows and is eventually delivered to the outside world. The necessary supplies of hormones are then altered qualitatively and increased quantities may also be required. These added needs have been met in various ways by the placental mammals and principally involve the function of the pituitary, the ovary, the placenta, and the uterus.

Progesterone, to use an oft-quoted phrase, is called "the hormone of pregnancy," but substantial, though usually smaller, amounts of estrogens are also used during gestation. These gonadal steroids maintain the endometrium and contribute to the considerable expansion that occurs in the myometrium during pregnancy. The hypertrophy of these muscles results from the stretching of the walls of the uterus and the induction, by estrogens, of new contractile proteins. Contractions of the uterus are not usually desirable during pregnancy, and the responsiveness of the myometrium to stimulation is reduced by progesterone. The contractile effects of oxytocin are, for instance, usually reduced by pretreatment of the uterus with progesterone, while estrogens have the opposite effect and enhance the responses to this neurohypophysial hormone. Such effects have not been demonstrated in all species but are very reproducible in some, like the rabbit. A most important role of progesterone in pregnancy in placental mammals is the inhibition of the estrous cycle and ovulation. This effect results from a negative-feedback inhibition of the release of gonadotropins from the pituitary and may be required when the periods of gestation exceed the length of the normal estrous cycle. Corpora lutea also persist in many viviparous and ovoviviparous nonmammals and although their precise role is uncertain it is suspected that they may also have a comparable role in these animals.

The problem of how to supply the added hormonal requirements of pregnancy has been met in various ways by different species of placental mammal. Estrogens and progesterone are typically secreted by the vertebrate ovary. The corpus luteum is usually the principal ovarian source of progesterone, but in nonpregnant animals, this structure does not usually persist longer than the estrous cycle. As we shall see, this situation even occurs in pregnant Australian marsupials. The period of gestation in these animals is similar to that of their estrous cycles so that a prolongation of the life of the corpus luteum is unnecessary. In the placental mammals, which have relatively longer periods of gestation, the corpus luteum persists for a much longer time and often remains functioning throughout the entire period of pregnancy. This extended survival is the result, in some species, of an inhibition of the effects of uterine luteolysins, caused by the presence of extra material in the uterus and the stimulating actions of mixtures of luteotropic hormones.

These hormonal combinations may consist of FSH, LH, prolactin, and gonadotropins that may be produced by the placenta. The precise hormonal content of this so-called *luteotropic complex* differs considerably from species to species. Its function is to extend the normal lifetime of the corpus luteum and to promote the secretion of progesterone.

The production of progesterone by the ovary may be supplemented in various ways. In some species (the horse), additional corpora lutea may form, but in others (humans and cattle) only a single corpus luteum is usually present. Animals that produce several young at a time have a correspondingly greater number of corpora lutea available for the production of progesterone. During pregnancy, the secretion of progesterone by individual corpora lutea may be increased by the action of the luteotropic complex. In addition, the amount of available progesterone is the net result of its rate of production and destruction. As described in Chapter 4, proteins that bind steroid hormones are present in the plasma and the rate of destruction of hormones is reduced when they are so bound. During pregnancy, the formation of such steroid hormone-binding proteins in the liver may be increased, probably as a result of stimulation by estrogens. There are considerable interspecific differences in the physiological patterns than ensure adequate progesterone in pregnancy.

In some species, such as the rabbit, ovariectomy during pregnancy always results in prompt abortion. In other species, such as the sheep and human, this operation does not necessarily result in a loss of the fetus. The placenta in these species produces sufficient gonadal steroids to support the uterus, though the supply may be inadequate during early pregnancy. There are considerable interspecific differences in the ability of the placenta to produce hormones. In sheep, humans, guinea pigs, and horses the placenta can provide sufficient progesterone for the needs of late pregnancy. However, in goats, pigs, cattle, dogs, rats, and rabbits the activity of the corpora lutea is necessary throughout pregnancy.

The fetus may also contribute hormones that are involved in gestation. The fetal adrenal cortex produces large amounts of two steroids that are substrates for the progesterone and estrogens formed by the placenta. Fetal pregnenolone sulfate can be converted to progesterone while dehydroepiandrosterone sulfate is used to synthesize estrogens. These steroids then pass into either the fetal or maternal circulation where they contribute to the maintenance of pregnancy and, at the appropriate time, to its termination.

The appearance of the fetoplacental unit as a temporary endocrine organ that helps to supply the hormonal requirements of pregnancy is a fascinating physiological adaptation. Such a role has not been described in nonmammals and is controversial in marsupial mammals. It is not possible to draw any orderly phyletic line as to the distribution of this hormone-secreting tissue in placental mammals, and it could have evolved separately on several occasions to suit the needs of each particular species. In recent years, it has become

apparent that tumorous tissues in mammals may produce a variety of hormones that normally arise from discrete endocrine glands. Perhaps there is
some analogy between such tumors and the evolution of an endocrine
placenta!

Parturition   The birth of the young is a precisely timed event about which
there has recently been an explosion of new information. The principal
experimental mammals have been sheep but others, such as goats, pigs, cattle,
rats, and rabbits, have also contributed information. Primates (humans and
monkeys) have also been studied (see Challis and Lye, 1994; Liggins and
Thorburn, 1994). Interspecific differences exist in the process of parturition
that probably reflect the mammal's reproductive strategy, such as single or
multiple young, the stage of maturity at birth, and the relative roles of the
corpus luteum and placenta in maintaining gestation. All mammals that have
been studied share a basic hormonal armamentarium related to parturition,
but it may be utilized in different ways. The hormones that may contribute to
this event are:

- progesterone, which has an inhibitory effect on the contractility of the
  uterus ("progesterone block"), an action which is usually alleviated
  during parturition
- estrogens, which promote the contractility of the uterus and increase its
  sensitivity to uterotropic substances; they can also modify placental
  activities
- oxytocin, which is secreted by the neurohypophysis and is the most
  potent uterotropic substance known; it is also formed by the ovary and
  the uterus itself
- relaxin formed by the ovary and placenta can inhibit uterine contractions and at birth can increase the flexibility of the cervix and pelvic
  ligaments
- prostaglandins, which are considered to be "local hormones"; they are
  derivatives of fatty acids and are formed locally at many tissue sites.
  Prostaglandins have a wide spectrum of action, which includes contraction of the uterus, a "ripening" of the cervix, and regression of the
  corpus luteum (luteolysis)
- cortisol from the fetal adrenal cortex, which can induce steroidogenic
  enzymes in the placenta and promote the maturation of fetal organs,
  especially the lungs.

Hippocrates in 460 BC is given the credit for the seminal pronouncement
that the fetus determines the time of its birth (Thorburn and Liggins, 1994).
This dictum was largely ignored until quite recently when contemporary
studies, mainly in the ewe, indicated that the maturation and stimulation of
the hypothalamo–pituitary–adrenal corex (HPA axis) in the fetus provided a

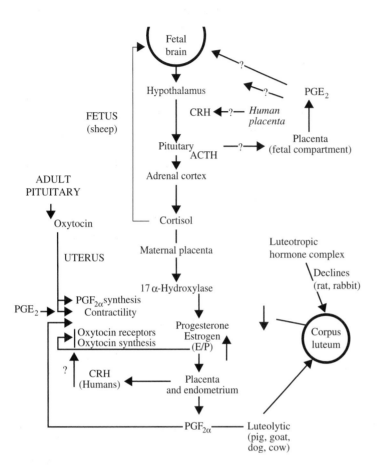

Fig. 9.13. Tentative summary of endocrine events that occur in mammalian parturition as mainly described in the ewe. For more details see the text.

major mechanism for the initiation of parturition. Electrocoagulation of the pituitary of the sheep fetus prolongs pregnancy while the infusion of cortisol or corticotropin promotes parturition. The same procedures performed on the ewe have no effects on the onset of parturition. A cascade of endocrine events can emanate from the activation of the fetal HPA axis in sheep. The primary trigger may be a genetic clock mechanism for that species, possibly facilitated by the fetal pituitary, which controls the synthesis of prostaglandin E_2. This substance appears to trigger a release of corticotropin via an action of CRH and, possibly, vasopressin (Fig. 9.13). The resulting surge of cortisol released from the fetal adrenal cortex passes to the placenta and induces the formation of 17α-hydroxylase. This enzyme diverts the placental ster-oidogenic pathway from progesterone to estrogens. The ratio estrogens to progesterone (E/P) then rises. This change has several effects that influence parturition. There is an increase in the contractility of the uterus, where oxytocin receptors are also induced. In humans and rats, mRNA for oxytocin has been identified in the uterus (Lefebve *et al.*, 1992). A local supplementation of the supply of this hormone is, therefore, available. Another important result of the rise in the E/P ratio is a facilitation of the synthesis of prostaglandin $F_{2\alpha}$ ($PGF_{2\alpha}$) by the placenta. This local hormone has a uterotonic action and its further synthesis in the uterus is promoted by oxytocin. In some species with corpora lutea that persist through pregnancy, such as pigs, goats, cattle, and dogs, $PGF_{2\alpha}$ can enter the uterine vein and pass into the ovario-uterine artery and exert a luteolytic effect in the ovary. In other species, such as rabbits, the timely regression of the corpora lutea may be caused by decline in the activity of the luteotropic complex of hormones.

In humans the initiation of parturition appears to utilize a unique endocrine mechanism, though it still shares some features with other mammals. The hypothalamic peptide CRH can be expressed in the human placenta, which starts to secrete it after the first trimester of pregnancy (Challis and Lye, 1994; McLean *et al.*,1995). Its action, is, however, normally blocked by the concomitant synthesis of a CRH-binding protein by the placenta and liver (Potter *et al.*, 1991). Commencing about 3 weeks prior to parturition, the synthesis of CRH exceeds that of its binding protein so that its actions may then become apparent (McLean *et al.*, 1995). These effects include local vasodilatation, a release of $PGF_{2\alpha}$, and an increased sensitivity of the uterus to oxytocin. Cortisol can promote the synthesis of placental CRH. It has been suggested such an effect could be part of a positive-feedback process following the passage of CRH into the fetal circulation where it could be stimulating the HPA axis (Challis and Lye, 1994). The timing of parturition in humans may thus be scheduled by the placental synthesis of CRH. This mechanism has been called a "placental clock." Its possible role in other primates remains to be investigated. The precise roles of relaxin in pregnancy and parturition have not been defined but differences between species appear to exist. Rises in its plasma levels have been observed prior to parturition in pigs, horses, guinea pigs, and rats (Liggins and Thorburn, 1994).

The fetus clearly gets considerable complex hormonal assistance in its progression through the "triumphal (pelvic) arch" to then be nurtured by expectant and hormonally primed parents.

Marsupial mammals

The marsupial and placental (eutherian) mammals appear to have originated from a common oviparous ancestor in the Cretaceous period about 100 million years ago. About 250 species of marsupials still exist, most of which live in Australia, with a few in the Americas and New Guinea. They occupy habitats ranging from tropical forests to open grasslands and dry desert regions. Such diversity is reflected by their varied reproductive strategies. (Tyndale-Biscoe and Renfree, 1987). The initiation of the study of the reproductive endocrinology of Australian marsupials was principally through the efforts of H. Waring on the occasion of his academic migration to Australia in 1948. He wisely recruited G. B. Sharman to perform the seminal experiments (Sharman, 1970). The first such observations were made on a small wallaby, the Rottnest Island quokka (*Setonix brachyurus*), which was historically appropriate as it was also the first Australian marsupial to be described by European explorers.

Systematically, the marsupials are distinguished from the placentals by the absence of a true placenta; only a yolk-sac placenta is present so that the physiological connection with the parent is more tenuous than in placentals. Young marsupials are born in a relatively immature state comparable, in some respects, to that of embryos in a quite early stage of the gestational period of placentals. The newborn young of marsupials are suckled on the teat, where they undergo a considerable part of the development that would normally occur within the uterus of placentals. In many, but not all, marsupials this takes place in an external pouch or marsupium. The period of development in a pouch far exceeds that of normal gestation and varies from about 60 days in bandicoots to 370 days in the koala. In the tammar wallaby it is about 250 days. As succinctly summarized by M. Renfree (1993) they have "exchanged the umbilical cord for the teat."

Australian marsupials are usually seasonal breeders who commonly give birth in midsummer and early winter. Such species can, therefore, be classified as short-day breeders, but some utilize long days for their reproductive timing and give birth in the spring (Gemmell, 1994). Various factors, especially rainfall and food supply, can impinge on their reproductive cycles so that the young are born and weaned at times that are most advantageous to their survival. Some species can breed continuously, as seen under domesticated conditions. The proximate physiological processes that influence such breeding behavior are uncertain but the pineal, and its secreted melatonin, relays photoperiodic information in some marsupials. A rhythmical release of prolactin can also function as a novel mediator in such reproductive processes.

While there are many similarities between the endocrine control of reproduction in placental and marsupial mammals, there are also some remarkable differences. The gonadal steroids, estrogens and progesterone, mediate the pre- and postovulatory changes in the reproductive tract during the estrous cycle of marsupials. Hypophysectomy of the tammar wallaby blocks the growth of the ovarian follicles and ovulation (Tyndale-Biscoe and Evans, 1980). This effect appears to reflect the absence of FSH and LH, which have been identified in extracts of the animal's pituitary gland. However, in contrast to eutherians, the marsupial corpus luteum develops normally following hypophysectomy and it is not dependent on the tropic action of LH. Instead, its activity appears to be controlled by prolactin, which has an inhibitory effect. Specific receptors for prolactin have been identified in the corpus luteum of the tammar (Sernia and Tyndale-Biscoe, 1979).

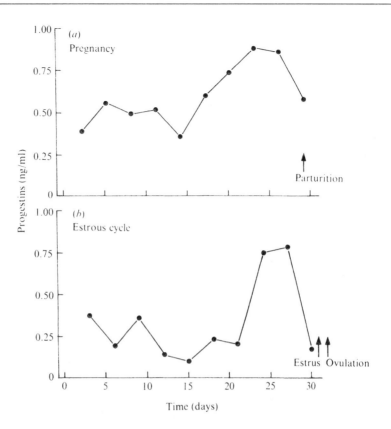

Fig. 9.14. Changes in the plasma progestin levels during pregnancy (*a*) and the estrous cycle (*b*) of a marsupial, the tammar wallaby *Macropus eugenii*. The commencement of pregnancy or the estrous cycle was initiated by removing the suckling young from the pouch, thereby initiating the development of the fertilized blastocyst (see text), or the next reproductive cycle. Progesterone levels commenced to rise on about day 15 and declined just prior to estrus and ovulation, or parturition. (From Lemon, 1972.)

In marked contrast to placentals, pregnancy in marsupials does not interfere with the concurrent estrous cycle and the maturation and ovulation of the egg. This egg is usually produced at the normal time. Certain marsupials also display an interesting form of delayed implantation or, more correctly called in this instance, an *embryonic diapause*, which differs from the process that is occasionally seen in placentals. It was first described by G. B. Sharman in the quokka, *Setonix brachyurus*, and has subsequently been observed most often in other macropodid marsupials (kangaroos). Embryonic diapause has also been observed in several groups of possum, including the pygmy possum, the feather-tail possum and the honey possum (Renfree, 1993).

The estrous cycle of marsupials ranges from 9.5 days to about 40 days, but it is usually about 28 days. As in placentals, it consists of an initial period of follicular growth that is accompanied by the development of the uterus and vagina, at first under the influence of estrogens and then, in the luteal phase, by progesterone. Estrus follows a sudden decline in the levels of progesterone (Fig. 9.14), reminiscent of that seen in some placentals, and lasts for several days during which ovulation occurs. This process, as in placentals, is initiated by a surge of LH in response to a rise in plasma estrogen levels, and it is usually spontaneous. Fertilization is followed by the development of the egg into a blastocyst, and if the animal is not already lactating, pregnancy will occur. A corpus luteum is formed in the ruptured follicle and persists for the period of time that is usual in the estrous cycle; its life is *not* prolonged by the pregnancy. An extended life for the corpus luteum is not necessary in marsupials as the period of gestation is usually nearly identical to the time of the normal estrous cycle (which continues to occur concurrently with the pregnancy!); nevertheless, the progesterone levels in pregnancy may be somewhat greater than those in the normal estrous cycle (Fig. 9.14), which seems to reflect a hypersecretion from the ovary. In the quokka, *Setonix brachyurus*, no difference in the levels of progesterone in the nonpregnant and pregnant animals could be detected (Cake, Owen, and Bradshaw, 1980). However, a sudden rise in the plasma progesterone concentration was detected between the third and fourth days of pregnancy. It was suggested that this "spike" may occur in response to a signal from the blastocyst and serve to establish a suitable secretory condition of the endometrium. Pregnancy varies from 9.5 days in the fat-tailed marsupial mouse, *Sminthopsis macroura*, to 38 days in the long-nosed rat kangaroo, *Potorous tridactylus*.

Most marsupials have a quite simple placenta that is formed by a vascularization of the chorion by blood vessels from the yolk sac (Renfree, 1980). Such a yolk-sac or choriovitelline placenta is quite small and, compared with that of eutherians, provides a rather tenuous connection between the mother and fetus. There is, however, considerable variation and the chorionic villi, which form the attachment, can be quite well developed in some species. In bandicoots, the allantois may also contribute to the placenta, as it does in eutherians. In the latter, the placenta plays a major role in furnishing the endocrine needs of pregnancy but it has usually been considered unlikely that this is so in marsupials. In several marsupials, ovariectomy in mid to late pregnancy does not result in the death of the fetus (though parturition does not occur) or in a regression of the uterine endometrium, which is supporting it (Renfree, 1980). This phenomenon is also seen in eutherians where survival of the fetus can, however, be attributed to the endocrine activities of the placenta. Fetal membranes collected from the quokka (*Setonix brachyurus*) and the tammar (*Macropus eugenii*) have been shown, *in vitro*, to be able to synthesize progesterone from pregnenolone (Bradshaw *et al.*, 1975; Heap, Renfree, and Burton, 1980). The tammar placenta, however, could not form estrogens. There is, therefore, some circumstantial and experimental evidence that the marsupial placenta may have an endocrine role during pregnancy.

Ovulation of the egg that ripens during pregnancy occurs at various times in relation to parturition. Ovulation may occur *prior* to parturition, as in the swamp wallaby (*Wallabia bicolor*) where the period of gestation is 35 days compared with only 32 days for the estrous cycle. In this marsupial, preparturition ovulation is followed by copulation. If fertilization takes place, a blastocyst develops, which, if lactation then occurs, lies dormant (see later). In other species like *Megaleia rufa*, parturition is closely succeeded by ovulation, postpartum copulation, and the formation of a blastocyst. In the grey kangaroo, *Macropus giganteus*, the period of gestation is much shorter than the estrous cycle, just as seen in the bush-tail possum, *Trichosurus*, and other nonkangaroos; prescheduled future ovulation is then inhibited by the suckling stimulus provided by the young. If, however, the young is removed, ovulation follows 9 days later. In the latter part of lactation of the grey kangaroo this inhibition may decline so that ovulation and fertilization may occur, though while the young is in the pouch, the fertilized egg does not develop further than the blastocyst stage.

When the young kangaroo leaves the pouch, the development of a dormant blastocyst can then proceed and pregnancy continues. The young kangaroos, however, remain with the mother and continue to suckle from outside the pouch; as a result, the female kangaroo may have one young in the pouch, and another, much older young, "at heel." The two young then feed from different teats and the composition of the milk that each feeds on is quite different, notwithstanding the fact that the endocrine secretions that are available to both glands are identical.

The delayed implantation in macropodid marsupials follows the division of the fertilized egg to a stage when 80 cells are present. This blastocyst, in contrast to the placental one, is surrounded by a shell membrane and a layer of albumin, in which state it can survive for several months. It lies in the uterus, in the branch opposite to that where the preceding pregnancy occurred. The temporary inhibition of the development of the blastocyst depends on the suckling stimulus from the young kangaroo in the pouch. Once suckling declines, the blastocyst then starts to develop further. Denervation of the mammary gland has the same effect even though suckling continues (Renfree, 1979). The nature of the inhibitory stimulus is thought to result from neural stimulation of the pituitary, as a result of suckling. Ovariectomy does not have any effect on the dormant blastocyst, but the injection of estrogen and, especially, progesterone can initiate its development. It appears that the corpus luteum of lactation, which is formed from the follicle that gave rise to the dormant blastocyst, is relatively quiescent during lactation and its subsequent development and rapid secretion initiates the succeeding pregnancy. A pulse of secreted progesterone appears to accomplish this change by initiating, and so synchronizing, both the further development of the blastocyst and the luteal phase of the uterus, the latter providing an environment that is necessary for the growth of the embryo. The nature of the inhibition of the further development of the corpus luteum during lactation was uncertain for several years. The injection of oxytocin, which is normally released by suckling, can prevent its activity in kangaroos but this hormone was found to be ineffective following hypophysectomy. Prolactin is also released in response to suckling and it is now accepted that it has an inhibitory luteostatic effect in kangaroos (Hinds, 1994; Renfree, 1994). This role for prolactin contrasts with its luteotropic one in the laboratory rat.

Some marsupials, the tammar (*Macropus eugenii*) and Bennett's wallaby (*Macropus rufogriseus*) undergo a seasonal quiescence during which time development of the dormant blastocyst is inhibited, even after the preceding joey leaves the pouch (Hinds, 1994). This period occurs between July and the summer solstice in late December, after which time an activation of the blastocyst may be precipitated and there is a continuation of the pregnancy (which may have already lasted 11 months, since the previous January). As with its role in the lactational delay in development of the corpus luteum, prolactin also functions as a luteostatic hormone during seasonal quiescence. It is released as a precise pulse each morning. This event is photoperiodically timed and appears to be related to the duration of the secretion of melatonin during the night. It can be overcome by denervation of the pineal gland, when an activation of the corpus luteum is initiated. Following a delay of 72 hours, when no further timed release of prolactin occurs, the corpus luteum secretes a pulse of progesterone that initiates the secretory phase of the uterine endometrium and further gestational development of the blastocyst and succeeding reproductive cycles. The dormant corpus luteum also appears to exert an inhibitory effect on the development of the Graafian follicles in these marsupials, which results in a suspension of the normal estrous cycle. The normal timing of the activation of the blastocyst in the tammar appears to be caused by a change in the photoperiod from the inhibitory threshold of 15 hours of light and 9 hours of dark to 12 hours of light and 12 hours of dark (12L/12D) (Hinds, 1994). This event can be mimicked by the appropriate administration of melatonin. In placental mammals, such as sheep, photoperiodic information influencing reproductive cycles is relayed to the GnRH pulse generator in the hypothalamus, which regulates the release of gonadotropins. In contrast, some marsupials have evolved a different mechanism: the utilization of prolactin to inhibit the activity of the corpus luteum. A tentative summary of the processes influencing the reproductive cycle in the tammar wallaby is shown in Fig. 9.15.

Parturition in marsupials appears to follow the general placental pattern and has been likened to that of the sheep (Renfree, 1993; 1994). There are, however, some interesting differences. The information that is available is largely confined to macropodids, especially the tammar wallaby. An intact pituitary and corpus luteum are necessary for normal parturition. The fetus has a sufficiently mature adrenal cortex to provide a cortisol signal, like in sheep, but there are no direct measurements that would provide confirmation of such a role. Plasma levels of several hormones change at the time of parturition. There is a decline in progesterone and a modest rise in estrogens, which do not, however, attain their maxima until the postpartum period when they initiate a release of LH and ovulation. Pulses of released prolactin

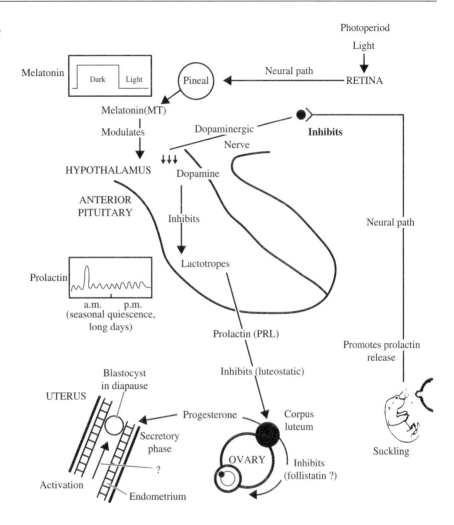

Fig. 9.15. Tentative summary of the factors affecting the reproductive cycle of the tammar wallaby during lactational and seasonal quiescence of breeding. For more details see the text. Based on information provided by Hinds (1994) and Renfree (1994).

and prostaglandins occur at the time of parturition. Prostaglandins ($PGF_{2\alpha}$ and PGE_2) appear to be formed by endometrium and this process could be a response to fetal cortisol. The prostaglandins may initiate a release of prolactin. Both of these excitants can exert a luteolytic action at this time, but blockade of prolactin release does not prevent parturition. However, antagonists of prostaglandin synthesis can delay birth. As term approaches, the uterus becomes more responsive to the contractile effects of oxytocin and prostaglandins. These responses can facilitate the passage of the fetus from the uterus. The administration of $PGF_{2\alpha}$ to nonpregnant tammar wallabies results in them adopting a "birth posture", when they sit on their lower backs with the tail extended between the legs and the body hunched forward. This response is also seen in immature females and even in males. Such a posture is vital for successfully directing the newborn towards the pouch.

Whether or not oxytocin is also involved in parturition, as in placentals, is unknown, but it is present in the marsupial pituitary and it contracts the uterus of the wallaby *Setonix brachyurus, in vitro* (Heller, 1973). This tissue is most sensitive in the late stages of pregnancy. Oxytocin also promotes milk letdown when injected into kangaroos, an effect shared with the placentals. This response may be of special importance for feeding the relatively undeveloped newly born young of marsupials. Indeed, it has been shown that the mammary gland of the agile wallaby (*Macropus agilis*) is highly sensitive to the action of oxytocin early in lactation, but this subsequently declines as the young grow larger (Lincoln and Renfree, 1981).

The reproductive pattern in marsupials shows distinct differences from that of placental mammals and appears to be well adapted to their manner of life. These patterns are quite varied and have been summarized and classified into four major groups (Tyndale-Biscoe and Renfree, 1987).

1. A group that have a relatively short pregnancy that is similar to the luteal phase of the estrous cycle. The scheduled postpartum estrus and ovulation are inhibited if lactation occurs. These animals exhibit polyestrus and they are polyovular. They include the larger possums (Phalangeridae) and some carnivorous species (Dasyuridae).

2. Species in which the period of gestation is much shorter than the luteal phase. The latter then extends into the period of lactation. The group includes bandicoots and bilbies (Peramelidae), which are polyestrus and polyovular.

3. Species in which gestation is about the same length as the estrous cycle so that ovulation occurs soon after parturition. If fertilization occurs then the development of the blastocyst is arrested as a result of the lactational stimulus. These marsupials are monoovular and polyestrus and include the Macropodidae (kangaroos). In two species, the tammar (*M. eugenii*) and Bennett's wallaby (*M. rufogriseus*), a seasonal photoperiodic stimulus can also result in embryonic diapause.

4. The last group are marsupials that have an unusually prolonged period of gestation that includes a time of embryonic diapause. These animals are polyestrus and polyovular and include pygmy possums, honey possums, and feather-tail gliders (Burramydidae, Tarsipedidae, and Acrobatidae).

Contrary to some popular opinion about the "lowly" state of development of these animals, their reproduction is an extremely efficient process, which many farmers and graziers will confirm. The embryonic diapause of the kangaroos constitutes an excellent "insurance" for continued reproduction, so that if a young is lost, or when it is weaned, another pregnancy follows with little delay.

Monotremes

These mammals are confined to Australia and New Guinea and are remarkable as they produce eggs with a keratinous shell that they care for. In the spiny anteater, *Tachyglossus*, the egg is lodged in a pouch for hatching, while platypuses lay their eggs and tend them in burrows. They are monoestrus (Griffiths, 1978; 1984). The gonads increase considerably in size prior to breeding in the late winter and spring. The platypus usually produces two eggs, and the echidna one. They are first incubated in the uterus for a period that, in the echidna, has been estimated at 9 to 27 days. Development equivalent to about 40 hours of incubation in chickens occurs at this time. The eggs are much smaller in size than those of birds and reptiles, reflecting the reduction of stored yolk proteins. Nutrients can, however, be accumulated from a secretory uterine endometrium. The processes of ovum maturation, fertilization, and subsequent development show many similarities to those of sauropsid reptiles (Austin, 1994). A sauropsid-like follicle, following ovulation, forms a mammalian-type corpus luteum that persists during pregnancy. In the platypus, the plasma concentrations of progesterone and estradiol rise during pregnancy to levels similar to those seen in placental mammals (Hughes and Carrick, 1978). However, it is uncertain whether these steroids are derived from the corpus luteum. Monotremes appear to have exchanged some storage of yolk proteins for a mammalian-type sustenance involving endometrial secretions, such as those controlled by progesterone in placentals. However, this is not a unique situation as it appears to also have been adopted by some elasmobranch fish (p. 439). It would be of special interest to know whether monotremes exhibit an hepatic vitellogenesis in response to estrogens, as seen in nonmammals but which is absent in the placentals.

In echidnas the period of incubation of the egg in the pouch is about 10 days. The young of monotremes are fed in the typical mammalian fashion on secreted milk. In the echidna it is released in greater amounts following injection of oxytocin. Development of the mammary glands of nonbreeding or ovariectomized echidnas can be promoted by the injection of estrogens. Relatively little is known about the endocrine processes that control reproduction in these very interesting and unique animals, but there appear to be a number of similarities to those of other mammals.

Nonmammals

Precise information about ovarian cycles in nonmammals, apart from birds, is meager compared with that in mammals. Much of the available knowledge is based on morphological and histological observations on the ovaries and the accessory and secondary sexual characters, especially the oviduct. Such information is related to endocrine changes on the basis of the abilities of injected, exogenous hormones to mimic or prevent such changes. These experimental approaches, while suffering from obvious limitations, have demonstrated that differences indeed exist between the ovarian cycles of different nonmammalian vertebrates. Using radioimmunoassay procedures for measuring hormone levels in the blood, the precise role of hormones in the reproductive life of nonmammals is now being investigated more rigorously. At the present time, the birds have received the most attention, but such studies are being extended to the reptiles, amphibians, and fishes.

It is usually somewhat difficult to make a strict comparison between the ovarian cycles of mammals and nonmammals. This partly reflects a lack of information about the latter, but the timing of the events also often differs rather radically. Birds usually take many weeks of preparation to come into breeding condition, when ovulation becomes possible. This latter process then occurs at regular intervals of about 24 hours, which can proceed for several weeks; in the domestic fowl it can occur for up to 300 days of the year with some minor breaks. This almost daily ovulation cannot be strictly compared with the estrous cycle of mammals but may be more synonymous with very short estrous cycles of rapidly successive periods of estrus. In other species of fish and amphibians, a single massive (sometimes referred to as "explosive") ovulation or oviposition may occur, but in the meantime the ovum may be held in readiness for some time; ovulation thus does not always appear to be an irrevocable event in a strictly pretimed program.

Birds   The ovarian cycle of most birds that live in temperate zones is under photoperiodic control. Species that live in equatorial and desert regions may utilize other cues to initiate their reproduction and some birds exhibit circannual endogenous reproductive rhythms (Follett, 1984). As described above, quail subjected to a long-day photoperiod exhibit a release of GnRH from the hypothalamus that results in a discharge of pituitary gonadotropins. Plasma LH and FSH levels have been shown to increase in Japanese quail that are photoperiodically stimulated by long daylengths (Nicholls, Scanes, and Follett, 1973; Follett, 1976). The changes in the levels of pituitary gonadotropins indicate that a release occurs at a precise time each day, which, in the quail, is in the evening after dusk. This daily rhythmical release of gonadotropins promotes the growth of the ovary and the maturation of the follicles. Estrogens are known to stimulate the growth of the avian oviduct (progesterone may also contribute to this increase) and secondary sexual characters and are released during such preparations for reproduction.

In most wild species of bird continuous stimulation by long spring days eventually results in a cessation of breeding activities and the release of gonadotropins and they become insensitive to further photoperiodic stimuli. This "refractoriness" is an important event that terminates reproduction and assures maturation of the young prior to inclement winter conditions. The process can be reversed by exposure of such birds to short days. The nature of the "clock" mechanism interpreting photoperiodic signals in birds is different to that in mammals (Follett, 1984). The pineal does not appear to be directly involved in such timing processes and photoreception does not necessarily depend on the eyes. As described earlier, photoreceptive extraretinal elements appear to be present in the brain but their precise location and nature are unknown.

There are some interesting interactions that occur between the thyroid gland and the photoperiodic mechanisms in birds (Follett, Nicholls, and Mayes, 1988; Follett and Nicholls, 1988). When T_4 is administered to quail hens maintained on short-day cycles of exposure to light, reproductive activity is initiated and they even lay eggs. However, if the birds are in a photorefractory state then this treatment results in a persistence of this condition even when they are transferred to a short-day cycle. Thyroidectomy performed during the refractory period then results in a restoration of photosensitivity. Thyroid hormones, therefore, appear to be able to mimic the effects of long days in initiating reproduction, but if the birds are in a refractory state the hormones will maintain them in this condition. It was suggested that the thyroid hormones may be influencing the functioning of photoperiodic mechanisms "high in the photoneuroendocrine system." A comparable interaction between T_4 administration and the photoperiodic control of reproduction has been observed in male American tree sparrows, *Spizella arborea* (Wilson and Reinert, 1993). Testicular growth was induced by T_4 in birds maintained on short days while refractory birds given this hormone and then moved to a short-day cycle failed to exhibit the expected restoration of photosensitivity. These observations have given rise to some interesting suggestions regarding the basic mediation of mechanisms for control of the photoperiodic processes. Such mechanisms appear to involve the brain and may reflect the initiation of genetic transcriptional processes such as those influenced by T_3 interacting with its ubiquitous nuclear receptors. Such a response could promote the induction of regulatory or morphogenetic proteins involved in the photoperiodic processes concerning both the initiation of reproduction and the maintenance of photorefractoriness.

When the bird is ready to breed, ovulation may begin. In domestic fowl, ducks, and geese, this is spontaneous but in other birds the presence of the male is usually necessary so that ovulation may then be said to be induced. There are several other factors that determine whether or not ovulation will occur in birds. If the newly laid eggs are continually removed from the nest some birds will continue to lay more eggs (*nondeterminate layers*). A house sparrow has thus been stimulated to produce 51 eggs in a season; ovulation apparently continued until the ovary was exhausted of suitable follicles. The

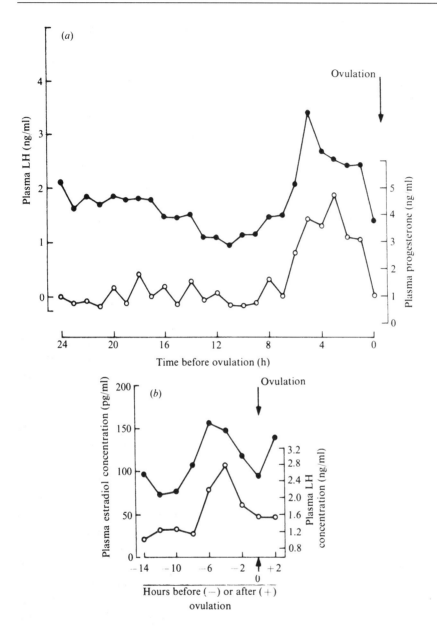

Fig. 9.16. The ovarian cycle of the domestic fowl. (*a*) Changes in the plasma levels of LH (●) and progesterone (○) during the ovulatory cycle. The relationship between the rises in the levels of progesterone and LH is not clear; the release of LH apparently does not initiate a release of progesterone as it never precedes it. (From Furr *et al.*, 1973.) (*b*) Changes in the levels of LH (○) and estradiol (●). The rise in the level of estradiol occurs about 2 hours before that of LH. (From Senior and Cunningham, 1974.)

domestic fowl, which can produce 300 eggs in a year, is an even more dramatic example of this phenomenon. How such birds recognize the number of eggs in the "clutch" is unknown, but it has been suggested that this may be the result of a tactile stimulus or sight. In cockatiels (*Nymphicus hollandicus*) it has been shown that rising plasma concentrations of prolactin may provide a proximate signal for the cessation of laying (Millam, Zhang, and El Halawani, 1996). Other types of bird (*determinate layers*) produce a set number of eggs, and changing the number in the nest does not modify ovulation.

In the domestic fowl, LH, as in mammals, initiates ovulation, but the mechanism controlling the release of this hormone is not clear. The injection of progesterone promotes ovulation while estrogens delay it, effects that are in direct contrast to those seen in mammals (Fraps, 1955). The normal rhythmical release of LH, which commences about 8 hours prior to ovulation is, however, preceded by an increase in plasma estradiol (Fig. 9.16b) (Follett and Davies,1979). Estrogens, however, do not directly stimulate the release of LH in birds, though they apparently serve to sensitize the hypothalamus to the action of progesterone. The surge of LH never *precedes* that of progesterone (Fig. 9.16a). Antisera to progesterone block ovulation in chickens, but antisera to estrogens do not (Furr and Smith, 1975). The mechanism of ovulation in birds is, therefore, clearly different from that in mammals.

It has been shown in the domestic fowl that as long as there is an egg in the oviduct further ovulation is inhibited. This effect can be mimicked by placing an irritant, such as a piece of thread, in the oviduct. Such a condition can be prolonged for about 3 weeks and as no regression in the ovary or oviduct occurs secretion of FSH and estrogens is thought to be unimpaired. The inhibition of ovulation by the presence of an egg in the oviduct is thought to be the result of a neural stimulus that may inhibit LH release. The injection of progesterone or LH into such birds overcomes this inhibition and promotes ovulation.

The precise site of origin of the circulating progesterone in birds is uncertain. Birds do not possess a corpus luteum but the ovary nevertheless secretes progesterone. This steroid may be formed by the follicles themselves, the interstitial tissue, or the corpora atretica. The role of progesterone in the ovarian cycle is also not clear. In mammals and other vertebrates, this steroid inhibits the release of LH and LH-like gonadotropins by its negative-feedback inhibition of the hypothalamus. Such an effect would be somewhat unexpected in birds if, as suspected, progesterone stimulates the release of LH. However, it will be recalled that estrogens can exert both positive- and negative-feedback effects on the release of LH in mammals. As birds lack a corpus luteum, LH also cannot exert its usual luteinizing and luteotropic actions though it is possible that it may have comparable effects at other sites in the ovary. There is need, at this stage, for a note of caution as it should be recalled that the endocrine observations on the avian ovarian cycle are nearly all confined to domestic species, especially *Gallus domesticus*.

The egg-laying cycle of the domestic fowl is thought to occur in the following manner (van Tienhoven and Planck, 1973). The eggs are laid in a clutch or "sequence" three to five in number, which are each produced at intervals of about 26 hours. Laying can only occur during a daylight period of about 8 hours called the "open period". Ovulation usually takes place about 2 hours after laying so that the time of the latter advances by a similar period each day. The open period thus limits the number of eggs that can be laid before an obligatory pause of 40 to 48 hours. Ovulation and oviposition are controlled by an endogenous rhythm lasting for 26 to 28 hours, the commencement of which is normally timed according to the photoperiod. Depending on such stimulation the length of the cycle can be retarded or advanced by about 2 hours and so may vary from 24 to about 30 hours. In continual light, other periodic events, such as the times of feeding and fluctuations in temperature, can be used to initiate the egg-laying cycle. Such external stimuli appear to sensitize regions of the hypothalamus, which, in response to stimulation by circulating progesterone, and possibly estrogens, secretes GnRH. LH is then released from the pituitary, which results in ovulation 4–8 hours later. Subsequent formation of the egg takes place in the oviduct, and the timing of the events there seems to depend on photoperiodic stimulation working in conjunction with the ruptured follicle. Removal of the latter tissue from the ovary results in a retention of the egg in the oviduct. Normally, oviposition occurs 13 to 14 hours after ovulation. It is possible that contractions of the oviduct that occur during oviposition are assisted by vasotocin, which is released from the neurohypophysis.

The production of eggs for human consumption is an important agricultural and economic industry that involves maintaining many millions of domestic hens. The costs of feeding them and providing light and heat are so large on a national basis that even quite small decreases in such costs may be of considerable economic importance. Therefore, any advance in the time of onset of the age of first laying or increases in the clutch size, by decreasing the interval between oviposition and the next ovulation, may potentially provide more and cheaper eggs. Artificial changes in the periods of light to which the birds are exposed each day have effected some important advances (Morris, 1979; Wilson and Cunningham, 1980). Sexual maturity in hens can be promoted by increasing the length of the daylight hours; for instance, pullets raised on a 14L/10D schedule mature and start laying 7–10 days earlier than those kept on a 6L/18D photoperiod. If pullets initially raised on a short-day protocol are subsequently changed to a long-day one, sexual maturation can be advanced by as much as 7 weeks. The particular decrement of time to maturity depends on when the change is made. Changing the lighting period of laying hens from 6L/18D to 14L/10D can enhance their egg laying compared with birds kept constantly on the latter schedule. This change is associated with an increase in the plasma LH concentrations. Light is power, and money, so that any reduction in its use may be an economic advantage. It has been found that periods of exposure to light need not be continuous and that intermittent exposures may be as effective as continual light. Thus, when hens on short days (8L/16D) are provided with a 1- to 10-minute pulse of light each hour for the first 8 hours of darkness, advances in sexual maturity

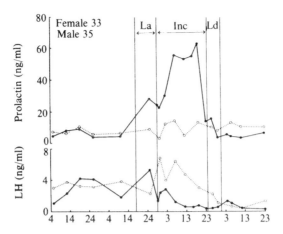

Fig. 9.17 Prolactin and LH concentrations in the plasma of mallards (*Anas platyrhynchos*) at different times of breeding. Values for a duck (●) and a drake (○) are shown. The vertical lines indicate the period of egg laying (La), incubation (Inc), and the leading of the young (Ld). Prolactin levels are greatly increased, whereas LH levels decline in the ducks when they are incubating the eggs. The drakes do not share this chore and do not exhibit these responses. (From Goldsmith and Williams, 1980.)

and onset of egg laying can occur. Such changes have been shown to increase plasma LH concentrations (Wilson and Cunningham, 1980).

Most birds incubate their eggs. In many instances both the male and female share periods on the nest and tend the young. The desire to incubate the eggs is called "broodiness" and is influenced by hormones. Preparations for incubating the eggs include building a nest and the development of a "brood patch" on the abdomen. In canaries and budgerigars, this morphological change includes a loss of feathers, which appears to involve the actions of estrogens and prolactin. Subsequently, there is an increase in the vascularity, which is estrogen dependent, and an increase in tactile sensitivity involving estrogen and, probably, progesterone (Hutchison, Hinde, and Steel, 1967; Hutchison, 1975). In ring doves, progesterone may play a more important role than other hormones in initiating broodiness (Stern and Lehrman, 1969).

Prolactin levels have been measured in the plasma of a wild population of mallards over the breeding season (Goldsmith and Williams, 1980) (Fig. 9.17). In these ducks, the prolactin concentrations increase about three-fold during incubation of the eggs and decline following the hatching of the ducklings. The drakes do not incubate the eggs, and relatively low levels of prolactin persist in the plasma over the breeding season. Plasma LH declines markedly in the ducks at the end of egg laying, but this change is not seen in the drakes until much later in the season. Female and male ring doves (*Streptopelia risoria*) share in the incubation of the eggs. Prolactin concentrations in the plasma then rise in both sexes (Lea, Vowles, and Dick, 1986). The release of this hormone is apparently facilitated by tactile stimuli from the nest and, subsequently, by the sight of the young. Macaroni and gentoo penguins from South Georgia share nesting duties, during which time plasma prolactin levels also increase in both sexes (Williams and Sharp, 1993). Wilson's phalarope (*Phalaropis tricolor*) exhibits a more complete sex role reversal as the male alone undertakes the incubation of the eggs. The prolactin concentrations in the plasma of the male bird are then even higher than in the female and his testosterone levels decline markedly (Oring *et al.*, 1988). Such hormonal changes appear to contribute to behavior appropriate for the hatching of the eggs and the care of the young. They may also have more physical effects such as the development of brood patches and the formation of crop-sac secretions to feed the young (Chapter 5). A release of prolactin has been observed in columbid birds following the exposure of either sex to the squabs (Buntin, 1979).

Birds show some interesting deviations and novelties in the use of hormones for integrating their reproductive processes. An evolution of the role of certain hormones has clearly occurred in this interesting offshoot from a reptilian stock.

Reptiles   Environmental temperature, as described earlier, appears to provide major proximate stimuli for the control of reproduction in reptiles (Licht, 1984; Whittier, 1994). Photoperiodic effects have been described but they are rare. Endogenous circannual rhythms appear to exist that may be influenced by external and internal stimuli, such as rain, the food supply, and the resulting somatic condition of the animals, as well as social interactions. The Reptilia contain oviparous, ovoviviparous, and viviparous species. Unlike birds, a distinct corpus luteum is formed following ovulation and although the available evidence indicates that it secretes progesterone (Highfill and Meade, 1975a,b; Colombo and Yaron, 1976) its role in the maintenance of pregnancy in viviparous reptiles is uncertain (Jones and Baxter, 1991). A pituitary LH and FSH have been identified in some reptiles, but in the snakes and lizards only a single gonadotropin appears to be present (Licht, 1979). The development of the ovary, its secretion and steroid hormones, and ovulation appear to be controlled by an FSH or a nonspecific gonadotropin. This is in contrast to mammals and birds, where there are two functioning gonadotropins, and amphibians, where an LH-like hormone appears to be predominant.

The changes that occur during the ovarian cycle of the ovoviviparous lizard *Sceloporus cyanogenys* are summarized in Fig. 9.18. The ovary starts to grow in October or November and this is accompanied by the development of the oviduct. These changes can be prevented by hypophysectomy. The gonadotropin stimulates gonadal growth and the secretion of estrogen. Implantation of small pellets of estrogens into the region of the median eminence reduces the growth of the oviduct. This effect is probably the result of a lower rate of ovarian estrogen secretion, caused by the inhibition of gonadotropin release by a negative-feedback inhibition. Ovulation is also prevented by such an estrogen implant. Mammalian FSH has been shown to promote ovulation in several species of lizards (Licht, 1970), and an endogenous gonadotropin no doubt also has this effect. In some reptiles, serial sampling of the plasma

Fig. 9.18. The annual ovarian cycle of the ovoviviparous lizard *Sceloporus cyanogenys*. The ovaries and oviducts start to develop (under the influence of gonadotropin) in October and ovulation may occur in December to January. Gestation lasts for about 12 weeks and the young are delivered in late March to mid-May. Corpora lutea persist during pregnancy and the plasma progesterone levels rise but then decline in late summer following parturition. (From Callard *et al.*, 1972.)

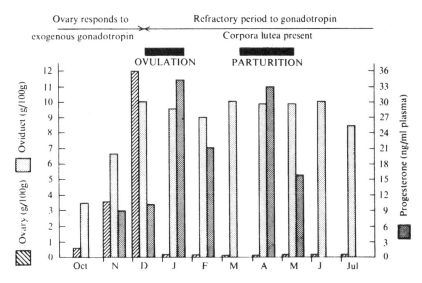

during the reproductive cycle indicates that progesterone levels peak just before ovulation (Jones and Baxter, 1991). It is, therefore, possible that reptiles may utilize a similar mechanism to induce ovulation as that observed in birds.

Following ovulation in *Sceloporus cyanogenys*, the corpus luteum develops and this is correlated with a three-fold increase in the circulating progesterone concentration. This elevated hormone level persists in the plasma until parturition, when it declines. A similar pattern in circulating progesterone levels has also been observed in the viviparous snake *Natrix sipedon* (Chan, Ziegel, and Callard, 1973). Direct evidence that reptilian corpora lutea can produce progesterone has been obtained following *in vitro* incubation of such tissue obtained from the snapping turtle, *Chelydra serpentina* (Klicka and Mahmoud, 1972). In *Sceloporus cyanogenys* the circulating progesterone levels are reduced following hypophysectomy, suggesting that there is some pituitary control over this hormone, but in pregnant lizards relatively high concentrations persist so that if a luteotropic effect is present it is apparently not vital (Callard *et al.*, 1972). In addition, the implantation of pellets of progesterone into the region of the median eminence results in a depression of the circulating progesterone concentration (Callard and Doolittle, 1973). This suggests the presence of a negative-feedback inhibition of the release of a tropic hormone and is accompanied by a decrease in the growth of the ovary and oviduct. The injection of progesterone has also been found to inhibit ovulation in the turtle *Chrysemys picta* (Klicka and Mahmoud, 1977). The hypothalamic control of gonadotropin release can, therefore, be influenced by estrogens and progesterone in a manner suggesting that a mechanism is present that is similar to that in mammals.

Progesterone, as we have seen, plays an important role in maintaining pregnancy in mammals, but there is no conclusive evidence to indicate that this occurs in reptiles. Ovariectomy or hypophysectomy does not affect the course of pregnancy in a variety of viviparous and ovoviviparous species of reptiles (Yaron, 1972). Progesterone, nevertheless, attains high concentrations during a reptilian pregnancy; so what is its function? Callard and Doolittle (1973) have suggested that its action in reptiles may be to inhibit gonadal growth during gestation, and this may represent a "more primitive" role than the regulation of the uterine environment that is seen in mammals. Corpora lutea are also formed following ovulation in oviparous reptiles, and it has been found that when this tissue is removed in gravid *Sceloporus undulatus* (an oviparous lizard) an earlier oviposition occurs, suggesting that progesterone may control the period of egg retention in such reptiles (Roth, Jones, and Gerrard, 1973). A viviparous snake, *Thamnophis elegans*, was also found to deliver its young somewhat earlier than expected following lutectomy (Highfill and Mead, 1975b).

In reptiles, as also in birds, estrogen stimulates the production of lipophosphoproteins by the liver and these are incorporated into the egg (vitellogenesis). Prolactin, when injected, has been shown to exert an antigonadal effect, but the significance of this inhibition is unknown. It is, nevertheless, interesting that these actions of estrogens and prolactin are shared with the birds.

A considerable amount of information is available about reproduction in native reptiles (Licht, 1984). A very varied picture emerges, partly reflecting their diverse phylogeny but also a paucity of controlled experiments.

Amphibians   Amphibians living in temperate zones exhibit a seasonal pattern of reproduction that appears to be mainly induced by climatic events especially temperature, but also rainfall (Lofts, 1984). Photoperiodic responses and endogenous rhythms appear to be rare. Most amphibians are oviparous though there are a few species that have ovoviviparous and even viviparous habits. Pseudoviviparity, when the young are incubated in dorsal cutaneous pouches and the stomach, occurs. An excellent account of the ovarian cycle in amphibians is given by Redshaw (1972) and Jørgenson, Hede, and Larsen (1978). Complete maturation of the oocytes usually takes several years while formation of the yolk, in species that live in temperate zones, commences in the summer preceding spawning (Fig. 9.19). The development of the ova is controlled by the adenohypophysis and the hypothalamus. The GnRH content of the hypothalamus is low in the nonbreeding season of the female African toad *Xenopus laevis*, but it rises in the spring when the ovaries enlarge (King and Millar, 1979b).

Hypophysectomy or transplantation of the pituitary, so that it is no longer in contact with the hypothalamus, interrupts oogenesis. Two distinct gonadotropins, an FSH and an LH, have been identified in amphibians, but their respective roles in ovarian development do not appear to parallel those

Fig. 9.19. The ovarian cycle of toads, *Bufo bufo*, in Denmark. The oocytes are recruited and the ova develop during the preceding summer and remain quiescent during hibernation in the winter. Ovulation occurs the following spring, in April. The ovarian cycle appears to be under the control of environmental factors such as temperature and light and requires adequate nutrition in the summer period. Arrow shown pointing backward from the larger to the smaller oocytes indicates an inhibiting effect of the former on further recruitment of oocytes for the final maturational process. (From Jørgenson *et al.*, 1978.)

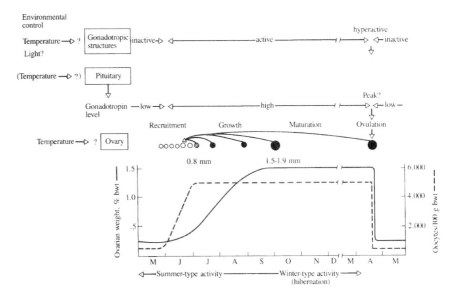

in mammals (Licht, 1979). The LH appears to have a predominant effect on the development of the oocytes, steriodogenesis, and ovulation.

Estrogens are produced by the ovarian follicles and these contribute to the development of other sexual characters, including the oviduct, as well as vitellogenesis. Progesterone may be formed by evanescent "corpora lutea" originating from ruptured follicles in oviparous species and more persistent corpora lutea in viviparous ones (Jones and Baxter, 1991). Progesterone may also be formed at other ovarian sites.

Ovulation can be readily promoted in amphibians by the injection of gonadotropin. This hormone may be obtained from amphibian pituitaries, but exogenous hormones from other species are also effective. The latter hormonal effects are particularly well known as they are the basis for a convenient pregnancy test for women. Human chorionic gonadotropin (hCG) that is secreted in the urine during pregnancy induces ovulation in frogs and toads. Both mammalian LH and amphibian LH are more effective than FSH in promoting the growth of the ovarian follicles and ovulation (Licht, 1979). The proximate cause of ovulation appears to be a surge of progesterone (Lofts, 1984). Ovulation is often, though not always, induced at the time of sexual pairing. *Xenopus* seems to be on the verge of ovulation for prolonged periods of time while in *Rana temporaria* the eggs are stored in the oviduct, from which they are expelled when mating occurs.

The oviduct undergoes a distinct annual cycle in *Bufo bufo* and attains its greatest size in the autumn (Jørgenson and Vijayakumar, 1970). A decline in the weight of the oviduct takes place during spawning in April. This decline results from the loss of secretory contents that coat the eggs with a "jelly." The

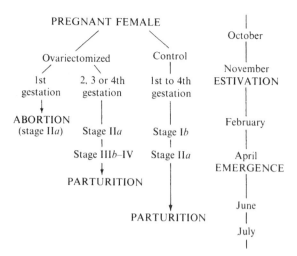

Fig. 9.20. Gestation in the viviparous frog *Nectophrynoides occidentalis* in relation to seasons. Normally these frogs ovulate and become pregnant in October. The dry season commences in November when they estivate. They emerge again, with the onset of rain, in April. The young are born in June. If these frogs are ovariectomized early in pregnancy, they may either abort, if the animals are young (and this is their first pregnancy) or, if they are large and it is the second, third, or fourth time of gestation, the development of the young is accelerated and they are born much earlier than usual. Ovarian progesterone is thought to delay the development of the young during the period of estivation. (From Zuber-Vogeli and Xavier, 1973.)

secretion of this jelly, like that of avidin in the fowl oviduct, is controlled by progesterone.

The pseudoviviparous South American tree frog *Gastrotheca rioambae* carries its young in pouches on its back. The development of these vascularized egg chambers is promoted by the synergistic actions of estrogens and progesterone (Jones and Baxter, 1991). Some very interesting observations have been made on the effects of progesterone on gestation in a viviparous frog *Nectophrynoides occidentalis* (Xavier and Ozon, 1971; Zuber-Vogeli and Xavier, 1973; Xavier, 1974). This frog lives in West Africa where it is subjected to periods of seasonal drought during which it estivates in burrows. Following ovulation in October the fertilized eggs are retained in the oviduct where development proceeds until parturition the following June. In November, these pregnant frogs estivate and do not emerge until April (Fig. 9.20). Corpora lutea are formed following ovulation that apparently secrete progesterone. When the ovaries from pregnant frogs are incubated *in vitro* with the progesterone substrate pregnenolone, they convert this steroid to progesterone. This ability to form progesterone declines as gestation progresses (Fig. 9.21). Following parturition, pregnenolone is converted to other steroids by the ovarian tissue and this process increases until ovulation again takes place. The preovulatory period is also the time when production of estrogens is thought to increase. If ovariectomy is performed early in gestation (see Fig. 9.20) of young frogs, during their first pregnancy, abortion occurs. In more mature frogs, however, development of the young is accelerated following ovariectomy and parturition takes place about 3 months earlier than usual. The implantation of progesterone into these frogs toward the end of gestation reduces the rate of growth of the embryos. Progesterone, therefore, appears to slow the rate of development of the embyros during the prolonged period of estivation so that the young are delivered at a more

Fig. 9.21. The ability of the ovarian tissue in the viviparous frog *Nectophrynoides occidentalis* to metabolize pregnenolone (*in vitro*) at different stages of its ovarian cycle. During gestation the principal steroid produced from the pregnenolone is progesterone but after parturition 17-hydroxyprogesterone, androstenedione, and testosterone are also formed. (From Xavier and Ozon, 1971.)

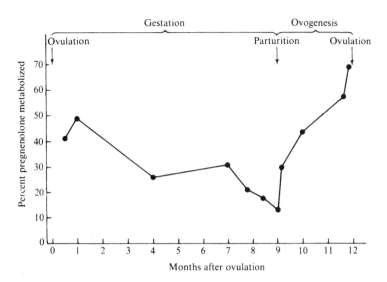

appropriate and favorable time of the year. This is indeed a novel role for progesterone.

Fishes The reproductive cycles of fish are varied, probably reflecting their phylogenetic diversity and wide geographical distribution. They frequently utilize environmental temperature as a cue, but also photoperiod, the availability of food, the proximity of a mate, and, in tropical species, even lunar cycles (Dodd and Sumpter, 1984; Wingfield and Kenagy, 1991). There may also be an underlying endogenous circannual rhythm upon which the factors just described may impinge. A role for the pineal gland in transducing photoperiodic stimuli probably varies among species. In some, such as the Atlantic salmon, a clear circadian rhythm in the formation of melatonin has been observed (Randall *et al.*, 1995). However, the possible relationship of such hormonal changes to the timing of reproduction has not been defined.

Most fishes are oviparous but some species have evolved ovoviviparous and viviparous methods of reproduction. In teleosts, the last two processes are usually rather different from the *in utero* development common to many other vertebrates. The young teleosts may develop *in situ* in the follicle or be incubated in the hollow central cavity of the ovary. Many elasmobranch fishes are ovoviviparous or viviparous and have long gestation periods during which time the young derive nutrients from the uterine wall. This process has distinct similarities to the processes maintaining embryos *in utero* in mammals.

In teleosts the development of the oocytes and vitellogenesis are dependent on the pituitary (Reinbloth, 1972). A single gonadotropin may be present in some species but two such hormones, GTH I and GTH II (Chapter 3), have

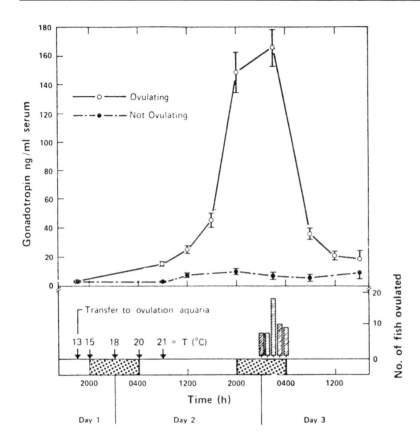

Fig. 9. 22. Changes in the serum gonadotropin concentrations at the time of ovulation in goldfish. The latter event was promoted by exposing the fish to a long day (16L/8D) and moving them to a warmer aquarium, at 20 °C. It can be seen that ovulation occurred in the dark early morning hours of the third day and in these fish (○ - ○) it was preceded by a large rise in serum gonadotropin concentration. No change in the level of this hormone was observed in the fish that did not ovulate (●··●). Each point represents a mean ± S.E. (From Stacey, Cook, and Peter, 1979.)

now been identified in an increasing number of teleosts. The injection of homologous gonadotropins can promote ovulation in teleosts. The roles of estrogens and progesterone in this process are, however, unknown. The endogenous release of gonadotropin in teleosts is under the influence of the hypothalamus, stimulation by GnRH and possibly, inhibition by dopamine occurs. Gonadotropin levels in the serum of the goldfish have been shown to rise at the time of ovulation (Fig. 9.22). This process can be promoted by transferring the fish on a photoperiod of 16 hours of light and 8 hours of dark from an aquarium at 13 °C to one at 20 °C. The gonadotropin level started to rise in the daylight period and reached a peak at the beginning of the dark period, during which ovulation occurred. Plasma cortisol concentrations also increased, succeeding the rise in gonadotropin (Cook, Stacey, and Peter, 1980). Such corticosteroids may be involved in the maturation process of the oocyte in some species, and possibly even in ovulation (Dodd and Sumpter, 1984).

The early development of the oocytes appears to be independent of hormones. Later development involves an accumulation of vitellogenins under the influence of GTH I. Meiosis can then continue but requires the action of the maturational gonadotropin GTH II, which promotes the synthesis of intermediary maturational hormones (maturation-inducing hormones, MIH) by the oocyte thecal and granulosa cells, and possibly the interrenals. Two such ovarian maturational steroids have been identified in different species: 17,20β-dihydroxy-4-pregnen-3-one (17,20β-P) and 17α20β,21- trihydoxy-4-pregnen-3-one (20βS) (Nagahama *et al.*, 1994; Thomas, 1994). These steroids appear to interact with receptors on the plasma membrane to induce the formation of a cytoplasmic maturation-promoting factor (MPF). This mechanism of action of a steroid hormone, involving a plasma membrane receptor, appears to be unique. Such a role for steroid hormones in maturation of the oocytes may be confined to teleosts, and possibly other fish, and amphibians. In mammals, a similar process of maturation occurs but it is directly promoted by LH (Tsafriri *et al.*, 1994).

In cyclostome and chondrichthyean fish, the pituitary gland appears to contribute to the control of the development of the ova and sperm (Dodd and Sumpter, 1984). However, the cyclostomes lack a hypophysial portal system, and in chondrichthyeans the ventral lobe of the pituitary, which is the site of synthesis of a GTH, has no direct vascular connection to the remainder of the gland. In lampreys, ovulation was found to occur as long as 2 weeks after hypophysectomy. Transplantation of the lamprey pituitary to another site in the body has no effect on ovarian development (Larsen, 1973; 1978). It is rather a mystery how cyclical control of reproduction is regulated in such fish. A GnRH has been identified in both groups of fish (Chapter 3) and it has been suggested that it may travel to the responsive gonadotrope cells by local diffusion in lampreys, and in the systemic circulation of both types of fish. It is also possible that the gonads exhibit a measure of internal, autonomous control over their reproductive activities (Fasano, Pierantoni, and Chieffi, 1989).

The role of corpora lutea in fish is contentious. Many teleosts possess so-called preovulatory corpora lutea, which are formed as a result of atresia of unovulated follicles and are more aptly called corpora atretica. Postovulatory corpora lutea, as well as corpora atretica, are present in many chondrichthyeans and there is considerable speculation as to whether they contribute to successful gestation in ovoviviparous and viviparous species (see Chieffi, 1967; Dodd, 1972a; Jones and Baxter, 1991). Hypophysectomy does not interrupt pregnancy, at least for the first 3 months, in the viviparous shark *Mustelus canis*, suggesting that an adenohypophysial control of progesterone secretion is not vital. It is not known whether progesterone contributes to gestation in ovoviviparous and viviparous fishes. Hisaw, in 1959, stated that "the elimination of yolk during follicular atresia and material from ruptured follicles at ovulation is a primitive function of corpora lutea and endocrine functions such as luteinization of the granulosa by pituitary luteinizing hormone and secretion of progesterone in response to pituitary luteotropic hormone as seen in mammals, are more recent adaptations."This interesting idea has, however, not been unquestionably accepted.

In many teleost and chondrichthyean fishes, the thyroid gland displays an increased activity during the breeding season. This season is associated with many physiological and environmental changes so that it is difficult to be certain whether the endocrine events are primarily related to reproduction. Sage (1973) considers that it is likely that the thyroid is involved in the reproductive process in fishes as it is necessary for gonadal maturation in some species. Consistent with such a suggestion is the observation that thyroidectomy in female dogfish results in their failure to develop mature oocytes compared with control fish (Dodd and Sumpter, 1984). It has also been found that the administration of T_4 to immature goldfish promotes the premature maturation of the oocytes (Hurlburt, 1977). One favored interpretation was that T_4 was acting synergistically with a gonadotropin to promote a response. A role for thyroid hormones in reproduction in fish is, however, still unresolved (Flett, Leatherland, and van der Kraak, 1994).

Oviposition and parturition in nonmammals–a role for the neurohypophysial hormones? The mammalian uterus is highly responsive to the contractile effects of oxytocin. This neurohypophysial hormone has been used to facilitate labor in humans and domestic mammals and it is involved physiologically in this process. The oviducts of birds, reptiles, amphibians, and fishes have been observed to contract when exposed to such neurohypophysial peptides (Heller, 1972). In nonmammals, arginine-vasotocin is usually the most active peptide though in elasmobranchs other oxytocin-like peptides may be involved (LaPointe, 1977). Measurements of the plasma concentrations of vasotocin at the time of oviposition have been made in birds (Koike, Shimada, and Cornett, 1988) and reptiles (Figler et al., 1989; Ferguson and Bradshaw, 1991; Guillette et al., 1991) and suggest that this peptide may be contributing to this event. Other factors could be involved, such as prostaglandins, relaxin, urotensin II in teleosts (Lederis, 1973), and galanin in birds (Li et al., 1996). The neurohypophysial peptides have assumed various roles during the evolution of the vertebrates, including regulation of water metabolism, milk let-down, and hypophysiotropic effects. They also have a special propensity to contract smooth muscle, and the effects on the contractility of the reproductive tract is a widespread and possibly useful manifestation of this ability.

Hormones and the evolution of viviparity: a summary

Viviparity, or giving birth to live young, has undoubtedly evolved separately in many different vertebrates (Sharman, 1976; Amoroso, Heap, and Renfree, 1979). Most people consider that this process has reached its highest level of physiological organization in eutherian mammals, but, apart from marsupials, viviparity is also found among the reptiles, amphibians, and fishes. In the last it is especially common in the Chondrichthyes, and it displays some interesting diversity in teleosts. The young in this group may, for instance, be retained not only in the oviducts but also in cavities in the ovaries and even in the ovarian follicles. In amphibians, they may be retained in enlarged vocal sacs of male frogs and special pouches in the skin. It is questionable, however, whether such refuges constitute true viviparity, though they do appear to conform to the definition in the *Oxford English Dictionary*. Suitable contact between the maternal and embryonic tissues can be made at a number of sites in the body; abnormal pregnancies in women have even occurred that involve the development of the fetus in the peritoneal cavity. The separate evolution of viviparity on so many occasions suggests that it can be advantageous to the survival of a species. A reduction in the number of eggs produced and the quantity of yolk proteins clearly may confer some saving in energy. The chances of survival of a particular young would appear to be enhanced though this is largely a "numbers game." Lots of eggs also favor the ultimate survival of more individuals. Speed of development may be increased through an assured and optimal supply of nutrients for a longer period of time, and the maintenance of a suitable temperature. In homeotherms, the role of the latter may be obligatory for survival, but even poikilotherms can, to some extent, regulate their body temperature, usually as a result of their behavior. Birds resort to incubating their eggs; however, this method of maintaining temperature usually restricts their movements.

Many contemporary elasmobranch fishes are viviparous, including about 55% of the sharks and rays (Wourms and Lombardi, 1992). In a remarkable example of an evolution that is convergent to that of the mammals, the elasmobranchs have developed endocrine strategies that utilize estrogens and progesterone to regulate the functioning of their reproductive tract and the development of viviparity (Callard and Koob, 1993). This path of evolution commenced about 400 million years ago, before the emergence of the mammals. They could, therefore, be considered to be an archetype for mammalian viviparity (Callard *et al.*, 1989). The viviparous elasmobranchs represent the first vertebrates to display a persistent corpus luteum during pregnancy. It appears to have an endocrine role and be able to synthesize progesterone and relaxin. A role for this corpus luteum in maintaining pregnancy in elasmobranchs is considered likely but it is still contentious (Jones and Baxter, 1991). The favored putative roles for the corpus luteum in elasmobranch viviparity are an ability of its hormones to limit uterine contractions, to promote its nutrient secretory activities, and to antagonize inappropriate vitellogenic effects of estrogens (Callard *et al.*, 1992).

Viviparity often involves a relatively long period of commitment on the part of the female parent, and also sometimes the male, to succor the young.

The period of pregnancy varies in different species; usually when it is prolonged the young are born in a better developed and more independent condition. This effect is especially apparent when one compares the newborn of marsupials with those of eutherians. Postnatal care of the young is not unique to viviparous species but it is a common accompaniment of this condition. It may involve the evolution of special arrangements for postnatal feeding, such as lactation by the mammary glands in mammals.

The morphological nature and degree of closeness of association of the embryo and parent vary in different species. In some, the egg may have stores of yolk for its development and it appears to be merely incubated inside the parent. However, the young even in these circumstances may still obtain nutrients through the secretory activities of the uterus. Such examples of ovoviviparity occur widely in nonmammals and in the monotremes. The sustenance of the young with true placental attachments, *in utero*, however, usually involves supplying the complete nutrients and providing a "sink" for the excretion of some products of metabolism. Therefore, blood systems show an association that allows exchanges to occur between mother and embryo.

Prolonged periods of gestation and care of the young require a number of physiological adjustments by the parent and even the offspring. Many of these require the intervention of hormones (it has been said that the embryo develops in "a sea of hormones").

1. The preparation and maintenance of the epithelium lining the oviduct and uterus for the reception and sustenance of the embryo are events that initially require estrogens, and then, mainly, progesterone.

2. Control of the motility of the muscle that envelops the gonoducts is hormonal; there is a relative immobility during gestation, but an increase in contractility at the time of birth. Progesterone reduces such activity while estrogens enhance it. At birth the latter steroids may enhance the formation of prostaglandins, which contract the uterus. Neurohypophysial hormones may also contribute to the initiation of contractions. A relaxation of, or softening of, the connective tissue of the cervix occurs under the influence of relaxin and prostaglandins.

3. Care of the young following their birth involves hormonal regulation of parental behavior. The development of the mammary glands and the initiation and maintenance of lactation take place under the influence of estrogens, progesterone, placental lactogen, prolactin, and oxytocin.

4. Concurrent reproductive activity is limited during gestation and lactation, this often involves the hormonal suppression of further reproductive cycles and ovulation or, in some species, a delay in the implantation of the blastocyst. These effects involve actions of progesterone and prolactin.

Viviparity has been accompanied by the evolution of several important endocrine mechanisms.

1. *Internal fertilization* is essential for viviparity to occur and involves the evolution of suitable morphological structures to aid the intromission of the sperm directly into the female genital tract. Such penile structures are usually secondary sex characters, which come under the control of androgens.

2. Changes in the *morphology of the female reproductive tract* may be necessary. For instance, the "plugging" role of the cervix is important in maintaining pregnancy and facilitating internal fertilization. Its activities are controlled by estrogens, progesterone, and relaxin.

3. The life of the *corpus luteum*, which is an important source of progesterone, can be prolonged during pregnancy in eutherians. In marsupials it stays in a quiescent condition during lactation. It is suspected that it may also have a role to play in pregnancy in nonmammals, especially reptiles and elasmobranchs.

4. The *placenta* may evolve endocrine mechanisms for the production of gonadotropins, which aid in the prolongation of the life of the corpus luteum, and progesterone, which supplements or even supplants its role. Estrogens are also produced that contribute to the process of birth. The endocrine association of the placenta and fetus (the fetoplacental unit) supplements the activities of the corpus luteum and facilitates the timing of parturition.

5. A prolonged contact with the maternal circulation may, potentially, be expected to influence the processes of *embryonic development*. The maternal plasma contains high concentrations of hormones, including sex steroids, which may affect the sexual differentiation of the reproductive apparatus and behavior. The mammalian fetus can, however, protect itself by forming a protein (α-fetoprotein) that binds, and so blocks, the actions of estrogens.

The control of metamorphosis in fish and amphibians

Many fishes and amphibians exist in two or more distinct morphological and physiological forms during their life cycles. The transformation from one type to another is called metamorphosis. This process can be defined as an irreversible change in the structure and physiology of an animal that is associated with an altered manner of life. The latter may involve movement to a different habitat, such as water to land, fresh water to the sea (or vice versa), as well as a drastically altered feeding pattern. Such transformations may be initiated by changes in environmental conditions such as light and temperature. It may occur at various stages of an animal's life cycle but the present discussion is confined to a so-called "first metamorphosis" that involves the development from a postembryonic larval form into a juvenile adult. "Second" and even "third" metamorphoses also occur, such as the transformation of juvenile salmon living in fresh water to a form that migrates to the sea (parr–smolt transformation, see Chapter 8).

The morphological and physiological changes that occur during metamorphosis are very complex and involve extensive reprograming of gene expression from a larval to an adult pattern (see, for instance, Kawahara, Baker, and Tata, 1991; Atkinson, Helbing, and Chen, 1994). It can involve both the activation and the repression of gene activity. Such genetic changes are triggered by specific nuclear transcription factors and interact with response elements on the promoter DNA of target organs. These transcriptional processes can be initiated by growth factors and hormones that combine with, and activate, the transcription factors (Chapter 4). A transcriptional response to growth factors or a hormone can occur within minutes. An activation of "early genes" may result in a downstream activation of further transcription factors ("third messengers") for "late" genes that are part of the overall reprograming process. A cascade of genetic responses can thus be initiated that may involve multitudes of genes. Amphibian metamorphosis is usually considered to be a multihormonal process in which thyroid hormone receptors play an especially important role. Depending on the tissue, activated thyroid hormone receptors may interact with numerous genes and so aid in the direction of normal development.

The mechanism of the metamorphosis of the ammocoete larvae of lampreys into their adult form has excited special interest, especially as initially it did not appear to involve thyroid hormones (Barrington, 1968; Youson, 1994). However, it was in the ammocoetes of *Petromyzon marinus* that the maturation of the endostyle into thyroid follicles was first observed (see Specker, 1988). Ammocoetes have a prolonged period of development taking 4 to 7 or more years as filter-feeding stream dwellers. Following their metamorphosis, some species migrate to large lakes or the sea where they parasitize other fish. The mature brook lampreys, however, subsist as free living individuals and stay closer to their original home. Measurements of the concentrations of thyroid hormones in the sea lamprey *Petromyzon marinus* indicated that both T_3 and T_4 are present in the ammocoetes and that they reach a peak concentration just before metamorphosis occurs (Youson, Plisetskaya, and Leatherland, 1994). Metamorphic climax is initiated in individuals that have attained an appropriate size and store of fat, and when the water temperature rises to a critical level. The thyroid hormone concentrations then decline drastically. Exposure of the ammocoetes to an anti-thyroid agent, potassium perchlorate, can promote metamorphosis (Holmes and Youson, 1993). It has, therefore, been suggested that metamorphosis in lampreys is initiated in suitably primed individuals that have attained high plasma levels of thyroid hormone, by the release of an anti-thyroid factor or hormone (Youson, 1994). The identity of such a factor is unknown but it is notable that in the Australian lamprey *Geotria australis* removal of the rostral region of the pituitary gland blocks metamorphosis (Joss, 1985). The pineal gland and melatonin could also be involved as removal of this gland has been shown to prevent metamorphosis in several species of lampreys (Joss, 1985).

Flounder (family Pleuronectidae) undergo a metamorphosis from a pelagic plankton-eating larva to an asymmetrical flat-fish that is a bottom-dwelling carnivore. The assumption of its asymmetrical appearance includes a migration of the right eye to the left side and a reduction of the dorsal fin rays. This metamorphosis has been extensively studied in the Japanese flounder *Paralichthys olivaceus* (Inui *et al.*, 1994). Whole body thyroid hormone (T_4 and T_3) levels were measured in these small larvae. A sudden surge of both hormones was observed at the onset of metamorphic climax and this peak was followed by a decline (Miwa *et al.*, 1988). Exogenous T_4 and T_3 both promoted metamorphosis, the latter being the more active, though the concentration of the former was much higher in the larvae under natural conditions. An anti-thyroid drug, thiourea, retarded metamorphosis while the administration of thyrotropin (TSH) promoted it. Histological observations suggested that there was an enhanced secretion of TSH at the time of metamorphic climax. There was also an increased concentration of tissue 5'- and 5-mono-deiodinase. It was, therefore, suggested that metamorphosis in the flounder is initiated by a maturation of the pituitary–thyroid axis, as well as by increased activity of enzymes that metabolize the thyroid hormones and convert T_4 to T_3. The effects of thyroid hormones on dorsal fin ray resorption was studied *in vitro*: their actions were increased by cortisol and antagonized by prolactin. The concentrations of growth hormone and prolactin in the pituitary gland increased during metamorphosis with a sudden rise at the climax. Metamorphosis of the leptocephali larvae of the conger eel, *Conger myriaster*, is associated with a similar pattern of thyroid hormone concentrations to those seen in the flounder (Yamano *et al.*, 1991). The roles of the endocrine glands in the metamorphosis of teleosts show many similarities to those observed in amphibian tadpoles (see below).

The possibility that changes in the endocrine glands, especially the thyroid, may initiate metamorphosis in fishes arose from the observation that thyroid hormones can initiate metamorphosis in many amphibians. The profound and dramatic morphological changes that accompany the metamorphosis of anuran tadpoles into adult frogs and toads have been a source of wonder to biologists for a long time. The transformation from a purely aquatic animal, with no limbs or lungs, into a terrestrial beast that breathes and hops about on four legs is also accompanied by many physiological and biochemical changes. The larval life and metamorphosis of amphibians may be relatively short, from several weeks in desert-dwelling species where water is available for only a short time to as long as 3 years in bullfrogs. The factors that determine the time of metamorphosis are not clear; they are partly genetic but they can also be modified by the environment. Bullfrog tadpoles from the tropical southern parts of the USA may metamorphose before the beginning of the first winter after hatching, while those in northern areas may endure three winters before this change occurs. Possible environmental factors that influence the time of metamorphosis include nutrition, temperature, the salinity and acidity of the water where they live, and the relative proximity of other tadpoles ("crowding"): some experiments even suggest that light may stimulate this process (Dent, 1968). One can foresee that such factors could exert their effects through the activation of endocrine glands, in this instance the thyroid through its hypothalamic and pituitary control mechanisms.

The feeding of thyroid gland extracts can produce metamorphosis in tadpoles far earlier than it would normally occur. Conversely, the administration of anti-thyroid drugs prolongs, or even prevents, metamorphosis. Natural metamorphosis in tadpoles is accompanied by a sudden increase in thyroid gland activity, as indicated by histological changes and an increase in the rate of uptake of radioactive iodine. There can be no doubt that the activity of the thyroid gland determines metamorphosis in tadpoles but this is only a part of the endocrine story. The thyroid in tadpoles, like that in other vertebrates, is under the control of TSH from the adenohypophysis. The injection of TSH into tadpoles also results in a premature metamorphosis. Hypophysectomized tadpoles do not metamorphose but grow larger and larger and attain "giant" proportions. The hypothalamus and median eminence, which are usually the next sites in the chain of thyroid control, are also involved in metamorphosis for when the tadpole pituitary is transplanted in the tail, metamorphosis is prevented and they grow more rapidly than usual.

The effect on growth is an important clue that may indicate (at least partly) a loss of the hypothalamic inhibition of the release of prolactin in this situation. Injections of prolactin into bullfrog tadpoles have been shown to antagonize tail resorption induced by T_4 and delay metamorphosis (Nicoll *et al.*, 1965; Etkin and Gona, 1967). The same effect has been achieved by grafts or extracts (Kikuyama, Yamamoto, and Mayumi, 1980) of tadpole pituitaries, so this effect probably also exists physiologically. Antisera prepared to amphibian prolactin can accelerate metamorphosis (Clemons and Nicoll, 1977). It would appear that prolactin opposes tadpole metamorphosis. Analogies between the actions of thyroid hormones and prolactin, respectively, with those of ecdysone and juvenile hormone in metamorphosis in insects were noted in the 1970s by H. A. Bern and W. Etkin (see Bern, 1992).

Etkin (1970), after a careful assessment, has provided a description of how metamorphosis is normally regulated in tadpoles. This is summarized in Fig. 9.23 and is based on the observations already described, which have been correlated with histological changes that occur in the hypothalamus, pitu-

Fig. 9.23. Role of hormones in growth and development of the tadpole. *Top*: development of the pituitary, especially in relation to establishing its connections to the hypothalamus. *Bottom*: changes in activity and hormone concentrations of the pituitary–prolactin and pituitary–thyroid axes. The roles and functioning of the pituitary–prolactin and pituitary–thyroid axes have not been clearly established (see text). Major events are acquisition by the hypothalamus of T_4 sensitivity, hastening its development, and onset of definitive roles of TRH and P-R-IH, so that TSH and T_4 increases while prolactin decreases. (Based on Etkin, 1970.)

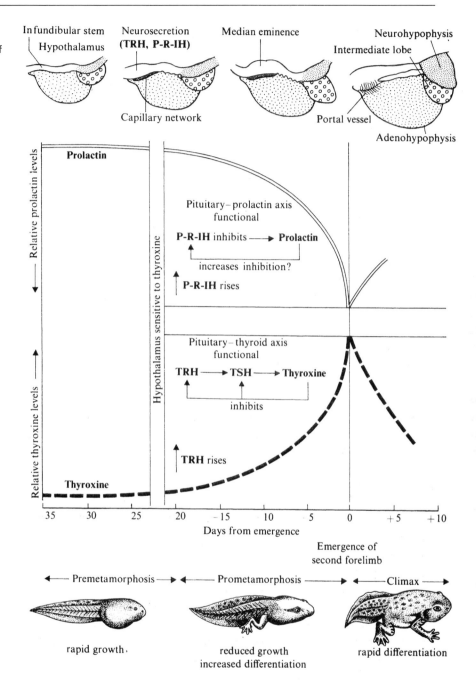

itary, and thyroid gland. Metamorphosis is divided into three stages: (*a*) a period of rapid growth called premetamorphosis; (*b*) a time of reduced growth but increased differentiation called prometamorphosis; and (*c*) metamorphic climax when there are "explosive changes", the tail is resorbed, and the frog emerges and takes up a terrestrial existence. Premetamorphosis is characterized endocrinologically as a time when thyroid hormone secretion is low, which reflects the anatomical immaturity of the hypothalamic–pituitary axis and a low rate of TSH secretion. This condition is stabilized further by the presence of large amounts of prolactin, possibly reflecting an immaturity of the inhibitory influence of the hypothalamus (P-R-IH), which further inhibits any progress toward metamorphic change. With the progressive maturation of the hypothalamic–pituitary axis toward the beginning of prometamorphosis, TRH possibly increases and releases TSH. Antisera to TSH slows metamorphosis (Eddy and Lipner, 1976). The T_4 concentration rises progressively during prometamorphosis. The hypothalamus continues to mature under the influence of T_4 (called a "positive feedback"), and when this is complete and the portal blood supply to the adenohypophysis is finally established, there is a massive stimulation of the thyroid, involving TSH, and metamorphic climax ensues. During this last period, the levels of T_4 are thought to be declining (owing to the operation of the adult negative-feedback inhibition of TSH release).

Many more details have emerged regarding tadpole metamorphosis since Etkin's prognostic summary (Fig. 9.23). The predicted increases in the levels of the plasma thyroid hormones (T_3 and T_4) have been observed in bullfrog tadpoles (Regard, Taurog, and Nakashima, 1978). This increase, which commences late in prometamorphosis, not only reflects a release of the hormones but also changes in their peripheral metabolism. Hepatic 5-monodeiodinase (Chapter 4) levels are initially high in late prometamorphosis but then progressively decline. The 5'-monodeiodinase in the skin and gut rises in a complementary manner (see Galton, 1988; 1992; Galton, Davey, and Schneider, 1994). 5-Monodeiodinase converts T_4 to reverse T_3, and T_3 diiodothyronine, which are both inactive. The 5'-monodeiodinase converts T_4 to T_3, which is more active, and this process increases as metamorphic climax approaches. The activity of 5-monodeiodinase limits the effects of T_4 and T_3 at a time when their actions would be inappropriate. The synthesis of TSH, as indicated by the presence of its mRNA, also rises at metamorphic climax. The levels of thyroid hormones subsequently decline, apparently reflecting the maturation of the negative-feedback loop of the pituitary (TSH)–thyroid gland axis (Buckbinder and Brown, 1993). TRH concentration also rises prior to metamorphosis (King and Millar, 1981), further completing the armamentarium for the control of thyroid hormone synthesis and release. However, the precise role of TRH in metamorphosis in uncertain (Dickoff, 1993).

Thyroid hormone receptors (TR) appear to arise early in the development of tadpoles. In *Xenopus laevis* their synthesis is promoted by thyroid hormone in a process of autoinduction (up-regulation), probably triggered initially by an interaction with maternal TR mRNA, which is present in the oocyte

(Kawahara *et al.*, 1991). Two isoforms of the receptors are formed: TRα and TRβ. Early in the development of bullfrog tadpoles TRα predominates but at metamorphic climax TRβ is more abundant in some tissues (Kawahara *et al.*, 1991; Atkinson *et al.*, 1994). The significance of this change in unknown.

Confirmation of the role of prolactin as a juvenile hormone has involved some controversies. An unexpected surge of prolactin has been observed late in metamorphic climax (Dent, 1988). The mRNA for this hormone also rises in the pituitary gland at this time (Takahashi *et al.*, 1990; Buckbinder and Brown, 1993). It is possible, however (Dent, 1988), that this surge occurs at a stage of development after the final programing of the adult gene expression has already been determined. It may, therefore, reflect the attainment of the adult pattern of prolactin synthesis.

Corticosterone concentrations rise in the plasma during prometamorphosis in bullfrog tadpoles and then decline in the late climax period (Dickoff, 1993). Such steroids appear to enhance the actions of thyroid hormones, possibly by stimulating peripheral conversion of T_4 to T_3 and also by reducing their inactivation (Galton, 1990).

Among amphibians, the phenomenon of metamorphosis from a larval to adult form is not confined to anurans but also occurs in urodeles. The latter group include species with permanent larval forms that are neotenous and can breed. Axolotls, *Ambystoma mexicanum*, are such urodeles. It was an event of some note among biologists when it was found that these larval forms could be induced to metamorphose into conventional adult-type salamanders by feeding them thyroid gland tissue. The same effect could be elicited by TSH, indicating that the natural disability resides in the pituitary and not the thyroid gland.

Studies of metamorphosis, especially in amphibians, have provided a most important model for understanding the process of development and its control by factors such as proto-oncogenes and hormones.

Sex determination and differentiation

The completion of reproduction involves the development and maturation of the young and the expression of its ability to continue the process of the proliferation of the species. In vertebrates, reproduction usually involves different sexes, though parthenogenesis does occur. The determination of the sex in the individual involves the early embryonic attainment of an ovary or a testis and the subsequent matching differentiation of the urogenital ducts, external genitalia, and diverse secondary sexual characters. In mammals, the predicted sex is displayed in the fertilized ovum by the disposition of its sex chromosomes: heterogametic (XY) and it becomes a male, homogametic (XX) and a female results. This chromosomal pattern is, however, not the same in all vertebrates. In birds, the female is heterogametic (ZW) and the male homogametic (ZZ). Other groups of vertebrates may display either pattern or in some there are no distinct sex chromosomes. The process of development as a male or female is ultimately under the control of genes that may be expressed in a cascading manner, the products of which initiate and control various morphogenetic and physiological processes. The trigger that initiates a male or female pattern is thought to reside in a single gene that acts as a "master switch" and is situated on one of the sex chromosomes. In mammals, this process directs the differentiation of an initially indeterminate, ambisexual gonad into a testis. The ovary appears to develop without such a positive stimulus so that the female is called the "default" sex and the male the "inducible" one. The crucial male-determining gene, which produces a testis-determining factor (TDF), is present on the Y chromosome and has been the subject of an intensive search. Former candidates were the HYA gene (*his*tocompatability *Y a*ntigen) and the ZFY gene (*z*inc *f*inger protein *Y*) neither of which were subsequently found to be invariably associated with the determination of the male sex in mammals. The current favored candidate is the SRY gene (*s*ex-determining *r*egion *Y*) (see Graves, 1994). Products of the SRY gene have been found to be consistent with activation or repression of "downstream" genes that are involved in the synthesis of hormones involved in sexual differentiation (Haqq *et al.*, 1993). The SRY (or *sry* in nonhumans) gene has been identified in humans, mice, rabbits, and two species of marsupials but it has, apparently not been successfully sought in nonmammals. When sex is determined by such a gene the process is known as *genotype sex determination* (GSD).

Studies on the sex ratios in hatchling reptiles have shown that in oviparous species sex determination may be influenced by the incubation temperature of the eggs. The first such observation appears to have been that of M. Charnier on an African lizard, *Agama agama* (Charnier, 1966). Subsequently, the effect was observed in a variety of other reptiles, including crododilians and some chelonians and lacertilians (Bull, 1980; Pieau *et al.*, 1994). It also appears to occur in amphibians and fishes. This phenomenon apparently does not occur in viviparous vertebrates. It may take place in species with or without sex chromosomes. Incubation of the eggs over a quite narrow range of temperature may result in the hatchings being all male, all female or a mixture of the sexes. In some species, the higher temperature may favor all males and the lower one all females with a gradation of distribution in between; in others, the high temperature may favor all females and the lower one all males. In the alligator *Alligator mississippiensis* a nest temperature of 30 °C or below results in all females, while at 33 °C or above all males result, with a gradated mixture of the sexes between 30° and 33 °C. Such a process of sex determination is called *temperature-dependent sex determination* (TSD). Its mechanism appears to involve a temperature-dependent gene expression (Crews, 1994; Pieau *et al.*, 1994).

A role of hormones in sexual differentiation has been suspected for at least 80 years. When cows bear twin fetuses of opposite sex, the bull calf usually develops normally but the female suffers a disordered sexual development and is sterile. It is then called a "freemartin". F. R. Lillie in 1917 described this condition, which was already well known to farmers, and suggested that as a result of anastomoses of their blood vessels the development of the cow calf was influenced by hormones passed to it from the bull calf. In mammals, it is now considered likely that at least one such hormone is antimullerian hormone (see below). Attempts to mimic the situation of the freemartin experimentally were performed by joining the blood vessels of pairs of larval salamanders (parabiosis) of the opposite sex. The results indicated that a sex reversal involving the gonads may be achieved in these amphibians (Burns, 1935). However, direct administration of sex steroid hormones to fetal mammals failed to produce such a sex reversal with the possible exception of a marsupial, the Virginia opossum (Burns, 1961). Such a sex reversal of the gonads has, however, been observed in nonmammals, including amphibians (Burns, 1938; 1939; Foote, 1938; Chang *et al.*, 1996) and teleost fish (Yamamoto, 1953; 1958).

Alfred Jost (see Jost, 1971) showed that when rabbit fetuses were gonadec-tomized at a stage when the gonads were still in an "indeterminant" sexual condition, the subsequent differentiation of the urogenital ducts all occurred in the female pattern. He identified two controlling hormones in the newly differentiated testis. Androgens from the Leydig cells, which promoted the growth of the Wolffian ducts (mesonephric ducts) to form the male ejacula-tory system (epididymis, vas deferens, etc.),and an antimullerian substance that promoted the regression of the Müllerian ducts (paramesonephric duct). In the female, the latter ducts give rise to the oviducts, uterus, and vagina: in fetal eutherian mammals, this process does not appear to involve hormones. The female is described as being the "neutral form." The antimullerian substance is now called antimullerian hormone (AMH, see Chapter 3). The initial transformation of the gonads of eutherian mammals into an ovary or testis does not appear to involve hormones (This view is, however, currently being questioned, see George and Wilson 1994). Jost's important observa-tions led to the paradigm: chromosomes determine gonadal sex and gonadal sex determines phenotypic sex.

Numerous apparent exceptions to this "rule" are seen in noneutherians, including fish, amphibians, reptiles, birds, and even marsupials (Wachtel, Wachtel, and Nakamura, 1991; Crews, 1994). Gonadal sex is influenced by incubation temperature and by the experimental manipulation of hormones in some species. Differentiation of the phenotypic sexual apparatus has not been thoroughly investigated in lower vertebrates but numerous observations suggest that male and female sex steroids may have quite diverse effects (Burns, 1961). The responses to such steroids can be influenced by such factors as the species, its stage of embryonic development, the particular hormone preparation used, and the dose given. In birds (the domestic fowl and ducks) gonadectomy of embryos (using X-rays) during their sexually indeterminant stage indicates that the normal regression of the Müllerian ducts in the male results (as in mammals) from the actions of hormones secreted by the testes (Wolff and Wolff, 1951). (In the female the right Müllerian duct also normally regresses and this process is under the control of the ovaries). However, in contrast to mammals, the Wolffian ducts persist following gonadectomy, and the external genitalia, in both sexes, differentiate in the male form (in this instance the male is the "neutral form"). In an amphibian, the newt *Triturus*, the ovaries and testes each participate in the appropriate differentiation of the gonoducts (Burns, 1961). Hence the precise roles of hormones in sexual differentiation and determination in mammals is not prototypical of all vertebrates.

The investigation of embryonic development has been facilitated by numerous modern research techniques, including those for identifying genes and their products. Inhibitors of specific steps in steroid hormone synthesis are also available. Such drugs include those that block the synthesis of estrogens by inhibiting the aromatase enzyme P450 aromatase (Chapter 4). Another group of such inhibitors are those that block the 5α-reductase that converts testosterone to its more active nonaromatizable metabolite 5α-dihydrotestosterone (DHT).

Genetically female chickens given an aromatase inhibitor while in the egg can be induced to produce testes that can even form sperm (Elbrecht and Smith, 1992). A similar sex-reversal effect has also been promoted in the young of the parthenogenetic whiptail lizards, *Cnemidopherus uniparens* (which normally only produces female young), and the red-eared turtle *Trachemys scripta*, which exhibits TSD (Wibbels and Crews, 1994; Wennstrom and Crews, 1995). Direct measurements of aromatase levels in the embryonic gonads of the European pond turtle *Emys orbicularis* show an increase at the female-favored incubation temperature but not at the male one (Pieau *et al.*, 1994). Similar changes were observed in aromatase concentrations in the gonad–adrenal–mesonephric duct complex of salt water crocodiles, *Crocodylus porosus* (Smith and Joss, 1994). It has, therefore, been suggested that the P450 aromatase gene may play a central role in sex determination in such species and possibly in other vertebrates (Fig. 9.24). Elevated concentrations of estrogens in ambisexual embryonic gonads may activate genes that promote the formation of ovaries, while a suppression of the formation of this steroid may favor testicular differentiation. The control of the expression of the P450 aromatase gene may depend on various factors. The effect of temperature in reptiles could be direct or via an action on another upstream gene. An SRY-like gene may inhibit the expression of the P450 aromatase gene. Such an inhibition would be consistent with the

Fig. 9.24. The possible roles of hormones in sexual differentiation of the gonads in vertebrates. A, androstenedione; AMH, antimullerian hormone; E_1 and E_2, estrogens; SRY, sex region of Y chromosome; T, testosterone; TDF, testis-determining factor. (From Pieau *et al.*, 1994.)

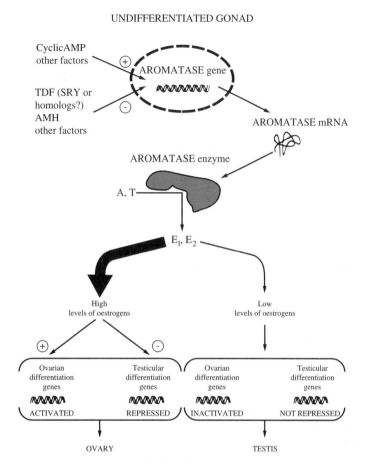

observation of the repressor properties of the SRY gene transcripts on the mammalian aromatase gene (Haqq *et al.*, 1993). A further effect of the SRY gene transcript in activating the antimullerian hormone gene was described. Antimullerian hormone, apart from suppressing the development of the Müllerian duct, has also been shown to induce sex reversal in the fetal ovaries of sheep and mice (Vigier *et al.*, 1989; Behringer *et al.*, 1990). It can block the synthesis of P450 aromatase. Antimullerian hormone may, therefore, also play an important role in sex determination.

A more detailed hypothesis regarding TSD in reptiles has been provided by Crews (1994). An inhibitor of 5α-reductase at a male-biased temperature has been shown to favor the development of female hatchlings in the red-eared turtle, suggesting that this enzyme may be contributing to the sex determination of the male young (Crews and Bergeron, 1994). Aromatase may mediate sex determination of the female young but instead of the male being merely in "default" owing to a repression of this enzyme, its sex determination may be enhanced by the synthesis of dihydrotestosterone by 5α-reductase (Crews,

Cantú, and Bergeron, 1996). Crews also suggests that an upregulation of either androgen or estrogen receptors at appropriate temperatures may contribute to the process of sex determination. Temperature-sensitive promotors could be involved in the activation of the P450 aromatase and 5α-reductase genes. Alternatively such an effect could be mediated by an upstream testis-determining factor-like (SRY-like) gene.

In some species of teleost fish, sex reversal is a normal part of their life cycle (see, for instance, Reinbloth, 1970; Shapiro, 1994). The gonads of such mature fish retain the potential to become ovaries or testes. They may commence life as either a female (protogyny) or a male (protandry) but subsequently if there is a dearth of the opposite sex they may then undergo a useful sex change. The gonads redifferentiate and produce the appropriate germ cells and hormones. Secondary sex characters such as integumental color schemes and behavior follow suit. The change of sex appears to be initiated behaviorally and reflects a novel social situation. A role for pheromones, such as mediate the hormonal control of spawning behavior (Stacey *et al.*, 1994), may be apt. Searches for possible genetic and endocrine mechanisms for sex reversal in teleosts has had inconsistent results. An obvious candidate would be a SRY-like gene "switch" that could respond to external or internal signals. A comprehensive understanding of this special process of sex reversal could illuminate our basic knowledge of sexual development.

Observations of endocrine influences on sex determination and differentiation in nonmammalian vertebrates has provided new insights about these important physiological processes. While the embryonic testis of eutherian mammals is known to have a steroidogenic ability, that of the early embryonic ovary has often been discounted. However, it appears that estrogens can be formed by the embryonic ovary of eutherians, though in relatively small amounts (Mittwoch and Burgess, 1991; George and Wilson, 1994). This ability may be a vestigial one but it could contribute to the processes of sexual development, as seen in birds and reptiles (Crews, 1994). The study of the roles of hormones in early sexual development of lower vertebrates may have initiated a reassessment of how such vital processes work in mammals.

Conclusions

Apart from nutrition, reproduction probably involves the most complex processes of humoral coordination that exist in vertebrates. It can also be considered as being under "multihormonal control." Reproduction and the preparations for this event involve a variety of tissues and organs, and several distinct, accurately timed physiological events (for instance, impregnation, ovulation, and parturition) and usually take a prolonged period of time to reach fruition. Hormones are especially suited to such needs for coordination.

The nature of the reproductive process displays considerable morphological and physiological variation and certain types of mechanism, such as viviparity in mammals and oviparity in birds, predominate in certain systematic groups, in which they are a feature and involve special endocrine mechanisms. There are, however, many examples of what may be parallel evolution, such as viviparity in fishes and reptiles in which the special role of hormones may have evolved independently but have a similar end result.

The diversity in reproductive processes does not appear to have involved many changes in the structure of the hormones themselves. The gonadal steroid hormones usually have the same structure in all vertebrates, though those from the pituitary show distinct differences that are probably of more consequence in limiting their action to a certain species than in reflecting any functional predilection to coordinate novel processes. It is, however, of special note that pituitary gonadotropins may exist as a single molecule in some vertebrates but in most tetrapods it now seems likely that two distinct hormones, one with LH and the other with FSH activity, have emerged. This dichotomy presumably allows for the operation of a more specific and precise control mechanism, but the physiological differences that, it would appear, must result are unknown. Another notable endocrine novelty that has emerged in eutherian mammals is the ability of the placenta to act as an additional site for steroid hormone synthesis and to form two pituitary-like hormones: chorionic gonadotropin and placental lactogen. There is, however, considerable interspecific variability in the endocrine function of the eutherian placenta, and it is suspected that this may have arisen on several separate evolutionary occasions and been perpetuated according to the requirements of the particular species involved.

The control of metamorphosis in amphibians and some fish is a unique and dramatic example of how endocrines can influence development, but they also contribute more ubiquitously, especially to sexual development and differentiation, in all groups of vertebrates. The phylogenetically novel process of lactation in mammals is a clear example of the evolution of endocrine function, though hormones may also contribute to the process of parental care of the young in a variety of other vertebrates.

References

Abdel-Meguid, S.S., Shieh, H.S., Smith, W.W., Dayringer, H.E., Violand, B.N., and Bentle, L.A. (1987). Three-dimensional structure of a genetically engineered variant of porcine growth hormone. *Proc. Natl. Acad. Sci. USA* **84**, 6434–6437.

Abe, K., Robinson, G.A., Liddle, G.W., Butcher, R.W., Nicholson, W.E., and Baird, C.E. (1969). Role of cyclic AMP in mediating the effects of MSH norepinephrine and melatonin on frog skin color. *Endocrinology* **85**, 674–682.

Achen, M.G., Duan, W., Pettersson, T.M., Harms, P.J., Richardson, S.J., Lawrence, M.C., Wettenhall, R.E.H., Aldred, A.R., and Schreiber, G. (1993). Transthyretin gene expression in the choroid plexus first evolved in reptiles. *Am. J. Physiol.* **265**, R982–R989.

Acher, R. (1978). Molecular evolution of neurohypophyseal hormones and neurophysins. In *Neurosecretion and neuroendocrine activity. Evolution, structure and function* (edited by W. Bargmann, A. Oksche, A. Polenov, and B. Scharrer). pp. 31–43, New York: Springer-Verlag.

Acher, R. (1980). Molecular evolution of biologically active polypeptides. *Proc. Roy. Soc., Lond., Ser. B* **210**, 21–43.

Acher, R. (1990). Structure evolution and processing adaptation of neurohypophysial–neurophysin precursors. In *Progress in Comparative Endocrinology* (edited by A. Epple, C.G. Scanes, and M.H. Stetson), pp. 1–9. New York: Wiley-Liss.

Acher, R., Chauvet, J., and Chauvet, M.T. (1972). Phylogeny of the neurohypophysial hormones. Two new active peptides isolated from a cartilaginous fish, *Squalus acanthias. Eur. J. Biochem.* **29**, 12–19.

Adelson, J.W. (1971). Enterosecretory proteins. *Nature, Lond.* **229**, 321–325.

Agellon, L.B., Davies, S.L., Chen, T.T., and Powers, D.A. (1988). Structure of a fish (rainbow trout) growth hormone gene and its evolutionary implications. *Proc. Natl. Acad. Sci. USA* **85**, 5136–5140.

Ahima, R.S., Prabakaran, D., Mantzoros, C., Qu, D., Lowell, B., Maratos-Flier, E., and Flier, J.S. (1996). Role of leptin in the neuroendocrine response to fasting. *Nature, Lond.* **382**, 250–252.

Akutsu, S., Takada, M., Ohki-Hamazaki, H., Murakami, S., and Arai, Y. (1992). Origin of luteinizing hormone-releasing hormone (LHRH) neurons in the chick embryo: effect of the olfactory placode ablation. *Neurosci. Lett.* **142**, 241–244.

Alvarado, R.H., and Johnson, S.R. (1966). The effects of neurohypophysial hormones on water and sodium balance in larval and adult bullfrogs (*Rana catesbeiana*). *Comp. Biochem. Physiol.* **18**, 549–561.

Amit, T., Hochberg, Z., Waters, M.J., and Barkey, R.J. (1992). Growth hormone- and prolactin-binding proteins in mammalian serum. *Endocrinology* **131**, 1793–1803.

Amoroso, E.C., Heap, A.R., and Renfree, M.B. (1979). Hormones and the evolution of viviparity. In *Hormones and Evolution* (edited by E.J.W. Barrington), pp. 925–989. New York: Academic Press.

Amoroso, E.C. and Marshall, F.H.A. (1960). External factors in sexual periodicity. In *Marshall's Physiology of Reproduction* (edited by A.S. Parkes), vol.1 (Pt. 2), pp. 707–831. London: Longmans.

Anderson, L.L. (1973). Effects of hysterectomy and other factors on luteal function. In *Handbook of Physiology*, Sect. 7 *Endocrinology*, vol. II. *Female reproductive system* (Pt. 2), pp. 69–86. Washington, DC: American Physiological Society.

Anderson, R.S. (1993). Transforming frogs and flies. *Nature, Lond.* **361**, 116–117.

Anderson, T.A., Bennett, L.R., Conlon, M.A., and Owens, P.C. (1993). Immunoreactive and receptor-active insulin-like growth factor-I (IGF-I) and IGF-binding protein in blood plasma of the freshwater fish *Macquaria ambigua* (golden perch). *J. Endocr.* **136**, 191–198.

Anderrsson, R.C.G., Karlsson, J.O., and Grundström, N. (1984). Adrenergic nerves and the alpha$_2$-adrenoreceptor system regulating melanosome aggregation within fish melanophores. *Acta Physiol. Scand.* **121**, 173–179.

Ando, M., Kondo, K., and Takei, Y. (1992). Effects of eel atrial natriuretic peptide on NaCl and water transport across the intestine of the seawater eel. *J. Comp. Physiol. B* **162**, 436–439.

Andrews, P.C., Brayton, K., and Dixon, J.E. (1987). Precursors of regulatory peptides: their proteolytic processing. *Experientia* **43**, 784–790.

Andrews, P.C., Pollock, H.G., Elliott, W.M., Youson, J.H., and Plisetskaya, E.M. (1988). Isolation and characterization of a variant somatostatin-14 and two related somatostatins of 34 and 37 residues from lamprey (*Petromyzon marinus*) *Proc. Natl. Acad. Sci. USA* **263**, 15 809–15 814.

Anthony, R.V., Liang, R., Kayl, E.P., and Pratt, S.L. (1995). The growth hormone/prolactin gene family in ruminant placentae. *J. Reprod. Fert. Suppl.* **49**, 83–95.

Arakawa, E., Hasegawa, S., Kaneko, T., and Hirano, T. (1993). Effects of changes in environmental calcium on prolactin secretion in Japanese eel, *Anguilla japonica. J. Comp. Physiol B* **163**, 99–106.

Ariëns Kappers, J. (1970). The pineal organ: An introduction. In *The Pineal Gland* (edited by G.E.W. Wolstenholme and J. Knight), pp. 3–25. London: Churchill Livingstone.

Argetsinger, L.S., Carter-Su, C. (1996). Mechanism of signalling by growth hormone receptor. *Physiol Rev.* **76**, 1098–1107.

Armour, K.J., O'Toole, L.B., and Hazon, N. (1993). The effect of dietary protein restriction on the secretory dynamics of 1α-hydroxycorticosterone and urea in the dogfish, *Scyliorhinus canicula*: a possible role for 1α-hydroxycorticosterone in sodium retention. *J. Endocr.* **138**, 275–282.

Armstrong, D.T. and Kennedy, T.G. (1972). Role of luteinizing hormones in regulation of the rat estrous cycle. *Am. Zool.* **12**, 245–255.

Arnason, S.S., Rice, G.E., Chadwick, A., and Skadhauge E. (1986). Plasma levels of arginine vasotocin, prolactin, aldosterone and corticosterone during prolonged dehydration in the domestic fowl: effect of dietary NaCl. *J. Comp. Physiol. B* **156**, 383–397.

Arnold-Reed, D.E. and Balment, R.J. (1989). Steroidogenic role of the caudal neurosecretory system in the flounder, *Platichthys flesus. Gen. Comp. Endocr.* **76**, 267–273.

Arnold-Reed, E.E. and Balment, R.J. (1991). Atrial natriuretic factor stimulates *in-vivo* and *in-vitro* secretion of cortisol in teleosts. *J. Endocr.* **128**, R17–R20.

Arnold-Reed, D.E. and Balment, R.J. (1994). Peptide hormones influence *in vitro* interrenal secretion of cortisol in the trout, *Oncorhynchus mykiss. Gen. Comp. Endocr.* **96**, 85–91.

Arnold-Reed, E.E., Balment, R.J., McCrohan, C.R., and Hackney, C.M. (1991). The caudal neurosecretory system of *Platichthys flesus*: general morphology and responses to altered salinity. *Comp. Biochem. Physiol. A* **99**, 137–143.

Arnold-Reed, D., Hazon, N. and Balment, R.J. (1991). Biological action of atrial natriuretic factor in flatfish. *Fish Physiol. Biochem.* **9**, 271–277.

Arsaky, A. (1813). *De piscium cerebro et medulla spinali*. Halae; Dissertatio inaugeralis.

Atkinson, B.G., Helbing, C. and Chen, Y. (1994). Reprogramming of gene expression in the liver of *Rana catesbeiana* tadpoles during spontaneous and thyroid hormone induced metamorphosis. In *Perspectives in Comparative Endocrinology* (edited by K.G. Davey, R.E. Peter, and S.S. Tobe), pp. 416–423. Ottawa: National Research Council of Canada.

Atlas, S.A. and Maack, T. (1992). Atrial natriuretic factor. In *Handbook of Physiology*, Sect. 8, *Renal Physiology*, vol. 2, pp. 1577–1672. Bethesda: American Physiological Society.

Augee, M.L. and McDonald, I.R. (1973). Role of the adrenal cortex in the adaptation of the monotreme *Tachyglossus aculeatus* to low environmental temperature. *J. Endocrinol.* **58**, 513–523.

Auperin, B., Rentier-Delrue, F., Martial, J.A., and Prunet, P. (1994a). Evidence that two tilapia (*Oreochromis niloticus*) prolactins have different osmoregulatory functions during adaptation to a hyperosmotic environment. *J. Mol. Endocr.* **12**, 13–24.

Auperin, B., Rentier-Delrue, F., Martial, J.A., and Prunet, P. (1994b). Characterization of a single prolactin (PRL) receptor in tilapia (*Oreochromis niloticus*) which binds both PRL₁ and PLR₁. *J. Mol. Endocr.* **13**, 241–251.

Auperin, B., Rentier-Delrue, F., Martial, J.A., and Prunet, P. (1995). Regulation of gill prolactin receptors in tilapia (*Oreochromis niloticus*) after a change in salinity or hypophysectomy. *J. Endocr.* **145**, 213–220.

Austin, C.R. (1994). Pre-implantation development. In *Marshall's Physiology of Reproduction*, 4th edn (edited by G.E. Lamming), vol. 3, *Pregnancy and Lactation*, Pt. 1, *Ovulation and Early Pregnancy*, pp. 93–155. London: Chapman & Hall.

Ayre, E.A., Yuan, H., and Pang, S.F. (1992). The identification of ¹²⁵I-labelled iodomelatonin-binding sites in the testes

and ovaries of the chicken (*Gallus domesticus*). *J. Endocr.* **133**, 5–11.

Baber, E.C. (1876). Contributions to the minute anatomy of the thyroid gland of the dog. *Proc. Roy. Soc.* **24**, 240–241.

Babiker, M.M. and Rankin, J.C. (1973). Effects of neurohypophysial hormones on renal function in the freshwater- and sea-water-adapted eel *Anguilla anguilla L. J. Endocr.* **57**, xi–xii.

Badura, L.L. and Goldman, B.D. (1992). Prolactin-dependent seasonal changes in pelage: role of the pineal gland and dopamine. *J. Exp. Zool.* **261**, 27–33.

Baeuerle, P.A. (1995). Enter a polypeptide messenger. *Nature, Lond.* **373**, 661–662.

Bagnara, J.T. (1969). Responses of pigment cells of amphibians to intermedin. *Colloq. Int. C.N.R.S.* **177**, 153–158.

Bagnara, J.T. and Hadley, M.E. (1970). Endocrinology of the amphibian pineal. *Am. Zool.* **10**, 201–216.

Bagnara, J.T. and Hadley, M.E. (1972). *Chromatophores and Color Change*. Englewood Cliffs, NJ: Prentice-Hall.

Bajaj, M., Blundell, T.L., Pitts, J.E., Wood, S.P., Tatnell, M.A., Falkmer, S., Emdin, S.O. Gowan, L.K., Crow, H., Schwabe, H., Wollmer, C., and Strassburger, W. (1983). Dogfish insulin: primary structure, conformation and biological properties of an elasmobranch insulin. *Eur. J. Biochem.* **135**, 535–542.

Baker, B.I. (1994). Melanin-concentrating hormone updated. Functional considerations. *Trends. Endocr. Metab.* **5**, 120–126.

Baker, B.I. and Ball, J.N. (1975). Evidence for a dual pituitary control of teleost melanophores. *Gen. Comp. Endocr.* **25**, 147–152.

Baker, B.I. and Buckingham, J.C. (1983). A study of corticotrophic and melanotrophic activities in the pituitary and brain of the lamprey *Lampetra fluviatilis. Gen. Comp. Endocr.* **52**, 283–290.

Baker, B.I. and Rance, T.A. (1983). Further observations on the distribution and properties of teleost melanin concentrating hormone. *Gen. Comp. Endocr.* **50**, 423–431.

Baker, B.I., Wilson, J.F., and Bowley, T.J. (1984). Changes in pituitary and plasma levels of MSH in teleosts during physiological colour change. *Gen. Comp. Endocr.* **55**, 142–149.

Baker. M.E. (1988). Is vitellogenin an ancestor of apolipoprotein B-100 of human low-density lipoprotein and human lipoprotein lipase? *Biochem. J.* **255**, 1057–1060.

Baksi, S.N. and Kenny, A.D. (1977). Vitamin D$_3$ metabolism in immature Japanese quail: effects of ovarian hormones. *Endocrinology* **101**, 1216–1220.

Baldwin, G.F. and Bentley, P.J. (1980). Calcium metabolism in bullfrog tadpoles (*Rana catesbeiana*). *J. Exp. Biol.* **88**, 357–365.

Baldwin, G.F. and Bentley, P.J. (1981). A role for skin in Ca metabolism of frogs? *Comp. Biochem, Physiol. A* **68**, 181–185.

Ball, J.N. and Baker, B.I. (1969). The pituitary gland: anatomy and histophysiology. In *Fish Physiology* (edited by W.S. Hoar and D.J. Randall), vol. II, *The Endocrine System*, pp. 1–110. New York: Academic Press.

Ball, J.N. and Ensor, D.M. (1965). Effect of prolactin on plasma sodium in the teleost. *Poecilia latipinna. J. Endocr.* **32**, 269–270.

Ball, J.N. and Ensor, D.M. (1969). Specific action of prolactin on plasma sodium levels in hypophysectomized *Poecilia latipinna* (Teleostei). *Gen. Comp. Endocr.* **8**, 432–440.

Ball, J.N. and Ingleton, P.M. (1973). Adaptive variations in prolactin secretion in relation to external salinity in the teleost *Poecilia latipinna. Gen. Comp. Endocr.* **20**, 312–325.

Balment, R.L., Hazon, N., and Perrott, M.N. (1987). Control of corticosteroid secretion and its relation to osmoregulation in lower vertebrates. In *Comparative Physiology of Environmental Adaptations* (edited by R. Kirsch and B. Lahlou), pp. 92–102. Basel: Karger.

Balment, R.J. and Lahlou B. (1987). Atrial natriuretic peptide (ANP) modulation of second messenger activity in the isolated trout gill cells. *J. Endocr.* **115** (Suppl.), 164.

Balment, R.J. and Loveridge, J.P. (1989). Endocrine and osmoregulatory mechanisms in the Nile crocodile, *Crocodylus niloticus. Gen. Comp. Endocr.*

73, 361–367.

Banerjee, P.P., Banerjee, S., Shen, S.-T., Kao, Y.-H., and Yu, J.Y.-L. (1993). Studies on isolation and properties of two isoforms of pituitary gonadotropin from black silver carp, *Aristichthys nobilis. Comp. Biochem. Physiol. B* **104**, 241–253.

Banerjee, S., Banerjee, P.P., Shen, S.-T., and Lu, J.Y.-L. (1994). Studies on purification and characterization of pituitary thyrotropin from black silver carp. *Aristichthys nobilis. Comp. Biochem, Physiol. B* **107**, 337–346.

Bar, A. and Norman, A.W. (1981). Studies on the mode of action of calciferol. XXXIV. Relationship of the distribution of 25-hydroxyvitamin D$_3$ metabolites to gonadal activity and egg shell formation in the quail. *Endocrinology* **109**, 950–955.

Bar, A., Striem, S., Vax, E., Talpaz, H., and Hurwitz, S. (1992). Regulation of mRNA and calbindin turnover in intestine and shell gland of the chicken. *Am. J. Physiol.* **262**, R800–R805.

Bar, A., Vax, E., Hunziker, W., Halevy, O., and Striem, S. (1996). The role of gonadal hormones in gene expression of callbindin (M_r 28 000) in the laying hen. *Gen. Comp. Endocrin.* **103**, 115–122.

Barajas, L. (1979). Anatomy of the juxtaglomerular apparatus. *Am. J. Physiol.* **237**, F333–F343.

Barash, I.A., Cheung, C.C., Weigle, D.S., Ren, H., Kabigting, E.B., Kuijper, J.L., Clifton, D.K., and Steiner, R.A. (1996). Leptin is a metabolic signal to the reproductive system. *Endocrinology* **137**, 3144–3147.

Bard, J., Walker, M.W., Branchek, T.A., and Weinshank, R.L. (1995). Cloning and functional expression of a human Y4 subtype on pancreatic polypeptide, neuropeptide Y, and peptide YY. *J. Biol. Chem.* **270**, 26 762–26 765.

Barrington, E.J.W. (1942). Blood sugar and the follicles of Langerhans in the ammocoete larva. *J. Exp. Biol.* **19**, 45–55.

Barrington, E.J.W. (1962). Hormones and vertebrate evolution. *Experientia* **18**, 201–210.

Barrington, E.J.W. (1968). Metamorphosis in lower chordates. In *Metamorphosis, A Problem in Developmental Biology* (edited by W. Etkin and L.I. Gilbert), pp. 223–270. New York:

Appleton-Century-Crofts.

Barrington, E.J.W. and Dockray, G.J. (1970). The effect of intestinal extracts of lampreys (*Lampetra fluviatilis* and *Petromyzon marinus*) on pancreatic secretion in the rat. *Gen. Comp. Endocr.* **14**, 170–177.

Barrington, E.J.W. and Dockray, G.J. (1972) Cholecystokinin-pancreozymin-like activity in the eel *Anguilla anguilla. Gen. Comp. Endocr.* **19**, 80–87.

Basu, S.L. (1969). Effects of hormones on the salientian spermatogenesis *in vivo* and *in vitro. Gen. Comp. Endocr.* **Suppl. 2**, 203–213.

Bataille, D. (1989). Gut glucagon. In *Handbook of Physiology*, Sect. 6 *The Gastrointestinal System*, vol. II, *Neural and Endocrine Biology*, pp. 455–474. Bethesda: American Physiological Society.

Baumann, H., Morella, K.K., White, D.W., Dembski, M., Bailon, P.S., Kim, H., Lai, C.-F., and Tartaglia, L.A. (1996). The full-length leptin receptor has signalling capabilities of interleukin 6-type cytokine receptors. *Proc. Natl. Acad. Sci.* **USA 93**, 8374–8378.

Baumbach, W.R., Horner, D.L., and Logan, J.S. (1989). The growth hormone-binding protein in rat serum is an alternatively spliced form of the rat growth hormone gene. *Genes Devel.* **3**, 1199–1205.

Bautista, C.M., Mohan, S., and Baylink, D.J. (1990). Insulin-like growth factors I and II are present in skeletal tissues of ten vertebrates. *Metabolism* **39**, 96–100.

Baxter, R.C. (1993). Circulating binding proteins for insulin-like growth factors. *Trend. Endocr. Metab.* **4**, 91–96.

Bayliss, W.M. and Starling, E.H. (1902). The mechanism of pancreatic secretion. *J. Physiol., Lond.* **28**, 325–353.

Bayliss, W.M. and Starling, E.H. (1903). On the uniformity of the pancreatic mechanism in vertebrata. *J. Physiol., Lond.* **29**, 174–180.

Bazan, J.F. (1990). Structural design and molecular evolution of a cytokine receptor superfamily. *Proc. Natl. Acad. Sci. USA* **87**, 6934–6938.

Beato, M. and Sánchez-Pacheco, A. (1996). Interaction of steroid hormone receptors with the transcription initiation complex. *Endocrine Rev.* **17**, 587–609.

Bedarkar, S., Turnell, W.G., Blundell, T.L., and Schwabe, C. (1977). Relaxin has conformational homology with insulin. *Science* **270**, 449–451.

Behringer, R.R., Cate, R.L., Froelick, G.J., Palmiter, R.D., and Brinster, R.L. (1990). Abnormal sexual development in transgenic mice chronically expressing Müllerian inhibiting substance. *Nature, Lond.* **345**, 167–170.

Bélanger, L.F., Dimond, M.T., and Copp, D.H. (1973). Histological observations on bone and cartilage of growing turtles treated with calcitonin. *Gen. Comp. Endocr.* **20**, 297–304.

Bellamy, D. and Leonard, R.A. (1965). Effect of cortisol on the growth of chicks. *Gen. Comp. Endocr.* **5**, 402–410.

Bennett, M.B. and Rankin, J.C. (1986). The effects of neurohypophysial hormones on the vascular resistance of the isolated perfused gill of the European eel, *Anguilla anguilla* L. *Gen. Comp. Endocr.* **64**, 60–66.

Bennett, M.B. and Rankin, J.C. (1987). The effects of catecholamines on tritiated water influx and the branchial vasculature of the European eel, *Anguilla anguilla* L. *J. Comp. Physiol. B* **157**, 327–333.

Bentley, P.J. (1962). Studies on the permeability of the large intestine and urinary bladder of the tortoise (*Testudo graeca*) with special reference to the effects of neurohypophysial and adrenocortical hormones. *Gen. Comp. Endocr.* **2**, 323–328.

Bentley, P.J. (1969). Neurohypophysial function in Amphibia: hormone activity in the plasma. *J. Endocr.* **43**, 359–369.

Bentley, P.J. (1971). *Endocrines and Osmoregulation. A Comparative Account of the Regulation of Water and Salt in Vertebrates.* New York: Springer-Verlag.

Bentley, P.J. (1972). Introductory remarks. Symposium on endocrinology and osmoregulation. *Fed. Proc.* **31**, 1583–1586.

Bentley, P.J. (1973). Osmoregulation in the aquatic urodeles *Amphiuma means* (the congo eel) and *Siren lacertina* (the mud eel). Effects of vasotocin. *Gen. Comp. Endocr.* **20**, 386–392.

Bentley, P.J. (1983). Urinary loss of calcium in an anuran amphibian (*Bufo marinus*) with a note on the effects of calcemic hormones. *Comp. Biochem. Physiol. B* **76**, 717–719.

Bentley, P.J. (1984). Calcium metabolism in the Amphibia. *Comp. Biochem. Physiol. A* **79**, 1–5.

Bentley, P.J. and Follett, B.K. (1963). Kidney function in a primitive vertebrate, the cyclostome *Lampetra fluviatilis*. *J. Physiol., Lond.* **169**, 902–918.

Bentley, P.J. and Follett, B.K. (1965). The effects of hormones on the carbohydrate metabolism of the lamprey *Lampetra fluviatilis*. *J. Endocr.* **31**, 127–137.

Bentley, P.J. and Grubb, B.R. (1989). Sodium and potassium concentrations in the intestinal fluids of the fowl. *Gallus domesticus. Comp. Biochem. Physiol. A* **92**, 577–579.

Bentley, P.J. and Schmidt-Nielsen, K. (1965). Permeability to water and sodium of the crocodilian, *Caiman sclerops*. *J. Cell Comp. Physiol.* **66**, 303–309.

Bentley, P.J. and Yorio, T. (1979). Do frogs drink? *J. Exp. Biol.* **79**, 41–46.

Bereiter-Hahn, J., Matoltsy, A.G., and Richards, K. (eds.) (1986). *Biology of the integument*, vol. 2, *Vertebrates*. Berlin: Springer-Verlag.

Berelowitz, M., Bruno, J.F., and White, J.D. (1992). Regulation of hypothalamic neuropeptide expression by peripheral metabolism. *Trend. Endocr. Metab.* **3**, 127–133.

Bergland, R.M. and Page, R.B. (1979). Pituitary–brain vascular relations: a new paradigm. *Science* **204**, 18–24.

Berkenbosch, F., Tilders, F.J.H., and Vermes, I. (1983). β-Adrenoreceptor activation mediates stress-related secretion of β-endorphin-related peptides from intermediate but not anterior pituitary. *Nature, Lond.* **305**, 237–239.

Bern, H.A. (1972). Comparative endocrinology – the state of the field and art. *Gen. Comp. Endocr.* **Suppl. 3**, 751–761.

Bern, H.A. (1992). The development of the role of hormones in development – a double remembrance. *Endocrinology* **131**, 2037–2038.

Bern, H.A. and Nicoll, C.S. (1968). The comparative endocrinology of prolactin. *Rec. Prog. Horm. Res.* **24**, 681–713.

Bern, H.A. and Nicoll, C.S. (1969). The zoological specificity of prolactin. *Colloq. Int. C.N.R.S.* **177**, 193–202.

Bern. H.A., Pearson, D., Larson, B.A., and Nishioka, R.S. (1985). Neurohormones from fish tails: the caudal neurosecretory system. I. "Urophysiology" and the caudal neurosecretory system of fishes. *Rec. Prog. Horm. Res.* **41**, 533–522.

Berridge, M.J. (1993). Inositol trisphosphate and calcium signalling. *Nature, Lond.* **361**, 315–325.

Bewley, T.A. and Li, C.H. (1970). Primary structures of human pituitary growth hormone and sheep pituitary lactogenic hormone compared. *Science* **168**, 1361–1362.

Bianco, A.C., Sheng, X., and Silva, J.E. (1988). Triiodothyronine amplifies norepinephrine stimulation of uncoupling protein gene transcription by a mechanism not requiring protein synthesis. *J. Biol. Chem.* **263**, 18 168–18 175.

Bianco, A.C. and Silva, J.E. (1987). Optimal response of key enzymes and uncoupling protein to cold in BAT depends on local T_3 generation. *Am. J. Physiol.* **253**, E255–E263.

Bicknell, R.J. (1985). Endogenous opioid peptides and hypothalamic neuroendocrine neurones. *J. Endocr.* **107**, 437–446.

Bikle, D.D., Spencer, E.M., Burke, W.H., and Rost, C.R. (1980). Prolactin but not growth hormone stimulates 1,25-dihydroxyvitamin D_3 production by chick renal preparations *in vitro*. *Endocrinology* **107**, 81–84.

Bilezikjian, L.M. and Vale, W.W. (1992). Local extragonadal roles for activins. *Trend. Endocr. Metab.* **3**, 218–223.

Billard, R., LeGac, F., and Loir, M. (1990). Hormonal control of sperm production in teleost fish. In *Progress in Comparative Endocrinology* (edited by A. Epple, C.G. Scanes, and M.H. Stetson), pp. 329–335. New York: Wiley-Liss.

Billington, C.J., Briggs, J.E., Link, J.G., and Levine A.S. (1991). Glucagon in physiological concentrations stimulates brown fat thermogenesis *in vivo*. *Am. J. Physiol.* **261**, R501–R507.

Bing, C., Pickavance L., Frankish, H.M., Keith, J., Trayhurn, P., and Williams, G.

(1996). The effects of cold exposure to *ob* (obese) gene expression in white adipose tissue of Zucker rats. *J. Endocr.* **151** (Suppl.), P48.

Binkley, S. (1988). *The Pineal: Endocrine and Nonendocrine Function*. Englewood Cliffs, NJ: Prentice-Hall.

Birnbaumer, M., Seibold, A., Gilbert, S., Ishido, M., Barberis, C., Antaramian, A., Brabet, P., and Rosenthal, W. (1992). Molecular cloning of the receptor for human antidiuretic hormone. *Nature, Lond.* **357**, 333–339.

Björnsson, B.Th., Young, G., Lin, R.J., Deftos, L.J., and Bern, H.A. (1989). Smoltification and seawater adaptation in coho salmon (*Oncorhynchus kisutch*): plasma calcium regulation, osmoregulation, and calcitonin. *Gen. Comp. Endocr.* **74**, 346–354.

Blair-West, J.R., Coghlan, J.P., Denton, D.A., Nelson, J.F., Orchard, E., Scoggins, B.A., Wright, R.D., Myers, K., and Junquera, C.L. (1968). Physiological, morphological and behavioural adaptation to a sodium deficient environment by native Australian and introduced species of animals. *Nature, Lond.* **217**, 922–928.

Blair-West, J.R., Coghlan, J.P., Denton, D.A., Funder, J.W., Scoggins, B.A., and Wright, R.D. (1971). The effect of the heptapeptide (2–8) and the hexapeptide (3–8) fragments of angiotensin II on aldosterone secretion. *J. Clin. Endocr. Metab.* **32**, 575–578.

Blomquist, A.G., Söderberg, C., Lundell, I., Milner, R.J. and Larhammar, D. (1992). Strong evolutionary conservation of neuropeptide Y: sequences of chicken, goldfish, and *Torpedo marmorata* DNA clones. *Proc. Natl. Acad. Sci. USA* **89**, 2350–2354.

Bloom, S.R. and Polak. J.M. (1978). Gut hormone overview. In *Gut hormones* (edited by S.R. Bloom), pp. 3–18. London: Churchill Livingstone.

Blum, J.J. (1967). An adrenergic control system in *Tetrahymena. Proc. Natl. Acad. Sci. USA* **58**, 81–88.

Boehlke, K.W., Church, R.L., Tiemeier, O.W., and Eleftheriou, B.E. (1966). Diurnal rhythm in plasma glucocorticoid levels in the channel catfish (*Ictalurus punctatus*). *Gen. Comp. Endocr.* **7**, 18–21.

Boelkins, J.N. and Kenny, A.D. (1973). Plasma calcitonin levels in Japanese quail. *Endocrinology* **92**, 1754–1760.

Bolton, J.P., Collie, N.L., Kawauchi, H., and Hirano, T. (1987). Osmoregulatory actions of growth hormone in rainbow trout (*Salmo gairdneri*). *J. Endocr.* **112**, 63–68.

Bona-Gallo, A., Licht, P., MacKenzie, D.S., and Lofts, B. (1980). Annual cycles in levels of pituitary and plasma gonadotropin, gonadal steroids, and thyroid activity in the chinese cobra (*Naja naja*). *Gen. Comp. Endocr.* **42**, 477–493.

Bootman, M.D. and Berridge, M.J. (1995). The elemental principles of calcium signalling. *Cell* **83**, 675–678.

Borski, R.J., Yoshikawa, J.S.M., Madsen, S.S., Nishioka, R.S., Zabetian, C., Bern, H.A., and Grau, E.G. (1994). Effects of environmental salinity on pituitary growth hormone content and cell activity in the euryhaline tilapia, *Oreochromis mossambicus. Gen. Comp. Endocr.* **95**, 483–494.

Boswell, T., Sharp, P.J., Hall, M.R., and Goldsmith, A.R. (1995). Migratory fat deposition in European quail: a role for prolactin? *J. Endocr.* **146**, 71–79.

Bouhtiauy, I., Lajeunesse, L., and Brunette, M.G. (1991). The mechanism of parathyroid hormone action on calcium reabsorption by the distal tubule. *Endocrinology* **128**, 251–258.

Bouillon, R., van Baelen, H., Rombauts, W., and de Moor, P. (1978). The isolation and characterization of the vitamin D-binding protein from rat serum. *J. Biol. Chem.* **253**, 4426–4431.

Boulanger, B.R., Lilly, M.P., Hamlyn, J.M., Laredo, J., Shurleff, D., and Gann, D.S. (1993). Ouabain is secreted by the adrenal gland of awake dogs. *Am. J. Physiol.* **264**, E413–E419.

Bourne, A.R. and Seamark, R.F. (1973). Seasonal changes in testicular function in the lizard *Tiliqua rugosa. J. Endocr.* **57**, x.

Bousfield, G.R., Perry, W.M., and Ward, D.N. (1994). Gonadotropins. Chemistry and biosynthesis. In *The Physiology of Reproduction*, 2nd edn (edited by E. Knobil and J.D. Neill), vol. 1, pp. 1749–1792. New York: Raven Press.

Boutin, J.-M., Jolicoeur, C. Okamura, H.,

Gagnon, J., Edery, M., Shirota, M., Banville, D., Dusanter-Fourt, I., Djiane, J., and Kelly, P.A. (1988). Cloning and expression of the rat prolactin receptor, a member of the growth hormone/ prolactin receptor gene family. *Cell* **53**, 69–77.

Bower, A., Hadley, M.E., and Hruby, V.J. (1974). Biogenic amines and control of melanophore stimulating hormone release. *Science* **184**, 70–72.

Boyle, M.R., Verchere, C.B., McKnight, G., Mathews, S., Walk, K., and Taborsky, G.J. (1994). Canine galanin: sequence, expression and pancreatic effects. *Regulatory Peptides* **50**, 1–11.

Brabant, G., Prank, K., Schöfl, C. (1992). Pulsatile patterns in hormone secretion. *Trend. Endocr. Metab.* **3**, 183–190.

Brackett, K.H., Fields, P.A., Dubois, W., Chang, S.-M.T., Mather, F.B., and Fields, M.J. (1997). Relaxin: an ovarian hormone in an avian species (*Gallus domesticus*). *Gen. Comp. Endcor.* **105**, 155–163.

Bradley, A.J., McDonald, I.R., and Lee, A.K. (1980). Stress and mortality in a small marsupial (*Antechinus stuartii*, Macleay). *Gen. Comp. Endocr.* **40**, 188–200.

Bradley, A.J. and Stoddart, D.M. (1990). Metabolic effects of cortisol, ACTH, adrenalin and insulin in the marsupial sugar glider, *Petaurus breviceps. J. Endocr.* **127**, 203–212.

Bradley, A.J. and Stoddart, D.M. (1992). Seasonal changes in plasma androgens, glucocorticoids and glucocorticoid-binding proteins in the marsupial sugar glider *Petaurus breviceps. J. Endocr.* **132**, 21–31.

Bradshaw, F.J. and Bradshaw S.D. (1996). Arginine vasotocin: locus of action along the nephron of the ornate dragon lizard, *Ctenophorous ornatus. Gen. Comp. Endocr.* **103**, 218–289.

Bradshaw, S.D. (1978). Volume regulation in desert reptiles and its control by pituitary and adrenal hormones. In *Osmotic and Volume Regulation. Alfred Benzon Symposium XI* (edited by C.B. Jørgensen and E. Skadhauge), pp. 38–53. Copenhagen: Munksgaard.

Bradshaw, S.D. (1986). Hormonal mechanisms and survival of desert reptiles. In *Endocrine Regulations as Adaptive Mechanisms to the Environment* (edited by I. Assenmacher and J. Boissin), pp. 415–438. Paris: Centre National de la Recherche Scientifique.

Bradshaw, S.D. (1997). *Homeostasis in Desert Reptiles*. Heidelberg: Springer.

Bradshaw, S.D., McDonald, I.R., Hahnel, R., and Heller, H. (1975). Synthesis of progesterone by the placenta of a marsupial. *J. Endocr.* **65**, 451–452.

Bradshaw, S.D., and Rice, G.E. (1981). The effects of pituitary and adrenal hormones on renal and postrenal reabsorption of water and electrolytes in the lizard, *Varanus gouldii* (Gray). *Gen. Comp. Endocr.* **44**, 82–93.

Bradshaw, S.D. and Waring, H. (1969). Comparative studies on the biological activity of melanin-dispersing hormone (MDH). *Colloq. Int. C.N.R.S.* **77**, 135–151.

Brain, S.D., Williams, T.J., Tippins, J.R., Morris, H.R., and MacIntyre, I. (1985). Calcitonin gene-related peptide is a potent vasodilator. *Nature, Lond.* **313**, 54–56.

Brenner, B.M., Ballermann, B.J., Gunning, M.E., and Zeidel, M.L. (1990). Diverse biological actions of atrial natriuretic peptide. *Physiol. Rev.* **70**, 665–699.

Bressler, R. and Bahl, J.J. (1990). Insulin regulation of metabolism relevant to gluconeogenesis. In *Insulin* (edited by P. Cuatrecasas and S. Jacobs). pp. 451–467. Berlin: Springer-Verlag.

Broadhead, C.L., O'Sullivan, U.T., Deacon, C.F., and Henderson, I.W. (1992). Atrial natriuretic peptide in the eel, *Anguilla anguilla L*: its cardiac distribution, receptors and actions on isolated branchial cells. *J. Mol. Endocr.* **9**, 103–114.

Bronner, F. (1996). Cytoplasmic transport of calcium and other inorganic ions. *Comp. Biochem. Physiol. B* **115**, 313–317.

Brown, E.M. (1992). Kidney and bone: physiological and pathophysiological relationships. In *Handbook of Physiology*, Sect. 8 *Renal Physiology*, vol. II, pp. 1841–1916. Bethesda: American Physiological Society.

Brown, E.M., Vassilev, P.M., and Hebert, S.C. (1995). Calcium ions as extracellular messengers. *Cell* **83**, 679–682.

Brown, J.A., Oliver, J.A., Henderson, I.W., and Jackson, B.A. (1980). Angiotensin and single nephron glomerular function in the trout *Salmo gairdneri*. *Am. J. Physiol.* **239**, R509–R514.

Brown, J.C., Buchan, A.M.J., McIntosh, C.H.S., and Pederson, R.A. (1989). Gastric inhibitory polypeptide. In *Handbook of Physiology*, Sect. 6 *The Gastrointestinal System*, vol. II, *Neural and Endocrine Biology*, pp. 403–430. Bethesda: American Physiological Society.

Brown, M.A., Cree, A., Daugherty, C.H., Dawkins, B.P., and Chambers, G.K. (1994). Plasma concentrations of vitellogenin and sex steroids in female tuatara (*Sphenodon punctatus punctatus*) from northern New Zealand. Comp. Endocr. **95**, 201–212.

Brown, P.B. and Robinson, E.H. (1992). Vitamin D studies with channel catfish (*Ictalurus punctatus*) reared in calcium-free water. *Comp. Biochem. Physiol. A* **103**, 213–219.

Brown, P.S. and Brown, S.C. (1973). Prolactin and thyroid hormone interactions in salt and water balance in the newt *Notophthalmus viridescens*. *Gen. Comp. Endocr.* **20**, 456–466.

Brown, P.S. and Brown, S.C. (1987). Osmoregulatory actions of prolactin and other adenohypophysial hormones. In *Vertebrate Endocrinology: Fundamentals and Biomedical Implications*, vol. 2, *Regulation of Water and Electrolytes* (edited by P.K.T. Pang and M.P. Schreibman), pp. 45–84. San Diego: Academic Press.

Brown, R.E. (1985). Hormones and paternal behavior in vertebrates. *Am. Zool.* **25**, 895–910.

Brown S.C. and Brown, P.S. (1980). Water balance in the California newt, *Taricha torosa*. *Am. J. Physiol.* **238**, R113–R118.

Brownstein, M. (1977). Neurotransmitters and hypothalamic hormones in the central nervous system. *Fedn. Proc.* **36**, 1960–1963.

Brunette, M.G., Chan, M., Ferriere, C., and Roberts, K.D. (1978). Site of 1,25(OH)$_2$vitamin D$_3$ synthesis in the kidney. *Nature, Lond.* **276**, 287–289.

Bryant, R.W., Epstein, A.N., Fitzsimmons, J.T., and Fluharty, S.J. (1980). Arousal of a specific and persistent sodium appetite in the rat with continuous intercerebroventricular infusion of angiotensin II. *J. Physiol., Lond.* **301**, 365–382.

Bubenik, G.A. (1986). Regulation of seasonal endocrine rhythms in male boreal cervids. In *Endocrine Regulations as Adaptive Mechanisms to the Environment* (edited by I. Assenmacher and J. Boissin), pp. 461–472. Paris: Centre National de la Recherche Scientifique.

Buckbinder, L. and Brown, D.D. (1993). Expression of *Xenopus laevis* prolactin and thyrotropin genes during metamorphosis. *Proc. Natl. Acad. Sci. USA* **90**, 3820–3824.

Budayr, A.A., Halloran, B.P., King, J.C., Kiep, D., Nissenson, R.A., and Strewler, G.H. (1989). High levels of a parathyroid hormone-like protein in milk. *Proc. Natl. Acad. Sci. USA* **86**, 7183–7185.

Buffenstein, R., Sergeev, I.N., and Pettifor, J.M. (1993). Vitamin D hydroxylase and their regulation in a naturally vitamin D-deficient subterranean mammal, naked mole rat (*Heterocephalus glaber*). *J. Endocr.* **138**, 59–64.

Buffenstein, R., Skinner, D.C., Yahav, S., Moodley, G.P., Cavaleros, M., Zachen, D., Ross, F.P., and Pettifor, J.M. (1991). Effect of oral cholecalciferol supplementation at physiological and supraphysiological doses in naturally vitamin D$_3$-deficient subterranean damara mole rats (*Cryptomys damarensis*). *J. Endocr.* **131**, 197–202.

Buffenstein, R. and Yahav, S. (1991). Cholecalciferol has no effect on calcium and inorganic phosphorus balance in a naturally cholecalciferol-deplete subterranean mammal, the naked mole rat (*Heterocephalus glaber*). *J. Endocr.* **129**, 21–26.

Bull, J.J. (1980). Sex determination in reptiles. *Quart. Rev. Biol.* **55**, 3–21.

Buntin, J.D. (1979). Prolactin release in parent ring doves after brief exposure to their young. *J. Endocr.* **82**, 127–130.

Burchill, S.A. and Thody, A.J. (1986). Melanocyte-stimulating hormone and the regulation of tyrosinase activity in hair follicle melanocytes of the mouse. *J. Endocr.* **111**, 225–232.

Burchill, S.A., Thody, A.J., and Ito, S.

(1986). Melanocyte-stimulating hormone, tyrosinase activity and the regulation of eumelanogenesis and phaeomelanogenesis in the hair follicular melanocytes of the mouse. *J. Endocr.* **109**, 15–21.

Burns, R.K. (1935). The process of sex transformation in parabiotic *Amblystoma. III.* Conversion of testis to ovary in heteroplastic airs of *A. tigrinum* and *A. punctatum. Anat. Rec.* **63**, 101–129.

Burns, R.K. (1938). The effects of crystalline sex hormones on sex differentiation in *Ambylstoma. I.* Estrone. *Anat. Rec.* **71**, 447–468.

Burns, R.K. (1939). The effects of crystalline sex hormones on sex differentiation in *Ambylstoma. II.* Testosterone propionate. *Anat. Rec.* **73**, 73–87.

Burns, R.K. (1961). Role of hormones in the differentiation of sex, In *Sex and Internal Secretions* (edited by W.C. Young), pp. 76–158. Baltimore, MD: Williams & Wilkins.

Burzawa-Gerard, E. and Fontaine, Y.A. (1972). The gonadotropins of lower vertebrates. *Gen. Comp. Endocr.* **Suppl. 3**, 715–728.

Burzawa-Gerard, E., Goncharov, B.F., and Fontaine, Y.-A. (1975). L'hormone gonadotrope hypophysaire d'un poisson Chondrostéen, L'Esturgeon (*Acipenser stellatus Pall.*). *Gen. Comp. Endocr.* **27**, 289–295.

Butkus, A., Roche, P.J., Fernley, R.T., Haralambidis, J., Penschow, J.D., Ryan, G.B., Trahair, J.F., Tregear, G.W., and Coghlan, J.P. (1987). Purification and cloning of a corpuscles of Stannius protein from *Anguilla australis. Mol. Cell. Endocr.* **54**, 123–133.

Butler, D.G., Oudit, G.Y. and Cadinouche, M.Z.A. (1995). Angiotensin I- and II- and norepinephrine-mediated pressor responses in an ancient holostean fish, the bowfin (*Amia calva*). *Gen Comp. Endocr.* **98**, 289–302.

Butler, D.G., Siwanowicz, H., and Puskas, D. (1989). A re-evaluation of experimental evidence for the hormonal control of avian nasal salt glands. In *Progress in Avian Osmoregulation* (edited by M.A. Hughes and A. Chadwick), pp. 127–141. Leeds: Leeds Philosophical and Literary Society.

Byamungu, N., Darras, V.M. and Kühn, E.R. (1991). Purification of tilapia thyrotropin from a crude pituitary homogenate by immunoaffinity chromatography using a matrix of antibodies against porcine follicle-stimulating hormone. *Gen. Comp. Endocr.* **84**, 183–191.

Byrne, B.M., Gruber, M., and Ab, G. (1989). The evolution of egg yolk proteins. *Prog. Biophys, Mol. Biol.* **53**, 33–69.

Cahill, G.F., Aoki, T.T., and Marliss, E.B. (1972). Insulin and muscle protein. In *Handbook of Physiology*, Sect. 7 *Endocrinology*, vol. I, *Endocrine Pancreas*, pp. 563–577, Washington, DC: American Physiological Society.

Cake, M.H., Owen, F.J., and Bradshaw, S.D. (1980). Difference in concentration of progesterone in plasma between pregnant and non-pregnant quokkas (*Setonix brachyurus*). *J. Endocr.* **84**, 153–158.

Callard, I.P., Callard, G.V., Lance, V., and Eccles, S. (1976). Seasonal changes in testicular structure and function and the effects of gonadotropins in the fresh-water turtle, *Chrysemis picta. Gen. Comp. Endocr.* **30**, 347–356.

Callard, I.P. and Doolittle, J.P. (1973). The influence of intrahypothalamic injections of progesterone on ovarian growth and function in the ovoviviparous iguanid lizard *Sceloporus cyanogenys. Comp. Biochem. Physiol. A* **44**, 625–629.

Callard, I.P., Doolittle, J., Banks, W.L., and Chan, S.W.C. (1972). Recent studies on the control of the reptilian ovarian cycle. *Gen. Comp. Endocr.* **Suppl. 3**, 65–75.

Callard, I.P., Fileti, L.A., Perez, L.E., Sorbera, L.A., Giannoukos, G. Klosterman, L.L., Tsang, P., and McCracken, J.A. (1992). Role of the corpus luteum and progesterone in the evolution of vertebrate viviparity. *Am. Zool.* **32**, 264–275.

Callard, I.P., Klosterman, L.L., Sorbera, L.A., Fileti, L.A., and Reese, J.C. (1989). Endocrine regulation of reproduction in elasmobranchs: archetype for terrestrial vertebrates. *J. Exp. Zool.* **Suppl, 2**, 12–22.

Callard, I.P. and Koob, T.J. (1993). Endocrine regulation of the elasmobranch reproductive tract. *J. Exp. Zool.* **266**, 368–377.

Callard, I.P., Riley, D., and Perez, L. (1990). Vertebrate vitellogenesis: molecular model for multihormonal control of gene regulation. In *Progress in Comparative Endocrinology* (edited by A. Epple, C.G. Scanes, and M.H. Stetson), pp. 343–348. New York: Wiley-Liss.

Campbell, R.R., Etches, R.J., and Leatherland, J.F. (1981). Seasonal changes in plasma prolactin concentration and carcass lipid levels in the lesser snow goose (*Anser caerulescens caerulescens*). *Comp. Biochem. Physiol. A* **68**, 653–657.

Campfield, L.A., Smith, F.J., Guisez, Y., Devos, R., and Burn, P. (1995). Recombinant mouse oB protein – evidence for a peripheral signal linking adiposity and central neural network. *Science* **269**, 546–549.

Capelli, J.P., Wesson, L.G., and Aponte, G.E. (1970). A phylogenetic study of the renin-angiotensin system. *Am. J. Physiol.* **218**, 1171–1178.

Cardwell, J.R., and Liley, N.R. (1991). Hormonal control of sex and color change in the stoplight parrotfish, *Sparisoma viride. Gen. Comp. Endocr.* **81**, 7–20.

Caron, M.G. and Lefkowitz, R.J. (1993). Catecholamine receptors: structure, function, and regulation. *Rec. Prog. Horm. Res.* **48**, 277–290.

Carrick, S. and Balment, R.J. (1983). The renin–angiotensin system and drinking in the euryhaline flounder, *Platichthys flesus. Gen. Comp. Endocr.* **51**, 423–433.

Cascone, O., Turyn, D., Dellacha, J.M., Machado, V.L.A., Marques, M., Vita, N., Cassan, C., Ferrara, P., and Guillemot, J.C. (1991). Isolation, purification, and primary structure of insulin from the turtle *Chrysemys dorbigni. Gen. Comp. Endocr.* **84**, 355–359.

Castrucci, A., Hadley, M.E., Wilkes, B.C., Hruby, V.J., and Sawyer, T.K. (1989). Melanotropin structure–activity studies on melanocytes of the teleost fish *Synbranchus marmoratus. Gen. Comp. Endocr.* **74**, 209–214.

Ceccatelli, S., Hulting, A., Zhang, X., Gustafsson, L., Villar, M., and Hökfelt, T. (1993). Nitric oxide synthase in the rat anterior pituitary gland and the role of nitric oxide in regulation of luteinizing hormone secretion. *Proc. Natl. Acad. Sci. USA* **90**, 11 292–11 296.

Chadwick, C.S. and Jackson, H.R. (1948). Acceleration of skin growth and molting in the red eft of *Triturus viridescens* by means of prolactin injections. *Anat. Rec.* **101**, 718.

Challis, J.R.G. and Lye, S.J. (1994). Parturition. In *The Physiology of Reproduction*, 4th edn (edited by E. Knobil and J.D. Neill), vol. 2, pp. 985–1032. New York: Raven Press.

Chan, D.K.O. (1972). Hormonal regulation of calcium balance in teleost fish. *Gen. Comp. Endocr.* **Suppl. 3**, 411–420.

Chan, D.K.O. (1975). Cardiovascular and renal effects of urotensins I and II in the eel, *Anguilla rostrata. Gen. Comp. Endocr.* **27**, 52–61.

Chan, D.K.O. and Chester Jones, I. (1969). Neurohypophysial peptide action in the eel. *J. Endocr.* **45**, 161–174.

Chan, D.K.O., Chester Jones, I., Henderson, I.W., and Rankin, J.C. (1967). Studies of the experimental alteration of water and electrolyte composition of the eel (*Anguilla anguilla* L.). *J. Endocr.* **37**, 297–317.

Chan, D.K.O. and Woo, N.Y.S. (1978). Effect of glucagon on the metabolism of the eel. *Anguilla japonica. Gen. Comp. Endocr.* **35**, 216–225.

Chan, S.-J., Cao, Q.-P., and Steiner D.F. (1990). Evolution of the insulin super-family: cloning of a hybrid insulin/insulin-like growth factor cDNA from amphioxus. *Proc. Natl. Acad. Sci. USA* **87**, 9319–9323.

Chan, S.J., Keim, P., and Steiner, D.F. (1976). Cell-free synthesis of rat preproinsulins; characterization and partial amino acid sequence determination. *Proc. Natl. Acad. Sci., USA* **73**, 1964–1968.

Chan, S.J., Nagamatus, S., Cao, Q.-P., and Steiner, D.F. (1992). Structure and evolution of insulin and insulin-like growth factors in chordates. *Prog. Brain Res.* **92**, 15–24.

Chan, S.T.H., and Phillips, J.G. (1969). The biosynthesis of steroids by the gonads of the ricefield eel, *Monopterus*

albus at various stages during natural sex-reversal. *Gen. Comp. Endocr.* **12**, 619–636.

Chan, S.W.C. and Phillips, J.G. (1971). Seasonal variations in production *in vitro* of corticosteroids by the frog (*Rana rugulosa*) adrenal. *J. Endocr.* **50**, 1–17.

Chan, S.W.C., Ziegel, S., and Callard, I.P. (1973). Plasma progesterone in snakes. *Comp. Biochem. Physiol. A* **44**, 631–637.

Chang, A.C., Janosi, J., Hulsbeek, M., Dejong, D., Jeffrey, K.J., Noble, J.R., and Reddel, R.R. (1995). A novel human cDNA highly homologous to the fish hormone stanniocalcin. *Mol. Cell Endocr.* **112**, 241–247.

Chang, C.Y., Liu, Y.-X., Zhu, T.-T., and Zhu, H.-H. (1984). The reproductive endocrinology of Amphioxus. In *Frontiers in Physiological Research* (edited by D. Carlick and P.I. Korner), pp. 79–86. Canberra: Australian Academy of Science.

Chang, L.-T., Yu, N.-W., Hsu, C.-Y., and Liu, H.-W. (1996). Gonadal transformation in male *Rana catesbeiana* tadpoles intraperitoneally implanted with estradiol capsules. *Gen. Comp. Endocr.* **102**, 299–306.

Channing, C.P., Licht, P., Papkoff, H., and Donaldson, E.M. (1974). Comparative activities of mammalian, reptilian and piscine gonadotropins in monkey granulosa cell cultures. *Gen. Comp. Endocr.* **22**, 137–145.

Charnier, M. (1966). Action de la température sur sex-ratio chez l'embryon *Agama agama* (Agamidae, Lacertilien). *Soc. Biol.Ouest Afr.* **160**, 620–622.

Chartrel, N., Conlon, J.M., Danger, J.-M., Fournier, A., Tonon, M.-C., and Vaudry, H. (1991). Characterization of melanotropin-release-inhibiting factor (melanostatin) from frog brain: homology with human neuropeptide Y. *Proc. Natl. Acad. Sci. USA* **88**, 3862–3866.

Chartrel, N., Tonon, M.-C., Vaudry, H., and Conlon, J.M. (1991). Primary structure of frog pituitary adenylate cyclase-activating polypeptide (PACAP) and effects of ovine PACAP on the frog pituitary. *Endocrinology* **129**, 3367–3371.

Chartrel, N., Wang, Y., Fournier, A., Vaudry, H., and Conlon, J.M. (1995). Frog vasoactive intestinal polypeptide and

galanin: primary structures and effects on pituitary adenylate cyclase. *Endocrinology* **136**, 3079–3086.

Chauvet, J., Rouille, Y., Chauveau, C., Chauvet, M.T., and Acher, R. (1994). Special evolution of neurohypophysial hormones in cartilaginous fishes: asvatocin and phasvatocin, two oxytocin-like peptides isolated from the spotted dogfish (*Scyliorhinus caniculus*). *Proc. Natl. Acad. Sci. USA* **91**, 11 266–11 270.

Chauvet, M.T., Hurpet, D., Chauvet, J., and Acher, R. (1980). Phenypressin (Phe^2-Arg^8-vasopressin), a new neurohypophysial peptide found in marsupials. *Nature, Lond.* **287**, 640–641.

Chavin, W. (1976). The thyroid of the Sarcopterygian fishes (Dipnoi and Crossopterygii) and the origin of the tetrapod thyroid. *Gen. Comp. Endocr.* **30**, 142–155.

Chavin, W., Kim, K., and Tchen, T.T. (1963). Endocrine control of pigmentation. *Ann. N.Y. Acad. Sci.* **100**, 678–685.

Chen, C.-L.C. (1993). Editorial: inhibin and activin as paracrine/autocrine factors. *Endocrinology* **132**, 4–5.

Chen, C., Jack, J., and Garofalo, R.S. (1996). The *Drosophila* insulin receptor is required for normal growth. *Endocrinology* **137**, 846–856.

Chen, H., Charlat, O., Tartaglia, L.A., Woolf, E.A., Weng, X., Ellis, S.J., Lakey, N.D., Culpepper, J., Moore, K.J., Breitbart, R.E., Duyk, G.M., Tepper, R.I. and Morgenstern, J.P. (1996). Evidence that the diabetes gene encodes the leptin receptor: identification of a mutation in the leptin receptor gene in *db/db* mice. *Cell* **84**, 491–495.

Chen, J.D. and Evans, R.M. (1995). A transcriptional co-repressor that interacts with nuclear hormone receptors. *Nature Lond.* **377**, 454–457.

Chen, T.T., Marsh, A., Shamblott, M., Chan, K.-M., Tang, Y.-L., Cheng, C.M., and Yang, B.-Y. (1994). Structure and evolution of fish growth hormone and insulin-like growth factor genes. In *Fish Physiology*, vol. XIII, *Molecular Endocrinology of Fish* (edited by N.M. Sherwood and C.Y Hew), pp. 179–209. San Diego: Academic Press.

Cherel, Y., Leloup, J., and Le Maho, Y. (1988a). Fasting in king penguin. II. Hormonal and metabolic changes during molt. *Am. J. Physiol.* **254**, R178–R184.

Cherel, Y., Robin, J.-P., Walch, O., Karmann, H., Netchitailo, P., and Le Maho, Y. (1988b). Fasting in king penguin. I. Hormonal and metabolic changes during breeding. *Am. J. Physiol.* **254**, R70–R177.

Chester Jones, I., Bellamy, D., Chan, D.K.O., Follett, B.K., Henderson, I.W., Phillips, J.G., and Snart, R.S. (1972). Biological actions of steroid hormones on nonmammalian vertebrates. In *Steroids in Nonmammalian Vertebrates* (edited by D.R. Idler), pp. 414–480. New York: Academic Press.

Chester Jones, I., Chan, D.K.O., and Rankin, J.C. (1969). Renal function in the European eel (*Anguilla anguilla* L.).–II. Effects of the caudal neurosecretory system, corpuscles of Stannius, neurohypophysial peptides and vasoactive substances. *J. Endocr.* **43**, 21–31.

Chester Jones, I., Henderson, I.W., Chan, D.K.O., Rankin, J.C., Mosley, W., Brown, J.J., Lever, A.F., Robertson, J.I.S., and Tree, M. (1966). Pressor activity in extracts of the corpuscles of Stannius from the European eel (*Anguilla anguilla* L.). *J. Endocr.* **34**, 393–408.

Chester Jones, I., and Phillips, J.G. (1986). The adrenal and interrenal glands. In *Vertebrate Endocrinology: Fundamentals and Biomedical Implications* vol. 1, *Morphological Implications* (edited by P.K.T. Pang and M.P. Schreibman), pp. 319–350. Orlando: Academic Press.

Chester Jones, I., Rodgers, G.R., Bradshaw, S.D., Bradshaw, F.J., and Stewart, T. (1994). Observations on the adrenocortical structure of the female brush-tail possum (*Trichosurus vulpecula*) and locations of cell division. *Gen. Comp. Endocr.* **93**, 163–176.

Cheung, C.C., Thornton, J.E., Kuijper, J.L., Weigle, D.S., Clifton, D.K. and Steiner, R.A. (1997). Leptin is a metabolic gate for the onset of puberty in the female rat. *Endocrinology* **138**, 855–858.

Cheung, W.Y. (1980). Calmodulin plays a pivotal role in cellular regulation. *Science* **207**, 19–27.

Chevalier, B., Anglade, P., Derouet, M. Mollé, D., and Simon, J. (1996). Isolation and characterization of muscovy (*Cairna moschata*) duck insulin. *Comp. Biochem. Physiol. B* **114**, 19–26.

Chieffi, G. (1967). The reproductive system of elasmobranchs: developmental and endocrinological aspects. In *Sharks, Skates and Rays* (edited by P.W. Gilbert, R.F. Mathewson, and D.P. Rall), pp. 553–580. Baltimore: Johns Hopkins University Press.

Chin, W., Kronenberg, H.M., Dee, P.C., Maloof, F., and Habener, J.F. (1981). Nucleotide sequence of the mRNA encoding the pre-α-subunit of mouse thyrotropin. *Proc. Natl. Acad. Sci. USA* **78**, 5329–5333.

Chinkers, M., Garbers, D.M., Chang, M.-S., Lowe, D.G., Chin, H., Goeddel, D.V., and Schulz, S. (1989). A membrane form of guanylate cyclase is an atrial natriuretic peptide receptor. *Nature, Lond.* **338**, 78–83.

Chiou, S. and Vesely D.L. (1995). Kaliuretic peptide: the most potent inhibitor of Na$^+$-K$^+$-ATPase of the atrial natriuretic peptides. *Endocrinology* **136**, 2033–2039.

Chiu, K.W. and Lynn, W.G. (1972). Observations on thyroidal control of sloughing in the garter snake, *Thamnophis sirtalis. Copeia* 1972 (no. 1), 158–163.

Chiu, K.W. and Phillips, J.G. (1971a) The effect of hypophysectomy and of injections of thyrotrophin into hypophysectomized animals on the sloughing cycle of the lizard *Gekko gecko* L. *J. Endocr.* **49**, 611–618.

Chiu, K.W. and Phillips, J.G. (1971b) The role of prolactin in the sloughing cycle in the lizard *Gekko gecko* L. *J. Endocr.* **49**, 625–634.

Chiu, K.W., Sham, J.S.K., Maderson, P.F.A., and Zucker, A.H. (1986). Interaction between thermal environments and hormones affecting skin-shedding frequency in the tokay (*Gekko gecko*) (Gekkonidae, Lacertilia), *Comp. Biochem. Physiol. A* **84**, 345–351.

Chiu, K.W., Wong, C.C., Lei, F.H., and Tam, V. (1975). The nature of thyroidal secretions in reptiles. *Gen. Comp. Endocr.* **25**, 74–82.

Choi, Y.-H., Furuse, M., Satoh, S., and Okumura, J. (1994). Endogenous cholecystokinin is not a major regulator of food intake in the chicken. *J. Comp. Physiol. B* **164**, 425–429.

Christakos, S., Brunette, M.G., and Norman, A.W. (1981). Localization of immunoreactive vitamin D-dependent calcium binding protein in chick nephron. *Endocrinology* **109**, 322–324.

Christakos, S. and Norman, A.W. (1978). Vitamin D$_3$ induced calcium binding protein in bone tissue. *Science* **202**, 70–71.

Churchill, P.C., Malvin, R.L., Churchill, M.C., and McDonald, F.D. (1979). Renal function in *Lophius americanus*: effects of angiotensin II. *Am. J. Physiol.* **236**, R297–R301.

Cinti, S., Frederich, R.C., Zingaretti, M.C., de Matteis, R., Flier, J.S., and Lowell, B.B. (1997). Immunohistochemical localization of leptin and uncoupling protein in white and brown fat tissue. *Endocrinology* **138**, 797–804.

Cioffi, J.A., Shafer, A.W., Zupancic, T.J., Smith-Gbur, M.J., Mikhail, A., Platika, D., and Snodgrass, H.R. (1996). Novel B219/OB receptor isoforms: possible role of leptin in hematopoiesis and reproduction. *Nature Med.* **2**, 585–588.

Citri, Y. and Schramm, M. (1980). Resolution, reconstitution and kinetics of the primary action of a hormone receptor. *Nature, Lond.* **287**, 297–300.

Clapham, D.E. (1995). Calcium signalling. *Cell* **80**, 259–268.

Clark, J.H. and Mani, S.K. (1994). Action of ovarian steroid hormones. In *The Physiology of Reproduction*, 2nd edn (edited by E. Knobil and J.D. Neill), vol. 1, pp. 1011–1962. New York: Raven Press.

Clark N.B. (1967). Influence of estrogens upon serum calcium, phosphate and protein concentrations of fresh-water turtles. *Comp. Biochem. Physiol.* **20**, 823–834.

Clark, N.B. (1972). Calcium regulation in reptiles. *Gen. Comp. Endocr. Supply* **3**, 430–440.

Clark, N.B. and Dantzler, W.H. (1972). Renal tubular transport of calcium and phosphate in snakes: role of parathyroid hormone. *Am. J. Physiol.* **223**, 1455–1464.

Clark, N.B., and Wideman, R.F. (1977). Renal excretion of phosphate and calcium in parathyroidectomized starlings. *Am. J. Physiol.* **223**, F138–F144.

Clark, N.B. and Wideman, R.F. (1980). Calcitonin stimulation of urine flow and sodium excretion in the starling. *Am. J. Physiol.* **238**, R406–R412.

Clarke, W.C., Farmer, S.W., and Hartwell, K.M. (1977). Effect of teleost pituitary growth hormone on growth of *Tilapia mossambica* and on growth and seawater adaptation of sockeye salmon (*Oncorhynchus nerka*). *Gen. Comp. Endocr.* **33**, 174–178.

Clemons, G.K. and Nicoll, C.S. (1977). Effects of antisera to bullfrog prolactin and growth hormone on metamorphosis on *Rana catesbeiana* tadpoles. *Gen. Comp. Endocr.* **31**, 495–497.

Cobb, C.S. and Brown, J.A. (1992). Angiotensin II binding to tissues of the rainbow trout, *Oncorhynchus mykiss*, studied by autoradiography. *J. Comp. Physiol. B* **162**, 197–202.

Cofré, G. and Crabbé, J. (1965). Stimulation by aldosterone of active sodium transport by the isolated colon of the toad *Bufo marinus*. *Nature, Lond.* **207**, 1299–1300.

Cohn, D.V., Smardo, F.L., and Morrissey, J.J. (1979). Evidence for internal homology in bovine preproparathyroid hormone. *Proc. Natl. Acad. Sci., USA* **76**, 1469–1471.

Collie, N.L. and Hirano, T. (1987). Mechanisms of hormone actions on intestinal transport. In *Vertebrate Endocrinology: Fundamentals and Biomedical Implications*, vol. 2, *Regulation of Water and Electrolyes* (edited by P.K.T. Pang and M.P. Schreibman), pp. 239–270. San Diego: Academic Press.

Collins, S., Kuhn, C.M., Petro, A.E., Swick, A.G., Chrunyk, B.A., and Surwit, R.S. (1996). Role of leptin in fat regulation. *Nature, Lond.* **380**, 677.

Colombo, L. and Yaron, Z. (1976). Steroid 21-hydroxylase activity in the ovary of the snake *Storeria dekayi* during pregnancy. *Gen. Comp. Endocr.* **28**, 403–412.

Conley, A.J. and Mason, J.I. (1994). Endocrine function of the placenta. In

Textbook of Fetal Physiology (edited by G.D. Thorburn and R. Harding), pp. 16–29. Oxford: Oxford University Press.

Conlon, J.M. (1989). Biosynthesis of regulatory peptides–evolutionary aspects. In *The Comparative Physiology of Regulatory Peptides* (edited by S. Holmgren) pp. 344–369. London: Chapman & Hall.

Conlon, J.M. (1990). Somatostatin: aspects of molecular evolution. In *Progress in Comparative Endocrinology* (edited by A. Epple, C.G. Scanes, and M.H. Stetson), pp. 10–15. New York: Wiley-Liss.

Conlon, J.M., Andrews, P.C., Falkmer, S., and Thim, L. (1988). Isolation and structural characterization of insulin from the holocephalan fish, *Chimaera monstrosa* (rabbit fish). *Gen. Comp. Endocr.* **72**, 154–160.

Conlon, J.M., Andrews, P.C., Thim, L., and Moon, T.W. (1991a). The primary structure of glucagon-like peptide but not insulin has been conserved between the American eel, *Anguilla rostrata* and the European eel *Anguilla anguilla*. *Gen. Comp. Endocr.* **82**, 23–32.

Conlon, M., Bjørnholm, B., Jørgenson, F.S., Yousin, J.H., and Schwartz, T.W. (1991b). Primary structure and conformational analysis of peptide methionine–tyrosine, a peptide related to neuropeptide Y and peptide YY isolated from lamprey intestine. *Eur. J. Biochem.* **199**, 293–298.

Conlon, J.M., Bondareva, V., Rusakov, Y., Plisetskaya, E.M., Mynarcik, D.C., and Whittaker, J. (1995a). Characterization of insulin, glucagon, and somatostatin from the river lamprey, *Lampetra fluviatilis*. *Gen. Comp. Endocr.* **100**, 96–105.

Conlon, J.M. Göke, R., Andrews, P.C., and Thim, L. (1989). Multiple molecular forms of insulin and glucagon-like peptide from the Pacific ratfish (*Hydrolagus colliei*). *Gen. Comp. Endocr.* **73**, 136–146.

Conlon, J.M., Nielsen, P.F., and Youson, J.H. (1993). Primary structures of glucagon and glucagon-like peptide isolated from intestine of parasitic phase lamprey, *Petromyzon marinus*. *Gen. Comp. Endocr.* **91**, 96–104.

Conlon, J.M., Nielsen, P.F., Youson, J.H.,

and Potter, I.C. (1995b). Proinsulin and somatostatin from the islet organ of the Southern-hemisphere lamprey, *Geotria australis*. *Gen. Comp. Endocr.* **100**, 413–422.

Conlon, J.M., O'Harte, F., Smith, D.D., Tonon, M.-C., and Vaudry, H. (1992). Isolation and primary structure of urotensin II from the brain of a tetrapod, the frog *Rana ridibunda*. *Biochem. Biophys. Res. Commun.* **188**, 578–583.

Conlon, J.M., O'Toole, L., and Thim, L. (1987). Primary structure of glucagon from the gut of the common dogfish (*Scyliorhinus caniculus*). *FEBS Lett.* **214**, 50–56.

Conlon, J.M., and Thim, L. (1985). Primary structure of glucagon from an elasmobranch fish, *Torpedo marmorata*. *Gen. Comp. Endocr.* **60**, 398–405.

Conlon, J.M., Yano, Y., Waugh, D., and Hazon, N. (1996). Distribution and molecular forms of urotensin II and its role in cardiovascular regulation in vertebrates. *J. Exp. Zool.* **275**, 226–238.

Constans, T., Chevalier, B., Derouet, M., and Simon, J. (1991). Insulin sensitivity and liver insulin receptor structure in ducks from two genera. *Am. J. Physiol.* **261**, R882–890.

Cook, A.F., Stacey, N.E., and Peter, R.E. (1980). Periovulatory changes in serum cortisol levels in the goldfish. *Gen. Comp. Endocr.* **40**, 507–510.

Cooper, D.M.F., Mons, N., and Karpen, J.W. (1995). Adenylyl cyclases and the interaction between calcium and cAMP signalling. *Nature, Lond.* **374**, 421–424.

Cooper, G.J.S. (1994). Amylin compared with calcitonin gene-related peptide: structure, biology, and relevance to metabolic disease. *Endocrine Rev.* **15**, 163–201.

Cooper, W.E. and Ferguson, G.W. (1972). Steroids and color change during gravidity in the lizard *Crotaphytus collaris*. *Gen. Comp. Endocr.* **18**, 69–72.

Coote, J.H., Johns, E.J., Macleod, V.H., and Singer, B. (1972). Effect of renal nerve stimulation, renal blood flow and adrenergic blockade on plasma renin activity in the cat. *J. Physiol., Lond.* **226**, 15–36.

Copp, D.H. (1969) The ultimobranchial glands and calcium regulation. In *Fish*

Physiology (edited by W.S. Hoar and D.J. Randall), vol. II *The Endocrine System*, pp. 377–398. New York: Academic Press.

Copp. D.H. (1972). Calcium regulation in birds. *Gen. Comp. Endocr.*, 441–447.

Copp. D.H. (1976). Comparative endocrinology of calcitonin. In *Handbook of Physiology*, sect. 7 *Endocrinology*, vol. 7, *Parathyroid Gland*, pp. 431–442. Washington, DC: American Physiological Society.

Copp, D. H., Cameron, E.C., Cheney, B.A., Davidson, A.G.F., and Henze, K.G. (1962). Evidence for calcitonin – a new hormone from the parathyroid that lowers blood calcium. *Endocrinology* 70, 638–649.

Copp, D.H., Cockcroft, D.W., and Keuk, Y. (1967a). Calcitonin from ultimobranchial glands of dogfish and chickens. *Science* 158, 924–926.

Copp, D.H., Cockcroft, D.W., and Keuk, Y. (1967b). Ultimobranchial origin of calcitonin, hypocalcemic effect of extracts from chicken glands. *Can. J. Physiol. Pharmacol.* 45, 1095–1099.

Copp, D.H. and Kline, I.W. (1989). Calcitonin. In *Vertebrate Endocrinology: Fundamentals and Biomedical Implications*, vol. 3, *Regulation of Calcium and Phosphate* (edited by P.K.T. Pang and M.P. Schreibman), pp. 79–103. San Diego: Academic Press.

Corradino, R.A. (1993). Calbindin D_{28K} regulation in precociously matured chick egg shell gland *in vitro*. *Gen. Comp. Endocr.* 91, 158–166.

Cortelyou, J.R. (1967). The effect of commercially prepared parathyroid extracts on plasma and urine calcium levels in *Rana pipiens*. *Gen. Comp. Endocr.* 9, 234–240.

Costa, A., Poma, A., Navarra, P., Forsling, M.L., and Grossman, A. (1996). Gaseous transmitters as new agents in neuroendocrine regulation. *J. Endocr.* 149, 199–207.

Costanzo, L.S. and Windhager, E.E. (1992). Renal tubular transport of calcium. In *Handbook of Physiology* Sect. 8, *Renal Physiology*, vol. II, pp. 1759–1783. Bethesda: American Physiological Society.

Coupland, R.E. (1968). Corticosterone and methylation of noradrenaline by extraadrenal chromaffin tissue. *J. Endocr.* 41, 487–490.

Courty, Y. and Dufaure, J.P. (1980). Levels of testosterone, dihydrotestosterone, and androstenedione in the plasma and testis of a lizard (*Lacerta vivipara* Jacquin) during the annual cycle. *Gen. Comp. Endocr.* 42, 325–333.

Cowie, A.T. (1972). Lactation and its hormonal control. In *Hormones in Reproduction* (edited by C.R. Austin and R.V. Short), pp. 106–143. Cambridge: Cambridge University Press.

Cowie, A.T., Forsyth, I.A., and Hart, I.C. (1980). *Hormonal Control of Lactation*. Berlin: Springer-Verlag.

Crabbé, J. and de Weer, P. (1964). Action of aldosterone on the bladder and skin of the toad. *Nature, Lond.* 202, 278–279.

Craik, J.C.A. (1978). Effects of hypophysectomy on vitellogenesis in the elasmobranch *Scyliorhinus canicula* L. *Gen. Comp. Endocr.* 36, 63–67.

Crews, D. (1994). Temperature, steroids and sex determination. *J. Endocr.* 142, 1–8.

Crews, D. and Bergeron, J.M. (1994). Role of reductase and aromatase in sex determination in the red-eared slider (*Trachemys scripta*), a turtle with temperature-dependent sex determination. *J. Endocr.* 143, 279–289.

Crews, D., Cantú, A.R., and Bergeron, J.M. (1996). Temperature and non-aromatizable androgens: a common pathway in male sex determination in a turtle with temperature-dependent sex determination? *J. Endocr.* 149, 457–463.

Csaba, G. (1986). Why do hormone receptors arise? *Experientia* 42, 715–718.

Currie, M.G., Fok, K.F., Kato, J., Moore, R.J., Hamra, F.K., Duffin, K.L., and Smith, C.E. (1992). Guanylin: an endogenous activator of intestinal guanylate cyclase. *Proc. Natl. Acad. Sci. USA* 89, 947–951.

Dacke, C.G. (1979). *Calcium Regulation in Sub-mammalian Vertebrates*. New York: Academic Press.

Dahl, G.E., Evans, N.P., Moenter, S.M., and Karsch, F.J. (1994). The thyroid gland is required for reproductive neuroendocrine responses to photoperiod in the ewe. *Endocrinology* 135, 10–15.

Dangé, A.D. (1986). Branchial

Na⁺–K⁺-ATPase activity in freshwater or salt-water acclimated tilapia, *Oreochromis* (Sarotheradon) *mossambicus*: effects of cortisol and thyroxine. *Gen. Comp. Endocr.* 62, 341–343.

Daniels-McQueen, S., McWilliams, D., Birken, S., Canfield, R., Landefeld, T., and Boime, I. (1978). Identification of mRNAs encoding the α and β subunits of human chorionic gonadotropin. *J. Biol. Chem.* 253, 7109–7114.

Danks, J.A., Devlin, A.J., Ho, P.M.W., Diefenbach-Jagger, H., Power, D.M., Canario, A., Martin, T.J., and Ingleton, P.M. (1993). Parathyroid hormone-related protein is a factor in normal fish pituitary. *Gen. Comp. Endocr.* 92, 201–212.

Dantzler, W.K. (1992). Comparative physiology of the kidney. In *Handbook of Physiology*, Sect. 8, *Renal Physiology*, vol. I, pp. 415–474. Bethesda: American Physiological Society.

Dark, J., Miller, D.R., and Zucker, I. (1996). Gonadectomy in the spring reinstates hibernation in male golden-mantled ground squirrels. *Am. J. Physiol.* 270, R1240–R1243.

Datta, H.K., Rathod, H., Manning, P., Turnbull, Y., and McNeil, C.J. (1996). Parathyroid hormone induces superoxide anion burst in the osteoclast: evidence for the direct instantaneous activation of the osteoclast by the hormone. *J. Endocr.* 149, 269–275.

Dauphin-Villemant, C., Tonon, M.C., and Vaudry, H. (1992). Lack of effect of TRH on α-MSH release from the neurointermediate lobe of the lizard *Lacerta vivipara*. *Gen. Com. Endocr.* 87, 183–188.

Davidson, M.B. (1987). Effect of growth hormone on carbohydrate and lipid metabolism. *Endocrin. Rev.* 8, 115–131.

Davies, N.T., Munday, K.A., and Parsons, B.J. (1970). The effect of angiotensin on rat intestinal fluid transfer. *J. Endocr.* 48, 39–46.

Davis, P.J., Gregerman, R.I. and Poole, W.E. (1969). Thyroxine-binding proteins in the serum of the grey kangaroo. *J. Endocr.* 45, 477–478.

Davis, W.L. and Jones, R.G. (1982). Lysosomal proliferation in rachitic avian intestinal absorptive cells following

1,25-dihydroxycholcalciferol. *Tissue Res.* **14**, 585–595.

Deavers, D.R. and Musacchia, X.J. (1979). The function of glucocorticoids in thermogenesis. *Fedn. Proc.* **38**, 2177–2181.

de Beer, G.R. (1928). *Vertebrate Zoology.* pp. 401. London: Sidgwick & Jackson.

de Boland, A.R. and Nemere, I. (1992). Rapid action with vitamin D compounds. *J. Cell. Mol. Biochem.* **49**, 32–36.

de Bold, A.J. (1979). Heart atria granularity effects of changes in water–electrolyte balance. *Proc. Soc. Exp. Biol. Med.* **161**, 508–511.

de Bold, A.J. (1985). Atrial natriuretic factor: a hormone produced by the heart. *Science* **230**, 767–770.

Deftos, L.J. (1997). Editorial: there's something fishy and perhaps even fowl about the mammalian calcitonin receptor and its ligand. *Endocrinology* **138**, 519–520.

Deftos, L.J., Burton, D.W., Watkins, W.B., and Catherwood, B.D. (1980). Immunohistological studies of artiodactyl and teleost pituitaries with antisera to calcitonin. *Gen. Comp. Endocr.* **42**, 9–18.

DeGroot, L.J. (1993). Editorial: TRAPS – thyroid receptor auxilary proteins. *Endocrinology* **133**, 963–964.

Deguchi, T. (1981). Rhodopsin-like photosensitivity of isolated chicken pineal gland. *Nature, Lond.* **290**, 706–707.

de Leeuw, R., Habibi, H.R., Nahorniak, C.S., and Peter, R.E. (1989). Dopaminergic regulation of pituitary gonadotrophin-releasing hormone receptor activity in the goldfish (*Carassius auratus*). *J. Endocr.* **121**, 239–247.

de Loof, A. and Schoops, L. (1990). Homologies between the amino acid sequences of some vertebrate peptide hormones and peptides isolated from invertebrate sources. *Comp. Biochem. Physiol. B* **95**, 459–468.

DeLuca, H.F. (1971). The role of vitamin D and its relationship to parathyroid hormone and calcitonin. *Rec. Prog. Horm. Res.* **27**, 479–510.

DeLuca, H.F. (1974). Vitamin D: the vitamin and the hormone. *Fedn. Proc.* **33**, 2211–2219.

DeLuise, M., Martin, T.J., Greenberg, P.B. and Michelangeli, V. (1972). Metabolism

of porcine, human and salmon calcitonin in the rat. *J. Endocr.* **53**, 475–482.

Denning-Kendall, P.A., Sumpter, J.P., and Lowry, P.J. (1982). Peptides derived from pro-opiocortin in the pituitary gland of the dogfish, *Squalus acanthias. J. Endocr.* **93**, 381–390.

Dent, J.N. (1968). Survey of amphibian metamorphosis. In *Metamorphosis, a Problem in Developmental Biology* (edited by W. Etkin and L.I. Gilbert), pp. 271–311, New York: Appleton-Century-Crofts.

Dent, J.N. (1975). Integumentary effects of prolactin in lower vertebrates. *Am. Zool.* **15**, 923–935.

Dent, J.N. (1988). Hormonal interaction in amphibian metamorphosis. *Am. Zool.* **28**, 297–308.

Denton, D.A., McKinley, M.J., and Weisinger, R.S. (1996). Hypothalamic integration of body fluid regulation. *Proc. Natl. Acad. Sci. USA* **93**, 7397–7404.

De Roos, R. and DeRoos, C.C. (1972). Comparative effects of the pituitary-adrenocortical axis and catecholamines on carbohydrate metabolism in elasmobranch fish. *Gen. Comp. Endocr.* **Suppl. 3**, 192–197.

DeRoos, R. and DeRoos, C.C. (1979). Severe insulin-induced hypoglycemia in the spiny dogfish shark (*Squalus acanthias*). *Gen. Comp. Endocr.* **37**, 186–191.

de Vos, A.M., Ultsch, M., and Kossiakoff, A.A. (1992). Human growth hormone and extracellular domain of its receptor: crystal structure of the complex. *Science* **255**, 306–312.

de Vos, P., Saladin, R., Auwerx, J., and Staels, B. (1995). Induction of *ob* gene expression by corticosteroids is accompanied by body weight loss and reduced food intake. *J. Biol. Chem.* **270**, 15 958–15 961.

Dharmamba, M., Mayer-Gostan, N., Maetz, J., and Bern, H.A. (1973). Effect of prolactin on sodium movement in *Tilapia mossambica* adapted to sea water. *Gen. Comp. Endocr.* **21**, 179–187.

Diaz, M. and Lorenzo A. (1991). Regulation of amiloride-sensitive Na$^+$ absorption in the lizard (*Gallotia galloti*) colon by aldosterone. *Comp. Biochem. Physiol. A* **100**, 63–68.

Diaz, M. and Lorenzo, A. (1992). Aldosterone regulation of active sodium transport in the lizard colon (*Gallotia galloti*). *J. Comp. Physiol. B* **162**, 189–196.

DiBattista, J.A., Mehdi, A.Z., and Sandor, T. (1984). A detailed investigation of the cytoplasmic cortisol-binding receptor of North American eel (*Anguilla rostrata*) tissues. *Can. J. Biochem. Cell Biol.* **62**, 991–997.

Dicker, S.E. and Elliot, A.B. (1973). Neurohypophysial hormones and homeostasis in the crab-eating frog, *Rana cancrivora. Horm. Res.* **4**, 224–260.

Dickoff, W.W. (1993). Hormones metamorphosis, and smolting. In *The Endocrinology of Growth, Development and Metabolism in Vertebrates* (edited by M.P. Schreibman, C.G. Scanes, and P.K.T. Pang), pp. 519–540. San Diego, CA: Academic Press.

Dimaline, R., Young, J., and Gregory, H. (1986). Isolation from chicken antrum, and primary amino acid sequence of a 36-residue peptide of the gastrin/CCK family. *FEBS Lett.* **205**, 318–322.

Divecha, N. and Irvine, R.F. (1995). Phospholipid signalling. *Cell* **80**, 269–278.

Dobson, S. and Dodd. J.M. (1977a). Endocrine control of the testis in the dogfish *Scyliorhinus canicula* L. I. Effects of partial hypophysectomy on gravimetric, hormonal and biochemical aspects of testis function. *Gen. Comp. Endocr.* **32**, 41–52.

Dobson, S. and Dodd, J.M. (1977b). Endocrine control of the testis in the dogfish *Scyliorhinus canicula* L. II. Histological and ultrastructural changes in the testis after partial hypophysectomy (ventral lobectomy). *Gen. Comp. Endocr.* **32**, 53–71.

Dockray, G.J. (1975). Comparative studies on secretin. *Gen. Comp. Endocr.* **25**, 203–210.

Dockray, G.J. (1978). Evolution of secretin-like hormones. In *Gut Hormones* (edited by S.R. Bloom), pp. 64–67. London: Churchill Livingstone.

Dockray, G.J. (1979). Comparative biochemistry and physiology of gut hormones. *Annu. Rev. Physiol.* **41**, 83–95.

Dockray, G.J. (1989). Comparative

neuroendocrinology of gut peptides. In *Handbook of Physiology*, Sect. 6, *The Gastrointestinal System*, vol. II, *Neural and Endocrine Biology*, pp. 133–170. Bethesda: American Physiological Society.

Dockray, G.J. and Gregory, R.A. (1989). Gastrin. In *Handbook of Physiology*, Sect. 6, *The Gastrointestinal System*, vol. II, *Neural and Endocrine Biology*, pp. 311–336. Bethesda: American Physiological Society.

Dodd, J.M. (1960). Gonadal and gonadotrophic hormones in lower vertebrates. In *Marshall's Physiology of Reproduction* (edited by A.S. Parkes), vol. I (Pt. 2), pp. 417–582. London: Longmans.

Dodd, J.M. (1972a). Ovarian control in cyclostomes and elasmobranchs. *Am. Zool.* **12**, 325–339.

Dodd, J.M. (1972b). The endocrine regulation of gametogenesis and gonad maturation in fishes. *Gen. Comp. Endocr.* **Suppl. 3**, 675–687.

Dodd, J.M. (1975). The hormones of sex and reproduction and their effects in fish and lower chordates: twenty years on. *Am. Zool* **15**, (Suppl.) 137–171.

Dodd, J.M. (1986). The ovary. In *Vertebrate Endocrinology: Fundamentals and Biomedical Implications*, vol. 1, *Morphological Implications*, (edited by P.K.T. Pang and M.P. Schreibman), pp. 351–397. Orlando, FL: Academic Press.

Dodd, J.M. and Dodd, M.H.I. (1969). Phylogenetic specificity of thyroid stimulating hormone with special reference to the Amphibia. *Colloq. Int. C.N.R.S.* **177**, 277–285.

Dodd, J.M., Dodd, M.H.I., Sumpter, J.P., and Jenkins, N. (1982). Gonadotrophic activity in the buccal lobe (Rachendachhypophyse) of the pituitary gland of the rabbit fish *Hydrolagus colliei* (Chondrichthyes: Holocephali). *Gen. Comp. Endocr.* **48**, 174–180.

Dodd, J.M. and Sumpter, J.P. (1984). Fishes. In *Marshall's Physiology of Reproduction*, 4th edn (edited by G.E. Lamming), vol. 1, *Reproductive Cycles of Vertebrates*, pp. 1–126. Edinburgh: Churchill Livingstone.

Dolman, D. and Edmonds, C.J. (1975). The effect of aldosterone and the renin-angiotensin system on sodium, potassium and chloride transport by proximal and distal rat colon *in vivo*. *J. Physiol., Lond.* **250**, 597–611.

Donaldson, E.M., Yamzaki, F., Dye, H.M., and Philleo, W.W. (1972). Preparation of gonadotrophin from salmon (*Onchorhynchus tshawytsha*) pituitary glands. *Gen. Comp. Endocr.* **18**, 469–481.

Donoso, A.O. and Segura, E.T. (1965). Seasonal variations of plasma adrenaline and noradrenaline in toads. *Gen. Comp. Endcor.* **5**, 440–443.

Doolittle, R.F. (1981). Similar amino acid sequences: chance or common ancestry. *Science* **214**, 149–159.

Doolittle, R.F. (1983). Angiotensinogen is related to the antitrypsin–antithrombin–ovalbumin family. *Science* **222**, 417–419.

Doolittle, R.F. (1989). Similar amino acid sequences revisited. *Trend. Biochem. Sci* **14**, 244–245.

Doolittle, R.F. (1990). *Molecular Evolution: Computer Analysis of Protein and Nucleic Acid Sequences*. San Diego, CA: Academic Press.

Dores, R.M., Adamczyk, D.L., and Joss, J.M.P. (1990). Analysis of ACTH-related and CLIP-related peptides partially purified from the pituitary of the Australian lungfish, *Neoceratodus forsteri*. *Gen. Comp. Endocr.* **79**, 64–73.

Dores, R.M., Finger, T.E., and Gold, M. (1984). Immunohistochemical localization of enkephalin- and ACTH-related substances in the pituitary of the lamprey. *Cell Tissue Res.*. **235**, 107–115.

Dores, R.M. and Gorbman, A. (1990) Detection of Met-enkephalin and Leu-enkephalin in the brain of the hagfish, *Eptatretus stoutii*, and the lamprey, *Petromyzon marinus. Gen. Comp. Endocr.* **77**, 489–499.

Dores, R.M., Hoffman, N.E., Chilcutt-Ruth, T., Lancha, A., Brown, C., Marra, L., and Youson, J. (1996). A comparative analysis of somatolactin-related immunoreactivity in the pituitaries of four neopterygian fishes and one chondrostean fish: an immunohistochemical study. *Gen. Comp. Endocr.* **102**, 79–87.

Dores, R.M., Kaneko, D.J., and Sandoval, F. (1993a). An anatomical and biochemical study of the pituitary proopiomelanocortin in the polypteriform fish *Calamoichthys calabaricus. Gen. Comp. Endocr.* **90**, 87–99.

Dores, R.M., Noso, T., Rand-Weaver, M., and Kawauchi, H. (1993b). Isolation of prolactin and growth hormone from the pituitary of the holostean fish *Amia calva. Gen. Comp. Endocr.* **90**, 346–354.

Dores, R.M., Rubin, D.A., and Quinn, T.W. (1996). Is it possible to construct phylogenetic trees using polypeptide hormone sequences? *Gen. Comp. Endocr.* **103**, 1–12.

Dorris, P.A. and Stocco, D.M. (1989). An endogenous digitalis-like factor derived from the adrenal gland: studies of adrenal tissue from various sources. *Endocrinology* **125**, 2573–2579.

Doty, S.B., Robinson, R.A., and Schofield, B. (1976). Morphology of bone and histochemical staining characteristics of bone cells. In *Handbook of Physiology*, Sect. 7 *Endocrinology*, vol. 7, *The Parathyroid Gland*, pp. 3–23. Washington DC: American Physiological Society.

Douglas, W.W. (1972). Secretomotor control of adrenal medullary secretion: synaptic, membrane and ionic events in stimulus-secretion coupling. In *Handbook of Physiology*, Sect. 7 *Endocrinology*, vol. 6, *Adrenal Gland*, pp. 367–388, Washington DC: American Physiological Society.

Douglas, W.W. (1974). Mechanisms of release of neurohypophysial hormones: stimulation secretion coupling. In *Handbook of Physiology*, Sect. 7 *Endocrinology*, vol. 4, *The Pituitary Gland and its Neuroendocrine Control* (P. 1) pp. 191–224, Washington DC: American Physiological Society.

Drakenberg, K., Sara, V.R., Lindahl, K.I., and Kewish, B. (1989). The study of insulin-like growth factors in tilapia, *Oreochromis mossambicus. Gen. Comp. Endocr.* **74**, 173–180.

Duchamp, C. and Barré, H. (1993). Skeletal muscle as the major site of nonshivering thermogenesis in cold-acclimated ducklings. *Am. J. Physiol.* **265**, R1076–R1083.

Duchamp, C., Chatonnet, J., Dittmar, A.,

and Barré, H. (1993). Increased role of skeletal muscle in the calorigenic response to glucagon in cold-acclimated ducklings. *Am. J. Physiol.* **265**, R1084–R1091.

Ducouret, B., Tujague, M., Ashraf, J., Mouchel, N., Servel, N., Valotaire, Y., and Thompson, E.B. (1995). Cloning of teleost fish glucocorticoid receptor shows that it contains a different deoxyribonucleic acid-binding domain from that of mammals. *Endocrinology* **136**, 3774–3783.

Duncan, M.J. and Goldman, B.D. (1984). Hormonal regulation of the annual pelage color cycle in the Djungarian hamster, *Phodopus sungorus*. I. Role of the gonads and pituitary. *J. Exp. Zool.* **230**, 89–95.

Duncan, M.J., Goldman, B.D., Di Pinto, M.N., and Stetson, M.H. (1985). Testicular function and pelage color have different critical daylengths in the Djungarian hamster, *Phodopus sungorus sungorus*. *Endocrinology* **116**, 424–430.

Dunn, A.D. (1980). Studies on iodoproteins and thyroid hormones in Ascidians. *Gen. Comp. Endocr.* **40**, 473–483.

Dunn, A.D. and Dunn, J.T. (1982a). Thyroglobulin degradation by thyroidal proteases: action of purified cathepsin D. *Endocrinology* **111**, 280–289.

Dunn, A.D. and Dunn, J.T. (1982b). Thyroglobulin degradation by thyroidal proteases: action of thiol endopeptidases *in vitro*. *Endocrinology* **111**, 290–298.

Dunn, J.T. Anderson, P.C., Fox, J.W., Fassler, C.A., Dunn, A.D., Hite, L.A., and Moore, R.C. (1987). The sites of thyroid hormone formation in rabbit thyroglobulin. *J. Biol. Chem.* **262**, 16948–16952.

Dupont, W., Leboulenger, F., Vaudry, H., and Vaillant, R. (1976). Regulation of aldosterone secretion in the frog *Rana esculenta* L. *Gen. Comp. Endocr.* **29**, 51–60.

Duve, H. and Thorpe, A. (1981). Gastrin/cholecystokinin (CCK)-like immunoreactive neurons in the brain of the blowfly, *Calliphora erythrocephala* (Diptera). *Gen. Comp. Endocr.* **43**, 381–391.

Dyer, R.G. and Robinson, J.E. (1989). The LHRH pulse generator. *J. Endocr.* **123**, 1–2.

Eagle, L.A., Tait, B., Harrison, J.C., and Morton, J.L. (1996). Effect of weight gain in male and female Wistar rats on the secretion of the insulin secretagogue beta-cell tropin. *J. Endocr.* **151** (Suppl.), P23.

Eales, J.G. (1990). Thyroid function in poikilotherms. In *Progress in Comparative Endocrinology* (edited by A. Epple, C.G. Scanes, and M.H. Stetson), pp. 415–420. New York: Wiley-Liss.

Eastman, J.T. and Portanova, R. (1982). ACTH activity in the pituitary and brain of the least brook lamprey, *Lampetra aepyptera*. *Gen. Comp. Endocr.* **47**, 346–350.

Ebberink, R.H.M., Smit, A.B., and van Minnen, J. (1989). The insulin family: evolution and structure and function in vertebrates and invertebrates. *Biol. Bull.* **177**, 176–182.

Ebenezer, I.S. (1996). Systemic administration of cholecystokinin (CCK) inhibits operant water intake in rats: implications for the CCK-satiety hypothesis. *Proc. Roy. Soc. Lond. Ser. B* **263**, 491–496.

Ebisawa, T., Karne, S., Lerner, M.R., and Reppert, S.M. (1994). Expression cloning of a high-affinity melatonin receptor from *Xenopus* dermal melanophores. *Proc. Natl. Acad. Sci. USA* **91**, 6133–6137.

Ebling, F.J. (1974). Hormonal control and methods measuring sebaceous gland activity. *J. Invest. Dermatol.* **62**, 161–171.

Ebling, F.J., Ebling, E., Randall, V., and Skinner, J. (1975). The synergistic action of α-melanocyte-stimulating hormone and testosterone on the sebaceous, prostate, preputial, Harderian and lachrymal glands, seminal vesicles and brown adipose tissue in the hypophysectomized-castrated rat. *J. Endocr.* **66**, 407–412.

Eddy, J.M.P. and Strahan, R. (1968). The role of the pineal complex in the pigmentary effector system of lampreys, *Mordacia mordax* (Richardson) and *Geotria australis* Gray. *Gen. Comp. Endocr.* **11**, 528–534.

Eddy, L. and Lipner, H. (1976). Amphibian metamorphosis: the role of thyrotropin-like hormone. *Gen. Comp. Endocr.* **29**, 333–336.

Edmonds, C.J. (1987). Peripheral metabolism of thyroxine. *J. Endocr.* **114**, 337–339.

Egami, N. and Ishii, S. (1962). Hypophysial control of reproductive functions in teleost fishes. *Gen. Comp. Endocr.* **Suppl. 1**, 248–253.

Elaroussi, M.A., Prahl, J.M., and De Luca, H.F. (1994). The vitamin D receptors: primary structures and their origins. *Proc. Natl. Acad. Sci. USA* **91**, 11 596–11 600.

Elbrecht, A. and Smith, R.G. (1992). Aromatase enzyme activity and sex determination in chickens. *Science* **255**, 467–470.

Elizondo, R.S. and LeBrie, S.J. (1969). Adrenal–renal function in water snakes, *Natrix cyclopion*. *Am. J. Physiol.* **217**, 419–425.

Elliot, A.B. (1968). Effects of adrenaline on water uptake in *Bufo melanostictus*. *J. Physiol., Lond.* **197**, 87P–88P.

Ellory, J.C., Lahlou, B., and Smith, M.W. (1972). Changes in the intestinal transport of sodium induced by exposure of goldfish to a saline environment. *J. Physiol. Lond.* **222**, 497–509.

Emerson, S.B. and Hess, D.L. (1996). The role of androgens in opportunistic breeding, tropical frogs. *Gen. Comp. Endocr.* **103**, 220–230.

Eng, J., Yu, J., Rattan, S., and Yalow, R.S. (1992). Isolation and amino acid sequences of opossum vasoactive intestinal polypeptide and cholecystokinin octapeptide. *Proc. Nat. Acad. Sci. USA* **89**, 1809–1811.

Ensor, D.M. (1978). *Comparative Endocrinology of Prolactin*. London: Chapman & Hall.

Ensor, D.M. and Ball, J.M. (1972). Prolactin and osmoregulation in fishes. *Fedn. Proc.* **31**, 1615–1623.

Epple, A. (1969). The endocrine pancreas. In *Fish Physiology* (edited by W.S. Hoar and D.J. Randall), vol. II, *The Endocrine System*, pp. 275–319. New York: Academic Press.

Epple, A. and Brinn, J.E. (1986).Pancreatic islets. In *Vertebrate Endocrinology: Fundamentals and Biomedical Implications*, vol. 1, *Morphological Considerations* (edited by P.K.T. Pang and M.P. Schreibman), pp. 279–317. Orlando, FL: Academic Press.

Epple, A., Cake, M.H., Potter, I.C., and Tajbakhsh, M. (1992). Impact of complete isletectomy on plasma glucose in the southern hemisphere lamprey, *Geotria australis*. *Gen. Comp. Endocr.* **86**, 284–288.

Epstein, F.H., Cynamon, M., and McKay, W. (1971). Endocrine control of Na–K-ATPase and seawater adaptation in *Anguilla rostrata*. *Gen. Comp. Endocr.* **16**, 323–328.

Epstein, F.H., Clark, B., Taylor, M., and Silva, P. (1989). Plasma levels of atrial natriuretic peptide in *Anguilla rostrata* during adaptation to freshwater (FW) and seawater (SW). *Bull. Mt Desert Isl. Biol. Lab.* **28**, 16.

Erickson, J.C., Clegg, K.E., and Palmiter, R.D. (1996). Sensitivity to leptin and susceptibility to seizures of mice lacking neuropeptide Y. *Nature, Lond.* **381**, 415–418.

Erikson, E.F., Colvard, D.S., Berg, N.J., Graham, M.L., Mann, K.G., Spelsberg, T.C. and Riggs, B.L. (1988). Evidence of estrogen receptors in normal osteoblast-like cells. **241**, 84–86.

Esch, F., Ling, N., Bohlen, P., Baird, A., Benoit, R., and Guillemin, R. (1984). Isolation and characterization of bovine hypothalamic corticotropin-releasing factor. *Biochem. Biophys. Res. Commun.* **122**, 899–905.

Esneu, M., Delarue, C., Remy-Jouet, I., Manzardo, E., Fasolo, A., Fournier, A., Saint-Pierre, S., Conlon, J.M., and Vaudry, H. (1994). Localization, identification, and action of calcitonin-gene related peptide in the frog adrenal gland. *Endocrinology* **135**, 423–430.

Estler, C.J. and Ammon, H.P.T. (1969). The importance of the adrenergic beta-receptors for thermogenesis and survival of acutely cold-exposed mice. *Can. J. Physiol. Pharmacol.* **47**, 427–434.

Etches, R.J. (1996). *Reproduction in Poultry* pp. 286–291. Wallingford: CAB International.

Etkin, W. (1970). The endocrine mechanism of amphibian metamorphosis, an evolutionary achievement. *Mem. Soc. Endocr.* **18**, 137–153.

Etkin, W. and Gona, A.G. (1967).

Antagonism between prolactin and thyroid hormone in amphibian development. *J. Exp. Zool.* **165**, 249–258.

Evans, D.H. (1990). An emerging role for a cardiac peptide hormone in fish osmoregulation. *Annu. Rev. Physiol.* **52**, 43–60.

Evans, D.H. (1991). Rat atriopeptin dilates vascular smooth muscle of the ventral aorta from the shark (*Squalus acanthias*) and the hagfish (*Myxine glutinosa*). *J. Exp. Biol.* **157**, 551–555.

Evans, D.H., Chipouras, E., and Payne, J.A. (1989). Immunoreactive atriopeptin in plasma of fishes: its potential role in gill hemodynamics. *Am. J. Physiol.* **257**, R939–R945.

Evans, R.M. (1988). The steroid and thyroid hormone receptor superfamily. *Science* **240**, 889–895.

Fagerlund, U.H.M. (1967). Plasma cortisol concentration in relation to stress in adult sockeye salmon during the freshwater state in their life cycle. *Gen. Comp. Endocr.* **8**, 197–207.

Fahrenkrug, J. (1989). Vasoactive intestinal peptide. In *Handbook of Physiology*, Sect. 6 *The Gastrointestinal System*, vol. II, *Neural and Endocrine Biology*, pp. 611–629. Bethesda: American Physiological Society.

Falkmer, S., and Ostberg, Y. (1977). Comparative morphology of pancreatic islets in animals. In *The Diabetic Pancreas* (edited by B.W. Volk and K.F. Wellmann), pp. 15–59. New York: Plenum.

Falkmer, S., Ostberg, Y., and van Noorden, S.V. (1978). Entero-insular endocrine systems of cyclostomes: a clue to hormone evolution. In *Gut Hormones* (edited by S.R. Bloom), pp. 57–63. London: Churchill Livingston.

Falkmer, S. and Patent, G.J. (1972). Comparative and embryological aspects of the pancreatic islets. In *Handbook of Physiology*, Sect. 7 *Endocrinology*, vol. I, *Endocrine pancreas*, pp. 1–23. Washington, DC: American Physiological Society.

Farer, L.S., Robbins, J., Blumberg, B.S., and Rall, J.E. (1962). Thyroxine–serum protein complexes in various animals. *Endocrinology* **70**, 686–696.

Farrar, E.S. and Frye, B.E.(1977). Seasonal

variation in the effects of adrenalin and glucagon in *Rana pipiens. Gen. Comp. Endocr.* **33**, 76–81.

Fasano, S., Pierantoni, R., and Chieffi, G. (1989). Reproductive biology of elasmobranchs with emphasis on endocrines. *J. Exp. Zool.* **Suppl. 2**, 53–61.

Feher, J.J. and Wasserman, R.H. (1979) Intestinal calcium-binding protein and calcium absorption in cortisol-treated chicks: effects of vitamin D_3. *Endocrinology* **104**, 547–551.

Fehmann, H.-C., Göke, R., and Göke, B. (1995). Cell and molecular biology of the incretin hormones glucagon-like peptide-I and glucose-dependent insulin releasing polypeptide. *Endocrin. Rev.* **16**, 390–369.

Feist, D., Florant, G., Greenwood, M.R.C., and Feist, K. (1986). Regulation of energy stores in Arctic ground squirrels: brown fat thermogenic capacity, lipoprotein lipase and pancreatic hormones during fat deposition. In *Living in the Cold. Physiological and Biochemical Adaptation* (edited by H.C. Heller, X.J. Musachia, and L.C.K. Wang), pp. 281–285. New York: Elsevier.

Fenwick, J.C. (1984). Effect of vitamin D_3 (cholecalciferol) on plasma calcium and intestinal ^{45}calcium absorption in goldfish, *Carassius auratis L. Can. J. Zool.* **62**, 34–36.

Fenwick, J.C. and Brasseur, J.G. (1991). Effects of stanniectomy and experimental hypercalcemia on plasma calcium levels and calcium influx in American eels, *Anguilla rostrata*, LeSueur. *Gen. Comp. Endocr.* **82**, 459–465.

Fenwick, J.C., Flik, G., and Verbost, P.M. (1995). A passive immunization technique against teleost hypocalcemic hormone stanniocalcin provides evidence for the cholinergic control of stanniocalcin release and the conserved nature of the hormone. *Gen. Comp. Endocr.* **98**, 202–210.

Fenwick, J.C. and Lam, T.J. (1988). Effects of calcitonin on plasma calcium and phosphate in the mudskipper, *Periophthalmodon schlosseri* (Teleostei), in water and during exposure to air. *Gen. Comp. Endocr.* **70**, 224–230.

Fenwick, J.C., Smith, K., Smith, J., and Flik, G. (1984). Effects of various vitamin D analogs on plasma calcium and phosphorus and intestinal calcium absorption in fed and unfed American eels, *Anguilla rostrata*. *Gen. Comp. Endocr.* **55**, 398–404.

Ferguson, D.R. and Heller, H. (1965). Distribution of neurohypophysial hormones in mammals. *J. Physiol., Lond.* **180**, 846–863.

Ferguson, G.W. and Chen, C.L. (1973). Steroid hormones, color change and ovarian cycling in free-living female collared lizard, *Crotaphytus collaris*. *Am. Zool.* **13**, 1277.

Fergusson, B. and Bradshaw, S.D. (1991). Plasma arginine vasotocin, progesterone, and luteal development during pregnancy in the viviparous lizards, *Tiliqua rugosa*. *Gen. Comp. Endocr.* **82**, 140–151.

Fernholm, B. (1972). Neurohypophysial-adenohypophysial regulations in hagfish (Myxinoidea, Cyclostomata). *Gen. Comp. Endocr.* **Suppl. 3**, 1–10.

Feyrter, F. (1938). *Über diffuse endokrine Epitheliale Organe*. Leipzig: Barth.

Fiddes, J.C. and Goodman, H.M. (1980). The cDNA for the β-subunit of human chorionic gonadotropin suggests evolution of a gene by readthrough into the 3'-untranslated region. *Nature, Lond.* **286**, 684–687.

Field, M., Graf, L.H., Laird, W.J., and Smith, P.L. (1978). Heat-stable endotoxin in *Escherichia coli*: in vitro effects on guanylate cyclase activity, cyclic GMP concentration, and ion transport in small intestine. *Proc. Natl. Acad. Sci. USA* **75**, 2800–2804.

Figler, R.A., MacKenzie, D.S., Owens, D.W., Licht, P., and Amoss, M.S. (1989). Increased levels of arginine vasotocin and neurophysin during nesting in sea turtles. *Gen. Comp. Endocr.* **73**, 223–232.

Finidori, J. and Kelly, P.A. (1995). Cytokine receptor signalling through two novel families of transducer molecules: janus kinases, and signal transducers and activators of transcription. *J. Endocr.* **147**, 11–23.

Firth, B.T., Kennaway, D.J., and Rozenbilds, M.A.M. (1979). Plasma melatonin in the scincid lizard,

Trachydosaurus rugosus: diel rhythm, seasonality, and the effect of constant light and constant darkness. *Gen. Comp. Endocr.* **37**, 493–500.

Fitzsimons, J.T. (1972). Thirst. *Physiol. Rev.* **52**, 468–561.

Fitzsimons, J.T. (1979). *The Physiology of Thirst and Sodium Appetite*. Monographs of the Physiological Society no. 35. Cambridge University Press.

Fitzsimons, J.T. and Kaufman, S. (1977). Cellular and extracellular dehydration, and angiotensin as stimuli to drinking in the common iguana, *Iguana iguana*. *J. Physiol., Lond.* **265**, 443–463.

Flett, P.A., Leatherland, J.F., and van der Kraak, G. (1994). Endocrine correlates a seasonal reproduction in fish: thyroid hormones and gonadal growth, steroidogenesis, and energy partitioning in Great Lakes salmon. In *Perspectives in Comparative Endocrinology* (edited by K.G. Davy, R.E. Peter, and S.S. Tobe), pp. 602–610. Ottawa: National Research Council of Canada.

Flik, G., Fenwick, J.C., Kolar, Z., Mayer-Gostan, N., and Wendelaar Bonga, S.E. (1986). Effects of ovine prolactin on calcium uptake and distribution in *Oreochromis mossambicus*. *Am. J. Physiol.* **250**, R161–R166.

Flik, G. and Perry, S.F. (1989). Cortisol stimulates whole body calcium uptake and the branchial calcium pump in freshwater rainbow trout. *J. Endocr.* **120**, 75–82.

Flik, G., Rentier-Delrue, F., and Wendelaar Bonga, S.E. (1994). Calcitropic effects of recombinant prolactins in *Oreochromis mossambicus*. *Am. J. Physiol.* **266**, R1302–R1308.

Flik, G., van der Velden, J.A., Dechering, K.J., Verbost, P.M., Schoenmakers, T.J.M., Kolar, Z.I., and Wendelaar Bonga, S.E. (1993). Ca^{2+} and Mg^{2+} transport in gut and gills of tilapia, *Oreochromis mossambicus*: a review. *J. Exp. Zool.* **265**, 356–365.

Flynn, T.G., de Bold, M., and de Bold, A.J. (1983). The amino acid sequence of atrial peptide with potent diuretic and natriuretic effects. *Biochem. Biophys. Res. Commun.* **117**, 859–865.

Foà, P.P. (1972). The secretion of glucagon. In *Handbook of Physiology*, Sect. 7

Endocrinology, vol. I, *Endocrine pancreas*, pp. 261–277. Washington, DC: American Physiological Society.

Follet, B.K. (1963). Mole ratios of the neurohypophysial hormones in the vertebrate neural lobe. *Nature, Lond.* **198**, 693–694.

Follet, B.K. (1976). Plasma follicle-stimulating hormone during photoperiodically induced sexual maturation in male Japanese quail. *J. Endocr.* **69**, 117–126.

Follet, B.K. (1984). Birds. In *Marshall's Physiology of Reproduction*, 4th edn (edited by G.E. Lamming), vol. 1, *Reproductive Cycles of Vertebrates*, pp. 283–350. Edinburgh: Churchill Livingstone.

Follet, B.K. and Davies, D.T. (1979). The endocrine control of ovulation in birds. In *Animal Reproduction* (Barc. Symposium 3 edited by H.W. Hawk), pp. 323–344. New York: Halsted Press.

Follet, B.K. and Maung, S.L. (1978). Rate of testicular maturation, in relation to gonadotrophin and testosterone levels, in quail exposed to various artificial photoperiods and to natural daylengths. *J. Endocr.* **78**, 267–280.

Follet, B.K. and Nicholls, T.J. (1988). Acute effect of thyroid hormones in mimicking photoperiodically induced release of gonadotropins in Japanese quail. *J. Comp. Physiol. B* **157**, 837–843.

Follet, B.K., Nicholls, T.J., and Mayes, C.R. (1988). Thyroxine can mimic photoperiodically induced gonadal growth in Japanese quail. *J. Comp. Physiol. B* **157**, 829–835.

Follet, B.K. and Redshaw, M.R. (1974). The physiology of vitellogenesis. In *Physiology of the Amphibia* (edited by B. Lofts), vol. II, pp. 219–298. New York: Academic Press.

Follet, B.K. and Riley, J. (1967). Effect of the length of the daily photoperiod on thyroid activity in the female Japanese quail (*Coturnix coturnix japonica*). *J. Endocr.* **39**, 615–616.

Follet, B.K. and Robinson, J.E. (1980). Photoperiod and gonadotrophin secretion in birds. *Prog. Reprod. Biol.* **5**, 39–51.

Fontaine, M. (1954). Du déterminisme physiologique des migrations. *Biol. Rev.* **29**, 390–418.

Fontaine, M. (1956). The hormonal control

of water and salt–electrolyte metabolism in fish. *Mem. Soc. Endocr.* **5**, 69–81.

Fontaine, M. (1964). Corpuscules de Stannius et régulation ionique (Ca, K. NA) du milieu interiéur de l'Anguille (*Anguilla anguilla* L.). *C.R. Acad. Sci., Paris* **259**, 875–878.

Fontaine, M., Callamand, O., and Olivereau, M. (1949). Hypophyse et euryhalinité chez l'anguille. *C.R. Acad. Sci., Paris* **228**, 513–514.

Fontaine, Y.-A (1969a). La spécificité zoologique des protéines hypophysaires capables de stimuler la thyroide. *Acta Endocrin.*, Kobn, **Suppl, 136**, 1–154.

Fontaine, Y.-A. (1969b). La spécificité zoologique d'action des hormones thyréotropes. *Colloq. Int. C.N.R.S.* **177**, 267–275.

Fontaine, Y.-A. and Burzawa-Gerard, E. (1977). Esquisse de l'evolution des hormones gonadotropes et thyreotropes des Vertébrés. *Gen. Comp. Endocr.* **32**, 341–347.

Fontaine, Y.-A., Leloup-Hatey, J., and Dufour, S. (1991). Endocrine control of reproduction: evolutionary considerations. In *Vertebrate Endocrinology: Fundamentals and Biomedical Implications*, vol. 4, Pt. B, *Reproduction* (edited by P.K.T. Pang and M.P. Schreibman), pp. 311–331. San Diego, CA: Academic Press.

Foote, C.L. (1938). Influence of sex hormones on sex differentiation in Amphibia (*Rana pipiens*). *Anat. Rec.* **72** (Suppl.) 120–121.

Forrest, J.N., Cohen, A.D., Schon, D.A., and Epstein, F.H. (1973a). Na transport and Na–K-ATPase in gills during adaptation to seawater: effects of cortisol. *Am. J. Physiol.* **224**, 709–713.

Forrest, J.N., MacKay, W.C., Gallagher, B., and Epstein, F.H. (1973b). Plasma cortisol response to saltwater adaptation in the American eel *Anguilla rostrata*. *Am. J. Physiol.* **224**, 714–717.

Forster, M.E. and Fenwick, J.C. (1994). Stimulation of calcium efflux from the hagfish, *Eptatretus cirrhatus*, gill pouch by an extract of corpuscles of Stannius from an eel (*Anguilla dieffenbachii*): Teleostii. *Gen. Comp. Endocr.* **94**, 92–103.

Forte, L.R. and Hamra, F.K. (1996). Guanylin and uroguanylin: intestinal

peptide hormones that regulate epithelial transport. *News Physiol. Sci.* **11**, 17–24.

Foskett, J.K. (1987). The chloride cell. In *Comparative Physiology of Environmental Adaptation* (edited by R. Kirsch and B. Lahlou), pp. 83–91. Basel: Karger.

Foskett, J.K., Bern, H.A., Machen, T.E., and Connor, M. (1983). Chloride cells and the hormonal control of teleost fish osmoregulation. *J. Exp. Biol.* **106**, 255–281.

Foster, D.O. (1986). Quantitative role of brown adipose tissue in thermogenesis. In *Brown Adipose Tissue* (edited by P. Trayhurn and D.G. Nicholls), pp. 31–51. London: Edward Arnold.

Fouchereau-Peron, M., Arlot-Bonnemains, Y., Moukhtar, M.S., and Milhaud, G. (1986). Adaptation of rainbow trout (*Salmo gairdnerii*) to sea water: changes in calcitonin levels. *Comp. Biochem. Physiol. A* **83**, 83–87.

Fouchereau-Peron, M., Arlot-Bonnemains, Y., Moukhtar, M.S., and Milhaud, G. (1987). Calcitonin induces hypercalcemia in grey mullet and immature freshwater and sea-water adapted rainbow trout. *Comp Biochem. Physiol. A* **87**, 1051–1053.

Fouchereau-Peron, M., Moukhtar, M.S., Le Gal, Y., and Milhaud, G. (1981). Demonstration of specific receptors for calcitonin in isolated trout gill cells. *Comp. Biochem, Physiol A* **68**, 417–421.

Foulkes, N.S. and Sassone-Corsi, P. (1992). More is better: activators and repressors of the same gene. *Cell* **68**, 411–414.

Fox, M.D., Hyde, J.F., Muse, K.N., Keeble, S.C., Howard, G., London, S.N., and Curry, T.E. (1994). Galanin: a novel intraovarian regulatory peptide. *Endocrinology* **135**, 636–641.

Francis, G.L., McNeil, K.A., Wallace, J.C., Ballard, F.J., and Owens, P.C. (1989a). Sheep insulin-like growth factors I and II: sequences, activities and assays. *Endocrinology* **124**, 1173–1183.

Francis, G.L., Owens, P.C., McNeil, K.A., Wallace, J.C., and Ballard, F.J. (1989b). Purification, amino acid sequences and assay cross-reactivities of porcine insulin-like growth factor-I and -II. *J. Endocr.* **122**, 681–687.

Franklin, C.E., Holmgren, S., and Taylor, G.C. (1996). A preliminary investigation

of the effects of vasoactive intestinal peptide on secretion from the lingual glands of *Crocodylus porosus*. *Gen. Comp. Endocr.* **102**, 74–78.

Frantz, A.G., Kleinberg, D.L., and Noel, G.L., (1972). Studies on prolactin in man. *Rec. Prog. Hormon. Res.* **28**, 527–573.

Fraps, R.M. (1955). Egg production and fertility in poultry. In *Progress in the Physiology of Farm Animals* (edited by J. Hammond), vol. II, pp. 661–740. London: Butterworths.

Fraser, D.R. (1980). Regulation of the metabolism of vitamin D. *Physiol. Rev.* **60**, 551–613.

Fraser, W.D. (1989). The structural and functional relationships between parathyroid hormone-related protein and parathyroid hormone. *J. Endocr.* **122**, 607–609.

Frawley, L.S. (1994). Role of the hypophysial neurointermediate lobe in the dynamic release of prolactin. *Trend. Endocr. Metab.* **5**, 107–112.

Freeman, H.C. and Idler, D.R. (1973). Effects of corticosteroids on liver transaminases in two salmonids, the rainbow trout (*Salmo gairdnerii*) and the brook trout (*Salvelinus fontinalis*). *Gen. Comp. Endocr.* **20**, 69–75.

Freeman, J.D. and Bernard, R.A. (1990). Atrial natriuretic peptide and salt adaptation in the sea lamprey *Petromyzon marinus*. *Physiologist* **33**, A38.

Freeman, M.E. (1993). Editorial: neuropeptide Y: a unique member of the constellation of gonadotropin-releasing hormones. *Endocrinology* **133**, 2411–2412.

Freeman, M.E. (1994). The neuroendocrine control of the ovarian cycle of the rat. In *The Physiology of Reproduction*, 2nd edn (edited by E. Knobil and J.D. Neill), vol. 2, pp. 613–658. New York: Raven Press.

Fridberg, G. and Bern, H.A. (1968). The urophysis and the caudal neurosecretory system of fishes. *Biol. Rev.* **43**, 175–199.

Friedman, J.M. and Leibel, R.L. (1992). Tackling a weighty problem. *Cell* **69**, 217–220.

Friedman, P.A., Coutermarsh, B.A., Kennedy, S.M., and Gesek, F.A. (1996). Parathyroid hormone stimulation of calcium transport is mediated by dual signalling mechanisms involving protein kinase A and protein kinase C. *Endocrinology* **137**, 13–20.

Fritsch, H.A.R. and Sprang, R. (1977). On the ultrastructure of polypeptide hormone-producing cells in the gut of the Ascidian, *Ciona intestinalis* L. and the Bivalve, *Mytilus edulus* L. *Cell Tiss. Res.* **177**, 407–413.

Fritz, I.B. (1972). Insulin actions on carbohydrate and lipid metabolism. In *Biochemical Actions of Hormones* (edited by G. Litwack), vol. II, pp. 165–214. New York: Academic Press.

Fritz, I.B. and Lee, L.P.K. (1972). Fat mobilization and ketogenesis. In *Handbook of Physiology*, Sect. 7 *Endocrinology*, vol. I, *Endocrine pancreas*, pp. 579–596. Washington, DC: American Physiological society.

Froesch, E.R., Bürgi, H., Ramseier, E.B., Bally, P., and Labhart, A. (1963). Antibody-suppressible and non-suppressible insulin-like activities in human serum and their physiologic significance. an insulin assay with adipose tissue of increased precision and specificity. *J. Clin. Invest.* **42**, 1816–1834.

Fryer, J., Lederis, K., and Rivier, J. (1983). Urotensin I, a CRF-like neuropeptide, stimulates ACTH release from the teleost pituitary. *Endocrinology* **113**, 2308–2310.

Fuji, R., Wakatabi, H., and Oshima, N. (1991). Inositol 1,4,5-trisphosphate signals the motile response of fish chromatophores. I. Aggregation of pigment in the tilapia melanophore. *J. Exp. Zool.* **259**, 9–17.

Fuller, B.B. and Viskochil, D.H. (1979). The role of RNA and protein synthesis in mediating the action of MSH on mouse melanoma cells. *Life Sci.* **24**, 2405–2416.

Funder, J.W. (1991). Steroids, receptors, and response elements: the limits of signal specificity. *Rec. Prog. Horm. Res.* **47**, 191–207.

Funder, J.W. (1993). Aldosterone action. *Annu. Rev. Physiol.* **55**, 115–130.

Funder, J.W., Feldman, D., and Edelman, I.S. (1973). The roles of plasma binding and receptor specificity in the mineralocorticoid action of aldosterone. *Endocrinology* **92**, 994–1004.

Funkenstein, B., Silbergeld, A., Cavari, B., and Laron, Z. (1989). Growth hormone increases plasma levels of insulin-like growth factor (IGF-I) in a teleost, the gilthead seabream (*Sparus aurata*). *J. Endocr.* **120**, R19–R21.

Furr, B.J.A., Bonney, R.C., England, R.J., and Cunningham, F.J. (1973). Luteinizing hormone and progesterone in peripheral blood during the ovulatory cycle in the hen. *Gallus domesticus. J. Endocr.* **57**, 159–169.

Furr, B.J.A. and Smith, G.K. (1975). Effects of antisera against gonadal steroids on ovulation in the hen *Gallus domesticus. J. Endocr.* **66**, 303–304.

Furutani, Y., Morimoto, Y., Shibahara, S., Noda, M., Takahashi, H., Hirose, T., Asai, M., Inayama, S., Hayashida, H., Miyata, T., and Numa, S. (1983). Cloning and sequence analysis of cDNA for ovine corticotropin-releasing factor precursor. *Nature, Lond.* **301**, 537–540.

Gabou, L., Boisnard, M., Gourdou, I., Jammes, H., Dulor, J.-P., and Djiane, J. (1996). Cloning a rabbit prolactin cDNA and prolactin gene expression in the rabbit mammary gland. *J. Mol. Endocr.* **16**, 27–37.

Galli-Gallardo, S.M., Pang, P.K.T., and Oguro, C. (1979). Renal responses of the Chilean toad, *Calyptocephalella caudiverbera*, and the mud puppy *Necturus maculosus*, to mesotocin. *Gen. Comp. Endocr.* **37**, 134–136.

Galloway, S. and Cutfield, J.F. (1988). Insulin-like material from the digestive tract of the tunicate *Pyura pachydermatina* (sea tulip). *Gen. Comp. Endocr.* **69**, 106–113.

Galton, V.A. (1988). Iodothyronine 5′-deiodinase activity in the amphibian *Rana catesbeiana* at different stages of the life cycle. *Endocrinology* **122**, 1746–1750.

Galton, V.A. (1990). Mechanisms underlying the acceleration of thyroid-induced tadpole metamorphosis by corticosterone. *Endocrinology* **127**, 2997–3002.

Galton, V.A. (1992). The role of thyroid hormone in amphibian metamorphosis. *Trend. Endocr. Metab.* **3**, 96–100.

Galton, V.A. Davey, J.C., and Schneider, M.H. (1994). Mechanisms of thyroid hormone action in developing *Rana catesbeiana* tadpoles. In *Perspectives in Comparative Endocrinology* (edited by K.G. Davey, R.E. Peter, and S.S. Tobe), pp. 412–415. Ottawa: National Research Council of Canada.

Ge, W., Chang, J.P., Peter R.E., Vaughan, J., Rivier, J., and Vale, W. (1992). Effects of porcine follicular fluid, inhibin-A, and activin-A on goldfish gonadotropin release *in vivo. Endocrinology* **131**, 1922–1929.

Ge, W., Cook, H., Peter, R.E., Vaughan, J., and Vale, W. (1993). Immunocytochemical evidence for the presence of inhibin and activin-like proteins and their localization in goldfish gonads. *Gen. Comp. Endocr.* **89**, 333–340.

Gemmell, R.T. (1994). Control of seasonal reproduction in Australian marsupials: effects of biotic factors. In *Perspectives in Comparative Endocrinology* (edited by K.G. Davey, R.E. Peter., and S.S. Tobe), pp. 568–573. Ottawa: National Research Council of Canada.

George, F.W. and Wilson, J.D. (1994). Sex determination and differentiation. In *The Physiology of Reproduction*, 2nd edn (edited by E. Knobil and J.D. Neill), vol. 1, pp. 3–28. New York: Raven Press.

Georges, D., Tashima L., Yamamoto, S., and Bryant-Greenwood, G.D. (1990). Relaxin-like peptides in ascidians.1. Identification of the peptide and its mRNA in ovary of *Herdmania momus. Gen. Comp. Endocr.* **79**, 423–428.

Gerald, C., Walker, M.W., Criscione, L., Gustafson, E.L., Batzi-Hartmann, C., Smith, K.E., Vaysse, P., Durkin, M.M., Laz, T.M., Linemeyer, D.L., Schaffhauser, A.O., Whitebread, S., Hofbauer, K.G., Taber, R.I., Branchek, T.A., and Weinshank, R.L. (1996). A receptor subtype involved in neuropeptide-Y-induced food intake. *Nature, Lond.* **382**, 168–171.

Gerald, C., Walker, M.W., Vaysse, P.J.-J., He, C., Branchek, T.A., and Weinshank, R.L. (1995). Expression cloning and pharmacological characterization of a human hippocampal neuropeptide Y/peptide YY Y$_2$ receptor subtype. *J. Biol. Chem.* **270**, 26758–26761.

Gern, W.N.A., Owens, D.W., and Ralph, C.L. (1978). Plasma melatonin in the trout: day-night change demonstrated by radioimmunoassay. *Gen. Comp. Endocr.* **34**, 453–458.

Gilham, I.D. and Baker, B.I. (1984). Evidence for the participation of melanin-concentrating hormone in physiological colour change in the eel. *J. Endocr.* **102**, 237–243.

Gilham, I.D. and Baker, B.I. (1985). A black background facilitates the response to stress in teleosts. *J. Endocr.* **105**, 99–105.

Ginsburg, M. (1968) Production, release, transportation and elimination of the neurohypophysial hormone. In *Neurohypophysial Hormones and Similar Polypeptides* (edited by B. Berde), pp. 286–371. Berlin: Springer-Verlag.

Girgis, S.I., Galan, F.G., Arnett, T.R., Rogers, R.M., Bone, Q., Ravazzola, M., and MacIntyre, I. (1980). Immunoreactive human calcitonin-like molecule in the nervous systems of protochordates and a cyclostome, *Myxine*. *J. Endocr.* **87**, 375–382.

Glass, C.K. (1996). Some new twists in the regulation of gene expression by thyroid hormone and retinoic acid receptors. *J. Endocr.* **150**, 349–357.

Glowacki, J., O'Sullivan, J., Miller, M., Wilkie, D.W., and Deftos, L.J. (1985). Calcitonin produces hypercalcemia in leopard sharks. *Endocrinology* **116**, 827–829.

Goddard, C., Butterwith, S.C., Roberts, R.D., and Duclos, M.J. (1993). Insulin-like growth factors and IGF binding proteins in *Avian Endocrinology* (edited by P.J. Sharp), pp. 275–284. Bristol: The Journal of Endocrinology.

Godet, M. (1961). Le problème hydrique et son controle hypophysaire chez le Protoptère. *Ann. Fac. Sci. Univ. Dakar* **6**, 183–201.

Godine, J.E., Chin, W.W., and Habener, J.F. (1980). Luteinizing and follicle-stimulating hormones. *J. Biol. Chem.* **255**, 8780–8783.

Goetz, K.L. (1991). Renal natriuretic peptide (urodilatin?) and atriopeptin: evolving concepts. *Am. J. Physiol.* **261**, F921–F932.

Goldman, J.M. and Hadley, M.E. (1969). The beta receptor and cyclic 3',-5'-adenosine monophosphate: possible roles in the regulation of melanophore

responses of the spadefoot toad, *Scaphiopus couchi. Gen. Comp. Endocr.* **13**, 151–163.

Goldsmith, A.R. and Williams, D.M. (1980). Incubation in mallards (*Anas platyrhynchos*): changes in plasma levels of prolactin and luteinizing hormone. *J. Endocr.* **86**, 371–379.

Gonzalez, G.C., Martinez, P.M., Lederis, K., and Lukowiak, K. (1992). Distribution and coexistence of urotensin I and urotensin II peptides in the cerebral ganglia of *Aplysia californica. Peptides* **13**, 695–703.

Goodman, J.M., Frick, G.P., and Souza, S. (1996). Species specificity of the primate growth hormone receptor. *News Physiol. Sci.* **11**, 157–161.

Goodman, R.H., Jacobs, J.W., Chin, W.E., Lund, P.K., Dee P., and Habener, J.F. (1980). Nucleotide sequence of a cloned structural gene coding for a precursor of pancreatic somatostatin. *Proc. Natl. Acad. Sci. USA* **77**, 5869–5873.

Goodman R.L. (1994). Neuroendocrine control of the ovine estrous cycle. In *The Physiology of Reproduction*, 2nd edn. (edited by E. Knobil and J.D. Neill), vol. 2, pp. 659–709. New York: Raven Press.

Goodrich, E.S. (1917). "Proboscis pores" in craniate vertebrates, a suggestion concerning the premandibular somites and the hypophysis. *Quart. J. Microscop. Sci.* **62**, 539–553.

Gorbman. A. (1940). Suitability of the common goldfish for assay of thyrotropic hormone. *Proc. Soc. Exp. Biol. Med., New York* **45**, 772–773.

Gorbman, A. (1995). Olfactory origins and evolution of the brain–endocrine system: facts and speculation. *Gen. Comp. Endocr.* **97**, 171–178.

Gorbman, A. and Bern, H.A. (1962). *A Textbook of Comparative Endocrinology*, p. 220. New York: Wiley.

Gotshall, R.W., Davies, J.O., Shade R.E., Spielman, W., Johnson, J.A., and Braverman, B. (1973). Effects of renal denervation on renin release in sodium-depleted dogs. *Am. J. Physiol.* **225**, 344–349.

Grau, E.G. (1987). Thyroid hormones. In *Vertebrate Endocrinology: Fundamentals and Biomedical Implications*, vol. 2, *Regulation of Water and Electrolytes*

(edited by P.K.T. Pang and M.P. Schreibman), pp. 85–102. San Diego, CA: Academic Press.

Grau, E.G., Nishioka, R.S., and Bern, H.A. (1982). Effects of somatostatin and urotensin II on tilapia pituitary prolactin release and interaction between somatostatin, osmotic pressure, Ca^{2+}, and adenosine 3', 5'-monophosphate in prolactin release *in vitro. Endocrinology* **110**, 910–915.

Grau, E.G., Richman, N.H., and Borski, R.J. (1994). Osmoreception and a simple endocrine reflex of the prolactin cell of the tilapia *Oreochromis mossambicus*. In *Perspectives in Comparative Endocrinology* (edited by K.G. Davy, R.E. Peter, and S.S. Tobe), pp. 251–256. Ottawa: National Research Council of Canada.

Graves, J.A.M. (1994). Mammalian sex-determining genes. In *The Differences Between the Sexes* (edited by R.V. Short and E. Balaban), pp. 397–418. Cambridge: Cambridge University Press.

Gray, D.A. (1993). Plasma atrial natriuretic factor concentrations and renal actions in the domestic fowl. *J. Comp. Physiol. B* **163**, 519–523.

Gray, D.A. (1994). Role of endogenous atrial natriuretic peptide in volume expansion diuresis and natriuresis of the Pekin duck. *J. Endocr.* **140**, 85–90.

Gray, D.A. and Erasmus, T. (1988). Glomerular filtration changes during vasotocin-induced antidiuresis in kelp gulls. *Am. J. Phyiol.* **255**, R936–R939.

Gray, D.A. and Erasmus, T. (1989a). Control of renal and extrarenal salt and water excretion by plasma agniotensin II in the kelp gull (*Larus dominicanus*). *J. Comp. Physiol. B* **158**, 651–660.

Gray, D.A. and Erasmus T. (1989b). Control of plasma arginine vasotocin in kelp gulls (*Larus dominicanus*): roles of osmolality, volume, and plasma angiotensin II. *Gen. Comp. Endocr.* **74**, 110–119.

Gray, D.H., Schütz, H., and Gerstberger, R. (1991). Interaction of atrial natriuretic factor and osmoregulatory hormones in the Pekin duck. *Gen. Comp. Endocr.* **81**, 246–255.

Gray, E.S. and Kelley K.M. (1991). Growth regulation in the gobiid teleost, *Gillichthys mirabilis*: roles of growth hormone, hepatic growth hormone receptors and insulin-like growth factor-I. *J. Endocr.* **131**, 57–66.

Green J.A. and Baker, B.I. (1989). Influence of nerves and hormones on the control of trout melanophores. *Life Sci.* **45**, 1127–1132.

Green, S. and Chambon, P. (1986). A superfamily of potentially oncogenic hormone receptors. *Nature, Lond.* **324**, 615–617.

Greenwood, A.W. and Blyth, J.S.S. (1935). Variation in plumage response of brown leghorn breast feather and its reaction to oestrone. *Proc. Zool. Soc. Lond. Ser. A* **109**, 247–288.

Greer, M.A. and Haibach, H. (1974). Thyroid secretion. In *Handbook of Physiology*, Sect. 7 *Endocrinology*, vol. III, *Thyroid*, pp. 135–146. Washington, DC: American Physiological Society.

Greger, R., Gögelein, H., and Schlatter, E. (1988). Stimulation of NaCl secretion in the rectal gland of the dogfish *Squalus acanthias*. *Comp. Biochem. Physiol. A* **90**, 733–737.

Gregory, R.A. (1962). *Secretory Mechanisms of the Gastrointestinal Tract*, p. 153. London: Edward Arnold.

Griffiths, M. (1978). *The Biology of the Monotremes*, pp. 209–254. New York: Academic Press.

Griffiths, M. (1984). Mammals: monotremes. In *Marhsall's Physiology of Reproduction*, 4th edn (edited by G.E. Lamming), vol. 1, *Reproductive Cycles of Vertebrates*, pp. 351–385. Edinburgh: Churchill Livingstone.

Grimm-Jørgensen, Y. (1983). Immunoreactive somatostatin in two pulmonate gastropods. *Gen. Comp. Endocr.* **49**, 108–114.

Grober, M.S., Myers, T.R., Marchaterre, M.A., Bass, A.H., and Myers, D.A. (1995). Structure, localization, and molecular phylogeny of a GnRH cDNA from a Paracanthopterygian fish, the plainfin midshipman (*Porichthys notatus*). *Gen. Comp. Endocr.* **99**, 85–99.

Groscolas, R. and Leloup, J. (1986). The endocrine control of reproduction and molt in male and female emperor (*Aptenodytes forsteri*) and Adelie (*Pygoscelis adeliae*) penguins. *Gen. Comp. Endocr.* **63**, 264–274.

Gross, M. and Kumar, R. (1992). Vitamin D endocrine system and calcium and phosphorous homeostasis. In *Handbook of Physiology*, Sect. 8, *Renal Physiology*, vol. II, pp. 1817–1839. Bethesda: American Physiological Society.

Grossman, A. (1994). Editorial: NO news is good news. *Endocrinology* **134**, 1003–1005.

Grubb, B.R. and Bentley, P.J. (1987). Aldosterone-induced, amiloride inhibitable short-circuit current in the avian ileum. *Am. J. Physiol.* **253**, G211–G216.

Grubb, B.R. and Bentley, P.J. (1989). Avian colonic ion transport: effects of corticosterone and dexamethasone. *J. Comp. Physiol. B* **159**, 131–138.

Grubb, B.R. and Bentley, P.J. (1992). Effects of corticosteroids on short-circuit current across the cecum of the domestic fowl, *Gallus domesticus*. *J. Comp. Physiol. B* **162**, 690–695.

Grubb, B.R., Driscoll, S.M., and Bentley, P.J. (1987). Electrical PD, short-circuit current and fluxes of Na and Cl across avian intestine. *J. Comp. Physiol. B* **157**, 181–186.

Guibbolini, M.E., Henderson, I.W., Mosley, W., and Lahlou, B. (1988). Arginine vasotocin binding to isolated branchial cells of the eel: effect of salinity. *J. Mol. Endocr.* **1**, 125–130.

Guillemin, R., Brazeau, P., Bohlen, P., Esch, F., Ling, N., and Wehrenberg, W.B. (1982). Growth hormone-releasing factor from a human pancreatic tumor that caused acromegaly. *Science* **218**, 585–587.

Guillemin, R. and Burgus, R. (1972). The hormones of the hypothalamus. *Sci. Am.* **227** (November), 24–33.

Guillette, L.J., Cox, M.C., and Crain, D.A. (1996). Plasma insulin-like growth factor-I concentration during the reproductive cycle of the American alligator (*Alligator mississippiensis*). *Gen. Comp. Endocr.* **104**, 116–122.

Guillette, L.J., Propper, C.R., Cree, A., and Dores, R.M. (1991). Endocrinology of oviposition in the tuatara (*Sphenodon punctatus*). II. Plasma arginine vasotocin concentrations during natural nesting. *Comp. Biochem. Physiol. A* **100**, 819–822.

Gurr, J.A., Catterall, J.F., and Kourides,

I.A. (1983). Cloning of cDNA encoding the pre-β subunit of mouse thyrotropin. *Proc. Natl. Acad. Sci. USA* **80**, 2122–2126.

Gwinner, E., Schwabl-Benzinger, I., Schabl, H., and Dittami, J. (1993). Twenty-four hour melatonin profiles in a nocturnally migrating bird during and between migratory seasons. *Gen. Comp. Endocr.* **90**, 119–124.

Habener, J.F., Singh, F.R., Deftos, L.J., Neer, R.M., and Potts, J.T. (1971). Explanation for unusual potency of salmon calcitonin. *Nature, New Biol.* **232**, 91–92.

Haddad, M., Roder, S., Olsen, H.S., and Wagner, G.F. (1996). Immunocytochemical localization of stanniocalcin cells in the rat kidney. *Endocrinology* **137**, 2113–2117.

Hadley, M.E. (1972). Functional significance of vertebrate integumental pigmentation. *Am. Zool.* **12**, 63–76.

Hadley, M.E. (1980). Control of release and mechanism of action of melanocyte stimulating hormone. *Gen. Comp. Endocr.* **40**, 311.

Halaas, J.L., Gajiwala, K.S., Maffei, M., Cohen, S.L., Chait, B.T., Rabinowitz, D., Lallone, R.L., Burley, S.K., and Friedman, J.M. (1995). Weight-reducing effects of the plasma protein encoded by the *obese* gene. *Science* **269**, 543–546.

Halaban, R., Pomerantz, S.H., Marshall, S., and Lerner, A.B. (1984). Tyrosinase activity and abundance of Cloudman melanoma cells. *Arch. Biochem. Biophys.* **230**, 383–387.

Hall, P.F. (1969). Hormonal control of melanin synthesis in birds. *Gen. Comp. Endocr.* **Suppl. 2**, 451–458.

Hall, T.R., Harvey, S., and Chadwick, A. (1986). Control of prolactin secretion in birds: a review. *Gen. Comp. Endocr.* **62**, 171–184.

Hall, V.D., Bartke, A., and Goldman, B.D. (1982). Role of the testis in regulating the duration of hibernation in the Turkish hamster, *Mesocricetus brandti*. *Biol. Reprod.* **27**, 802–810.

Hall, V. and Goldman, B. (1980). Effects of gonadal steroid hormones on hibernation in the Turkish hamster (*Mesicricetus brandti*). *J. Comp. Physiol.* **135**, 107–114.

Hamra, F.K., Eber, S.L., Chin, D.T.,

Currie, M.G., and Forte, L.R. (1997). Regulation of intestinal uroguanylin/guanylin receptor-mediated responses by mucosal acidity. *Proc. Natl. Acad. Sci. USA* **94**, 2705–2710.

Hamra, F.K., Forte, L.R., Eber, S.L., Pidhorodeckyj, N.V., Krause, W.J., Freeman, R.H., Chin, D.T., Tompkins, J.A., Fok, K.F., Smith, C.E., Duffin, K.L., Siegel, N.R., and Currie, M.G. (1993). Uroguanylin: structure and activity of a second endogenous peptide that stimulates intestinal guanylate cyclase. *Proc. Natl. Acad. Sci. USA* **90**, 10464–10468.

Handa, R.K., Ferrario, C.M., and Strandhoy, J.W. (1996). Renal actions of angiotensin (1–7): *in vivo* and *in vitro* studies. *Am. J. Physiol.* **270**, F141–F147.

Handler, J.S., Bensinger, R., and Orloff, J. (1968). Effects of adrenergic agents on toad bladder response to ADH, 3', 5'-AMP, and theophyline. *Am. J. Physiol.* **215**, 1024–1031.

Hansel, W. and Echternkamp, S.E. (1972). Control of ovarian function in domestic animals. *Am. Zool.* **12**, 225–243.

Hanssen, R.G.J.M., Lafeber, F.P.J.G., Flik, G., and Wendelaar Bonga, S.E. (1989). Ionic and total calcium levels in the blood of the European eel (*Anguilla anguilla*): effects of stanniectomy and hypocalcin replacement therapy. *J. Exp. Biol.* **141**, 177–186.

Hanstrom, B. (1966). Gross anatomy of the hypophysis in mammals. In *The Pituitary Gland* (edited by G.W. Harris and B.T. Donovan), vol. 1, pp. 1–57. Berkeley, CA: University of California Press.

Haqq, C.M., King, C.-Y., Donahoe, P.K., and Weiss, M.A. (1993). SRY recognizes conserved DNA sites in sex-specific promotors. *Proc. Natl. Acad. Sci. USA* **90**, 1097–1101.

Harmeyer, J. and DeLuca, H.F. (1969). Calcium-binding protein and calcium absorption after vitamin D administration. *Arch. Biochem. Biophys.* **133**, 247–254.

Harmon, J.S. and Sheridan, M.A. (1992). Effects of nutritional state, insulin, and glucagon on lipid mobilization in rainbow trout, *Oncorhynchus mykiss*, *Gen. Comp. Endocr.* **87**, 214–221.

Harms, P.J., Tu, G.-F., Richardson, S.J.,

Aldred, A.R., Jaworowski, A., and Schreiber, G. (1991). Transthyretin (prealbumin) gene expression in choroid plexus is strongly conserved during evolution of vertebrates. *Comp. Biochem. Physiol. B* **99**, 239–249.

Harper, A.A. and Raper, H.S. (1943). Pancreozymin, a stimulant of the secretion of pancreatic enzymes in extracts of the small intestine. *J. Physiol., Lond.* **102**, 115–125.

Harper, C. and Toverud, S.U. (1973). Ability of thyrocalcitonin to protect against hypercalcemia in adult rats. *Endocrinology* **93**, 1354–1359.

Harri, M.N.E. (1972). Effect of season and temperature acclimation on the tissue catecholamine level and utilization in the frog *Rana temporaria*. *Comp. Gen. Pharmacol.* **3**, 101–112.

Harri, M. and Hedenstam, R. (1972). Calorigenic effect of adrenaline and noradrenaline in the frog, *Rana temporaria. Comp. Biochem. Physiol. A* **41**, 409–419.

Harrington, R.W. (1968). Delimitation of the thermolabile phenocritical period of sex determination and differentiation in the ontogeny of the normally hermaphroditic fish *Rivulus marmoratus*, Poey. *Physiol. Zool.* **41**, 447–459.

Hartman, F.A. and Brownell, K.A. (1949). *The Adrenal Gland*. London: Henry Kimpton.

Haruta, K., Yamashita, T., Kawashima, S. (1991). Changes in arginine vasotocin content in the pituitary of the medaka (*Oryzias latipes*) during osmotic stress. *Gen. Comp. Endocr.* **83**, 327–336.

Harvey, S. (1990). Thyrotropin-releasing hormone: a growth hormone-releasing factor. *J. Endocr.* **125**, 345–358.

Harvey, S. (1993). Growth hormone secretion in poikilotherms and homeotherms. In *The Endocrinology of Growth, Development and Metabolism in Vertebrates* (edited by M.P. Schreibman, C.G. Scanes, and P.K.T. Pang), pp. 151–162. San Diego, CA: Academic Press.

Harvey, S., Hall, T.R., Chadwick, A., and Ensor, D.M. (1989). Osmoregulation and the physiology of prolactin. In *Progress in Avian Osmoregulation* (edited by M.A. Hughes and A. Chadwick), pp.

81–109. Leeds: Leeds Philosophical and Literary Society.

Harvey, S., Trudeau, V.L., Ashworth, R.J., and Cockle, S.M. (1993). *p* Glutamylglutamylprolineamide modulation of growth hormone secretion in domestic fowl: antagonism of thyrotropin-releasing hormone action? *J. Endocr.* **138**, 137–147.

Harvey, S., Zeng, Y.-Y., and Pang, P.K.T. (1987). Parathyroid hormone-like immunoreactivity in fish plasma and tissues. *Gen. Comp. Endocr.* **68**, 136–146.

Hassin, S., Elizur, A., and Zohar, Y. (1995). Molecular cloning and sequence analysis of stripped bass (*Morone saxatilis*) gonadotropin-I and -II subunits. *J. Mol. Endocr.* **15**, 23–35.

Havel, P. and Taborsky, G.J. (1994). The contribution of the autonomic nervous system to changes in glucagon and insulin secretion during hypoglycemic stress. In *Endocrine Reviews Monographs, 2. The Endocrine Pancreas, Insulin Action, and Diabetes* (edited by L.E. Underwood), pp. 182–200. Bristol: The Endocrine Society.

Hawa, N.S., Hewison, M., Farrow, S.M., and O'Riordan, J.L.H. (1994). Editorial: cell mates in the superfamily. *Endocrinology* **135**, 1–3.

Hayashida, T. (1970). Immunological studies with rat pituitary growth hormone (RGH). II. Comparative immunochemical investigation of GH from representatives of various vertebrate classes with monkey antiserum to RGH. *Gen. Comp. Endocr.* **15**, 432–452.

Hayashida, T. (1971). Biological and immunochemical studies with growth hormone in pituitary extracts of holostean and chondrostean fishes. *Gen. Comp. Endocr.* **17**, 278–280.

Hazelwood, R.L. (1973). The avian endocrine pancreas. *Am. Zool.* **13**, 699–709.

Hazelwood, R.L. (1990). Pancreatic polypeptide (PP) and its relevant relatives. In *Progress in Comparative Endocrinology* (edited by A. Epple, C.G. Scanes, and M.H. Stetson), pp. 250–256. New York: Wiley-Liss.

Hazelwood, R.L. (1993). Pancreatic hormones and metabolism. In *The Endocrinology of Growth, Development, and Metabolism in Vertebrates* (edited by

M.P. Schreibman, C.G. Scanes, and P.K.T. Pang), pp. 289–325. San Diego, CA: Academic Press.

Hazon, N., Balment, R.J. Perrott, M., and O'Toole, L.B. (1989). The renin–angiotensin system and vascular and dipsogenic regulation in elasmobranchs. *Gen. Comp. Endocr.* **74**, 230–236.

Hazon, N. Bjenning, C., and Conlon, J.M. (1993). Cardiovascular actions of dogfish urotensin II in the dogfish *Scyliorhinus canicula. Am. J. Physiol.* **265**, R573–R576.

Hazon, N. and Henderson, I.W. (1985). Factors affecting the secretory dynamics of 1α-hydroxycorticosterone in the dogfish, *Scyliorhnus canicula. J. Endocr.* **59**, 50–55.

He, J. and Furmanski, P. (1995). Sequence specificity and transcriptional activation in the binding of lactoferrin to DNA. *Nature, Lond.* **373**, 721–724.

Heap, R.B. (1972). Role of hormones in pregnancy. In *Hormones in Reproduction* (edited by C.R. Austin and R. V. Short), pp. 72–105. Cambridge University Press.

Heap, R.B., Flint, A.P., and Gadsby, J.E. (1979). Role of embryonic signals in the establishment of pregnancy. *Brit. Med. Bull.* 35, No. 2, 129–135.

Heap, R.B., Perry, J.S., and Challis, J.R.G. (1973). Hormonal maintenance of pregnancy. In *Handbook of Physiology*, Sect. 7 *Endocrinology*, vol. II, *Female reproductive system* (Pt. 2), pp. 217–260. Washington, DC: American Physiological Society.

Heap, R.B., Renfree, M.B., and Burton, R.D. (1980). Steroid metabolism in the yolk sac placenta and endometrium of the tammar wallaby, *Macropus eugenii. J. Endocr.* **87**, 339–349.

Hedges, S.B., Moberg, K.D., and Maxson, L.R. (1990). Tetrapod phylogeny inferred from 18S and 28S ribosomal RNA sequences and a review of the evidence for Amniote relationships. *Mol. Biol. Evol.* **7**, 607–633.

Heierhorst, J., Mahlmann, S., Morley, S.D., Coe, I.R., Sherwood, N.M., and Richter, D. (1990). Molecular cloning of two distinct precursor cDNAs from chum salmon (*Onchorhynchus keta*) suggests an ancient gene duplication. *FEBS Lett.* **260**,

301–304.

Heierhorst, J., Morley, S.D., Figueroa, J., Krentler, C., Lederis, K., and Richter, D. (1989). Vasotocin and isotocin precursors from the white sucker, *Catostomus commersoni*: cloning and sequence analysis of the cDNAs. *Proc. Natl. Acad. Sci. USA* **86**, 5242–5246.

Heinig, J.A., Keeley, F.W., Robson, P., Sower, S.A., and Youson, J.H. (1995). The appearance of proopiomelanocortin early in vertebrate evolution: cloning and sequencing of POMC from a lamprey cDNA library. *Gen. Comp. Endocr.* **99**, 137–144.

Heller, H. (1972). The effect of neurohypophysial hormones on the female reproductive tract of lower vertebrates. *Gen. Comp. Endocr.* **Suppl. 3**, 703–714.

Heller, H. (1973). The effects of oxytocin and vasopressin during the oestrous cycle and pregnancy on the uterus of a marsupial species, the quokka (*Setonix brachyurus*). *J. Endocr.* **58**, 657–671.

Heller, H. (1974). Molecular aspects of comparative endocrinology. *Gen. Comp. Endocr.* **22**, 315–332.

Henderson, I.W. and Chester Jones, I. (1967). Endocrine influences on the net extrarenal fluxes of sodium and potassium in the European eel (*Anguilla anguilla* L.) *J. Endocr.* **37**, 319–325.

Henderson, I.W., Jotisankasa, V., Mosley, W., and Oguri, M. (1976). Endocrine and environmental influences upon plasma cortisol concentrations and plasma renin activity in the eel, *Anguilla anguilla* L. *J. Endocr.* **70**, 81–95.

Henderson, I.W., Oliver, J.A., McKeever, A., and Hazon, N. (1981). Phylogenetic aspects of the renin–angiotensin system. In *Advances in Animal and Comparative Physiology* (edited by G. Pethes and V.L. Frenyo), pp. 355–363. New York: Pergamon Press.

Henderson, J.R. (1969). Why are the islets of Langerhans? *Lancet* **ii**, 469–470.

Henry, H.L., Dutta, C., Cunningham, N., Blanchard, R., Penny, R., Tang, C., Marchetto, G., and Chou, S.-Y. (1992). The cellular and molecular regulation of 1,25(OH)$_2$D$_3$ production. *J. Steroid Biochem. Mol. Biol.* **41**, 401–407.

Henry, H. and Norman, A.W. (1975).

Presence of renal 25-hydroxyvitamin-D-1-hydroxylase in species of all vertebrate classes. *Comp. Biochem. Physiol. B* **50**, 431–434.

Henry, H.L. and Norman. A.W. (1978). Vitamin D: two dihydroxylated metabolites are required for normal chicken egg hatchability *Science* **201**, 835–837.

Herbert, J. (1972). Behavioural patterns. In *Reproduction in Mammals*, 4. *Reproduction Patterns* (edited by C.R. Austin and R.V. Short), pp. 34–68. Cambridge University Press.

Herrmann-Erlee, M.P.M. and Flik, G. (1989). Bone: comparative studies on endocrine involvement in bone metabolism. In *Vertebrate Endocrinology: Fundamentals and Biomedical Implications*, vol. 3, *Regulation of Calcium and Phosphate* (edited by P.K.T. Pang and M.P. Schreibman), pp.211–242. San Diego, CA: Academic Press.

Hewison, M. (1992). Vitamin D and the immune system. *J. Endocr.* **132**, 171–175.

Higgs, D.A. and Eales, J.G. (1973). Measurement of circulating thyroxine in several freshwater teleosts by competitive binding analysis. *Can. J. Zool.* **51**, 49–53.

Highfill, D.R. and Mead, R.A. (1975a). Function of corpora lutea of pregnancy in the viviparous snake, *Thamnophis elegans. Gen. Comp. Endocr.* **27**, 401–407.

Highfill, D.R. and Mead. R.A. (1975b). Sources and levels of progesterone during pregnancy in the garter snake *Thamnophis elegans. Gen. Comp. Endocr.* **27**, 389–400.

Hill, C.W. and Fromm, P.O. (1968). Response of the interrenal gland of rainbow trout (*Salmo gairdneri*) to stress. *Gen. Comp. Endocr.* **11**, 69–77.

Hill, P.A., Reynolds, J.J., and Meikle, M.C. (1995). Osteoblasts mediate insulin-like growth factor-I and -II stimulation of osteoclast formation and function. *Endocrinology* **136**, 124–131.

Hilliard, R.W., Epple, A., and Potter, I.C. (1985). The morphology and histology of the endocrine pancreas of the southern hemisphere lamprey, *Geotria australis* Gray. *J. Morphol.* **184**, 253–261.

Himick, B.A., and Peter, R.E. (1994). CCK/gastrin-like immunoreactivity in brain and gut, and CCK suppression of feeding in goldfish. *Am. J. Physiol.* **267**, R841–R851.

Himick, B.A., Vigna, S.R., and Peter, R.E. (1996). Characterization of cholecystokinin binding sites in goldfish brain and pituitary. *Am. J. Physiol.* **271**, R137–R143.

Hinde, R.A., Steel, E., and Follett, B.K. (1974). Effect of photoperiod on oestrogen-induced nest-building in ovariectomized or refractory female canaries (*Serinus canarius*). *J. Reprod. Fert.* **40**, 383–399.

Hinds, L.A. (1994). Prolactin, a hormone for all seasons: endocrine regulation of seasonal breeding in the Macropodidae. In *Oxford Reviews of Reproductive Biology* (edited by H.M. Charlton), vol. 16, pp. 249–301. Oxford University Press.

Hinson, J.P. (1990). Paracrine control of adrenocortical function: a new role for the medulla? *J. Endocr.* **124**, 7–9.

Hinson, J.P., Dawnay, A.B., and Raven, P.W. (1995). Why we should give a qualified welcome to ouabain: a whole new family of adrenal steroid hormones? *J. Endocr.* **146**, 369–372.

Hirano, T., Hasegawa, S., Yamauchi, H., and Orimo, H. (1981). Further studies on the absence of hypocalcemic effects of eel calcitonin in the eel, *Anguilla japonica*. *Gen. Comp. Endocr.* **43**, 42–50.

Hiraoka, S., Suzuki, M., Yanagisawa, T., Iwata, M., and Urano, A. (1993). Divergence of gene expression in neurohypophysial hormone precursors among salmonids. *Gen. Comp. Endocr.* **92**, 292–301.

Hirsch, P.F. and Munson, P.L. (1969). Thyrocalcitonin. *Physiol. Rev.* **49**, 548–622.

Hisaw, F.L. (1926). Experimental relation of the pubic ligament of the guinea pig. *Proc. Soc. Exp. Biol. Med., New York* **23**, 661–663.

Hisaw, F.L. (1959). The corpora lutea of elasmobranch fishes. *Anat. Rec.* **133**, 289.

Ho, S.-M. (1991). Vitellogenesis. In *Vertebrate Endocrinology: Fundamentals and Biomedical Implications*, vol. 4, Pt. A, *Reproduction* (edited by P.K.T. Pang and M.P. Schreibman), pp. 91–126. San Diego, CA: Academic Press.

Ho, S.-M, L'Italien, J., and Callard, I.P. (1980). Studies on reptilian yolk:

Chrysemys vitellogenin and phosvitin. *Comp. Biochem. Physiol. B* **65**, 139–144.

Hobart, P., Crawford, R., Shen, L.-P., Pictet, R., and Rutter, W.J. (1980). Cloning and sequence analysis of cDNAs encoding two distinct somatostatin precursors found in the endocrine pancreas of anglerfish. *Nature, Lond.* **288**, 137–141.

Hodgins, M.B., (1989). Peripheral oestrogen synthesis in man: its regulation and possible physiological significance. *J. Endocr.* **121**, 1–3.

Hoffman, C.W. and Dent, J.N. (1977). Hormonal regulation of cellular proliferation in the epidermis of the red-spotted newt. *Gen. Comp. Endocr.* **32**, 522–530.

Hogben, L.T. (1924). *The Pigmentary Effector System.* p. 67. Edinburgh: Oliver & Boyd.

Hogben, L.T. (1942). Chromatic behaviour. *Proc. Roy. Soc. Lond., Series B* **131**, 111–136.

Holick, M.F. (1989). Phylogenetic and evolutionary aspects of vitamin D from phytoplankton to humans. In *Vertebrate Endocrinology: Fundamentals and Biomedical Implications*, vol. 3, *Regulation and Calcium and Phosphate* (edited by P.K.T. Pang and M.P. Schreibman), pp. 7–43. San Diego, CA: Academic Press.

Holmes, J.A. and Youson, J.H. (1993). Induction of metamorphosis in landlocked sea lampreys, *Petromyzon marinus*. *J. Exp. Zool.* **267**, 598–604.

Holmes, R.L. and Ball, J.N. (1974). *The Pituitary Gland. A Comparative Account.* Cambridge University Press.

Holmes, W.N., Butler, D.G., and Phillips J.G. (1961). Observations on the effects of maintaining glaucous-winged gulls (*Larus glaucescens*) on fresh water and sea water for long periods. *J. Endocr.* **23**, 53–61.

Holmquist, A.L., Dockray, G.J., Rosenquist, G.L., and Walsh, J.H. (1979). Immunochemical characterization of cholecystokinin-like peptides in lamprey gut and brain. *Gen. Comp. Endocr.* **37**, 474–481.

Holstein B. (1982). Inhibition of gastric acid secretion in the Atlantic cod, *Gadus morhua*, by sulphated and desulphated gastrin, caerulein, and CCK-like

octapeptide. *Acta Physiol. Scand.* **114**, 453–459.

Holt, W.F. and Idler, D.R. (1975). Influence of the interrenal gland on the rectal gland of the skate. *Comp. Biochem. Physiol. C* **50**, 111–119.

Honma, Y. and Chiba, A. (1996). Immunohistochemical and ultrastructural characterization of the hypophyseal ventral lobe in the cloudy dogfish. In *Environmental & Conservation Endocrinology* (edited by J. Joss), pp. 177–178. Sydney: Macquarie University.

Horikawa, S., Takai, T., Toyosato, M., Takahashi, H., Noda, M., Kakidani, H., Kubo, T., Hirose, T., Inayama, S., Hayashida, H., Miyata, T., and Numa, S. (1983). Isolation and structural organization of the human preproenkephalin B gene. *Nature, Lond.* **306**, 611–614.

Hornsby, P.J. (1995). Current challenges for DHEA research. *Ann. N.Y. Acad. Sci.* **774**, xiii–xiv.

Hornsey, D.J. (1977). Triiodothyronine and thyroxine levels in the thyroid and serum of the sea lamprey *Petromyzon marinus* L. *Gen. Comp. Endocr.* **31**, 381–383.

Horseman, N.D. (1994). Editorial: famine to feast – growth hormone and prolactin signal transducers. *Endocrinology* **135**, 1289–1291.

Horwitz, B.A. (1989). Biochemical mechanisms and control of cold-induced cellular thermogenesis in placental mammals. In *Advances in Comparative and Environmental Physiology*, vol. 4, *Animal Adaptation to Cold* (edited by L.C.H. Wang), pp. 83–116. Berlin: Springer-Verlag.

Hoshino, S., Suzuki, M., Kakegawa, T., Imai, K., Wakita, M., Kobayashi, Y., and Yamada, Y. (1988). Changes in plasma thyroid hormone, luteinizing hormone (LH), estradiol, progesterone and corticosterone of laying hens during a forced molt. *Comp. Biochem, Physiol. A* **90**, 355–359.

Hotchiss, J. and Knobil, E. (1994). The menstrual cycle and its neuroendocrine control. In *The Physiology of Reproduction*, 2nd edn (edited by E. Knobil and J.D. Neill), vol. 2, pp. 711–749. New York: Raven Press.

Howe, A. (1973). The mammalian pars intermedia: a review of its structure and function. *J. Endocr.* **59**, 385–409.

Hughes, M.A. and Chadwick, A. (eds.) (1989). *Progress in Avian Osmoregulation* Leeds: Leeds Philosophical and Literary Society.

Hughes, R.L. and Carrick, F.N. (1978). Reproduction in female montremes. *Aust. Zool.* **20**, 233–253.

Hulbert, A.J. and Hudson, J.W. (1976). Thyroid function in a hibernator. *Spermophilus trideceneatus. Am. J. Physiol.* **230**, 1211–1216.

Hulbert, A.J. and Williams, C.A. (1988). Thyroid function in a lizard, a tortoise, and a crocodile, compared with mammals. *Comp. Biochem. Physiol. A* **90**, 41–48.

Humbel, R.E., Bosshard, H.R., and Zahn, H. (1972). Chemistry of insulin. In *Handbook of Physiology*, Sect. 7 *Endocrinology*, vol. 1, *Endocrine pancreas*, pp. 111–132. Washington, DC: American Physiological Society.

Hunter, T. (1995). Protein kinases and phosphatases: the Yin and Yan of protein phosphorylation and signalling. *Cell* **80**, 225–236.

Hurlburt, M.E. (1977). Role of the thyroid gland in ovarian maturation of the goldfish. *Carassius auratus L. Can. J. Zool.* **55**, 1906–1913.

Hurwitz, S. (1989). Calcium homeostasis in birds. *Vit. Horm.* **45**, 173–221.

Hutchison, R.E. (1975). Effects of ovarian steroids and prolactin on the sequential development of nesting behaviour in female budgerigars. *J. Endocr.*. **67**, 29–39.

Hutchison, R.E., Hinde, R.A., and Steel, E. (1967). The effects of oestrogen progesterone and prolactin on brood patch formation in ovariectomized canaries. *J. Endocr.* **39**, 379–385.

Hyodo, S., Ishii, S., and Joss, J.M.P. (1996). Neurohypophyseal hormone precursors in Australian lungfish. In *Environmental & Conservation Endocrinology* (edited by J. Joss), pp. 266–267. Sydney: Macquarie University.

Hyodo, S. and Urano, A. (1991). Changes in expression of provasotocin and proisotocin genes during adaptation to hyper- and hypo-osmotic environments

in rainbow trout. *J. Comp. Physiol. B* **161**, 549–556.

Idelman, S. (1979). The structure of the mammalian adrenal cortex. In *General, Comparative and Clinical Endocrinology of the Adrenal Cortex* (edited by I. Chester Jones and I.W. Henderson), vol. 2, pp. 1–199. London: Academic Press.

Idler, D.R. (ed.) (1972). *Steroids in Nonmammalian Vertebrates.* New York: Academic Press.

Idler, D.R. and Burton, M.P.M. (1976). The pronephroi as the site of presumptive interrenal cells in the hagfish *Myxine glutinosa* L. *Comp. Biochem, Physiol. A* **53**, 73–77.

Idler, D.R. and Kane, K.M. (1976). Interrenalectomy and Na–K ATPase activity in the rectal gland of the skate *Raja ocellata. Gen. Comp. Endocr.* **28**, 100–102.

Idler, D.R. and Ng, T.B. (1979). Studies of two types of gonadotropins from both salmon and carp pituitaries. *Gen Comp. Endocr.* **38**, 421–440.

Idler, D.R., Sangalang, G.B., and Truscott, B. (1972). Corticosteroids in the South American lungfish. *Gen. Comp. Endocr.* **Suppl. 3**, 238–244.

Idler, D.R. and Truscott, B. (1972). Corticosteroids in fish. In *Steroids in Nonmammalian Vertebrates* (edited by D.R. Idler), pp. 126–252. New York: Academic Press.

Ingalls, A.M., Dickie, M.M., and Snell, G.D. (1950). *Obese*, a new mutation in the house mouse. *J. Heredity* **41**, 317–318.

Ingleton, P.M. Hazon, N., Ho, P.M.W., Martin, T.J. and Danks, J.A. (1995). Immunodetection of parathyroid hormone-related protein in plasma and tissues of an elasmobranch (*Scyliorhinus canicula*). *Gen. Comp. Endocr.* **98**, 211–218.

Innis, R.B. and Snyder, S.H. (1980). Distinct cholecystokinin receptors in brain and pancreas. *Proc. Natl. Acad. Sci. USA* **77**, 6917–6921.

Inui, Y., Miwa, S., Yamano, K., and Hirano, T. (1994). Hormonal control of flounder metamorphosis. In *Perspectives in Comparative Endocrinology* (edited by K.G. Davey, R.E. Peter, and S.S. Tobe), pp. 408–411. Ottawa: National Research

Council of Canada.

Ireland, M.P. (1973). Effects of arginine vasotocin on sodium and potassium metabolism in *Xenopus laevis* after skin gland stimulation and sympathetic blockade. *Comp. Biochem. Physiol. A* **44**, 487–493.

Isaia, J. (1984). Water and nonelectrolyte permeation. In *Fish Physiology* (edited by W.S. Hoar and D.J. Randall), vol. X, *Gills*, Pt. B, *Ion and Water Transfer*, pp. 1–38. Orlando, FL: Academic Press.

Isaia, J., Maetz, J., and Haywood, G.P. (1978). Effects of epinephrine on branchial non-electrolyte permeability in rainbow trout. *J. Exp. Biol.* **74**, 227–237.

Ishida, I., Ichikawa, T., and Deguchi, T. (1986). Cloning and sequence analysis of cDNA encoding urotensin I precursor. *J. Biol. Chem.* **83**, 308–312.

Ishii, S., Ando, H., Wako, H., and Kubota, Y. (1993). Molecular biology of gonadotrophins. In *Avian Endocrinology* (edited by P.J. Sharp), pp. 123–134. Bristol: Journal of Endocrinology.

Ismail-Beigi, F. (1993). Thyroid hormone regulation of Na,K-ATPase expression. *Trend. Endocr. Metab.* **4**, 152–155.

Ito, H., Koide, Y., Takamatsu, N. Kawauchi, H., and Shiba, T. (1993). cDNA cloning of the β subunit of teleost thryotropin. *Proc. Natl. Acad. Sci. USA* **90**, 6052–6055.

Itoh, H., Suzuki K., and Kawauchi, H. (1988). The complete amino acid sequences of β-subunits of two distinct chum salmon GTHs. *Gen. Comp. Endocr.* **71**, 438–451.

Iturriza, F.C. (1969). Further evidences for the blocking effect of catecholamines on the secretion of melanocyte-stimulating hormone in toads. *Gen. Comp. Endocr.* **12**, 417–426.

Iturriza, F., Verzi, D.H., and Di Maggio, L.A. (1995). Glucagon of caviomorphs and other tetrapods immunohistochemically investigated with two antisera against the N- and C-terminal portions of the molecule. *Gen. Comp. Endocr.* **99**, 178–184.

Ivell, R., and Richter, D. (1984). Structure and comparison of the oxytocin and vasopressin genes from rat. *Proc. Natl. Acad. Sci. USA* **81**, 2006–2010.

Ivy, A.C. and Oldberg, E. (1928). A

hormone mechanism for gall bladder contraction and evacuation. *Am. J. Physiol.* **86**, 599–613.

Jackson, I.M.D. (1981a). Abundance of immunoreactive thyrotropin-releasing hormone-like material in the alfalfa plant. *Endocrinology* **108**, 344–346.

Jackson, I.M.D. (1981b). Evolutionary significance of the phylogenetic distribution of the mammalian hypothalamic releasing hormones. *Fedn. Proc.* **40**, 2545–2552.

Jackson, I.M.D. and Reichlin, S. (1974). Thyrotropin-releasing hormone distribution in hypothalamic and extrahypothalamic brain tissues of mammalian and submammalian chordates. *Endocrinology* **95**, 854–862.

Jackson, I.M.D. and Reichlin, S. (1977). Thyrotropin-releasing hormone: abundance in the skin of the frog. *Rana pipiens. Science* **198**, 414–415.

Jackson, R.G. and Sage, M. (1973). A comparison of the effects of mammalian TSH on the thyroid glands of the teleost *Galeichthys felis* and the elasmobranch *Dasyatis sabina. Comp. Biochem. Physiol. A* **44**, 867–870.

Jallageas, M. and Assenmacher, I. (1986). Endocrine correlates of hibernation in the edible dormouse (*Glis glis*). In *Living in the Cold. Physiological and Biochemical Adaptations* (edited by H.C. Heller, X.J. Musacchia, and L.C.H. Wang), pp. 265–271. New York: Elsevier.

Janský, L. (1995). Humoral thermogenesis and its role in maintaining energy balance. *Physiol. Rev.* **75**, 237–257.

Janssens, P.A. (1964). The metabolism of the aestivating African lungfish. *Comp. Biochem. Physiol.* **11**, 105–117.

Janssens, P.A. (1967). Interference of metyrapone with the actions of cortisol in *Xenopus laevis* Daudin and the laboratory rat. *Gen. Comp. Endocr.* **8**, 94–100.

Janssens, P.A., Vinson, G.P., Chester Jones, I., and Mosley, W. (1965). Amphibian characteristics of the adrenal cortex of the African lungfish (*Protopterus* sp.) *J. Endocr.* **32**, 373–382.

Janssens, P.M.W. (1987). Did vertebrate signal transduction mechanisms originate in eukaryotic microorganisms. *Trend. Biochem. Sci.* **12**, 456–459.

Jansz, H.S. and Zandberg, J. (1992).

Identification and partial characterization of the salmon calcitonin/CGRP by polymerase chain reaction. *Ann. N.Y. Acad. Sci.* **657**, 63–69.

Jensen, J. and Conlon, J.M. (1992). Characterization of peptides related to neuropeptide tyrosine and peptide tyrosine–tyrosine from the brain and gastrointestinal tract of teleost fish. *Eur. J. Biochem.* **210**, 405–410.

Jia, L., Canny, B.J., Orth, D.N. and Leong, D.A. (1991). Distinct classes of corticotropes mediate corticotropin-releasing hormone- and arginine vasopressin-stimulated adrenocorticotropin release. *Endocrinology* **128**, 197–203.

John-Alder, H.B. (1984). Seasonal variations in activity, aerobic energetic capacities, and plasma thyroid hormones (T_3 and T_4) in an iguanid lizard. *J. Comp. Physiol. B* **154**, 409–419.

Johnsen, A.H., and Rehfeld, J.F. (1990). Cionin: a disulfotyrosyl hybrid of cholecystokinin and gastrin from the neural ganglion of the protochordate *Ciona intestinalis. J. Biol. Chem.* **265**, 3054–3058.

Johnson, P.A., and Wang, S.-Y. (1993). The molecular biology and endocrinology of inhibin in the domestic hen. In *Avian Endocrinology* (edited by P.J. Sharp), pp. 297–308. Bristol: Journal of Endocrinology.

Johnston, C.I., Davis, J.O., Wright, F.S., and Howards, S.S. (1967). Effects of renin and ACTH on adrenal steroid secretion in the American bullfrog. *Am. J. Physiol.* **213**, 393–399.

Jones, J.I. and Clemmons, D.R. (1995). Insulin-like growth factors and their binding proteins: biological actions. *Endocrine Rev.* **16**, 3–34.

Jones, M.E.E., Bradshaw, S.D., Fergusson, B., and Watts, R. (1990). Effects of available surface water on levels of antidiuretic hormone (lysine-vasopressin) and water and electrolyte metabolism of the Rottnest Island quokka (*Setonix brachyurus*). *Gen. Comp. Endocr.* **77**, 75–87.

Jones, R.E. and Baxter, D.C. (1991). Gestation, with emphasis on corpus luteum biology, placentation, and parturition. In *Vertebrate Endocrinology:*

Fundamentals and Biomedical Implications, vol. 4. Pt. A, *Reproduction* (edited by P.K.T. Pang and M.P. Schreibman), pp. 205–302. San Diego, CA: Academic Press.

Jönsson, A.-C. (1989). Gastrin/cholecystokinin-related peptides–comparative aspects. In *The Comparative Physiology of Regulatory Peptides* (edited by S. Holmgren), pp. 61–86. London: Chapman & Hall.

Jönsson, A.-C. and Hölmgren, S. (1989). Gut secretion. In *The Comparative Physiology of Regulatory Peptides* (edited by S. Hölmgren), pp. 256–271. London: Chapman & Hall.

Jørgenson, C.B. (1988). Nature of moulting control in amphibians: effects of cortisol implants in toads *Bufo bufo. Gen. Comp. Endocr.* **71**, 29–35.

Jørgenson, C.B., Hede, K.-E., and Larsen, L.O. (1978). Environmental control of annual ovarian cycle in the toad *Bufo bufo* L.: role of temperature. In *Environmental Physiology* (edited by I. Assenmacher and D.S. Farner), pp. 28–36. Berlin: Springer-Verlag.

Jørgenson, C.B., Larsen, L.O., and Rosenkilde, P. (1965). Hormonal dependency of molting in amphibians: effect of radiothyroidectomy in the toad *Bufo bufo* L. *Gen. Comp. Endocr.* **5**, 248–251.

Jørgenson, C.B. and Vijayakumar, S. (1970). annual oviduct cycle and its control in the toad *Bufo bufo* L. *Gen. Comp. Endocr.* **14**, 404–411.

Jörnvall, H., Carlström, A., Pettersson, T., Jacobssen, B., Persson, M., and Mutt, V. (1981). Structural homologies between prealbumin, gastrointestinal prohormones and other proteins. *Nature, Lond.* **291**, 261–263.

Jorpes, E. and Mutt, V. (1966). Cholecystokinin and pancreozymin, one single molecule? *Acta Physiol. Scand.* **66**, 196–202.

Joss, J.M.P. (1985). Pituitary control of metamorphosis in the southern hemisphere lamprey, *Geotria australis. Gen. Comp. Endocr.* **60**, 58–62.

Joss, J.M.P., Arnold-Reed, D.E., and Balment, R.J. (1994). The steroidogenic response to angiotensin II in the Australian lungfish. *Neoceratodus forsteri.*

J. Comp. Physiol. B. **164**, 378–382.

Joss, J.M.P., Edwards, A., and Kime, D.E. (1996). *In vitro* biosynthesis of androgens in the Australian lungfish, *Neoceratodus forsteri. Gen. Comp. endocr.* **101**, 256–263.

Josso, N. (1994). Anti-mullerian hormone: a masculinizing relative of TGF-β. In *Oxford Reviews of Reproductive Biology* (edited by H.M. Charlton), vol. 16, pp. 139–163. Oxford University Press.

Jost, A. (1971). Hormones in development; past and present prospects. In *Hormones and Development* (edited by M. Hamburgh and E.J.W. Barrington), pp. 1–18. New York: Apple-Century-Crofts.

Jüppner, H., Abou-Samra, A.-B., Uneno, S., Gu, W.-X., Potts, J.T., and Segre, G.V. (1988). The parathyroid hormone-like peptide associated with humoral hypercalcemia of malignant and parathyroid hormone bind to the same receptor on the plasma membrane of ROS 17/2.8 cells. *J. Biol. Chem.* **263**, 8557–8560.

Kadowaki, K., Kishimoto, J., Leng, G., and Emson, P.C. (1994). Up-regulation of nitric oxide synthase (NOS) gene expression together with NOS activity in the rat hypothalamo–hypophysial system after chronic salt loading: evidence of a neuromodulatory role of nitric oxide in arginine vasopressin and oxytocin secretion. *Endocrinology* **134**, 1011–1017.

Kahn, C.R., White, M.F., Shoelson, S.E., Backer, J.M., Araki, E., Cheatham, B., Csermely, P., Folli, F., Goldstein, B.J., Huertas, P., Rothenberg, P.L., Saad, M.J.A., Siddle, K., Sun, X.-J., Wilden, P.A., Yamada, K, and Kahn, S.A. (1993). The insulin receptor and its substrate: molecular determinants of early events in insulin action. *Rec. Prog. Horm. Res.* **48**, 291–339.

Kaiya, H. and Takei, Y. (1996a). Osmotic and volaemic regulation of atrial and ventricular natriuretic peptide secretion in concious eels. *J. Endocr.* **149**, 441–447.

Kaiya, H. and Takei, Y. (1996b) Atrial and ventricular natriuretic peptide concentrations in plasma of freshwater- and seawater-adapted eels. *Gen. Comp. Endocr.* **102**, 183–190.

Kaji, H., Sugimoto, T., Kanatani, M., Nasu, M., and Chihara, K. (1996).

Estrogen blocks parathyroid hormone (PTH)-stimulated osteoclast-like cell formation by selectively affecting PTH-responsive cyclic-adenosine monophosphate pathway. *Endocrinology* **137**, 2217–2224.

Kakizawa, S., Kaneko, T., Hasegawa, S., and Hirano, T. (1993). Activation of somatolactin cells in the pituitary of the rainbow trout. *Oncorhynchus mykiss* by low environmental calcium. *Gen. Comp. Endocr.* **91**, 298–306.

Kakizawa, S., Kaneko, T., and Hirano, T. (1996). Elevation of plasma somatolactin concentrations during acidosis in rainbow trout (*Oncorhynchus mykiss*). *J. Exp. Biol.* **199**, 1043–1051.

Kaneko, T. and Pang, P.K.T. (1987). Immunocytochemical detection of parathyroid hormone-like substance in the goldfish brain and pituitary gland. *Gen. Comp. Endocr.* **68**, 147–152.

Kangawa, K., Tawaragi, Y., Oikawa, S., Mizuno, A., Sakuragawa, Y., Nakazato, H., Fukudu, A., Minamino, N., and Matsuo, H. (1984). Identification of γ atrial natriuretic polypeptide and characterization of the cDNA encoding its precursor. *Nature, Lond.* **312**, 152–155.

Kamiya, M. (1972). Sodium-potassium-activated adenosinetriphosphatase in isolated chloride cells from eel gills. *Comp. Biochem. Physiol. B* **43**, 611–617.

Kanis, J.A., Cundy T., Bartlett, M., Smith, R., Heynen, G., Warner, G.T., and Russell, R.G.G. (1978). Is 24,25-dihydroxycholecalciferol a calcium-regulating hormone in man? *Br. Med. J.* **1**, 1382–1386.

Karnaky, K.J. (1986). Structure and function of the chloride cell of *Fundulus heteroclitus* and other teleosts. *Am. Zool.* **26**, 209–224

Karsch, F.J. (1980). Seasonal reproduction: a saga of reversible fertility. *The Physiologist* **23**, 29–38.

Karsch, F.J. and Dahl, G.E. (1994). Role of the thyroid gland in seasonal reproduction in the ewe. In *Perspectives in Comparative Endocrinology* (edited by K.G. Davey, R.E. Peter, and S.S. Tobe), pp. 594–601. Ottawa: National Research Council of Canada.

Karsch, F.J., Dahl, G.E., Hachigian, T.M., and Thrun, L.A. (1995). Involvement of thyroid hormones in seasonal reproduction. *J. Reprod. Fertility* **Suppl. 49**, 409–422.

Kasson, B.G., Adashi, E.Y., and Hsueh, A.J.W. (1986). Arginine vasopressin in the testis: an intragonadal peptide control system. *Endocrine Rev.* **7**, 156–168.

Kastin, A.J., Schally, A.V., and Kostrzewa, R.M. (1980). Possible aminergic mediation of MSH release and of the CNS effects of MSH and MIF-I. *Fedn. Proc.* **39**, 2931–2936.

Kawahara, A., Baker, B.S., and Tata, J.R. (1991). Developmental and regional expression of thyroid hormone receptor genes during *Xenopus* metamorphosis. *Development* **112**, 933–943.

Kawase, T., Howard, G.A., Roos, B.A., and Burns, D.M. (1996). Calcitonin gene-related peptide rapidly inhibits calcium uptake in osteoblastic cell lines via activation of adenosine triphosphate-sensitive potassium channels. *Endocrinology* **137**, 984–990.

Kawauchi, H. (1983). Chemistry of proopiocortin-related peptides in the salmon pituitary. *Arch. Biochem. Biophys.* **227**, 343–350.

Kawauchi, H.C. (1989). Structure and biosynthesis of melanin-concentrating hormone. *Life Sci.* **45**, 1132–1140.

Kawauchi, H., Kawazoe, I., Tsubokawa, M., Kishida, M., and Baker, B.I. (1983). Characterization of melanin-concentrating hormone in chum salmon pituitaries. *Nature, Lond.* **305**, 321–323.

Kawauchi, H., Yasuda, A., and Rand-Weaver, M. (1990). Evolution of prolactin and growth hormone family. In *Progress in Comparative Endocrinology* (edited by A. Epple, C.G. Scanes, and M.H. Stetson), pp. 47–53. New York: Wiley-Liss.

Keller, H.K., Redding, J.M., Moberg, G., and Dores, R.M. (1994). Analysis of the post-translational processing of α-MSH in the pituitaries of the chondrichthyean fishes, *Acipenser transmontanus* and *Polyodon spathula. Gen. Comp. Endocr.* **94**, 159–165.

Keller, N., Richardson, I.U., and Yates, F.E. (1969). Protein binding and the biological activity of corticosteroids: *in vivo* induction of hepatic and pancreatic alanine amino-transferases by corticosteroids in normal and estrogen-treated rats. *Endocrinology* **84**, 49–62.

Kelley, D.B. (1978). Neuroanatomical correlates of hormone sensitive behaviors in frogs and birds. *Am. Zool.* **18**, 477–488.

Kelley, K.M. (1993). Experimental diabetes mellitus in a teleost fish. I. Effect of complete isletectomy and subsequent hormonal treatment on metabolism in the goby, *Gillichthys mirabilis. Endocrinolgy* **132**, 2689–2695.

Kelley, K.M., Gray, E.S., Siharath, K., Nicoll, C.S., and Bern, H.A. (1993). Experimental diabetes mellitus in a teleost fish. II. Roles of insulin, growth hormone (GH), insulin-like growth factor-I, and hepatic GH receptors in diabetic growth inhibition in the goby, *Gillichthys mirabilis. Endocrinology* **132**, 2696–2702.

Kennaway, D.J. and Rowe, S.A. (1995). Melatonin binding sites and their role in seasonal reproduction. *J. Reprod. Fertility* **Suppl. 49**, 423–435.

Kenny, A.D. (1971). Determination of calcitonin in plasma by bioassay. *Endocrinology* **89**, 1005–1013.

Kenny, A.D. (1976). Vitamin D metabolism: physiological regulation in egg-laying Japanese quail. *Am. J. Physiol.* **230**, 1609–1615.

Kenny, A.D. and Dacke, C.G. (1974). The hypercalcaemic response to parathyroid hormone in Japanese quail. *J. Endocr.* **62**, 15–23.

Kenyon, C.J., McKeever, A., Oliver, J.A., and Henderson, I.W. (1985). Control of renal and adrenocortical function by the renin–angiotensin system in two euryhaline teleost fishes. *Gen. Comp. Endocr.* **58**, 93–100.

Kerkof, P.R., Boschwitz, D., and Gorbman, A. (1973). The response of hagfish thyroid tissue to thyroid inhibitors and to mammalian thyroid-stimulating hormone. *Gen. Comp. Endocr.* **21**, 231–240.

Kesner, J.S., Convey, E.M., and Anderson, C.R. (1981). Evidence that estradiol induces the preovulatory LH surge in cattle by increasing pituitary sensitivity to LHRH and then increasing LHRH

release. *Endocrinology* **108**, 1386–1391.

Keutmann, H.T., Sauer, M.M., Hendy, G.N., O'Riordan, J.L.H., and Potts, J.T. (1978). Complete amino acid sequence of human parathyroid hormone. *Biochemistry* **17**, 5723–5729.

Keys, A. and Bateman, J.B. (1932). Branchial responses to adrenaline and pitressin in the eel. *Biol. Bull.* **63**, 327–336.

Khosla, S., Demay, M., Pines, M., Hurwitz, S., Potts, J.T. and Kronenberg, H.M. (1988). Nucleotide sequence of cloned cDNAs encoding chicken preproparathyroid hormone. *J. Bone Mineral. Res.* **3**, 689–6698.

Kiebzak, G.M. and Minnich, J.E. (1982). Effects of calcitonin on electrolyte excretion in the lizard *Dipsosaurus dorsalis. Gen. Comp. Endocr.* **48**, 232–238.

Kikuyama, S., Yamamoto, K., and Mayumi, M. (1980). Growth-promoting and antimetamorphic hormone in pituitary glands of bullfrogs. *Gen. Comp. Endocr.* **41**, 212–216.

Kimmel, J.R., Maher, M.J., Pollock, H.G., and Vensel, W.H. (1976). Isolation and characterization of reptilian insulin: partial amino acid sequence of rattlesnake (*Crotalus atrox*) insulin. *Gen. Com. Endocr.* **28**, 320–333.

Kimura, T., Tanizawa, O., Mori, K., Brownstein, M.J., and Okayama, H. (1992). Structure and expression of a human oxytocin receptor. *Nature, Lond.* **356**, 526–529.

King. J.A. and Millar, R.P. (1979a). Phylogenetic and anatomical distribution of somatostatin in vertebrates. *Endocrinology* **105**, 1322–1329.

King, J.A. and Millar, R.P. (1979b). Hypothalamic luteinizing hormone-releasing hormone content in relation to the seasonal reproductive cycle of *Xenopos laevis. Gen. Comp. Endocr.* **39**, 309–312.

King, J.A. and Millar, R.P. (1980). Radioimmunoassay of methionine[5]-enkephalin sulphoxide: phylogenetic and anatomical distribution. *Peptides* **1**, 211–216.

King, J.A. and Millar, R.P. (1981). TRH, GH-RIH, and LH-RH in metamorphosing *Xenopus laevis. Gen.*

Comp. Endocr. **44**, 20–27.

King, J.A. and Millar, R.P. (1992). Evolution of gonadotropin-releasing hormones. *Trends Endocr. Metab.* **3**, 339–346.

King, J.R. and Farner, D.S. (1965). Studies of fat deposition in migratory birds. *Ann. N.Y. Acad. Sci.* **131**, 422–440.

King, W.J. and Greene, G.L. (1984). Monoclonal antibodies localize oestrogen receptor in the nuclei of target cells. *Nature, Lond.* **307**, 745–747.

Kirschner, L.B. (1980). Comparison of vertebrate salt-excreting organs. *Am. J. Physiol.* **238**, R219–R223.

Kishida, M., Baker, B.I., and Eberle, A.N. (1989). The measurement of melanin-concentrating hormone in trout blood. *Gen. Comp. Endocr.* **74**, 221–229.

Kita, T., Smith, C.E., Fok, K.F., Duffin, K.L., Moore, W.M., Karabatsos, P.J., Kachur, J.F., Hamra, F.K., Pidhorodeckyj, N.V., Forte, L.R., and Currie, M.G. (1994). Characterization of human uroguanylin: a member of the guanylin peptide family. *Am. J.Physiol.* **266**, F342–F348.

Kitamura, K., Kangawa, K., Kawamoto, M., Ichiki, Y., Nakamura, S., Matsuo, H., and Eto, T. (1993a). Adrenomedullin: a novel hypotensive peptide isolated from human pheochromocytoma. *Biochem. Biophys. Res. Commun.* **192**, 553–560.

Kitamura, K., Kangawa, K., Matsuo, H., and Eto, T. (1995). Adrenomedullin. Implications for hypertension research. *Drugs* **49**, 485–495.

Kitamura, K. Sakata, J., Kangawa, K., Kojima, M., Matsuo, H., and Eto, T. (1993b). Cloning and characterization of cDNA encoding a precursor of human adrenomedullin. *Biochem. Biophys. Res. Commun.* **194**, 720–725.

Kjeld, J.M., Sigurjónsson, J., and Árnason, A. (1992). Sex hormone concentrations in blood serum from North Atlantic fin whale (*Balaenoptera physalus*). *J. Endocr.* **134**, 405–413.

Klapper, D.G., Svoboda, M.E., and van Wyk, J.J. (1983). Sequence analysis of somatomedin-C: confirmation of identity with insulin-like growth factor I. *Endocrinology* **112**, 2215–2217.

Kleiber, M. (1961). *The Fire of Life. An Introduction to Animal Energetics*, p. 312.

New York: Wiley.

Klicka, J. and Mahmoud, I.Y. (1972). Conversion of pregenolone-4^{14}C to progesterone-4^{14}C by turtle corpus luteum. *Gen. Comp. Endocr.* **19**, 367–369.

Klicka, J. and Mahmoud, I.Y. (1977). The effects of hormones on the reproductive physiology of the painted turtle. *Chrysemys picta. Gen. Comp. Endocr.* **31**, 407–413.

Kline, L.W. (1981). A hypocalcemic response to synthetic salmon calcitonin in the green iguana, *Iguana iguana. Gen. Comp. Endocr.* **44**, 476–479.

Kline, L.W. (1982). An age-dependent response to synthetic salmon calcitonin in the chuckwalla, *Sauromalus obesus. Can. J. Zool.* **60**, 1359–1361.

Kline, L.W. and Longmore, G.A. (1986). Determination of calcitonin in reptilian serum by heterologous radioimmunoassay. *Gen. Comp. Endocr.* **61**, 1–4.

Kloas, W., Flugge, G., Fuchs, E., and Stolte, H. (1988). Binding sites for atrial natriuretic peptide in the kidney and aorta of the hagfish (*Myxine glutinosa*). *Comp. Biochem. Physiol. A* **91**, 685–688.

Klosterman, L.L., Mural, M.T., and Siiteri, P.K. (1986). Cortisol levels, binding, and properties of corticosteroid-binding globulin in the serum of primates. *Endocrinology* **118**, 424–434.

Knight, T.W., Tervit, H.R., and Lynch, P.R. (1983). Effects of boar pheromones, ram's wool and presence of bucks on ovarian activity in anovular ewes early in the breeding season. *Anim. Reprod. Sci.* **6**, 129–134.

Knobil, E. (1992). Remembrance: the discovery of the hypothalamic gonadotropin-releasing hormone pulse generator and of its physiological significance. *Endocrinology* **131**, 1005–1006.

Kobayashi, H. (1981). Angiotensin-induced drinking in parrots. *Gen. Comp. Endocr.* **43**, 399–401.

Kobayashi, H., Owada, K., Yamada, C., and Okawara, Y. (1986). The caudal neurosecretory system in fishes. In *Vertebrate Endocrinology: Fundamentals and Biomedical Implications*, vol.1, *Morphological Considerations* (edited by

P.K.T. Pang and M.P Schreibman), pp. 147–174. Orlando, FL: Academic Press,

Kobayashi, K., Kitamura, K., Etoh, T., Nagatomo, Y., Takenaga, M., Ishikawa, T., Imamura, T., Koiwaya, Y., and Eto, T. (1996). Increased plasma adrenomedullin levels in chronic congestive heart failure. *Am. Heart J.* **131**, 994–998.

Kobayashi, T., Sakai, N., Adachi, S., Asahina, K., Iwasawa, H., and Nagahama, Y. (1993). 17α, 20α-dihydroxy-4-pregnen-3-one is the naturally occurring spermiation-inducing hormone in the testis of a frog, *Rana nigromaculata. Endocrinology* **133**, 321–327.

Kobayashi, H., Uemura, H., Takei, Y., Itatsu, N., Ozawa, M., and Ichinohe, K. (1983). Drinking induced by angiotensin II in fishes. *Gen. Comp. Endocr.* **49**, 295–306.

Kobayashi, H., Uemura, H., Wada, M., and Takei, Y. (1979). Ecological adaptation of angiotensin-induced thirst mechanism in tetrapods. *Gen. Comp. Endocr.* **38**, 93–104.

Kocsis, J.F., McIlroy, P.J., and Carsia, R.V. (1995). Atrial natriuretic peptide stimulates aldosterone production by turkey (*Meleagris gallopava*) adrenal steroidogenic cells. *Gen. Comp. Endocr.* **99**, 364–372.

Kodicek, E. (1974). The story of vitamin D from vitamin to hormone. *Lancet* **i**, 325–329.

Koike, T.I., Pryor, L.R., Neldon, H.L., and Venable, R.S. (1977). Effect of water deprivation of plasma radioimmunoassayable arginine vasotocin in conscious chickens (*Gallus domesticus*). *Gen. Comp. Endocr.* **33**, 359–364.

Koike, T.I., Shimada, K., and Cornett, L.E. (1988). Plasma levels of immunoreactive mesotocin and vasotocin during oviposition in chickens: relationship to oxytocic action of the peptides *in vitro* and peptide interaction with myometrial membrane binding sites. *Gen. Comp. Endocr.* **70**, 119–126.

Komourdjian, M.P., Saunders, R.L., and Fenwick, J.C. (1976). The effect of porcine somatotrophin on growth, and survival in seawater of Atlantic salmon (*Salmo salar*) parr. *Can. J. Zool.* **54**,

531–535.

Korf, H.-W. and Oksche, A. (1986). The pineal organ. In *Vertebrate Endocrinology: Fundamentals and Biomedical Implications*, vol. 1, *Morphological Considerations* (edited by P.K.T. Pang and M.P. Schreibman), pp. 105–145. Orlando, FL: Academic Press.

Kourides, I.A. and Weintraub, B.D. (1979). mRNA-directed biosynthesis of α-subunit of thyrotropin: translation in cell-free and whole-cell systems. *Proc. Natl. Acad. Sci. USA* **76**, 298–302.

Krebs, E.G. (1972). Protein kinases. *Curr. Top. Cell Regul.* **5**, 99–133.

Krieger, D.T. (1971). The hypothalamus and neuroendocrinology. *Hosp. Practice* September, 87-99.

Krieger, D.T. (1972). Circadian corticosteroid periodicity: critical period for abolition by neonatal injection of corticosteroid. *Science* **178**, 1205–1207.

Krieger, D.T. and Liotta, A.S. (1979). Pituitary hormones in brain: where, how and why? *Science* **205**, 366–372.

Krishna, G., Hynie, S., and Brodie, B.B. (1968). Effects of thyroid hormones on adenyl cyclase in adipose tissue and on free fatty acid mobilization. *Proc. Natl. Acad. Sci., USA* **59**, 884–889.

Krishnamurthy, V.G. and Bern, H.A. (1969). Correlative histological study of the corpuscles of Stannius and the juxtaglomerular cells of teleost fishes. *Gen. Comp. Endocr.* **13**, 313–335.

Krishnan, A.V., Cramer, S.D., Bringhurst, F.R., and Feldman, D. (1995). Regulation of 1,25-dihydroxyvitamin D$_3$ receptors by parathyroid hormone in osteoblastic cells: role of second messenger pathways. *Endocrinology* **136**, 705–712.

Kühn, E.R. (1990). Hormonal control of peripheral monodeiodination in vertebrates. In *Progress in Comparative Endocrinology* (edited by A. Epple, C.G. Scanes, and M.H. Stetson), pp. 421–426. New York: Wiley-Liss.

Kühn, E.R., Berghman, L.R., Moons, L., Vandesande, F., Decuypere, E., and Darras, V.M. (1993). Hypothalamic and peripheral control of thyroid function during the life cycle of the chicken. In *Avian Endocrinology* (edited by P.J. Sharp), pp. 29–46. Bristol: Journal of endocrinology.

Kuhn, M., Raida, M., Adermann, K., Schulz-Knappe, P., Gerzer, R., Heim. J.-M., and Forssmann, W.-G. (1993). The circulating bioactive form of human guanylin is a higher molecular weight peptide (10.3 kDa). *FEBS Lett.* **318**, 205–209.

Kumar, M.A. and Sturtridge, W.C. (1973). The physiological role of calcitonin assessed through chronic calcitonin deficiency in rats. *J. Physiol., Lond.* **233**, 33–43.

Kuno, S., Kubokawa, K., Nagasawa, H., Urano, A., and Ishii, S. (1996). Cloning and sequence analysis of isotocin genes in chum salmon, *Oncorhynchus keta*. In *Environmental & Conservation Endocrinology* (edited by J. Joss), pp. 211–212. Sydney: Macquarie University.

Kvetnoy, I., Sandvik, A.K., and Waldum, H.L. (1997). The diffuse neuroendocrine system and extrapineal melatonin. *J. Endocr.* **18**, 1–3.

Lacy, E.R. and Reale, E. (1990). The presence of a juxtaglomerular apparatus in elasmobranch fish. *Anat. Embryol.* **182**, 249–262.

Lafeber, F.P.J.G., Hanssen, R.G.J.M., Choy, Y.M., Flik, G., Herrmann-Erlee, M.P.M., Pang, P.K.T., and Wendelaar Bonga, S.E. (1988). Identification of hypocalcin (teleocalcin) isolated from trout Stannius corpuscles. *Gen. Comp. Endocr.* **69**, 19–30.

Lafeber, F.P.J.G., Flik, G., Wendelaar Bonga, S.E., and Perry, S.F. (1988). Hypocalcin from Stannius corpuscles inhibits gill calcium uptake in trout. *Am. J. Physiol.* **254**, R891–R896.

Lafeber, F.P.J.G. and Perry, S.F. (1988). Experimental hypercalcemia induces hypocalcin release and inhibits branchial Ca^{2+} influx in freshwater trout. *Gen. Comp. Endocr.* **72**, 136–143.

Lagios, M.D. (1975). The pituitary gland of the coelacanth *Latimeria chalumnae* Smith. *Gen. Comp. Endocr.* **25**, 126–146.

Lagios, M.D. and Stasko-Concannon, S. (1979). Presumptive interrenal tissue (adrenocortical homolog) in the coelacanth *Latimeria chalumnae. Gen. Comp. Endocr.* **37**, 404–406.

Lam, T.J. (1972). Prolactin and hydromineral metabolism in fishes. *Gen.*

Comp. Endocr. **Suppl. 3**, 328–338.

Lamharzi, N. and Fouchereau-Peron, M. (1996). Adaptation of rainbow trout to seawater: changes in calcitonin gene-related peptide levels are associated with an increase in hormone-receptor interaction in gill membranes. *Gen. Comp. Endocr.* **102**, 274–280.

Lance, V.A., Elsey, R.M., and Coulson, R.A. (1993). Biological activity of alligator, avian, and mammalian insulins in juvenile alligators: plasma glucose and amino acids. *Gen. Comp. Endocr.* **89**, 267–275.

Lance, V., Hamilton, J.W., Rouse, J.B., Kimmel, J.R., and Pollock, H.G. (1984). Isolation and characterization of reptilian insulin, glucagon, and pancreatic polypeptide: complete amino acid sequence of alligator (*Alligator mississipiensis*) insulin and pancreatic polypeptide. *Gen. Comp. Endocr.* **55**, 112–124.

Land, H., Schütz, G., Schmale, H., and Richter, D. (1982). Nucleotide sequence of cloned cDNA encoding bovine arginine vasopressin-neurophysin II precursor. *Nature, Lond.* **295**, 299–303.

Landas, S., Phillips, M.I., Stamler, J.F., and Raizada, M.K. (1980). Visualization of specific angiotensin II binding sites in the brain by fluorescent microscopy. *Science* **210**, 791–793.

Langehorne, P. and Simpson, T.H. (1986). The interrelationship of cortisol, gill (Na+K) ATPase, and homeostasis during the parr–smolt transformation of Atlantic salmon (*Salmo salar* L.). *Gen. Comp. Endocr.* **61**, 203–213.

LaPointe, J.I. (1977). Comparative physiology of neurohypophysial hormone action on the vertebrate oviduct–uterus. *Am. Zool.* **17**, 763–773.

Larhammar, D., Blomquist, A.G., Yee, F., Jazin, E., Yoo, H., and Wahlested, C. (1992). Cloning and functional expression of a human neuropeptide Y/peptide YY receptor of the Y_1 type. *J. Biol. Chem.* **267**, 10 935–19 938.

Larner, J. (1990). Effects of insulin on glycogen metabolism. In *Insulin* (edited by P. Cuatrecasas and S. Jacobs), pp. 367–384. Berlin: Springer-Verlag.

Larsen, L.O. (1965). Effects of hypophysectomy in the cyclostome,

Lampetra fluviatilis (L) Gray. *Gen. Comp. Endocr.* **5**, 16–30.

Larsen, L.O. (1969). Effects of gonadectomy in the cyclostome, *Lampetra fluviatilis. Gen. Comp. Endocr.* **13**, 516–517.

Larsen, L.O. (1973). Development in adult, freshwater river lampreys and its hormonal control. Thesis: University of Copenhagen.

Larsen, L.O. (1978). Hormonal control of sexual maturation in lampreys. In *Comparative Endocrinology* (edited by F.J. Gaillard and H.H. Boer), pp. 105–108. Amsterdam: Elsevier North-Holland.

Larsen, L.O. (1980). Physiology of adult lampreys, with special regard to natural starvation, reproduction, and death after spawning. *Can. J. Fish. Aquat. Sci.* **37**, 1762–1779.

Larsen, L.O. and Rosenkilde, P. (1971). Iodine metabolism in normal, hypophysectomized, and thyrotropin-treated river lampreys, *Lampetra fluviatilis* (Gray) L. (Cyclostomata). *Gen. Comp. Endocr.* **17**, 94–104.

Larsen, P.J., Jukes, K.E., Chowdrey, H.S., Lightman, S.L., and Jessop, D.S. (1994). Neuropeptide-Y potentiates the secretion of vasopressin from the neurointermediate lobe of the rat pituitary gland. *Endocrinology* **134**, 1635–1639.

Larson, B. and Bern, H.A. (1987). The urophysis and osmoregulation. In *Vertebrate Endocrinology: Fundamentals and Biomedical Implications*, vol. 2, *Regulation of Water and Electrolytes* (edited by P.K.T. Pang and M.P. Schreibman), pp. 143–156. San Diego, CA: Academic Press.

Larson, B.A. and Madani, Z. (1991). Increased urotensin I and II immunoreactivity in the urophysis of *Gillichthys mirabilis* transferred to low salinity water. *Gen. Comp. Endocr.* **83**, 379–387.

Larsson, A. and Lewander, K. (1972). Effects of glucagon administration to eels (*Anguilla anguilla* L.). *Comp. Biochem. Physiol. A* **43**, 831–836.

Larsson, A.L. (1973). Metabolic effects of epinephrine and norepinephrine in the eel *Anguilla anguilla* L. *Gen. Comp.*

Endocr. **20**, 155–167.

Larsson, D., Björnsson, B.T., and Sundell, K. (1995). Physiological concentrations of 24,25-dihydroxyvitamin D$_3$ rapidly decrease the *in vitro* intestinal calcium uptake in the Atlantic cod. *Gadus morhua. Gen. Comp. Endocr.* **100**, 211–217.

Larsson, L.-I. (1980). Peptide secretory pathways in GI tract: cytochemical contributions to regulatory physiology of the gut. *Am. J. Physiol.* **239**, G237–G246.

Larsson, L.-I. and Rehfeld, J.F. (1977). Evidence for a common evolutionary origin of gastrin and cholecystokinin. *Nature, Lond.* **269**, 335–338.

Larsson, M., Pettersson, T., and Carlström, A. (1985). Thyroid hormone binding in serum of 15 vertebrate species: isolation of thyroxine-binding globulin and prealbumin analogs. *Gen. Comp. Endocr.* **58**, 360–375.

Lasak, R., Wiborg, O., Richter, K., Burgschwaiger, S., Vuust, J., and Kreil, G. (1987). Conserved exon–intron organization in two different caerulein precursor genes. *Eur. J. Biochem.* **169**, 53–58.

Lasmoles, F., Jullienne, A., Desplan, C., Milhaud, G., and Moukhtar, M.S. (1985). Structure of chicken calcitonin predicted by partial nucleotide sequence of its precursor. *FEBS Lett.* **180**, 113–116.

Laurent, F., Karmann, H., Harvey, S., Foltzer, C., and Mialhe, P. (1989). Growth hormone and somatostatin in the plasma of transiently diabetic ducks: basal variation and response to glucose. *Gen. Comp. Endocr.* **74**, 181–189.

Laurentie, M.P., Barenton, B., Charrier, J., Garcia-Villar, R., Marnet, P.G., Blanchard, M., and Toutain, P.L. (1989). Instantaneous secretion rate of growth hormone in lambs: relationships with sleep, food intake, and posture. *Endocrinology* **125**, 642–651.

Laverty, G. and Clark, N.B. (1981). Renal clearance of phosphate and calcium in the fresh-water turtle: effects of parathyroid hormone. *J. Comp. Physiol. B* **141**, 463–469.

Laverty, G. and Clark, N.B. (1989). The kidney. In *Vertebrate Endocrinology:*

Fundamentals and Biomedical Implications, vol. 3, *Regulation of Calcium and Phosphate* (edited by P.K.T. Pang and M.P. Schreibman), pp. 277–317. San Diego, CA: Academic Press.

Lavin, J.H., Wittert, G., Sun, W.-M., Horowitz, M., Morley, J.E., and Read, N.W. (1996). Appetite regulation by carbohydrate: role of blood glucose and gastrointestinal hormones. *Am. J. Physiol.* **271**, E209–E214.

Lawson, D.E.M., Fraser, D.R., Kodicek, E., Morris, H.R., and Williams, D.H. (1971). Identification of 1,25-dihydroxycholecalciferol, a new kidney hormone controlling calcium metabolism. *Nature, Lond.* **230**, 228–230.

Lea, R.W., Klandorf, H., Harvey, S., and Hall, T.R. (1992). Thyroid and adrenal function in the ring dove (*Streptopelia risoria*) during food deprivation and a breeding cycle. *Gen. Comp. Endocr.* **86**, 138–146.

Lea, R.W., Vowles, D.M., and Dick, H.R. (1986). Factors affecting prolactin secretion during the breeding cycle of the ring dove (*Streptopelia risoria*) and its possible role in incubation. *J. Endocr.* **110**, 447–458.

Lederis K. (1973). Current studies on urotensin. *Am. Zool.* **13**, 771–773.

Lederis, K. (1987). Non-mammalian corticotropin release-stimulating peptides. *Ann. New York Acad. Sci.* **512**, 129–138.

Lederis, K., Fryer, J.N., Okawara, Y., Schönrock, C.H.R., and Richter, D. (1994). Corticotropin-releasing factors acting on the fish pituitary: experimental and molecular analysis. In *Fish Physiology*, Vol. XIII, *Molecular Endocrinology of Fish* (edited by N.M. Sherwood and C.L. Hew), pp. 67–100. San Diego, CA: Academic Press.

Lederis, K., Fryer, J., Rivier, J., MacCannell, K.L., Kobayashi, Y., Woo, N., and Wong, K.L. (1985). Neurohormones from fish tails. II: actions of urotensin I in mammals and fishes. *Res. Prog. Horm. Res.* **41**, 553–573.

Lee, A.K. and Mercer, E.H. (1967). Cocoon surrounding desert-dwelling frogs. *Science* **157**, 87–88.

Lee, J. and Malvin, R.L. (1987). Natriuretic response to homologous heart extract in

aglomerular toadfish. *Am. J. Physiol.* **252**, R1055–R1058.

Lee, M.M., and Donahoe, P.K. (1993). Mullerian inhibiting substance: a gonadal hormone with multiple functions. *Endocrine Rev.*. **14**, 152–164.

Lee, T.M., Pelz, K., Licht, P., and Zucker, I. (1990). Testosterone influences hibernation in golden-mantled ground squirrels. *Am. J. Physiol.* **259**, R760–R767.

Lee, Y.-M., Beinborn, M., McBride, E.W., Lu, M., Kolakowski, L.F., and Kopin, A.S. (1993). The human brain cholecystokinin/gastrin receptor. Cloning and characterization. *J. Biol. Chem.* **268**, 8164–8169.

Lefebvre, D.L., Giaid, A., Bennett, H., Larivière, R., and Zingg, H.H. (1992). Oxytocin gene expression in the rat uterus. *Science* **256**, 1553–1555.

Le Fevre, M.D. (1973). Effects of aldosterone on the isolated substrate-depleted turtle bladder. *Am. J. Physiol.* **225**, 1252–1256.

Legan, S.J., Karsch, F.J., and Foster, D.L. (1977). The endocrine control of seasonal reproductive function in the ewe: a marked change in response to negative feedback action of estradiol on luteinizing hormone secretion. *Endocrinology* **101**, 818–823.

Leibson, L. and Plisetskaya, E.M. (1968). Effect of insulin on blood sugar level and glycogen content in organs in some cyclostomes and fish. *Gen. Comp. Endocr.* **11**, 318–392.

Leloup, J. and Fontaine, M. (1960). Iodine metabolism in lower vertebrates. *Ann. N.Y. Acad. Sci.* **86**, 316–353.

Le Magnen, J. (1992). *Neurobiology of Feeding and Nutrition.* San Diego, CA: Academic Press.

Lemon, M. (1972). Peripheral plasma progesterone during pregnancy and the oestrous cycle in the tammar wallaby, *Macropus eugenii. J. Endocr.* **55**, 63–71.

Le Ninan, F., Cherel, Y., Robin, J.-P., Leloup J., and Le Maho, Y. (1988). Early changes in plasma hormones and metabolites during fasting in king penguin chicks. *J. Comp. Physiol. B* **158**, 395–401.

Leopold, A.S., Erwin, M., Oh, J.C., and Browning, B. (1976). Phytoestrogens:

adverse effects on reproduction of California quail. *Science* **191**, 98–100.

LeRoith, D. (1995). Editorial: a novel *Drosophila* insulin receptor : fly in the ointment or evolutionary conservation. *Endocrinology* **136**, 2355–2356.

Le Roith, D., Delahunty, G., Wilson, G.L., Roberts, C.T., Shemer, J., Hart, C., Lesniak, M.A., Shiloach, J., and Roth, J. (1986). Evolutionary aspects of the endocrine and nervous systems. *Rec. Prog. Horm. Res.* **42**. 549–582.

Le Roith, D., Pickens, W., Wilson, G.L., Miller, B., Berelowitz, M., Vinik, A.I., Collier, E., Cleland, C.F. (1985). Somatostatin-like material is present in flowering plants. *Endocrinology* **117**, 2093–2097.

Leung, D.W., Spencer, S.A., Cachianes, G., Hammonds, R.G., Collins, C., Henzel, W.J., Barnard, R., Waters, M.J., and Wood, W.I. (1987). Growth hormone receptor and serum binding protein: purification, cloning and expression. *Nature, Lond.* **330**, 537–543.

Levin, E.R. (1993). Natriuretic peptide C-receptor: more than a clearance receptor. *Am. J. Physiol.* **264**, E483–E489.

Levitin, H.P. (1980). Monaminergic control of MSH release in the lizards *Anolis carolinensis. Gen. Comp. Endocr.* **41**, 279–286.

Levy, A. and Lightman, S.L. (1989). Posterior pituitary vasopressin mRNA in salt-loaded rats: effects of pituitary stalk section. *J. Physiol. Lond.* **418**, 140P.

Lewander, K., Dave, G., Johansson-Sjöbeck, M.L., Larsson, A., and Lidman, U. (1976). Metabolic effects of insulin in the European eel *Anguilla anguilla* L. *Gen. Comp. Endocr.* **29**, 455–467.

Lewin, R. (1988). Molecular clocks turn a quarter of a century. *Science*, **239**, 561–563.

Lewis, R.V., and Erickson, B.W. (1986). Evolution of proenkephalin and prodynorphin. *Am. Zool.* **26**, 1027–1032.

Lewis, S.A. and Diamond, J.M. (1976). Na$^+$ transport by rabbit urinary bladder, a tight epithelium. *J. Memb. Biol.* **28**, 1–40.

L'Horset, F., Perret, C., Brehier, A., and Thomasset, D. (1990). 17β-Estradiol stimulates the calbindin-D$_{9K}$ (CaBP9k)

gene expression at the transcriptional and posttranscriptional levels in the rat uterus. *Endocrinology* **127**, 2891–2897.

Li, C.H. (1969). Recent studies on the chemistry of human growth hormone. *Colloq. Int. C.N.R.S.* **177**, 175–179.

Li, C.H. (1972). Recent knowledge of the chemistry of lactogenic hormones. In *Lactogenic Hormones* (edited by G.E.W. Wolstenholme and J. Knight), pp. 7–22. London: Churchill.

Li, C.H. and Chung, D. (1976). Primary structure of human β-lipotrophin. *Nature, Lond.* **260**, 622–624.

Li, D., Tsutsui, K., Muneoka, Y., Minakata, H., and Nomoto, K. (1996). An oviposition-inducing peptide: isolation, localization, and function of avian galanin in the quail oviduct. *Endocrinology* **137**, 1618–1626.

Li, X.-J., Wu, Y.-N., North, R.A., and Forte, M. (1992). Cloning, functional expression, and developmental regulation of a neuropeptide Y receptor from *Drosophila melanogaster. J. Biol. Chem.* **267**, 9–12.

Licht, P. (1970). Effects of mammalian gonadotropins (ovine FSH and LH) in female lizards. *Gen. Comp. Endocr.* **14**, 98–106.

Licht, P. (1972). Environmental physiology of reptilian breeding cycles: role of temperature. *Gen. Comp. Endocr.* **Suppl. 3**, 477–487.

Licht, P. (1979). Reproductive endocrinology of reptiles and amphibians: gonadotropins. *Annu. Rev. Physiol.* **41**, 337–351.

Licht, P. (1983). Evolutionary divergence in the structure and function of pituitary gonadotropins of tetrapod vertebrates. *Am. Zool.* **23**, 673–683.

Licht, P. (1984). Reptiles. In *Marshall's Physiology of Reproduction*, 4th edn (edited by G.E. Lamming), vol. 1, *Reproductive Cycles of Vertebrates*, pp. 206–282. Edinburgh: Churchill Livingstone.

Licht, P. and Bona Gallo, A.B. (1978). Immunochemical relatedness among pituitary follicle-stimulating hormones of tetrapod vertebrates. *Gen. Comp. Endocr.* **36**, 575–584.

Licht, P., Farmer, S.W., Gallo, A.B., and Papkoff, H. (1979). Pituitary gonadotropins in snakes. *Gen. Comp.*

Endocr. **39**, 34–52.

Licht, P., Farmer, S.W., and Papkoff, H. (1978). Biological activity of hybrid combinations of ovine and sea turtle subunits. *Gen. Comp. Endocr.* **35**, 289–294.

Licht, P. and Papkoff, H. (1974). Separation of two distinct gonadotropins from the pituitary gland of the snapping turtle (*Chelydra serpentina*). *Gen. Comp. Endocr.* **22**, 218–237.

Licht, P., Papkoff, H., Farmer, S.W., Muller, C.H., Tsui, H.W., and Crews, D. (1977). Evolution of gonadotropin structure and function. *Rec. Progr. Horm. Res.* **33**, 169–243.

Licht, P. and Pavgi, S. (1992). Identification and purification of a high-affinity thyroxine binding protein from albumin and prealbumin in the blood of a turtle, *Trachemys scripta. Gen. Comp. Endocr.* **85**, 179–192.

Licht, P. and Stockell-Hartree, A. (1971). Actions of mammalian, avian and piscine gonadotrophins in the lizard. *J. Endocr.* **49**, 113–124.

Liel, Y., Kraus, S., Levy, J., and Shany, S. (1992). Evidence that estrogens modulate activity and increase the number of 1,25-dihydroxyvitamin D receptors in osteoblast-like cells (ROS 17/2,8). *Endocrinology* **130**, 2597–2601.

Lieverse, R.J., Jansen, J.B.M.J., Masclee, A.M. and Lamers, C.B.H.W. (1994). Role of cholecystokinin in the regulation of satiation and satiety in humans. *Ann. N.Y. Acad. Sci.* **713**, 268–272.

Liggins, G.C. and Thorburn, G.D. (1994). Initiation of parturition. In *Marshall's Physiology and Reproduction*, 4th edn (edited by G.E. Lamming), vol. 3, *Pregnancy and Lactation*, Pt. 2, *Fetal Physiology, Parturition and Lactation*, pp. 862–1002. London: Chapman & Hall.

Lillie, F.R. (1917). The free-martin; a study of the actions of sex hormones in the foetal life of cattle. *J. Exp. Biol.* **23**, 371–452.

Lin, X.-W., Lin, H.-R., and Peter, R.E. (1996). Direct influences of temperature on gonadotropin-II release from perifused pituitary fragments of common carp (*Cyprinus carpio* L.) *in vitro. Comp. Biochem. Physiol. A* **114**, 341–347.

Lincoln, D.W., McNeilly, A.S., and Sharpe,

R.M. (1989). Reproductive physiology of inhibin and related peptides. In *Recent Advances in Endocrinology and Metabolism* (edited by C.R.W. Edwards and D.W. Lincoln), No. 3, pp. 77–107. Edinburgh: Churchill Livingstone.

Lincoln, D.W. and Renfree, M.B. (1981). Milk ejection in a marsupial, *Macropus agilis*. *Nature, Lond.* **289**, 504–506.

Lincoln, G.A. (1979). Use of pulsed infusion of luteinizing hormone releasing hormone to mimic seasonally induced endocrine changes in the ram. *J. Endocr.* **83**, 251–260.

Lincoln, G.A. and Baker, B.I. (1995). Seasonal and photoperiod-induced changes in the secretion of α-melanocyte-stimulating hormone in Soay sheep: temporal relationships with changes in β-endorphin, prolactin, follicle-stimulating hormone, activity of the gonads and growth of the wool and horns. *J. Endocr.* **144**, 471–481.

Lindberg, I. and White, L. (1986). Reptilian enkephalins: implications for the evolution of proenkephalin. *Arch. Biochem. Biophys.* **245**, 1–7.

Ling, J.K. (1972). Adaptive functions of vertebrate molting cycles. *Am. Zool.* **12**, 77–93.

Liposits, Z., Merchenthaler, I., Reid, J.J., and Negro-Vilar, A. (1993). Galanin-immunoreactive axons innervate somatostatin-synthesizing neurons in the anterior periventricular nucleus of the rat. *Endocrinology* **132**, 917–923.

Lofts, B. (1964). Seasonal changes in the functional activity of the interstitial and spermatogenetic tissues of the green frog *Rana esculenta*. *Gen. Comp. Endocr.* **4**, 550–562.

Lofts, B. (1968). Patterns of testicular activity. In *Perspectives in Endocrinology, Hormones in the Lives of Lower Vertebrates* (edited by E.J.W. Barrington and C.B. Jorgenson), pp. 239–304. New York: Academic Press.

Lofts, B. (1969). Seasonal cycles in reptilian testes. *Gen. Comp. Endocr.* **Suppl. 2**, 147–155.

Lofts, B. (1984). Amphibians. In *Marshall's Physiology of Reproduction*, 4th edn (edited by G.E. Lamming), vol. 1, *Reproductive Cycles of Vertebrates*, pp. 127–205. Edinburgh: Churchill Livingstone.

Lofts, B. and Bern, H.A. (1972). The functional morphology of steroidogenic tissues. In *Steroids in Nonmammalian Vertebrates* (edited by D.R. Idler), pp. 37–125. New York: Academic Press.

Lofts, B., Follett, B.K., and Murton, R.K. (1970). Temporal changes in the pituitary gonadal axis. *Mem. Soc. Endocr.* **18**, 545–575.

Lofts, B., Murton, R.K., and Thearle, R.J.P. (1973). The effects of testosterone propionate and gonadotropins on the bill pigmentation and testes of the house sparrow (*Passer domesticus*). *Gen. Comp. Endocr.* **21**, 202–209.

Lofts, B., Phillips, J.G., and Tam, W.H. (1971). Seasonal changes in the histology of the adrenal gland of the cobra, *Naja naja*. *Gen. Comp. Endocr.* **16**, 121–131.

Logan, A.G. Moriarty, A.J., and Rankin, J.C. (1980). A micropuncture study of kidney function in the river lamprey, *Lampetra fluviatilis*, adapted to fresh water. *J. Exp. Biol.* **85**, 137–147.

Lolait, S.J., O'Carroll, A.-M., McBride, O.W., Konig, M., Morel, A., and Brownstein, M.J. (1992). Cloning and characterization of a vasopressin V$_2$ receptor and possible link to nephrogenic diabetes insipidus. *Nature, Lond.* **357**, 336–339.

Lomedico, P.T., Chan, S.J., Steiner, D.F., and Saunders, G.F. (1977). Immunological and chemical characteristics of bovine prepro-insulin. *J. Biol. Chem.* **252**, 7971–7978.

Loretz, C.A. and Bern, H.A. (1981). Stimulation of sodium transport across the teleost urinary bladder by urotensin II. *Gen. Comp. Endocr.* **43**, 325–330.

Loretz, C.A., Howard, M.E., and Seigel. A.J. (1985). Ion transport in goby intestine: cellular mechanism of urotensin II stimulation. *Am. J. Physiol.* **249**, G284–G293.

Loudon, A., Rothwell, N., and Stock, M. (1985). Brown fat, thermogenesis and physiological birth in a marsupial. *Comp. Biochem. Physiol. A* **81**, 815–819.

Loumaye, E., Thorner, J., and Catt, K.T. (1982). Yeast mating pheromone activates mammalian gonadotrophs: evolutionary conservation of a reproductive hormone? *Science* **218**, 1323–1325.

Lovegrove, B.G. and Knight-Eloff, A. (1988). Soil and burrow temperatures, and the resource characteristics of the social mole-rat *Cryptomys damarensis* (Bathyergidae) in the Kalahari Desert. *J. Zool. Lond.* **216**, 403–416.

Lowry, P.J. and Scott, A.P. (1975). The evolution of vertebrate corticotrophin and melanocyte stimulating hormone. *Gen. Comp. Endocr.* **26**, 16–23.

Lowry, P.J. and Scott, A.P. (1975). The evolution of vertebrate corticotrophin and melanocyte stimulating hormone. *Gen. Comp. Endocr.* **26**, 16–23.

Lu, M., Wagner, G.F., and Renfro, J.L. (1994). Stanniocalcin stimulates phosphate reabsorption by flounder renal proximal tubule in primary culture. *Am. J. Physiol.* **267**, R1356–R1362.

Lyons, W.R. (1958). Hormonal synergism in mammary growth. *Proc. Roy. Soc. Lond. Series B* **149**, 303–325.

Ma, S.W.Y. and Copp, D.H. (1978). Purification, properties and action of a glycopeptide from the corpuscles of Stannius which affects calcium metabolism in the teleost. In *Comparative Endocrinology* (edited by P.J. Gaillard and H.H. Boer), pp. 283–286, Amsterdam: Elsevier.

MacDonald, O.A., Hwang, C.-S., Fan, H., and Lane, M.D. (1995). Regulated expression of the obese gene product (leptin) in white adipose tissue and 3T3-L1 adipocytes. *Proc. Natl. Acad. Sci. USA* **92**, 9034–9037.

MacKenzie, D.S. and Licht, P. (1984). Studies on the specificity of thyroid response to pituitary glycoprotein hormones. *Gen. Comp. Endocr.* **56**, 156–166.

MacKenzie, D.S., Licht, P., and Papkoff, H. (1981). Purification of thyrotropin from the pituitaries of two turtles: the green sea turtle and the snapping turtle. *Gen. Comp. Endocr.* **45**, 39–48.

MacLaughlin, D.T., Hudson, P.L., Graciano, A.L., Kenneally, M.K., Ragin, R.C., Manganaro, T.F., and Donahoe, P.K. (1992). Mullerian duct regression and antiproliferative bioactivities of Mullerian inhibiting substance reside in its carboxy-terminal domain. *Endocrinology* **131**, 291–296.

MacLeod, R.M. and Lehmeyer, J.E. (1974).

Studies on the mechanism of the dopamine-mediated inhibition of prolactin secretion. *Endocrinology* **94**, 1077–1085.

Madej, T., Boguski, M.S., and Bryant, S.H. (1995). Threading analysis suggests that the obese gene product may be a helical cytokine. *FEBS Lett.* **373**, 13–18.

Maderson, P.F.A., Chiu, K.W., and Phillips, J.G. (1970). Endocrine–epidermal relationships in squamate reptiles. *Mem. Soc. Endocr.* **18**, 259–284.

Maderson, P.F.A. and Licht, P. (1967). Epidermal morphology and sloughing frequency in normal and prolactin treated *Anolis carolinensis* (Iguanidae, Lacertilia). *J. Morphol.* **123**, 157–172.

Madsen, S.S. (1990a). Effects of repetitive cortisol and thyroxine injections on chloride cell number and Na$^+$/K$^+$-ATPase activity in gills of freshwater acclimated rainbow trout. *Salmo. gairdneri. Comp. Biochem. Physiol. A* **95**, 171–175.

Madsen, S.S. (1990b). The role of cortisol and growth hormone in seawater adaptation and development of hypoosmoregulatory mechanisms in sea trout parr (*Salmo trutta trutta*). *Gen. Comp. Endocr.* **79**, 1–11.

Maetz, J., Bourguet, J., Lahlou, B., and Houdry, J. (1964). Peptides neurohypophysaires et osmorégulation chez *Carassius auratus. Gen. Comp. Endocr.* **4**, 508–522.

Maetz, J., Mayer, N., and Chartier-Baraduc, M.M. (1967). La balance minérale du sodium chez *Anguilla anguilla* en eau de mer, en eau douce et au cours de transfert d'un milieu à l'autre: effets de l'hypophysectomie et de la prolactine. *Gen. Comp. Endocr.* **8**, 177–188.

Maetz, J., and Rankin, J.C. (1969). Quelques aspects du rôle biologique des hormones neurohypophysaires chez les poissons. *Colloq. Int. C.N.R.S.* **177**, 45–54.

Maher, M.J. (1965). The role of the thyroid gland in the oxygen consumption of lizards. *Gen. Comp. Endocr.* **5**, 320–325.

Mahlmann, S., Meyerhof, W., Hausman, H., Heierhorst, J., Schönrock, C., Zwiers, H., Lederis, J., and Richter, D. (1994). Structure, function, and phylogeny of [Arg^8]vasotocin receptors from teleost fish and toad. *Proc. Natl. Acad. Sci. USA* **91**,

1342–1345.

Malaise. W.J. (1990). Regulation of insulin release by the intracellular mediators cyclic AMP, Ca^{2+}, inositol 1,4,5-trisphosphate, and diacylglycerol. In *Insulin* (edited by P. Cuatrecasas and S. Jacobs), pp. 113–124. Berlin: Springer-Verlag.

Malpaux, B., Robinson, J.E., Wayne, N.L., and Karsch, F.J. (1989). Regulation of the onset of the breeding season of the ewe: importance of long days and of an endogenous reproductive rhythm. *J. Endocr.* **122**, 269–278.

Mangelsdorf, D.J. and Evans, R.M. (1995). The RXR heterodimers and orphan receptors. *Cell* **83**, 841–850.

Manns, J.G., Boda, J.M., and Willes, R.F. (1967). Probable role of propionate and butyrate in control of insulin secretion in sheep. *Am. J. Physiol.* **212**, 756–764.

Marcinkiewicz, M., Day, R., Seidah, N.G., and Chrétien, M. (1993). Ontogeny of the prohormone convertases PC1 and PC2 in the mouse hypophysis and their colocalization with corticotropin and α-melanotropin. *Devel. Biol.* **90**, 4922–4926.

Marcocci, C., Freake, H.C., Iwasaki, J., Lopez, E., and MacIntyre, I. (1982). Demonstration and organ distribution of the 1,25-dihydroxyvitamin D$_3$-binding protein in fish (*A. anguilla*). *Endocrinology* **110**, 1347–1354.

Marshall, F.H.A. (1956). The breeding season. In *Marshall's Physiology of Reproduction* (edited by A.S. Parkes), vol. 1 (Pt. 1), pp. 1–42. London: Longmans.

Marshall, W.S. and Bern, H.A. (1980). Ion transport across the isolated skin of the teleost *Gillichthys mirabilis*. In *Epithelial Transport in Lower Vertebrates* (edited by B. Lahlou), pp. 337–350. Cambridge University Press.

Marshall, W.S. and Bern, H.A. (1981). Active chloride transport by the skin of a marine teleost is stimulated by urotensin I and inhibited by urotensin II. *Gen. Comp. Endocr.* **43**, 484–491.

Marsigliante, S., Muscella, A. Vilella, S., Nicolardi, G., Ingrosso, L., Ciardo, V., Zonno, V., Vinson, G.P. Ho, M.M., and Storelli, C. (1996). A monoclonal antibody to mammalian angiotensin II AT$_1$ receptor recognizes one of the

angiotensin II receptor isoforms expressed by the eel (*Anguilla anguilla*). *J. Mol. Endocr.* **16**, 45–56.

Martens, G.J.M. (1992). Molecular biology of G-protein-coupled receptors. *Prog. Brain Res.* **92**, 201–214.

Martens, G.J.M., Civelli, O., and Herbert, E. (1985). Nucleotide sequence of cloned cDNA for pro-opiomelanocortin in the amphibian *Xenopus laevis. J. Biol. Chem.* **260**, 13685–13689.

Martens, G.J.M., and Herbert, E. (1984). Polymorphism and absence of Leu-enkephalin sequences in proenkephalin genes in *Xenopus laevis. Nature, Lond.* **310**, 251–254.

Martens, G.J.M., Molhuizen, H.O.F., Gröneveld, D., and Roubos, E.W. (1991). Cloning and sequence analysis of brain cDNA encoding a *Xenopus* D$_2$ dopamine receptor. *FEBS Lett.* **281**, 85–89.

Martin, B. (1975). Steroid–protein interactions in nonmammalian vertebrates. *Gen. Comp. Endocr.* **25**, 42–51.

Martin, D.R., Pevahouse, J.B., Trigg, D.J., Vesely, D.L., and Buerkert, J.E. (1990). Three peptides from the ANF prohormone NH$_2$-terminus are natriuretic and/or kaliuretic. *Am. J. Physiol.* **258**, F1401–F1408.

Martin, G.B. and Walkden-Brown, S.W. (1995). Nutritional influences on reproduction in mature male sheep and goats. *J. Reprod. Fertility* **Suppl. 49**, 437–449.

Martin, G.B., Walkden-Brown, S.W., Boukhliq, R., Tjondronegoro, S., Miller, D.W., Fisher, J.S., Hötzel, M.J., Restall, B.J., and Adams, N.R. (1994). Non-photoperiodic inputs into seasonal breeding in male ruminants. In *Perspectives in Comparative Endocrinology* (edited by K.G. Davey, R.E. Peter, and S.S. Tobe), pp. 574–585. Ottawa: National Research Council of Canada.

Martin, I.K. and McDonald, I.R. (1986). Adrenocortical functions in a macropodid marsupial *Thylogale billardierii. J. Endocr.* **110**, 471–480.

Martin, P.A. and Faulkner, A. (1996). Effects of somatostatin-28 on circulating concentrations of insulin and gut hormones in sheep. *J. Endocr.* **151**,

107–112.

Martin, T.J., Moseley, J.M. and Gillespie, M.T. (1991). Parathyroid hormone-related protein: biochemistry and molecular biology. *Crit. Rev. Biochem. Mol. Biol.* **26**, 377–395.

Martinez, A. Unsworth, E.J., and Cuttitta, F. (1996). Adrenomedullin-like immunoreactivity in the nervous system of the starfish *Marthasterias glacialis. Cell. Tissue Res.* **283**, 169–172.

Marx, S.J., Woodward, C.J., and Aurbach, G.D. (1972). Calcitonin receptors in kidney and bone. *Science* **178**, 999–1001.

Mason, A.J., Hayflick, J.S., Ling, N., Esch, F., Ueno, N., Ying, S.-Y., Guillemin, R., Niall, H., and Seeburg, P.H. (1985). Complementary DNA sequences of ovarian follicular fluid inhibin show precursor structure and homology with transforming growth factor-*β. Nature, Lond.* **318**, 659–663.

Matsumoto, A. and Ishii, S. (eds.) (1992). *Atlas of Endocrine Organs. Vertebrates and Invertebrates.* English edition. Berlin: Springer-Verlag.

Maurel, D. and Coutant, C. (1986). Comparative aspects of molt in relation to the thyroidal and gonadal cycles in the European badger (*Meles meles L*), the red fox (*Vulpes vulpes L.*) and the mink (*Mustela vison*): the role of the testis. In *Endocrine Regulations as Adaptive Mechanisms to the Environment* (edited by I. Assenmacher and J. Boissin), pp. 277–309. Paris: Centre Nationale de la Recherche Scientifique.

Maurel, D., Coutant, D., and Boissin, J. (1987). Thyroid and gonadal regulation of hair growth during seasonal molt in the male European badger, *Meles meles L. Gen. Comp. Endocr.* **65**, 317–327.

Maurel, D., Coutant, C., and Boissin, J. (1989). Effects of photoperiod, melatonin implants and castration on molting and on plasma thyroxine, testosterone and prolactin levels in the European badger (*Meles meles*). *Comp. Biochem. Physiol. A* **93**, 791–797.

Mayer, N., Maetz, J., Chan, D.K.O., Forster, M., and Chester Jones, I. (1967). Cortisol, a sodium excreting factor in the eel (*Anguilla anguilla* L.) adapted to sea water. *Nature, Lond.* **214**, 1118–1120.

Mayer-Gostan, N., Flik, G., and Pang, P.T. (1992). An enzyme-linked immuno-absorbant assay for stanniocalcin, a major hypocalcemic hormone in teleost. *Gen. Comp. Endocr.* **86**, 10–19.

Mayer-Gostan, N., Wendelaar Bonga, S.E., and Balm, P.H.M. (1987). Mechanisms of hormone actions on gill transport. In *Vertebrate Endocrinology: Fundamentals and Biomedical Implications,* vol. 2, *Regulation of Water and Electrolytes* (edited by P.K.T. Pang and M.P. Schreibman), pp. 211–238. San Diego, CA: Academic Press.

Mayo, K.E., Vale, W., Rivier, J., Rosenfeld, M.G., and Evans, R.M. (1983). Expression-cloning and sequence of cDNA encoding human growth hormone-releasing factor. *Nature, Lond.* **306**, 86–88.

Maywood, E.S. and Hastings, M.H. (1995). Lesions of the iodomelatonin-binding sites of the mediobasal hypothalamus spare the lactotropic, but block the gonadotropic response of male Syrian hamsters to short photoperiod and to melatonin. *Endocrinology* **136**, 144–153.

Mazzocchi, G., Rebuffat, P., Gottardo, G., and Nussdorfer, G.G. (1996). Adrenomedullin and calcitonin gene-related peptide inhibit aldosterone secretion in rats, acting via a common receptor. *Life Sci.* **58**, 839–844.

McClanahan, L. (1967). Adaptations of the spadefoot toad. *Scaphiopus couchi* to desert environments. *Comp. Biochem. Physiol.* **20**, 73–99.

McCormick, S.D. (1996). Effects of growth hormone and insulin-like growth factor I on salinity tolerance and gill Na$^+$, K$^+$-AtPase in Atlantic salmon (*Salmo salar*): interaction with cortisol. *Gen. Comp. Endocr.* **101**, 3–11.

McCormick, S.D., Sakamoto, T., Hasegawa, S., and Hirano, T. (1991). Osmoregulatory actions of insulin-like growth factor-I in the rainbow trout (*Oncorhynchus mykiss*). *J. Endocr.* **130**, 87–90.

McCullagh, D.R. (1932). Dual endocrine activity of the testes. *Science* **76**, 19–20.

McDonald, L.R. (1980). Physiology of the adrenal cortex in mammals. In *Comparative Physiology: Primitive Mammals* (edited by K. Schmidt-Nielsen, L. Bolis, and C.R. Taylor), pp. 316–323. Cambridge University Press.

McDonald, L.K., Joss, J.M.P., and Dores, R.M. (1991). The phylogeny of Met-enkephalin and Leu-enkephalin: studies on the holostean fish *Lepidosteus platyrhincus* and the Australian lungfish, *Neoceratodus forsteri. Gen. Comp. Endocr.* **84**, 228–236.

McGregor, G.P., Desaga, J.F., Ehlenz, K., Fischer, A., Heese, F., Hegele, A., Lämmer, C., Peiser, C., and Lang, R.E. (1996). Radioimmunological measurement of leptin in plasma of obese and diabetic human subjects. *Endocrinology* **137**, 1501–1504.

McLachlan, R.I., Wreford, N.G., O'Donnell, L., de Kretser, D.M., and Robertson, D.M. (1996). The endocrine regulation of spermatogenesis: independent roles for testosterone and FSH. *J. Endocr.* **148**, 1–9.

McLean, E. and Donaldson, E.M. (1993). The role of growth hormone in the growth of poikilotherms. In *The Endocrinology of Growth, Development, and Metabolism in Vertebrates* (edited by M.P. Schreibman, C.G. Scanes, and P.K.T. Pang), pp. 43–72. San Diego, CA: Academic Press.

McLean, M., Bisits, A., Davies, J., Woods, R., Lowry, P., and Smith, R. (1995). A placental clock controlling the length of human pregnancy. *Nature Medicine* **1**, 460–463.

McMaster, D., Belenky, M.A., Polenov, A.L., and Lederis K. (1992). Isolation and amino acid sequence of urotensin II from the sturgeon *Acipenser ruthenus. Gen. Comp. Endocr.* **87**, 275–285.

McMillan, J.E. and Wilkinson, R.F. (1972). The effect of pancreatic hormones on blood glucose in *Ambystoma annulatum. Copeia* 1972, 664–668.

McNabb, F.M.A. and Freeman, T.B. (1990). Comparative studies of thyroid hormone deiodinase systems. In *Progress in Comparative Endocrinology* (edited by A. Epple, C.G. Scanes, and M.H. Stetson), pp. 433–438. New York: Wiley-Liss.

McNabb, R.A. (1969). The effects of thyroxine on glycogen stores and oxygen consumption in the leopard frog, *Rana pipiens. Gen Comp. Endocr.* **12**, 276–281.

McNatty, K.P., Cashmore, M., and Young, A. (1972). Diurnal variation in plasma cortisol levels in sheep. J. Endocr. **54**, 361–362.

McNeilly, A.S., Forsyth, I.A., and McNeilly, J.R. (1994). Regulation of post-partum fertility in lactating mammals. In *Marshall's Physiology of Reproduction*, 4th edn (edited by G.E. Lamming), vol. 3, Pt. 2, *Fetal Physiology, Parturition and Lactation*, pp. 1037–1101. London: Chapman & Hall.

Means, A.R. and Dedman, J.R. (1980). Calmodulin–an intracellular calcium receptor. *Nature, Lond.* **285**, 73–77.

Medica, P.A., Turner, F.B., and Smith, D.D. (1973). Hormonal induction of color change in female leopard lizards *Crotaphytus wislizenii*. *Copeia* 1973 (no. 4), 658–661.

Meier, A.H. and Farner, D.S. (1964). A possible endocrine basis for premigratory fattening in the white-crowned sparrow, *Zonotrichia leucophrys gambelli* (Nuttall). *Gen. Comp. Endocr.* **4**, 584–595.

Meister, B. and Hokfelt, T. (1988). Peptide- and transmitter-containing neurons in the mediobasal hypothalamus and their relationship to GABAergic systems: possible roles in control of prolactin and growth hormone secretion. *Synapse* **2**, 585–605.

Meltzer, V., Weinreb, S., Bellorin-Font, E., and Hruska, K.A. (1982). Parathyroid hormone stimulation of renal phosphoinositide metabolism in a cyclic nucleotide-independent effect. *Biochem, Biophys. Acta* **712**, 258–267.

Mepham, T.B. and Kuhn, N.J. (1994). Physiology and biochemistry of lactation. In *Marshall's Physiology of Reproduction*, 4th end, vol. 3, *Pregnancy and Lactation*, Part 2 *Fetal Physiology, Parturition and Lactation* (edited by G.E. Lamming), pp. 1103–1186. London: Chapman & Hall.

Mercken, L., Simons, M.-J., Swillens, S., Massaer, M., and Vassart, G. (1985). Primary structure of bovine thyroglobulin deduced from the sequence of its 8,431-base complementary DNA. *Nature, Lond.* **316**, 647–651.

Meyer, M., Richter, R., Brunkhorst, R., Wrenger, E., Schulz-Knappe, P., Kist, A., Mentz, P., Brabant, E.G., Koch, K.M., Rechkemmer, G., and Forssmann, W.-G.

(1996). Urodilatin is involved in sodium homeostasis and exerts sodium-state-dependent natriuretic and diuretic effects. *Am. J. Physiol.* **271**, F489–F497.

Migitaka, H., Hattori, A., Itoh, M., and Suzuki, T. (1996). Identification and changes of melatonin in edible plants. In *Environmental & Conservation Endocrinology* (edited by J. Joss) pp. 163–164. Sydney: Macquarie University.

Mikami, S.-I. (1992). 3. Hypophysis. In *Atlas of Endocrine Organs, Vertebrates and Invertebrates* (edited by A. Matsumoto and S. Ishii), pp. 39–62. English Edition. Berlin: Springer-Verlag.

Milet, C., Peignoux-Deville, J., and Martelly, E. (1979). Gill calcium fluxes in the eel, *Anguilla anguilla* (L). Effects of Stannius corpuscles and ultimobranchial body. *Comp. Biochem. Physiol. A* **63**, 63–70.

Milhaud, G., Rankin, J.C., Bolis, L., and Benson, A.A. (1977). Calcitonin: its hormonal action on the gill. *Proc. Natl. Acad. Sci., USA* **74**, 4693–4696.

Millam, J.R., Zhang, B., and El Halawani, M.E. (1996). Egg production in cockatiels (*Nymphicus hollandicus*) is influenced by number of eggs in nest after incubation begins. *Gen. Comp. Endocr.* **101**, 205–210.

Millar, R.P., Davidson, J.S., Flanagan, C., Illing, N., Becker, I., Jacobs, G., Wakefield, I., Zhou, W., Chi, L., and Sealfon, S.C. (1994). Gonadotropin-releasing hormone receptor structure and function. In *Perspectives in Comparative Endocrinology* (edited by K.G. Davey, R.E. Peter, and S.S. Tobe), pp. 264–268. Ottawa: National Research Council of Canada.

Millar, R., Illing, N., Hapgood, J., Rumbak, E., Flanagan, C., Davidson, J., Blackman B., Sun, Y., Sealfon, S., Weinstein, B.H., Konvickab, K., Guanierib, F., and King, J. (1996). Co-ordinated structural evolution of GnRHs and their receptors–insights from cooperative studies. In *Environmental & Conservation Endocrinology* (edited by J. Joss), pp. 13–17. Sydney: Macquarie University.

Millar, R.P. and King, J.A. (1994). Plasticity and conservation in gonadotropin-releasing hormone structure and function. In *Perspectives in*

Comparative Endocrinology (edited by K.G. Davey, R.E. Peter, and S.S. Tobe), pp. 129–136. Ottawa: National Research Council of Canada.

Miller, J.D. (1993). On the nature of the circadian clock in mammals. *Am. J. Physiol.* **264**, R821–R832.

Milliken, C.E., Fargher, R.C., Butkus, A., McDonald, M. and Copp, D.H. (1990). Effects of synthetic peptide fragments of teleocalcin (hypocalcin) on calcium uptake in juvenile rainbow trout (*Salmo gairdneri*). *Gen. Comp. Endocr.* **77**, 416–422.

Mills, I., Barge, R.M., Silva, J.E., and Larsen, P.R. (1987). Insulin stimulation of iodothronine 5'-deiodinase in rat brown fat adipocytes. *Biochem. Biophys. Res. Commun.* **143**, 81–86.

Minghetti, P.P. and Norman, A.W. (1988). 1,25 $(OH)_2$-vitaminD$_3$ receptors: gene regulation and genetic circuitry. *FASEB J.* **2**, 3043–3053.

Minick, M.C. and Chavin, W. (1973). Effects of catecholamines upon serum FFA levels in normal and diabetic goldfish, *Carassius auratus* L. *Comp. Biochem. Physiol. A* **44**, 1003–1008.

Mittwoch, U. and Burgess, A.M.C. (1991). How do you get sex? *J. Endocr.* **128**, 329–331.

Miura, T., Kobayashi, T., and Nagahama, Y. (1994). Hormonal regulation of spermatogenesis in the Japanese eel (*Anguilla japonica*). In *Perspectives in Comparative Endocrinology* (edited by K.G. Davey, R.E. Peter, and S.S. Tobe), pp. 631–635. Ottawa: National Research Council of Canada.

Miwa, S., Tagawa, M., Inui, Y., and Hirano, T. (1988). Thyroxine surge in metamorphosing flounder larvae. *Gen. Comp. Endocr.* **70**, 158–163.

Miyata, A., Arimura, A., Dahl, R.R., Minamino, N., Uehara, A., Jiang, L., Culler, M.D., and Coy, D.H. (1989). Isolation of a novel 38 residue–hypothalamic polypeptide which stimulates adenylate cyclase in pituitary cells. *Biochem. Biophys. Res. Commun.* **164**, 567–574.

Miyata, A., Minamino, N., Kangawa, K., and Matsuo, H. (1988). Identification of a 29-amino acid natriuretic peptide in chicken heart. *Biochem. Biophys. Res.*

Commun. **155**, 1330–1337.

Mizuno, J., and Takeda, N. (1988). Phylogenetic study of the arginine-vasotocin/arginine-vasopressin-like immunoreactive system in invertebrates. *Comp. Biochem. Physiol. A* **91**, 739–747.

Mojsov, S., Heinrich, G., Wilson, I.B., Ravazzola, M., Orci, L., and Habener, J.F. (1986). Preproglucagon gene expression in pancreas and intestine diversifies at the level of post-translational processing. *J. Biol. Chem.* **261**, 11 880–11 889.

Mommsen, T.P. and Moon, T.W. (1990). Metabolic response of teleost hepatocytes to glucagon-like peptide and glucagon. *J. Endocr.* **126**, 109–118.

Monaco, F., Dominici, R., Andreoli, M., de Pirro, R., and Roche, J. (1981). Thyroid hormone formation in thyroglobulin synthesized in the amphioxus (*Branchiostoma lanceolatum* PALLAS). *Comp. Biochem. Physiol. B* **70**, 341–343.

Moore, W.T. and Ward, D.N. (1980). Pregnant mare serum gonadotropin. *J. Biol. Chem.* **255**, 6930–3936.

Mor, A., Chartrel, N., Vaudry, H., and Nicolas, P. (1994). Skin peptide tyrosine–tyrosine, a member of the pancreatic polypeptide family: isolation, structure, synthesis, and endocrine activity. *Proc. Natl. Acad. Sci. USA* **91**, 10 295–10 299.

Moran, T.H., Ameglio, P.J., Schwartz, G.J. and McHugh, P.R. (1992). Blockade of type A, but not type B, CCK receptors attenuates satiety actions of exogenous and endogenous CCK. *Am. J. Physiol.* **262**, R46–R50.

Moreau, R., Raoelison, C., and Sutter, B.Ch.J. (1981). An intestinal insulin-like molecule in *Apis mellifica* L. (Hymenoptera). *Comp. Biochem. Physiol. A* **69**, 79–83.

Morel, A., O'Carrol, A.-M., Brownstein, M.J., and Lolait, S.J. (1992). Molecular cloning and expression of a rat Vla arginine vasopressin receptor. *Nature, Lond.* **356**, 523–526.

Morgan, F.J., Birken, S., and Canfield, R.E. (1975). The amino acid sequence of human chorionic gonadotropin. *J. Biol. Chem.* **250**, 5247–5258.

Morishita, F., Shimada, A., Fujimoto, M., Katayama, H., and Yamada, K. (1993).

Inhibition of adenylate cyclase activity in the goldfish melanophore is mediated by α_2-adrenoreceptors and a pertussis toxin-sensitive GTP-binding protein. *J. Comp. Physiol. B* **163**, 533–540.

Morisset, J., Levenez, F., Corring, T., Benrezzak, O., Pelletier, G., and Calvo, E. (1996). Pig pancreatic acinar cells possess predominantly the CCK-B receptor type. *Am. J. Physiol.* **271**, E397–E402.

Morley, M., Chadwick, A., and El Tounsy, E.M. (1981). The effect of prolactin on water absorption by the intestine of the trout (*Salmo gairdneri*). *Gen. Comp. Endocr.* **44**, 64–68.

Morley, S.D., Schönrock, C., Heierhorst, J., Figueroa, J., Lederis, K., and Richter, D. (1990). Vasotocin genes of the teleost fish *Catostomus commersoni*: gene structure, exon–intron boundary, and hormone precursor organization. *Biochemistry* **29**, 2506–2511.

Morris, T.R. (1979). The influence of light on ovulation in domestic birds. In *Animal Reproduction* (edited by H.W. Hawk). Beltsville Symposia in Agricultural Research, 3, pp. 307–322. Montclair, NJ: Allenheld Osmun.

Mosimann, R., Imboden, H., and Felix, D. (1996). The neuronal role of angiotensin II in thirst, sodium appetite, cognition and memory. *Biol. Rev.* **71**, 545–559.

Mountjoy, K.G., Robbins, L.S., Mortrud, M.T., and Cone, R.D. (1992). The cloning of a family of genes that encode the melanocortin receptors. *Science*, **257**, 1248–1251.

Mounzih, K., Lu, R., and Chehab, F.F. (1997). Leptin treatment rescues the sterility of genetically obese *ob/ob* males. *Endocrinology* **138**, 1190–1193.

Mrosovsky, N. and Sherry, D.F. (1980). Animal anorexia. *Science* **207**, 837–842.

Mueller, W.J., Brubaker, R.L., Gay, C.V., and Boelkins, J.N. (1973a). Mechanisms of bone resorption in laying hens. *Fedn. Proc.* **32**, 1951–1954.

Mueller, W.J., Hall, K.L., Maurer, C.A., and Joshua, I.G. (1973b). Plasma calcium and inorganic phosphate response of laying hens to parathyroid hormone. *Endocrinology* **92**, 853–856.

Munday, K.A., Parsons, B.J., and Poat, J.A. (1971). The effect of angiotensin on

cation transport by rat kidney cortex slices. *J. Physiol., Lond.* **215**, 269–282.

Murat, J.C., Plisetskaya, E.M., and Woo, N.Y.S. (1981). Endocrine control of nutrition in cyclostomes and fish. *Comp. Biochem. Physiol. A* **68**, 149–158.

Murphy, T.J., Nakamura, Y., Takeuchi, K., and Alexander, R.W. (1993). A cloned angiotensin receptor isoform from the turkey adrenal gland is pharmacologically distinct from mammalian angiotensin receptors. *Mol. Pharmacol.* **44**, 1–7.

Mutt, V. (1994). Historical perspectives on cholecystokinin research. *Ann. N.Y. Acad. Sci.* **713**, 1–10.

Nachman, R.J., Holman, G.M., Haddon, W.F., and Ling, N. (1986). Leucosulfakinin, a sulfated insect neuropeptide with homology to gastrin and cholecystokinin. *Science* **234**, 71–73.

Nagahama, Y. (1986). Testis. In *Vertebrate Endocrinology: Fundamentals and Biomedical Implications*, vol. 1, *Morphological Considerations* (edited by P.K.T. Pang and M.P. Schreibman), pp. 399–437. Orlando, FL: Academic Press.

Nagahama, Y., Yoshikuni, M., Yamashita, M., and Tanaka, M. (1994). Regulation of oocyte maturation in fish. In *Fish Physiology*, vol. XIII, *Molecular Endocrinology of Fish* (edited by N.M. Sherwood and C.L. Hew), pp. 393–439. San Diego, CA: Academic Press.

Nagamatsu, S., Chan, S.J., Falkmer, S., and Steiner, D.F. (1991). Evolution of the insulin gene superfamily. Sequence of a preproinsulin-like growth factor cDNA from the Atlantic hagfish. *J. Biol. Chem.* **266**, 2397–2402.

Nagy, G., Mulchahey, J.J., Smyth, D.G., and Neill, J.D. (1988). The glycopeptide moiety of vasopressin–neurophysin precursor is neurohypophysial prolactin releasing factor. *Biochem. Biophys. Res. Commun.* **151**, 524–529.

Nakagawa, Y., Kosugi, H., Miyajima, A., Arai, K.-I., and Yokota, T. (1994). Structure of the gene encoding the α subunit of the human granulocyte–macrophage colony stimulating factor receptor. Implications for the evolution of the cytokine receptor superfamily. *J. Biol. Chem.* **269**, 10 905–10 912.

Nakamura, M., Yamada, K., and Yokote,

M. (1971). Ultrastructural aspects of the pancreatic islets in carp of spontaneous diabetes mellitus. *Experientia* **27**, 75–76.

Nakamura, T., Takio, K., Eto, Y., Shibai, H., Titani, K., and Sugino, H. (1990). Activin-binding protein from rat ovary is follistatin. *Science* **247**, 863–838.

Nakanishi, S., Inoue, A., Kita, T., Nakamura, M., Chang, A.C.Y., Cohen, S.N., and Numa, S. (1979). Nucleotide sequence of cloned cDNA for bovine corticotropin–β-lipotropin precursor. *Nature, Lond.* **278**, 423–427.

Nalbandov, A.V. (1969). Specificity of action of gonadotrophic hormones. *Colloq. Int. C.N.R.S.* **177**, 335–342.

Navarro, I., Gutierrez, J., Caixach, J., Rivera, J., and Planas, J. (1991). Isolation and primary structure of glucagon from the endocrine pancreas of *Thunnus obesus*. *Gen. Comp. Endocr.* **83**, 227–232.

Nebert, D.W. and Gonzalez, F.J. (1987). P450 genes: structure, evolution, and regulation. *Annu. Rev. Biochem.*, **56**, 945–993.

Néchard, M. (1986). Structure and development of brown adipose tissue. In *Brown Adipose Tissue* (edited by P. Trayhurn and D.G. Nicholls), pp. 1–30. London: Edward Arnold.

Neer, E.J. (1995). Heterotrimeric G proteins: organizers of transmembrane signals. *Cell* **80**, 249–257.

Neill, J.D. and Nagy, G.M. (1994). Prolactin secretion and its control. In *The Physiology of Reproduction*, 2nd edn (edited by K. Knobil and J.D. Neill), vol. 1, pp. 1833–1860. New York: Raven Press.

Nelson, R.A. (1980). Protein and fat metabolism in hibernating bears. *Fedn. Proc.* **39**, 2955–2958.

Nemere, I. and Norman A.W. (1991). Transport of calcium. In *Handbook of Physiology*, Sect. 6, *The Gastrointestinal System* vol. IV, *Intestinal Absorption and Secretion*, pp. 337–360. Bethesda: American Physiological Society.

Ng, T.B. and Idler, D.R. (1980). Gonadotropic regulation of androgen production in flounder and salmonids. *Gen. Comp. Endocr.* **42**, 25–38.

Ng, T.B., Idler, D.R., and Burton, M.P. (1980). Effects of teleost gonadotropins and their antibodies on gonadal histology in winter flounder. *Gen. Comp. Endocr.* **42**, 355–364.

Ng, T.B., Idler, D.R., and Eales, J.G. (1982). Pituitary hormones in the thyroidal system in teleost fishes. *Gen. Comp. Endocr.* **48**, 372–389.

Ng, T.B., Idler, D.R., and Eales, J.G. (1991). Pituitary hormones that stimulate the thyroidal system in teleost fishes. *Gen. Comp. Endocr.* **48**, 372–389.

Niall, H. D. (1982). The evolution of peptide hormones. *Annu. Rev. Physiol.* **44**, 615–624.

Niall, H.D., Hogan, M.L., Sauer, R., Rosenblum, I.Y., and Greenwood, F. (1971). Sequences of pituitary and placental lactogenic and growth hormones: evolution from a primordial peptide by gene reduplication. *Proc. Natl. Acad. Sci., USA* **68**, 866–869.

Nicholls, D.G., Cunningham, S.A., and Riall, E. (1986). The bioenergetic mechanisms of brown adipose tissue thermogenesis. In *Brown Adipose Tissue* (edited by T. Trayhurn and D.G. Nicholls), pp. 52–85. London: Edward Arnold.

Nicholls, T.J., Scanes, C.J., and Follett, B.K. (1973). Plasma pituitary luteinizing hormone in Japanese quail during photoperiodically induced gonadal growth and regression. *Gen. Comp. Endocr.* **21**, 84–98.

Nichols, R. Schneuwly, S.A., and Dixon, J.E. (1988). Identification and characterization of a *Drosophila* homologue of the vertebrate neuropeptide cholecystokinin. *J. Biol. Chem.* **263**, 12167–12170.

Nicholson, H.D., Guldenaar, S.E.F., Boer, G.J., and Pickering, B.T. (1991). Testicular oxytocin: effects of intratesticular oxytocin in the rat. *J. Endocr.* **130**, 231–238.

Nicoll, C.S., Bern, H.A., Dunlop, D., and Strohman, R.C. (1965). Prolactin, growth hormone, thyroxine and growth in tadpoles of *Rana catesbeiana*. *Am. Zool.* **5**, 738–739.

Nicoll, C.S., Mayer, G.L., and Russell, S.M. (1986). Structural features of prolactins and growth hormones that can be related to their biological activities. *Endocrine Rev.* **7**, 169–203.

Nilsson, A. (1970). Gastrointestinal

hormones in the holocephalian fish *Chimaera monstrosa* (L). *Comp. Biochem. Physiol.* **32**, 387–390.

Nishida, Y., Hata, M., Nishizuka, Y., Rutter, W.J., and Ebina, Y. (1986). Cloning of a *Drosophila* cDNA encoding a polypeptide similar to the human insulin receptor precursor. *Biochem. Biophys. Res. Commun.* **141**, 474–481.

Nishii, M., Movérus, B., Bukovskaya, O.S., Takahashi, A., and Kawauchi, H. (1995). Isolation and characterization of [Pro²] somatostatin-14 and melanotropins from Russian sturgeon, *Acipenser gueldenstaedti* Brandt. *Gen. Comp. Endocr.* **99**, 6–12.

Nishimura, H. (1978). Physiological evolution of the renin-angiotensin system. *Jap. Heart J.* **19**, 806–822.

Nishimura, H. (1987). Role of the renin–angiotensin system in osmoregulation. In *Vertebrate Endocrinology: Fundamentals and Biomedical Implications*, vol. 2, *Regulation of Water and Electrolytes* (edited by P.K.T. Pang and M.P. Schreibman), pp. 157–187. San Diego, CA: Academic Press.

Nishimura, H. and Imai, M. (1982). Control of renal function in freshwater and marine teleosts. *Fedn. Proc.* **41**, 2355–2360.

Nishimura, H., Ogawa, M., and Sawyer, W.H. (1973). Renin-angiotensin system in primitive bony fishes and a holocephalian. *Am. J. Physiol.* **224**, 950–956.

Nishimura, H. and Sawyer, W.H. (1976). Vasopressor, diuretic, and natriuretic responses to angiotensins by the American eel, *Anguilla rostrata*. *Gen. Comp. Endocr.* **29**, 337–348.

Nishimura, H., Sawyer, W.H., and Nigrelli, R.F. (1976). Renin, cortisol and plasma volume in marine teleost fishes adapted to dilute media. *J. Endocr.* **70**, 47–59.

Nishizuka, Y. (1988). The molecular heterogeneity of protein kinase C and its implications for cellular regulation. *Nature, Lond.* **334**, 661–665.

Nojiri, H., Ishida, I., Miyashita, E., Sato, M., Urano, A., and Deguchi, T. (1987). Cloning and sequence analysis of cDNAs for neurohypophysial hormones vasotocin and mesotocin for the hypothalamus of the toad. *Bufo japonicus. Proc. Natl. Acad.*

Sci. USA **84**, 3043–3046.

Nolly, H.L., and Fasciola, J.C. (1973). The specificity of the renin-angiotensinogen reaction through the phylogenetic scale. *Comp. Biochem. Physiol. A* **44**, 639–645.

Norgren, R.B., and Lehman, M.N. (1991). Neurons that migrate from the olfactory epithelium in the chick express luteinizing hormone-releasing hormone. *Endocrinology* **128**, 1676–1678.

Norman, A.W. (1994). Editorial: the vitamin D endocrine system: identification of another piece of the puzzle. *Endocrinology* **134**, 1601A–1601B.

Norman, A.W. and Henry, H. (1974). 1,25-dihydroxycholecalciferol–a hormonally active form of vitamin D₃. *Rec. Prog. Horm. Res.* **30**, 431–473.

Norman, R.L. and Spies, H.G. (1986). Cyclic ovarian function in a male macaque: additional evidence for a lack of sexual differentiation in the physiological mechanisms that regulate the cyclic release of gonadotropins in primates. *Endocrinology* **118**, 2608–2610.

Noso, T., Lance, V.A., and Kawauchi, H. (1995). Complete amino acid sequence of crocodile growth hormone. *Gen. Comp. Endocr.* **98**, 244–252.

Noso, T., Nicoll, C.S., and Kawauchi, H. (1993a). Lungfish prolactin exhibits close tetrapod relationships. *Biochim. Biophys. Acta* **1164**, 159–165.

Noso, T., Nicoll, C.S., Polenov, A.L., and Kawauchi, H. (1993b). The primary structure of sturgeon prolactin: phylogenetic implication. *Gen.Comp. Endocr.* **91**, 90–95.

Novales, R.R. (1972). Recent studies of the melanin-dispersing effect of MSH on melanophores. *Gen. Comp. Endocr.* **Suppl. 3**, 125–135.

Novales, R.R. (1973). Discussion of "Endocrine regulation of pigmentation" by Frank S. Abbott. *Am. Zool.* **13**, 895–897.

Nozaki, M. and Gorbman, A. (1984). Distribution of immunoreactive sites for several components in the pituitary and brain of adult lampreys, *Petromyzon marinus* and *Entosphenus tridentatus*. *Gen. Comp. Endocr.* **53**, 335–352.

Nozaki, M., Gorbman, A., and Sower, S.A. (1994). Diffusion between the

neurohypophysis and the adenohypophysis of lampreys, *Petromyzon marinus*. *Gen. Comp. Endocr.* **96**, 385–391.

Nozaki, M., Takahashi, A., Amemiya, Y., Kawauchi, H., and Sower, S.A. (1995). Distribution of lamprey adrenocorticotropin and melanotropins in the pituitary of the adult sea lamprey, *Petromyzon marinus*. *Gen. Comp. Endocr.* **98**, 147–156.

Odum, E.P. (1965) Adipose tissue in migratory birds. In *Handbook of Physiology*, Sect, 5 *Adipose Tissue*, pp. 37–43. Washington, DC: American Physiological Society.

Ogawa, M. Yagasaki, M., and Yamazaki, J. (1973). The effect of prolactin on water influx in isolated gills of the goldfish *Carassius auratus* L. *Comp. Biochem. Physiol. A* **44**, 1177–1183.

O'Grady, S.M., Field, M., Nash, N.T., and Rao. M.C. (1985). Atrial natriuretic factor inhibits Na–K–Cl cotransport in teleost intestine. *Am. J. Physiol.* **249**, C531-C534.

Oguro, C. (1973). Parathyroid gland and serum calcium concentration in the giant salamander, *Megalobatrachus davidianus*. *Gen. Comp. Endocr.* **21**, 565–568.

Oguro, C. and Sasayama, Y. (1976). Morphology and function of the parathyroid gland of the caiman, *Caiman crocodilus*. *Gen. Comp. Endocr.* **29**, 161–169.

Oguru, C. and Sasayama, Y. (1978). Function of the parathyroid gland in serum calcium regulation in the newt, *Tylotriton andersoni* Boulenger. *Gen. Comp. Endocr.* **35**, 10–15.

Oguro, C. and Tomisawa, A. (1972). Effects of parathyroidectomy on serum calcium concentration of the turtle *Geoclemys reevesii*. *Gen. Comp. Endocr.* **19**, 587–588.

Oguro, C., Tomisawa, A., and Matuoka, N. (1974). Effect of parathyroidectomy on the serum calcium and phosphorous concentrations in the tortoise, *Testudo graeca*. *Zool. Magazine* **83**, 201–202.

Ohning, G.V., Wong, H.C., Lloyd, K.C.K., and Walsh, J.H. (1996). Gastrin mediates the gastric mucosal proliferative response to feeding. *Am. J. Physiol.* **271**, G470-G476.

Ohsako, S., Ishida, I., Ichikawa, T., and Deguchi, T. (1986). Cloning and sequence analysis of cDNAs encoding precursors of urotensin II-α and-γ. *J. Neurosci.* **6**, 2730–2735.

Okelo, O. (1986). Neuroendocrine control of physiological color change in *Chameleo gracilis*. *Gen. Comp. Endocr.* **64**, 305–311.

Oliva, A.A., Steiner, D.F., Chan, S.J. (1995). Proprotein convertases in amphioxus: predicted structure and expression of proteases SPC2 and SPC3. *Proc. Natl. Acad. Sci. USA* **92**, 3591–3595.

Oliver, C. and Porter, J.C. (1978). Distribution and characterization of α-melanocyte-stimulating hormone in the rat brain. *Endocrinology* **102**, 697–705.

Olivereau, M. (1967). Observations sur l'hypophyse de l'anguille femelle en particulier lors de la maturation sexuelle, *Z. Zellforsch. mikrosk. Anat.* **80**, 286–306.

Olivereau, M. (1978). Serotonin and MSH secretion: effect of parachlorophenylalanine on the pituitary cytology of the eel. *Cell Tiss. Res.* **191**, 83–92.

Olivereau, M. and Olivereau, J. (1978). Prolactin, hypercalcemia and corpuscles of Stannius in seawater eels. *Cell Tiss. Res.* **186**, 81–96.

Olsen, H.S., Cepeda, M.A., Zhang, Q.Q., Rosen, C.A., Vozzolo, B.L., and Wagner, G.F. (1996). Human stanniocalcin – a possible hormonal regulator of mineral metabolism. *Proc. Natl. Acad. Sci. USA* **93**, 1792–1796.

Olson, K.R. and Duff, D.W. (1992). Cardiovascular and renal effects of eel and rat atrial natriuretic peptide in rainbow trout, *Salmo gairdneri*. *J. Comp. Physiol B* **162**, 408–425.

Olsson, R. (1990). Evolution of chordate endocrine organs. In *Progress in Comparative Endocrinology* (edited by A. Epple, C.G. Scanes, and M.H. Stetson), pp. 272–281. New York: Wiley-Liss.

O'Malley, B.W. (1989). Editorial: did eucaryotic steroid receptors evolve from intracrine gene regulators. *Endocrinology* **125**, 1119–1120.

O'Malley, B.W., Tsai, S.Y., Bagchi, M. Weigel, N.L., Schrader, W.T., and Tsai, M.-J. (1991). Molecular mechanism of action of a steroid hormone receptor. *Rec. Prog. Horm. Res.* **47**, 1–68.

Ono, M. and Kawauchi, H. (1994). The somatolactin gene. In *Fish Physiology*, vol. XIII, *Molecular Endocrinology of Fish* (edited by N.M. Sherwood and C.Y. Hew), pp. 159–177. San Diego, CA: Academic Press.

Ono, M., Takayama, Y., Rand-Weaver, M., Sakata, S., Yasunaga, T., Noso, T., and Kawauchi, H. (1990). cDNA cloning of somatolactin, a pituitary protein related growth hormone and prolactin. *Proc. Natl. Acad. Sci. USA* **87**, 4330–4334.

Onstott, D. and Elde, R. (1986). Immunohistochemical localization of urotensin I/corticotropin-releasing factor, urotensin II, and serotonin immunoreactivities in the caudal spinal cord of non teleost fishes. *J. Comp. Neurol.* **249**, 205–225.

Orci, L. and Unger, R.H. (1975). Functional subdivision of the islets of Langerhans and possible role of D cells. *Lancet* **ii**, 1243–1244.

Oring, L.W., Fivizzani, A.J., Colwell, M.A., and El Halawani, M.E. (1988). Hormonal changes associated with natural and manipulated incubation in the sex-role reversed Wilson's phalarope. *Gen. Comp. Endocr.* **72**, 247–256.

Ormseth, O.A., Nicolson, M., Pelleymounter, M.A., and Boyer, B.B. (1996). Leptin inhibits prehibernation hyperphagia and reduces body weight in arctic ground squirrels. *Am. J. Physiol.* **271**, R1775–R1779.

Ornoy, A., Goodwin, D., Noff, D., and Edlestein, S. (1978). 24,25-diydroxyvitamin D is a metabolite of vitamin D essential for bone formation. *Nature, Lond.* **276**, 517–519.

Orozco, A., Silva, J.E., and Valverde-R.C. (1997). Rainbow trout liver expresses two iodothyronine phenolic deiodinase pathways with the characteristics of mammalian Types I and II 5'-deiodinases. *Endocrinology* **138**, 254–258.

O'Shea, D., Morgan, D.G.A., Meeran, K., Edwards, C.M.B., Turton, M.D., Choi, S.J., Heath, M.M., Gunn, I., Taylor, G.M., Howard, J.K., Bloom, C.I., Small, C.J., Haddo, O., Ma, J.J., Callinan, W., Smith, D.M., Ghatei, M.A., and Bloom,

S.R. (1997). Neuropeptide Y induced feeding in the rat is mediated by a novel receptor. *Endocrinology* **138**, 196–202.

O'Toole, L.B., Armour, K.J., Decourt, C., Hazon, N., Lahlou, B., and Henderson, I.W. (1990). Secretory patterns of 1α-hyydroxycorticosterone in the isolated perifused interrenal gland of the dogfish, *Scyliorhinus canicula. J. Mol. Endocr.* **5**, 55–60.

Ottlecz, A., Snyder, G.D., and McCann, A.M. (1988). Regulatory role of galanin in control of hypothalamic–pituitary function. *Proc. Natl. Acad. Sci. USA* **85**, 9861–9865.

Oughterson, S.M., Munoz-Chapuli, R., de Andres, V., Lawson, R., Heath, S., and Davies, D.H. (1995). The effects of calcitonin on serum calcium levels in immature brown trout, *Salmo trutta. Gen. Comp. Endocr.* **97**, 42–48.

Owada, K., Kawata, M., Akaji, K., Takagi, A., Moriga, M., and Kobayashi, H. (1985). Urotensin II-immunoreactive neurons in the caudal neurosecretory system of freshwater and seawater fish. *Cell Tissue Res.* **239**, 349–354.

Owada, K., Yamada, C., and Kobayashi, H. (1985). Immunohistochemical investigation of urotensins in the caudal spinal cord of four species of elasmobranchs and the lamprey, *Lampetra japonica. Cell Tissue Res.* **242**, 527–530.

Owyang, C. (1996). Physiological mechanisms of cholecystokinin action on pancreatic secretion. *Am. J. Physiol.* **271**, G1–G7.

Oyer, P.E., Cho, S., Peterson, J.D., and Steiner, D.F. (1971). Studies on human proinsulin. *J. Biol. Chem.* **246**, 1375–1386.

Ozon, R. (1972). Androgens in fishes, amphibians, reptiles and birds. In *Steroids in Nonmammalian Vertebrates* (edited by D.R. Idler), pp. 329–389. New York: Academic Press.

Packard, G.C., Packard, M.J., and Gorbman, A. (1976). Serum thyroxine concentrations in the Pacific hagfish and lamprey and in the leopard frog. *Gen. Comp. Endocr.* **28**, 365–367.

Palmer, R.M.J., Ferrige, A.G., and Moncada, S. (1987). Nitric oxide release accounts for the biological activity of endothelium-derived relaxing factor.

Nature, Lond. **327**, 524–526.

Pang, P.K.T. (1973). Endocrine control of calcium metabolism in teleosts. *Am. Zool.* **13**, 775–792.

Pang, P.K.T., Furspan, P.B., and Sawyer, W.H. (1983). Evolution of neurohypophyseal hormone actions in vertebrates. *Am. Zool.* **23**, 655–662.

Pang, P.K.T., Pang, R.K., and Sawyer, W.H. (1973). Effects of environmental calcium and replacement therapy on the killifish, *Fundulus heteroclitus*, after the surgical removal of the corpuscles of Stannius. *Endocrinology* **93**, 705–710.

Pang, P.K.T., and Sawyer, W.H. (1974). Effects of prolactin on hypophysectomized mud puppies *Necturus maculosus. Am. J. Physiol.* **226**, 458–462.

Pang, P.K.T. and Sawyer, W.H. (1975). Parathyroid hormone preparations, salmon calcitonin, and urine flow in the South American lungfish, *Lepidosiren paradoxa. J. Exp. Zool.* **193**, 407–412.

Pang, P.K.T. and Sawyer, W.H. (1978). Renal and vascular responses of the bullfrog (*Rana catesbeiana*) to mesotocin. *Am. J. Physiol.* **235**, F151–F155.

Pang, P.K.T. and Schreiban, M.P. (eds.) (1986). *Vertebrate Endocrinology: Fundamentals and Biomedical Implications*, vol. 1, *Morphological Considerations*. Orlando, FL: Academic Press.

Pang, P.K.T. and Schreibman, M.P. (eds.) (1987). *Vertebrate Endocrinology: Fundamentals and Biomedical Implications*, vol. 2, *Regulation of Water and Electrolytes*. San Diego, CA: Academic Press.

Pang, P.K.T. and Schreibman, M.P. (eds.) (1989). *Vertebrate Endocrinology: Fundamentals and Biomedical Implications*, vol. 3, *Regulation of Calcium and Phosphate*. San Diego, CA: Academic Press.

Pang, P.K.T., Schreibman, M.P., Balbontin, F., and Pang, R.K. (1978). Prolactin and pituitary control of calcium regulation in the killifish, *Fundulus heteroclitus. Gen. Comp. Endocr.* **36**, 306–316.

Pang, P.K.T., Uchiyama, M., and Sawyer, W.H. (1982). Endocrine and neural control of amphibian renal function. *Fedn. Proc.* **41**, 2365–2370.

Pant, K. and Chandola-Saklani, A. (1993). A role for thyroid hormones in the development of premigratory disposition in the redheaded bunting, *Emberiza bruniceps. J. Comp. Physiol. B* **163**, 389–394.

Papkoff, H. (1972). Subunit interrelationships among the pituitary glycoprotein hormones. *Gen. Comp. Endocr.* **Suppl. 3**, 609–616.

Parker, K.L. and Schimmer, B.P. (1993). Transcriptional regulation of the adrenal steroidogenic enzymes. *Trend. Endocr. Metab.* **4**, 46–50.

Parker, S. (1988). The expanding family of nuclear hormone receptors. *J. Endocr.* **119**, 175–177.

Parkes, A.S. and Marshall, A.J. (1960) The reproductive hormones in birds. In *Marshall's Physiology of Reproduction* (edited by A.S. Parkes), vol. 1 (pt. 2), pp. 583–706. London: Longmans.

Parkes, D. and Vale, W. (1992). Secretion of melanin-concentrating hormone and neuropeptide-EI from cultured rat hypothalamic cells. *Endocrinology* **131**, 1826–1831.

Parkinson, T.J. and Follett, B.K. (1995). Thyroidectomy abolishes seasonal testicular cycles of Soay rams. *Proc. Roy. Soc. Lond. Ser. B* **259**, 1–6.

Parmentier, M., Ghysens, M., Rypens, F., Lawson, D.E.M., Pasteels, J.L., and Pochet, R. (1987). Calbindin in vertebrate classes: immunohistochemical localization and Western blot analysis. *Gen. Comp. Endocr.* **65**, 399–407.

Patent, C.J. (1970). Comparison of some hormonal effects on carbohydrate metabolism in an elasmobranch (*Squalus acanthias*) and a holecephalan (*Hydrolagus collei*). *Gen. Comp. Endocr.* **14**, 215–242.

Pathak, V.K. and Chandola, A. (1982). Seasonal variation in extrathyroidal conversion of thyroxine to tri-iodothyronine and migratory disposition in redheaded bunting. *Gen. Comp. Endocr.* **47**, 433–439.

Patthy, L. (1990). Homology of a domain of the growth hormone/prolactin receptor family with type III modules of fibronectin. *Cell* **61**, 13–14.

Patzner, R.A. and Ichikawa, T. (1977). Effects of hypophysectomy on the testis of the hagfish, *Eptatretus burgeri* Girard (Cyclostomata). *Zool. Anz.*, Jena **199**, 371–380.

Peach, M.J. (1977). Renin–angiotensin system: biochemistry and mechanism of action. *Physiol. Rev.* **57**, 313–369.

Peaker, M. (1995). Endocrine signals from the mammary gland. *J. Endocr.* **147**, 189–193.

Peaker, M. and Linzell, J.L. (1975). *Salt Glands in Birds and Reptiles*. Cambridge University Press.

Pearse, A.G.E. (1968). Common cytochemical and ultrastructural characteristics of cells producing polypeptide hormones (the APUD series) and their relevance to thyroid and ultimobranchial C cells and calcitonin. *Proc. Roy. Soc. Lond., Ser. B* **170**, 71–80.

Pearse, A.G.E. (1976). Morphology and cytochemistry of thyroid and ultimobranchial C cells. In *Handbook of Physiology*, Sect. 7 Endocrinology, vol. 7, *Parathyroid Gland*, pp. 411–421. Washington, DC: American Physiological Society.

Peart, W.S. (1977). The kidney as an endocrine organ. *Lancet* **ii**, 543–548.

Pelleymounter, M.A., Cullen, M.J., Baker, M.B., Hecht, R., Winters, D., Boone, T., and Collins, F. (1995). Effects of the obese gene product on body weight regulation in *ob/ob* mice. *Science* **269**, 540–543.

Penhos, J.C. and Ramey, E. (1973). Studies on the endocrine pancreas of amphibians and reptiles. *Am. Zool.* **12**, 667–698.

Penny, R.J., Tilders, F.J.H., and Thody, A.J. (1979). The effect of hypothalamic lesions on immuno-reactive α-melanocyte stimulating hormone secretion in the rat. *J. Physiol. Lond.* **292**, 59–67.

Perez, J., Gutierrez, J., Carrillo, M., Zanuy, S., and Fernandez, J. (1989). Effect of bonito insulin on plasma immunoreactive glucagon levels and carbohydrate and lipid metabolism of sea bass (*Dicentrarchus labrax*). *Comp. Biochem. Physiol. A* **94**, 33–36.

Perrott, M.N. and Balment, R.J. (1990). The renin–angiotensin system and the regulation of plasma cortisol in the flounder. *Platichthys flesus. Gen. Comp. Endocr.* **78**, 414,–420.

Perrott, M.N., Carrick, S., and Balment, R.J. (1991). Pituitary and plasma arginine vasotocin levels in teleost fish. *Gen. Comp. Endocr.* **83**, 68–74.

Pertseva, M. (1991). The evolution of hormone signalling systems. *Comp. Biochem. Physiol. A* **100**, 775–787.

Peter, M.C.S. and Oommen, O.V. (1993). Stimulation of oxidative metabolism by thyroid hormones in propranolol/alloxan-treated bony fish, *Anabas testudineus* (Bloch). *J. Exp. Zool.* **266**, 85–91.

Peter, R.E. (1971). Feedback effects of thyroxine on the hypothalamus and pituitary of the goldfish. *Carassius auratus. J. Endocr.* **51**, 31–39.

Peter, R.E. and Crim. L.A. (1979). Reproductive endocrinology of fishes: gonadal cycles and gonadotropin in teleosts. *Annu. Rev. Physiol.* **41**, 323–335.

Peter, R.E., Yu, K.-L., Marchant, T.A., and Rosenblum, P.M. (1990). Direct neural regulation of the teleost adenohypophysis. *J. Exp. Zool.* **4** (Suppl.), 84–89.

Peterson, J.D., Steiner, D.F., Emdin, S.O., and Falkmer, S. (1975). The amino acid sequence of the insulin from a primitive vertebrate, the Atlantic hagfish (*Myxine glutinosa*). *J. Biol. Chem.* **250**, 5183–5191.

Petruzzelli, L., Herrera, R., Arenas-Garcia, R., Fernandez, R., Birbaum, M.J., and Rosen, O.M. (1986). Isolation of a *Drosophila* genomic sequence homologous to the kinase domain of the human insulin receptor and detection of the phosphorylated *Drosophila* receptor with an anti-peptide antibody. *Proc. Natl. Acad. Sci. USA* **83**, 4710–4714.

Philbrick, W.M., Wysolmerski, J.J., Galbraith, S., Holt, E., Orloff, J.J., Yang, K.H., Vasavada, R.C., Weir, E.C., Broadus, A.E., and Stewart, A.F. (1996). Defining the roles of parathyroid hormone-related protein in normal physiology. *Physiol. Rev.* **76**, 127–173.

Pic, P., Mayer-Gostan, N., and Maetz, J. (1973). Sea-water teleosts: Presence of α- and β-adrenergic receptors in the gill regulating salt extrusion and water permeability. In *Comparative Physiology* (edited by L. Bolis, K. Schmidt-Nielsen, and S.H.P. Maddrell), pp. 292–322. Amsterdam: Elsevier North-Holland.

Pickering, A.D. (1972). Effects of hypophysectomy on the activity of the endostyle and thyroid gland in the larval and adult river lamprey, *Lampetra fluviatilis* L. *Gen. Comp. Endocr.* **18**, 335–343.

Pickering, A.D. (1976). Effects of gonadectomy, oestradiol and testosterone on the migrating river lamprey, *Lampetra fluviatilis* L. *Gen. Comp. Endocr.* **28**, 473–480.

Pickford, G.E. and Kosto, B. (1957). Hormonal induction of melanogenesis in hypophysectomized killifish (*Fundulus heterolitus*). *Endocrinology* **61**, 177–196.

Pickford, G.E., Pang, P.K.T., Weinstein, E., Torretti, J., Hendler, E., and Epstein, F.H. (1970). The response of the hypophysectomized Cyprinodont, *Fundulus heteroclitus*, to replacement therapy with cortisol: Effects on blood serum and sodium-potassium activated adenosine triphosphatase in the gills, kidney, and intestinal mucosa. *Gen. Comp. Endocr.*, **14**, 524–534.

Pickford, G.E. and Phillips, J.G. (1959). Prolactin, a factor in promoting survival of hypophysectomized killifish in fresh water. *Science* **130**, 454–455.

Pictet, R. and Rutter, W.J. (1972). Development of the embryonic endocrine pancreas. In *Handbook of Physiology*, Sect. 7 *Endocrinology*, vol. 1, *Endocrine pancreas*, pp. 25–66. Washington, DC: American Physiological Society.

Pieau, C., Girondot, M., Desvages, G., Dorizzi, M., Richard-Mercier, N., and Zaborski, P. (1994). Environmental control of gonadal differentiation. In *The Differences Between the Sexes* (edited by R.V. Short and E. Balaban), pp. 433–448. Cambridge University Press.

Pierson, P.M., Guibbolini, M.E., and Lahlou, B. (1996). A V_1-type receptor for mediating the neurohypophysial hormone-induced ACTH release in trout pituitary. *J. Endocr.* **149**, 109–115.

Pike, J.W., Spanos, E., Colston, K.W., MacIntyre, I., and Haussler, M.R. (1978). Influence of estrogen on renal vitamin D hydroxylase and serum 1α, $25-(OH)_2D_3$ in chicks. *Am. J. Physiol.* **235**, E338–E343.

Pitcher, T., Buffenstein, R., Keegan, J.D., Moodley, G.P., and Yahav, S. (1992). Dietary calcium content, calcium balance and mode of uptake in a subterranean mammal, the Damara mole-rat. *J. Nutrit.* **122**, 108–114.

Pitcher, T., Sergeev, I.N., and Buffenstein, R. (1994). Vitamin D metabolism in the Damara mole-rat is altered by exposure to sunlight yet mineral metabolism is unaffected. *J. Endocr.* **143**, 367–374.

Plisetskaya, E., Kazakov, V.K., Soltitskaya, L., and Leibson, L.G. (1978). Insulin-producing cells in the gut of freshwater bivalve molluscs *Anodonta cygnea* and *Unio picotrum* and the role of insulin in the regulation of their carbohydrate metabolism. *Gen. Comp. Endocr.* **35**, 133–145.

Plisetskaya, E.M., Ottolenghi, C., Sheridan, M.A., Mommsen, T.P., and Gorbman, A. (1989). Metabolic effects of salmon glucagon and glucagon-like peptide in coho and chinook salmon. *Gen. Comp. Endocr.* **73**, 205–216.

Plisetskaya, E.M., Pollock, H.G., Elliott, W.M. Youson, J.H., and Andrews, P.C. (1988). Isolation and structure of lamprey (*Petromyzon marinus*) insulin. *Gen. Comp. Endocr.* **69**, 46–55.

Plisetskaya, E.M., Pollock H.G., Rouse, J.B., Hamilton, J.W., Kimmel, J.R., and Gorbman, A. (1986). Isolation and structures of coho salmon (*Oncorhynchus kisutch*) glucagon and glucagon-like peptide. *Reg. Peptides* **14**, 57–67.

Poffenbarger, P.L., Burns, R., and Bennett-Novak, A. (1976). A phylogenetic study of serum nonsuppressible insulin-like activity (NSILA), *Comp. Biochem. Physiol A* **52**, 223–226.

Pohorecky, L.A. and Wurtman, R.J. (1971). Adrenocortical control of epinephrine synthesis. *Pharmacol. Rev.* **23**, 1–35.

Pollock, H.G., Hamilton, J.W., Rouse, J.B., Ebner, K.E., and Rawitch, A.B. (1988a). Isolation of peptide hormones from the pancreas of the bullfrog. (*Rana catesbeiana*). *J. Biol. Chem.* **263**, 9746–9751.

Pollock, H.G., and Kimmel, J.R. (1975). Chicken glucagon. Isolation and amino acid sequence studies. *J. Biol. Chem* **250**, 9377–9380.

Pollock, H.G., Kimmel, J.R., Ebner, K.E., Hamilton, J.W., Rouse, J.B., Lance, V., and Rawitch, A.B. (1988b). Isolation of alligator gar (*Lepidosteus spatula*)

glucagon, oxyntomodulin, and glucagon-like peptide: amino acid sequences of oxyntomodulin and glucagon-like peptide. *Gen. Comp. Endocr.*, **69**, 133–140.

Porter, T., Hargis, B.M., Silsby, J.L., and El Halawani, M. (1989). Differential steroid production between theca interna and theca externa cells: a three-cell model for follicular steroidogenesis in avian species. *Endocrinology* **125**, 109–116.

Potter, E., Behan, D.P., Fischer, W.H., Linton, E.A., Lowry, P.J. and Vale, W.W. (1991). Cloning and characterization of the cDNAs for human and rat corticotropin releasing factor-binding proteins. *Nature, Lond.* **349**, 423–426.

Potts, J.T., Keutmann, H.T., Niall, H.D., Habener, J.F., and Tregear, G.W. (1972). Comparative biochemistry of parathyroid hormone. *Gen. Comp. Endocr.* **Suppl. 3**, 405–410.

Powell, J.F.F., Krueckl, S.L., Collins, P.M., and Sherwood, N.M. (1996). Molecular forms of GnRH in three model fishes: rockfish, medaka and zebrafish. *J. Endocr.* **150**, 17–23.

Powell, J.F.F., Zohar, Y., Elizur, A., Park, M., Fischer, W.H., Craig, A.G., Rivier, J.E., Lovejoy, D.A., and Sherwood, N.M. (1994). Three forms of gonadotropin-releasing hormone characterized from brains of one species. *Proc. Natl. Acad. Sci. USA* **91**, 12081–12085.

Prager, D. and Melmed, S. (1993). Editorial: insulin and insulin-like growth factor I receptors: are there functional distinctions. *Endocrinology* **132**, 1419–1420.

Pratt, D.S., Beinborn, M., Langhans, N., McBride, E.W., Kolakowski, L.F., and Kopin, A.S. (1994). The *Xenopus* brain cholecystokinin receptor: an evolutionary precursor of the human cholecystokinin receptor subtypes. (abstract). *Dig. Dis. Sci.* **39**, 1765.

Price, C.A. (1991). The control of FSH secretion in larger domestic species. *J. Endocr.* **131**, 177–184.

Prigge, W.F. and Grande, F. (1971). Effects of glucagon, epinephrine and insulin on *in vitro* lipolysis of adipose tissue from mammals and birds. *Comp. Biochem.*

Physiol. B **39**, 69–82.

Proux, J. and Rougon-Rapuzzi, G. (1980). Evidence for a vasopressin-like molecule in migratory locust. Radioimmunological measurements in different tissues; correlation with various states of hydration. *Gen. Comp. Endocr.* **42**, 378–383.

Prunet, P., Boeuf, G., Bolton, J.P., and Young, G. (1989). Smoltification and seawater adaptation in Atlantic salmon (*Salmo salar*): plasma prolactin, growth hormone, and thyroid hormones. *Gen. Comp. Endocr.* **74**, 355–364.

Prunet, P., Pisam, M., Claireaux, J.P., Boeuf, G., and Rambourg, A. (1994). Effects of growth hormone on gill chloride cells in juvenile Atlantic salmon (*Salmo salar*). *Am. J. Physiol.* **266**, R850–R857.

Putney, J.W. (1987). Calcium-mobilizing receptors. *Trend. Pharmacol. Sci* **8**, 481–486.

Qu, D.Q., Kudwig, D.S., Gammeltoft, S., Piper, M., Pelleymounter, M.A., Cullen, M.J., Mathes, W.F., Przypek, J., Kannarek, R., and Maratosfliier, E. (1996). A role for melanin-concentrating hormone in the central regulation of feeding behaviour. *Nature, Lond.* **380**, 243–247.

Quay, W.B. (1972). Integument and the environment: glandular composition, function and evolution. *Am. Zool.* **12**, 95–108.

Quevedo, W.C. (1972). Epidermal melanin units: melanocyte-keratinocyte interactions. *Am. Zool.* **12**, 35–41.

Rall, J.E., Robbins, J., and Lewallen, C.G. (1964). The thyroid. In *The Hormones* (edited by G. Pincus, K.V. Thimann, and E.B. Astwood), vol. V, pp. 159–439. New York: Academic Press.

Ramenofsky, M. (1990). Fat storage and fat metabolism in relation to migration. In *Bird Migration* (edited by E. Gwinner), pp. 214–231. Berlin: Springer-Verlag.

Rance, T. and Baker, B.I. (1979). The teleost melanin-concentrating hormone – a pituitary hormone of hypothalamic origin. *Gen. Comp. Endocr.* **37**, 64–73.

Randall, C.F., Bromage, N.R., Thorpe, J.E., Miles, M.S., and Muir, J.S. (1995). Melatonin rhythms in Atlantic salmon (*Salmo salar*) maintained under natural and out-of-phase photoperiods. *Gen. Comp. Endocr.* **98**, 73–86.

Rand-Weaver, M., Pottinger, T.G., and Sumpter, J.P. (1995). Pronounced seasonal rhythms in plasma somatolactin levels in rainbow trout. *J. Endocr.* **146**, 113–119.

Rand-Weaver, M., Swanson, P., Kawauchi, H., and Dickoff, W.W. (1992). Somatolactin, a novel pituitary protein: purification and plasma levels during reproductive maturation of coho salmon. *J. Endocr.* **133**, 393–403.

Rankin, J.C. and Bolis, L. (1984). Hormonal control of water movement across the gills. In *Fish Physiology* (edited by W.S. Hoar and D.J. Randall), vol. X, *Gills*, Pt B, *Ion and Water Transfer*, pp. 177–201. Orlando, FL: Academic Press.

Rankin, J.C. and Maetz, J. (1971). A perfused teleostean gill preparation: vascular actions of neurohypophysial hormones and catecholamines. *J. Endocr.* **51**, 621–635.

Rankin, M.W. (1991). Endocrine effects on migration. *Am. Zool.* **31**, 217–230.

Rao, D.S. and Raghuramulu, N. (1996). Food chain as origin of vitamin D in fish. *Comp. Biochem. Physiol. A* **114**, 15–19.

Rasmussen, H. and Goodman, D.B.P. (1977). Relationships between calcium and cyclic nucleotides in cell activation. *Physiol. Rev.* **57**, 421–509.

Rasquin, P. and Rosenbloom, L. (1954). Endocrine imbalance and tissue hyperplasia in teleosts maintained in darkness. *Bull. Am. Mus. Nat. Hist.* **104**, 359–420.

Rawding, R.S. and Hutchison, V.H. (1992). Influence of temperature and photoperiod on plasma melatonin in the mudpuppy, *Necturus maculosus. Gen. Comp. Endocr.* **88**, 364–374.

Rawdon, B.B. and Andrew, A. (1990). Vertebrate gut endocrine cells: comparative and developmental aspects. In *Progress in Comparative Endocrinology* (edited by A. Epple, C.G. Scanes, and M.H. Stetson), pp. 504–509. New York: Wiley-Liss.

Rawlings, S.R. and Hezareh, M. (1996). Pituitary adenylate cyclase-activating polypeptide (PACAP) and PACAP/vasoactive intestinal polypeptide receptors: actions on the anterior pituitary gland. *Endocrine Rev.* **17**, 4–29.

Redei, E., Hilderbrand, H., and Aird, F. (1995). Corticotropin-release-inhibiting factor is preprothyrotropin-releasing hormone-(178–199). *Endocrinology* **136**, 3557–3563.

Redshaw, M.R. (1972). The hormonal control of the amphibian ovary. *Am. Zool.* **12**, 289–306.

Regard, E., Taurog, A., and Nakashima, T. (1978). Plasma thyroxine and triiodothyronine levels in spontaneously metamorphosing *Rana catesbeiana* tadpoles and in adult anuran amphibia. *Endocrinology* **102**, 674–684.

Reinbloth, R. (1970). Intersexuality in fishes. *Mem. Soc. Endocr.* **18**, 515–541.

Reinbloth, R. (1972). Hormonal control of the teleost ovary. *Am. Zool.* **12**, 307–324.

Reinecke, M. (1989). Atrial natriuretic peptides – localization, structure, function, and phylogeny. In *The Comparative Physiology of Regulatory Peptides* (edited by S. Holmgren), pp. 3–33. London: Chapman & Hall.

Reinhart, G.A. and Zehr, J.E. (1994). Atrial natriuretic factor in the freshwater turtle *Pseudemys scripta:* a partial characterization. *Gen. Comp. Endocr.* **96**, 259–269.

Reiter, R.J. (1980). The pineal and its hormones in the control of reproduction in mammals. *Endocrine Rev.* **1**, 109–131.

Reiter, R.J. (1991a). Pineal melatonin: cell biology of its synthesis and of its physiological interactions. *Endocrine Rev.* **12** 151–180.

Reiter, R.J. (1991b). The pineal gland: reproductive interactions. In *Vertebrate Endocrinology: Fundamentals and Biomedical Implications*, vol. 4, Pt B, *Reproduction* (edited by P.K.T. Pang and M.P. Schreibman), pp. 269–310. San Diego, CA: Academic Press.

Reiter, R.J. and Sorrentino, S. (1970). Reproductive effects of the mammalian pineal. *Am. Zool.* **10**, 247–258.

Renaud, F.L., Chiesa, R., De Jesús, J.M., López, A., Miranda, J., and Tomassini, N. (1991). *Comp. Biochem. Physiol. A* **100**, 41–45.

Renfree, M.B. (1979). Initiation of development of diapausing embryo by mammary denervation during lactation in a marsupial. *Nature, Lond.* **278**,

549–551.

Renfree, M.B. (1980). Placental function and embryonic development in marsupials. In *Comparative Physiology: Primitive Mammals* (edited by K. Schmidt-Nielsen, L. Bolis, and C.R. Taylor), pp. 269–284, Cambridge University Press.

Renfree, M.B. (1993). Diapause, pregnancy, and parturition in Australian marsupials. *J. Exp. Zool.* **266**, 450–462.

Renfree, M.B. (1994). Endocrinology of pregnancy, parturition and lactation in marsupials. In *Marshall's Physiology of Reproduction*, 4th edn (edited by G.E. Lamming), vol. 3, *Pregnancy and Lactation*, Pt 2, *Fetal Physiology, Parturition and Lactation*, pp. 677–766. London: Chapman & Hall.

Renoir, J.-M, Mercier-Bodard, C., and Baulieu, E.M. (1980). Hormonal and immunological aspects of the phylogeny of sex steroid binding plasma protein. *Proc. Natl. Acad. Sci. USA* **77**, 4578–4582.

Reppert, S.M. and Weaver, D.R. (1995). Melatonin madness. *Cell* **83**, 1059–1062.

Reppert, S.M. and Weaver, D.R. (1997). Forward genetic approach strikes gold: cloning of a mammalian *clock* gene. *Cell* **89**, 487–490.

Reppert, S.M., Weaver, D.R., and Ebisawa, T. (1994). Cloning and characterization of a mammalian melatonin receptor that mediates reproductive and circadian responses. *Neuron* **13**, 1177–1185.

Rice, G.E. (1982). Plasma arginine vasotocin concentrations in the lizard *Varanus gouldii* (Gray) following water loading, salt loading, and dehydration. *Gen. Comp. Endocr.* **47**, 1–6.

Rice, G.E., Arnason, S.S., Arad, Z., and Skadhauge, E. (1985). Plasma concentrations of arginine vasotocin, prolactin, aldosterone and corticosterone in relation to oviposition and dietary NaCl in the domestic fowl. *Comp. Biochem. Physiol. A* **81**, 769–777.

Rice, G.E., Bradshaw, S.D., and Prendergast F.J. (1982). The effects of bilateral adrenalectomy on renal function in the lizard *Varanus gouldii* (Gray). *Gen. Comp. Endocr.* **47**, 182–189.

Richardson, S.J., Bradley, A.J., Duan, W., Wettenhall, R.E.H., Harms, P.J., Babon, J.J., Southwell, B.R., Nicol, S., Donnellan, S.C., and Schreiber, G. (1994). Evolution of marsupial and other vertebrate thyroxine-binding plasma proteins. *Am. J. Physiol.* **266**, R1359–R1370.

Ricquier, D. and Bouillard, F. (1986). The brown adipose tissue mitochondrial uncoupling protein. In *Brown Adipose Tissue* (edited by P. Trayhurn and D.G. Nicholls), pp. 86–104. London: Edward Arnold.

Rinderknecht, E. and Humbel, R.E. (1976a). Polypeptides with nonsuppressible insulin-like and cell-growth promoting activities in human serum: isolation, chemical characterization, and some biological properties of forms I and II. *Proc. Natl. Acad. Sci., USA* **73**, 2365–2369.

Rinderknecht, E. and Humbel, R.E. (1976b). Amino-terminal sequences of two polypeptides from human serum with nonsuppressible insulin-like and cell-growth promoting activities; evidence for structural homology with insulin B chain. *Proc. Natl. Acad. Sci., USA* **73**, 4379–4381.

Rindernecht, E. and Humbel, R.E. (1978). The amino acid sequence of human insulin-like growth factor I and its structural homology with proinsulin. *J. Biol. Chem.* **253**, 2769–2776.

Ritter, R.C., Brenner, L.A., and Tamura, C.S. (1994). Endogenous CCK and the peripheral neural substrates in intestinal satiety. *Ann. N.Y. Acad. Sci.* **713**, 255–267.

Rivier, J., Spiess, J., Thorner, M., and Vale, W. (1982). Characterization of a growth hormone-releasing factor from a human pancreatic islet tumour. *Nature, Lond.* **300**, 276–278.

Rizzo, A.J. and Goltzman, D. (1981). Calcitonin receptors in the central nervous system of the rat. *Endocrinology* **108**, 1672–1677.

Robbins, L.S., Nadeau, J.H., Johnson, K.R., Kelly, M.A., Roselli-Rehfuss, L., Baack, E., Mountjoy, K.G., and Cone, R.D. (1993). Pigmentation phenotype of variant extension locus alleles result from point mutations that alter MSH receptor function. *Cell,* **72**, 827–834.

Roberts, J.R. (1991). Renal function and plasma arginine vasotocin during water deprivation in an Australian parrot, the galah (*Cacatua roseicapilla*). *J. Comp. Physiol. B* **161**, 620–625.

Robertshaw, D., Taylor, C.R., and Mazzia, L.M. (1973). Sweating in primates: role of secretion of the adrenal medulla during exercise. *Am. J. Physiol.* **224**, 678–681.

Robertson, D.R. (1968). The ultimobranchial gland in *Rana pipiens*. IV. Hypercalcemia and glandular hypertrophy. *Z. Zellforsch. mikroskop. Anat.* **85**, 441–542.

Robertson, D.R. (1969a). The ultimobranchial body of *Rana pipiens*. VIII. Effects of extirpation upon calcium distribution and bone cell types. *Gen. Comp. Endocr.* **12**, 479–490.

Robertson, D.R. (1969b). The ultimobranchial body in *Rana pipiens*. IX. Effects of extirpation and transplantation on urinary calcium excretion. *Endocrinology.* **84**, 1174–1178.

Robertson, D.R. (1971). Cytological and physiological activity of ultimobranchial gland in the premetamorphic anuran *Rana catesbeiana. Gen. Comp. Physiol.* **16**, 329–341.

Robertson, D.R. (1975). Effects of ultimobranchial and parathyroid glands and vitamins D_2, D_3 and dihydrotachysterol$_2$ on blood calcium and intestinal calcium transport in the frog. *Endocrinology* **96**, 934–940.

Robertson, D.R. (1987). Plasma immunoreactive calcitonin in the frog (*Rana pipiens*). *Comp. Biochem. Physiol. A* **88**, 701–705.

Robertson, O.H., Krupp, M.A., Thomas, S.F., Favour, C.B., Hane, S., and Wexler, B.C. (1961). Hyperadrenocorticoidism in spawning migratory and non-migratory rainbow trout (*Salmo gairdnerii*): comparison with pacific salmon (Genus *Oncorhynchus*). *Gen. Comp. Endocr.* **1**, 473–484.

Robertson, O.H. and Wexler, B.C. (1959). Histological changes in the organs and tissues of migrating and spawning Pacific salmon (Genus *Oncorhynchus*). *Endocrinology* **66**, 222–239.

Robin, J.-P, Frain, M., Sardet, C., Groscolas, R., and Le Maho, Y. (1988). Protein and lipid utilization during long-term fasting in emperor penguins.

Am. J. Physiol. **254**, R61–R68.

Robinson, P.A., Hawkey, C., and Hammond, G.L. (1985). A phylogenetic study of the structural and functional characteristics of corticosteroid binding globulin in primates. *J. Endocr.* **104**, 251–257.

Rodbell, M., Lin, M.C., Salomon, Y., Londos, C., Harwood, J.P., Martin, B.R., Rendell, M., and Berman, M. (1975). Role of adenine and guanine nucleotides in the activity and responses of adenylate cyclase systems to hormones: evidence for multisite transition sites. In *Advances in Nucleotide Research* (edited by G.I. Drummond, P. Greengard, and G.A. Robison), vol. 5, pp. 3–29. New York: Raven Press.

Rodríguez-Sinovas, A., Fernández, E., Manteca, X., Fernandez, A.G., and Goñalons, E. (1997). CCK is involved in both peripheral and central mechanisms controlling food intake in chicken. *Am. J. Physiol.* **272**, R334–R340.

Roe, M.T., Anderson, P.C., Dunn, A.D., and Dunn, J.T. (1989). The hormonogenic sites of turtle thyroglobulin and their homology with those of mammals. *Endocrinology* **124**, 1327–1332.

Rohner-Jeanrenaud, F. and Jeanrenaud, B. (1996). Obesity, leptin, and the brain. *New Eng. J. Med.* **334**, 324–325.

Roos, B.A., Yoon, M., Cutshaw, S.V., and Kalu, D.N. (1980). Calcium regulatory action of endogenous calcitonin demonstrated by passive immunization with calcitonin antibodies. *Endocrinology* **107**, 1320–1326.

Rosenberg, J., Pines, M., and Hurwitz, S. (1988). Regulation of aldosterone secretion by avian adrenocortical cells. *J. Endocr.* **118**, 447–453.

Rosenfeld, M.G., Mermod, J.-J., Amara, S.G., Swanson, L.W., Sawchenko, P.E., Rivier, J., Vale, W.W., and Evans, R.M. (1983). Production of a novel neuropeptide encoded by the calcitonin gene via tissue-specific RNA processing. *Nature, Lond.* **304**, 129–135.

Rosenzweig, A., and Seidman, C.E. (1991). Atrial natriuretic factor and related peptide hormones. *Annu. Rev. Biochem.* **60**, 229–255.

Rossi, M., Choi, S.J., O'Shea, D., Miyoshi, T., Ghatei, M.A. and Bloom, S.R. (1997). Melanin-concentrating hormone acutely stimulates feeding but chronic administration has no effect on body weight. *Endocrinology* **138**, 351–355.

Roth, J. (1990). Insulin receptor structure. In *Insulin* (edited by P. Cuatrecasas and S. Jacobs), pp. 169–181. Berlin: Springer-Verlag.

Roth, J.J., Jones, R.E., and Gerrard, A.M. (1973). Corpora lutea and oviposition in the lizards *Sceloporus undulatus. Gen. Comp. Endocr.* **21**, 569–572.

Roth, S.I. and Schiller, A.L. (1976). Comparative anatomy of the parathyroid glands. In *Handbook of Physiology*, Sect. 7 *Endocrinology*, vol. 7, *Parathyroid Gland*, pp. 281–311. Washington, DC: American Physiological Society.

Rouille, Y., Chauvet, M.T., Chauvet, J., Acher, R., and Hadley, M.E. (1988). The distribution of lysine vasopressin (lysipressin) in placental mammals: a reinvestigation of the Hippopotamidae (*Hippopotamus amphibius*) and Tayassuidae (*Tayassu angulatus*) families. *Gen. Comp. Endocr.* **71**, 475–483.

Roupas, P. and Herington, A.C. (1994). Postreceptor signaling mechanisms for growth hormone. *Trend. Endocr. Metab.* **5**, 154–158.

Rowan, W. (1925). Relation of light to bird migration and developmental changes. *Nature, Lond.* **115**, 494–495.

Rowlands, I.W. and Weir, B.J. (1984). Mammals: Non-primate eutherians. In *Marhsall's Physiology of Reproduction*, 4th edn (edited by G.E. Lamming), vol. 1, *Reproductive Cycles of Vertebrates* pp. 455–659. Edinburgh: Churchill Livingstone.

Rubin D., and Dores, R.M. (1994). Cloning of a growth hormone from a primitive bony fish and its phylogenetic relationships. *Gen. Comp. Endocr.* **95**, 71–83.

Rubin, D.A., Youson, J.H., Marra, L.E., and Dores, R.M. (1996). Cloning of a gar (*Lepidosteus osseus*) GH cDNA: trends in actinopterygian GH structure. *J. Mol. Endocr.* **16**. 73–80.

Rudinger, J. (1968). Synthetic analogues of oxytocin: an approach to problems of hormone action. *Proc. Roy, Soc., Lond. Ser. B* **170**, 17–26.

Rushakoff, R.J., Liddle, R.A., Williams, J.A., and Goldfine, I.D. (1990). The role of cholecystokinin and other gut peptides on regulation of postprandial glucose and insulin levels. In *Insulin* (edited by P. Cuatrecasas and S. Jacobs), pp. 124–142. Berlin: Springer-Verlag.

Russell, S.M. and Nicoll, C.S. (1990). Evolution of growth hormone and prolactin receptors and effectors. In *Progress in Comparative Endocrinology* (edited by A. Epple, C.G. Scanes, and M.H. Stetson), pp. 168–173. New York: Wiley-Liss.

Rust, C.C. and Meyer, R.K. (1968). Effects of pituitary autografts on hair color in the short-tailed weasel. *Gen. Comp. Endocr.* **11**, 548–551.

Rust, C.C. and Meyer, R.K. (1969). Hair color, molt, and testis size in male, short-tailed weasels treated with melatonin. *Science* **165**, 921–922.

Ryan, G.B., Coghlan, J.P., and Scoggins, B.A. (1979). The granulated peripolar epithelial cell: a potential secretory component of the renal juxtaglomerular complex. *Nature, Lond.* **277**, 655–656.

Sage, M. (1973). The evolution of thyroidal function in fishes. *Am. Zool.* **13**, 899–905.

Sahlin, K. (1988). Gastrin/CCK-like immunoreactivity in Hatschek's groove of *Branchiostoma lanceolatum* (Cephalochordata). *Gen. Comp. Endocr.* **70**, 436–441.

Sahu, A. and Kalra, S.P. (1993). Neuropeptidergic regulation of feeding behavior. Neuropeptide Y. *Trend. Endocr. Metab.* **4**, 217–224.

Sairam, M.R., Papkoff, H., and Li, C.H., (1972). Human pituitary interstitial stimulating hormone: primary structure of the α subunit. *Biochem. Biophys. Res. Commun.* **48**, 530–537.

Saito, N., Kinzler, S., and Koike, T.I. (1990). Arginine vasotocin and mesotocin levels in theca and granulosa layers of the ovary during oviposition cycle in hens (*Gallus domesticus*). *Gen. Comp. Endocr.* **79**, 54–63.

Sakamoto, T. and Hirano, T. (1993). Expression of insulin-like growth factor I gene in osmoregulatory organs during seawater adaptation in the salmonid fish: possible mode of osmoregulatory action of growth hormone. *Proc. Natl. Acad. Sci. USA* **90**, 1912–1916.

Sakamoto, T., Ogasawara, T., and Hirano, T. (1990). Growth hormone kinetics during adaptation to hyperosmotic environment in rainbow trout. *J. Comp. Physiol. B* **160**, 1–6.

Sakamoto, T., Shepherd, B.S., Madsen, S.S., Nishioka, R.S., Siharath, K., Rickman, N.H., Bern, H.A., and Grau, E.G. (1997). Osmoregulatory actions of growth hormone and prolactin in an advanced teleost. *Gen. Comp. Endocr.* **106**, 95–101.

Sakata, J.-I., Kangawa, K., and Matsuo, H. (1988). Identification of new atrial natriuretic peptides in frog heart. *Biochem. Biophys. Res. Commun.* **155**, 1338–1345.

Saladin, R., de Vos, P., Guerre-Millo, M., Leturque, A., Girard, J., Staels, B., and Auwerx, J. (1995). Transient increase in *obese* gene expression after food intake or insulin administration. *Nature, Lond.* **377**, 527–529.

Salmon, W.D. and Daughaday, W.H. (1957). A. hormonally controlled serum factor which stimulates sulfate incorporation by cartilage *in vitro*. *J. Lab. Clin. Med.* **49**, 825–836.

Saltiel, A.R. (1996). Diverse signalling pathways in the cellular actions of insulin. *Am. J. Physiol.* **270** E375–E385.

Sammak, P.J., Adams, S.R., Harootunian, A.T., Schliwa, M., and Tsien, R.Y. (1992). Intracellular cyclic AMP, not calcium, determines the direction of vesicle movement in melanophores: direct measurement by fluorescence ratio imaging. *J. Cell. Biol.* **117**, 57–72.

Samson, W.K. (1992). Natriuretic peptides. A family of hormones. *Trend. Endocr. Metab.* **3**, 86–90.

Samson, W.K. and Murphy, T.C. (1997). Adrenomedullin inhibits salt appetite. *Endocrinology* **138**, 613–616.

Samon, W.K., Murphy, T.C., Robison, D., Vargas, T., Tau, E., and Chang, J.-K. (1996). A 35 amino acid fragment of leptin inhibits feeding in the rat. *Endocrinology* **137**, 5182–5185.

Sandberg, K., Ji, H., Millan, M.A., and Catt, K.J. (1991). Amphibian myocardial angiotensin II receptors are distinct from mammalian AT_1 and AT_2 receptor

subtypes. *FEBS Lett.* **284**, 281–284.

Sandor, T. (1969). A comparative survey of steroids and steroidogenic pathways throughout the vertebrates. *Gen. Comp. Endocr.* **Suppl. 2**, 284–298.

Sandor, T., DiBattista, J.A., and Mehdi, A.Z. (1984). Glucocorticoid receptors in the gill tissue of fish. *Gen. Comp. Endocr.* **53**, 353–364.

Santini, F., Chopra, I.J., Hurd, R.E., and Hua Checo, G.N. (1992). A study of the characteristics of hepatic iodothyronine 5'-monodeiodinase in various vertebrate species. *Endocrinology* **131**, 830–834.

Sasayama, Y. and Clark, N.B. (1984). Renal handling of phosphate, calcium sodium, and potassium in intact and parathyroidectomized *Rana pipiens*. *J. Exp. Zool.* **229**, 197–203.

Sasayama, Y. and Oguro, C. (1976). Effects of ultimobranchialectomy on calcium and sodium concentrations of serum and coelomic fluid in bullfrog tadpoles under high calcium and high sodium environment. *Comp. Biochem. Physiol. A* **55**, 35–37.

Sasayama, Y., Suzuki, N., Oguro, C., Takei, Y., Takahashi, A., Watanabe, T.X., Nakajima, K., and Sakakibara, S. (1992). Calcitonin of the stingray: comparison of the hypocalcemic activity with other calcitonins. *Gen. Comp. Endocr.* **86**, 269–274.

Sassin, J.F., Frantz, A.G., Weizman, E.D., and Kapen, S. (1972). Human prolactin: 24-hour pattern with increased release during sleep. *Science* **177**, 1205–1207.

Sato, T. and Wake, K. (1992). 1. Pineal Organ. In *Atlas of Endocrine Organs. Vertebrates and Invertebrates* (edited by A. Matsumoto and S. Ishii), pp. 11–24. English Edition. Berlin: Springer-Verlag.

Sawyer, W.H. (1970). Vasopressor, diuretic, and natriuretic responses by lungfish to arginine vasotocin. *Am. J. Physiol.* **218**, 1789–1794.

Sawyer, W.H. (1972a). Lungfishes and amphibians: endocrine adaptation and the transition from aquatic to terrestrial life. *Fedn. Proc.* **31**, 1609–1614.

Sawyer, W.H. (1972b). Neurohypophysial hormones and water and sodium excretion in African lungfish. *Gen. Comp. Endocr.* **Suppl. 3**, 345–349.

Sawyer, W.H. (1987). Neurohypophysial

hormones and osmoregulatory adaptations. In *Comparative Physiology of Environmental Adaptations* (edited by R. Kirsch and B. Lahlou). pp. 77–82. Basel: Karger.

Sawyer, W.H., Blair-West, J.R., Simpson, P.A., and Sawyer, M.K. (1976). Renal responses of Australian lungfish to vasotocin, angiotensin II, and NaCl infusion. *Am. J. Physiol.* **231**, 593–602.

Sawyer, W.H., Uchiyama, M., and Pang, P.K.T. (1982). Control of renal functions in lungfishes. *Fedn. Proc.* **41**, 2361–2364.

Saxena, B.B. and Rathnam, P. (1976). Amino acid sequence of the β subunit of follicle-stimulating hormone from human pituitary glands. *J. Biol. Chem.* **251**, 993–1005.

Scanes, C.G. and Campbell, R.M. (1993). Evolution of growth-related hormones and factors: the insulin and insulin-like growth factor family. In *The Endocrinology of Growth, Development, and Metabolism in Vertebrates* (edited by M.P. Schreibman, C.G. Scanes, and P.K.T. Pang), pp. 559–588. San Diego, CA: Academic Press.

Scanes, C.G., Dobson, S., Follett, B.K., and Dodd, J.M. (1972). Gonadotrophic activity in the pituitary gland of the dogfish (*Scyliorhinus canicula*). *J. Endocr.* **54**, 343–344.

Scanes, C.G., Follett, B.K., and Goos, H.J.Th. (1972). Cross-reaction in a chicken LH radioimmunoassay with plasma and pituitary extracts from various species. *Gen. Comp. Endocr.* **19**, 596–600.

Schally, A.V., Arimura, A., and Kastin, A.J. (1973). Hypothalamic regulatory hormones. *Science* **179**, 341–350.

Scheide, J.I. and Zadunaisky, J.A. (1988). Effect of atriopeptin II on isolated opercular epithelium of *Fundulus heteroclitus*. *Am. J. Physiol.* **254**, R27–R32.

Schell, D.A., Vari, R.C., and Samson, W.K. (1996). Adrenomedullin– a newly discovered hormone controlling fluid and electrolyte homeostasis. *Trend. Endocr. Metab.* **7**, 7–13.

Schlinger, B.A. and Arnold, A.P. (1993). Estrogen synthesis *in vivo* in the adult zebra finch: additional evidence that circulating estrogens can originate in the brain. *Endocrinology* **133**, 2610–2616.

Schmidt-Nielsen, K., Borut, A., Lee, P., and Crawford, E.C. (1963). Nasal salt secretion and possible function of the cloaca in water conservation. *Science* **142**, 1300–1301.

Schnoes, H.K. and DeLuca, H.F. (1980). Recent progress in vitamin D metabolism and the chemistry of vitamin D metabolites. *Fedn. Proc.* **39**, 2723–2729.

Schreibman, M.P. and Kallman, K.D. (1969). The effect of hypophysectomy on freshwater survival in teleosts of the order Antheriniformes. *Gen. Comp. Endocr.* **13**, 27–38.

Schultz, R.W., Bosma, P.T., Zandbergen, M.A., van der Sanden, M.C.A., van Dijk, W., Peute, J., Bogerd, J., and Goos, H.J.Th. (1993). Two gonadotropin-releasing hormones in the African catfish, *Clarias gariepinus*: localization, pituitary receptor binding, and gonadotropin release activity. *Endocrinology* **133**, 1569–1577.

Schussler, G.C. and Orlando, J. (1978). Fasting decreases triiodothyronine receptor capacity. *Science* **199**, 686–688.

Schütz, H. and Gerstberger R. (1990). Atrial natriuretic factor controls salt gland secretion in the Pekin duck (*Anas platyrhynchos*) through interaction with high affinity receptors. *Endocrinology* **127**, 1718–1726.

Schwabe, C. (1994). Theoretical limitations of molecular phylogenetics and the evolution of relaxins. *Comp. Biochem, Physiol. B* **107**, 167–177.

Schwabe, C. and McDonald, J.K. (1977). Relaxin: a disulfide homolog of insulin. *Science* **197**, 914–915.

Schwabl, H. (1993). Yolk is a source of maternal testosterone for developing birds. *Proc. Natl. Acad. Sci. USA* **90**, 11446–11450.

Schwabl, H. (1996). Maternal testosterone in the avian egg enhances postnatal growth. *Comp. Biochem. Physiol. A* **114**, 271–276.

Schwabl, H., Bairlein, F., and Gwinner, E. (1991). Basal and stress-induced corticosterone levels of garden warblers, *Sylvia borin*, during migration. *J. Comp. Physiol. B* **161**, 576–580.

Schwabl, H., Mock, D.W., and Gieg, J.A. (1997). A hormonal mechanism for parental favouritism. *Nature, Lond.* **386**, 231.

Schwanzel-Fukuda, M. and Pfaff, D.W. (1989). Origin of luteinizing hormone-releasing hormone neurons. *Nature, Lond.* **338**, 161–164.

Schwartz, M.W., Dallman, M.F., and Woods, S.C. (1995). Hypothalamic response to starvation: implications for the study of wasting disorders. *Am. J. Physiol.* **269**, R949–R957.

Schwartz, M.W., Figlewicz, D.P., Baskin, D.G., Woods, S.C., and Porte, D. (1992). Insulin in the brain: a hormonal regulator of energy balance. *Endocrine Rev.* **13**, 387–414.

Schwartz, M.W., Peskind, E., Raskind, M., Boyko, E.J., and Porte, D. (1996). Cerebrospinal fluid leptin levels: relationship to plasma levels and to adiposity in humans. *Nature Med.* **2**, 589–593.

Schwartz, N.B. (1973). Mechanisms controlling ovulation in small mammals. In *Handbook of Physiology*, Sect. 7 *Endocrinology*, vol. II, *Female reproductive system* (Pt. 1), pp. 125–141. Washington, DC: American Physiological Society.

Schwerdtfeger, W.K. (1979). Qualitative and quantitative date on the fine structure of the guppy (*Poecilia reticulata* Peters) epidermis following treatment with thyroxine and testosterone. *Gen. Comp. Endocr.* **38**, 484–490.

Scott, A.P., Besser, G.M., and Ratcliffe, J.G. (1971). A phylogenetic study of pituitary corticotrophic activity. *J. Endocr.* **51**, i–ii.

Scott, A.P. and Vermeirissen, E.L.M. (1994). Production of conjugated steroids by teleost gonads and their role as pheromones. In *Perspectives in Comparative Endocrinology* (edited by K.G. Davey, R.E. Peter, and S.S. Tobe), pp. 645–654. Ottawa: National Research Council of Canada.

Seal, U.S. and Doe, R.P. (1963). Corticosteroid-binding globulin, species distribution and small-scale purification. *Endocrinology*, **73**, 371–376.

Seeley, R.J., Matson, C.A., Chavez, M., Woods, S.C., Dallman, M.F., and Schwartz, M.W. (1996). Behavioral, endocrine, and hypothalamic responses to involuntary feeding. *Am. J. Physiol.* **271**, R819–R823.

Sefkow, A.J., DiStefano, J.J., Himick, B.A., Brown, S.B., and Eales, J.G. (1996). Kinetic analysis of thyroid hormone secretion and interconversion of the 5-day-fasted rainbow trout, *Oncorhynchus mykiss. Gen. Comp. Endocr.* **101**, 123–138.

Segre, G.V. and Goldring, S.R. (1993). Receptors for secretin, calcitonin, parathyroid hormone (PTH)/PTH-related peptide, vasoactive intestinal peptide, glucagonlike peptide I, growth hormone-releasing hormone, and glucagon belong to a newly discovered G-protein-linked receptor family. *Trend. Endocr. Metab.* **4**, 309–314.

Senior, B.E. and Cunningham, F.J. (1974). Oestradiol and luteinizing hormone during the ovulatory cycle of the hen. *J. Endocr.* **60**, 201–202.

Sergeev, I.N., Buffenstein, R., and Pettifor, J.M. (1993). Vitamin D receptors in a naturally vitamin D-deficient subterranean mammal, the naked mole rat (*Heterocephalus glaber*): biochemical characterization. *Gen. Comp. Endocr.* **90**, 338–345.

Sernia, C. (1980). Physiology of the adrenal cortex in monotremes. In *Comparative physiology: primitive mammals* (edited by K. Schmidt-Nielsen, L. Bolis, and C.R. Taylor), pp. 308–315. Cambridge University Press.

Sernia, C., Bradley, A.J., and McDonald, I.R. (1979). High affinity binding of adrenocortical and gonadal steroids by plasma proteins of Australian marsupials. *Gen. Comp. Endocr.* **38**, 496–503.

Sernia, C. and McDonald, I.R. (1977). Metabolic effects of cortisol and adrenocorticotrophin in a prototherian mammal *Tachyglossus aculeatus* (SHAW). *J. Endocr.* **75**, 261–269.

Sernia, C. and McDonald, I.R. (1993). Changes in the activity of the adrenal cortex in fasted echidnas (*Tachyglossus aculeatus*) exposed to low ambient temperatures. *Comp. Biochem. Physiol. A* **106**, 87–90.

Sernia C. and Tyndale-Biscoe, C.H. (1979). Prolactin receptors in the mammary gland, corpus luteum and other tissues of the tammar wallaby, *Macropus eugenii. J. Endocr.* **83**, 79–89.

Shafrir, E. and Wertheimer, E. (1965).

Comparative physiology of adipose tissue in different sites and in different species. In *Handbook of Physiology*, Sect. 5 *Adipose Tissue*, pp. 417–429. Washington, DC: American Physiological Society.

Shapiro, B. and Pimstone, B.L. (1977). A phylogenetic study of sulphation factor activity in 26 species. *J. Endocr.* **74**, 129–135.

Shapiro, D., (1994). Sex changes in fishes – how and why? In *The Differences Between the Sexes* (edited by R.V. Short and E. Balaban), pp. 105–130. Cambridge: Cambridge University Press.

Shapiro, M., Nicholson, W.E., Orth, D.N., Mitchel, W.M., Island, D.P., and Liddle, G.W. (1972). Preliminary characterization of the pituitary melanocyte stimulating hormones of several vertebrate species. *Endocrinology* **90**, 249–256.

Sharif, M., and Hanley M.R. (1992). Stepping up the pressure. *Nature, Lond.* **357**, 279–280.

Sharman, G.B. (1970). Reproductive physiology of marsupials. *Science*, **167**, 1221–1228.

Sharman, G.B. (1976). Evolution of viviparity in mammals. In *Reproduction in Mammals*, 6. *The Evolution of Reproduction* (edited by C.R. Austin and R.V. Short), pp. 32–70. Cambridge University Press.

Sheridan, M.A. (1994). Regulation of lipid metabolism in poikilothermic vertebrates. *Comp. Biochem. Physiol. B* **107**, 495–508.

Sherwood, N.M., Lovejoy, D.A., and Coe, I.R. (1993). Origin of mammalian gonadotropin-releasing hormones. *Endocr. Rev.* **14**, 241–254.

Sherwood, N.M., Parker, D.B., McRory, J.E., and Lescheid, D.W. (1994). Molecular evolution of growth hormone-releasing hormone and gonadotropin-releasing hormone. In *Fish Physiology*, Vol. XIII, *Molecular Endocrinology of Fish* (edited by N.M. Sherwood and C.L. Hew), pp. 3–66. Sand Diego, CA: Academic Press.

Sherwood, O.D. (1994). Relaxin. In *The Physiology of Reproduction*, 2nd edn (edited by E. Knobil and J.D. Neill), vol. 1, pp. 861–1010. New York: Raven Press.

Shine, J. (1994). Editorial: structural conservation and functional diversity – profile of a political peptide. *Endocrinology* **134**, 1989–1990.

Shughrue, P.J., Lane, M.V., and Merchenthaler, I. (1996). Glucagon-like peptide I receptor (GLP1-R) mRNA in the rat hypothalamus. *Endocrinology* **137**, 5159–5162.

Shuldiner, A.R., Phillips, S., Roberts, C.T., LeRoith, D., and Roth, J. (1989). *Xenopus laevis* contains two nonallelic preproinsulin genes, cDNA cloning and evolutionary perspective. *J. Biol. Chem.* **264**, 9428–9432.

Shuster, S., Burton, J.L., Thody, A.J., Plummer, N., Goolamali, S.K., and Bates, D. (1973). Melanocyte-stimulating hormone and Parkinsonism. *Lancet* **i**, 463–465.

Shuster, S. and Thody A.J. (1974). The control and measurement of sebum secretion. *J. Invest. Dermatol.* **62**, 172–190.

Shuttleworth, T.J. (1978). The effect of adrenaline on potentials in the isolated gills of the flounder (*Platichthys flesus* L.). *J. Comp. Physiol. B* **124**, 129–136.

Siberian correspondent (1973). Centenarian triton. *Nature (Lond.)* **242**, 369.

Sikkel, P.C. (1993). Changes in plasma androgen levels associated with changes in male reproductive behavior in brood cycling marine fish. *Gen. Comp. Endocr.* **89**, 229–237.

Silva, J.E. (1993). Hormonal control of thermogenesis and energy dissipation. *Trend. Endocr. Metab.* **4**, 25–32.

Silva, J.E. and Larsen, P.R. (1986). Hormonal regulation of iodothyronine 5'-deiodinase in rat brown adipose tissue. *Am. J. Physiol.* **251**, E639–E643.

Silva, P., Solomon, R.J., and Epstein, F.H. (1996). The rectal gland of *Squalus acanthias*: a model for the transport of chloride. *Kidney Int.* **49**, 1552–1556.

Silveira, P.F., Schiripa, L.N., Carmona, E., and Picarelli, Z.P. (1992). Circulating vasotocin in the snake *Bothrops jararaca*. *Comp. Biochem. Physiol. A* **103**, 59–64.

Simmons, D.J. (1971). Calcium and skeletal tissue physiology in teleost fishes. *Clin. Orthopaedics* **76**, 244–280.

Simon, J. and Taouis, M. (1993). The insulin receptor in chicken tissues. In *Avian Endocrinology* (edited by P.J.

Sharp), pp. 275–285. Bristol: The Journal of Endocrinology.

Singh, S. and Srivastav, A.K. (1993). Effects of calcitonin administration on serum calcium and inorganic phosphate levels of the fish, *Heteropneustes fossilis*, maintained either in artificial freshwater, calcium-rich freshwater, or calcium-deficient freshwater. *J. Exp. Zool.* **265**, 35–39.

Sinha, Y.N. (1995). Structural variants of prolactin: occurrence and physiological significance. *Endocrine Rev.* **16**, 354–369.

Siris E.S., Nisula, B.C., Catt, K.J., Horner, K., Birkin, S., Canfield, R.A., and Ross, G.T. (1978). New evidence for intrinsic follicle-stimulating hormone-like activity in human gonadotropin and luteinizing hormone. *Endocrinology* **102**, 1356–1361.

Skadhauge, E. (1969). Activités biologique des hormones neurohypophysaires chez les oiseaux et les reptiles. *Colloq. Int. C.N.R.S.* **177**, 63–68.

Skadhauge, E. (1981). *Osmoregulation in Birds* Berlin: Spring-Verlag.

Skadhauge, E. (1989). An overview of the interaction of kidney, cloaca, lower intestine, and salt gland in avian osmoregulation. In *Progress in Avian Osmoregulation* (edited by M.A. Hughes and A. Chadwick), pp. 333–346. Leeds: Leeds Philosophical and Literary Society.

Slootweg, M.C., Salles, J.P., Ohlsson, C. de Vries, C.P., Engelbregt, M.J.E., and Netelenbos, J.C. (1996). Growth hormone binds to single high affinity receptor site on mouse osteoblasts: modulation by retinoic acid and cell differentiation. *J. Endocr.* **150**, 465–472.

Smit, A.B., Geraerts, W.P.M., Meester, I., van Heerikhuizen, H., and Joosse, J. (1991). Characterization of cDNA clone encoding molluscan insulin-related peptide II of *Lymnaea stagnalis*. *Eur. J. Biochem.* **199**, 699–703.

Smit, A.B., van Marle, A., van Elk, R., Bogerd, J., van Heerikhuizen, H., Geraerts, W.P.M. (1993). Evolutionary conservation of the insulin gene structure in invertebrates: cloning of the gene encoding molluscan insulin-related peptide III from *Lymnaea stagnalis*. *J. Mol. Endocr.* **11**, 103–113.

Smith, A.I., and Funder, J.W. (1988). Proopiomelanocortin processing in the pituitary, central nervous system, and peripheral tissues. *Endocrine Rev.* **9**, 159–179.

Smith, C.A. and Joss, J.M.P. (1994). Steroidogenic enzyme activity and ovarian differentiation in the saltwater crocodile. *Crocodylus porosus. Gen. Comp. Endocr.* **93**, 232–245.

Smith, D.C.W. (1956). The role of the endocrine organs in the salinity tolerance of trout. *Mem. Soc. Endocr.* **5**, 83–98.

Smith, F.J., Campfield, L.A., Moschera, J.A., Bailon, P.S., and Burn, P. (1996). Feeding inhibition by neuropeptide Y. *Nature, Lond.* **382**, 307.

Smith, G.P. and Gibbs, J. (1994). Satiating effect of cholecystokinin. *Ann. N.Y. Acad. Sci.* **713**, 236–241.

Smith, H.W. (1930). Metabolism of the lungfish. *Protopterus aethiopicus. J. Biol. Chem.* **88**, 97–130.

Smith, J.P. (1982). Changes in blood levels of thyroid hormones in two species of passerine birds. *Condor*, **84**, 160–167.

Smith, N.F., Eddy, F.B., Struthers, A.D., and Talbot, C. (1991). Renin, atrial natriuretic peptide and blood plasma ions in parr and smolts of Atlantic salmon *Salmo salar* L. and rainbow trout *Oncorhynchus mykiss* (WALBAUM) in fresh water and after short-term exposure to sea water. *J. Exp. Biol.* **157**, 63–74.

So, Y.P., and Fenwick, J.C. (1979). *In vivo* and *in vitro* effects of Stannius corpuscle extract on the branchial uptake of ^{45}Ca in stanniectomized North American eels (*Anguilla rostrata*). *Gen. Comp. Endocr.* **37**, 143–149.

Sokabe, H., Nishimura, H., Ogawa, M., and Oguri, M. (1970). Determination of renin in the corpuscles of Stannius of the teleost. *Gen. Comp. Endocr.* **14**, 510–516.

Sokabe, H., Ogawa, M., Oguri, M., and Nishimura, H. (1969). Evolution of the juxtaglomerular apparatus in the vertebrate kidneys. *Texas Reports Biol. Med.* **27**, 867–885.

Solomon, R., Protter, A., McEnroe, G., Porter, J.G., and Silva, P. (1992). C-type natriuretic peptides stimulate chloride secretion in the rectal gland of *Squalus acanthias. Am. J. Physiol.* **262**, R707–R711.

Solomon, R., Taylor, M., Dorsey, D., Silva, P., and Epstein, F.H. (1985). Atriopeptin

stimulation of rectal gland function in *Squalus acanthias. Am. J. Physiol.* **249**, R348–R354.

Somoza, G.M., Stéfano, A., D'Eramo, J.L., Canosa, L.F., and Fridman, O. (1994). Immunoreactive GnRH suggesting a third form of GnRH in addition to cIIGnRH and sGnRH in the brain and pituitary gland of *Prochilodus lineatus* (Caraciformes). *Gen. Compt. Endocr.* **94**, 44–52.

Sonnemans, M.A.F., Evans, D.A.P., Burbach, J.P.H., and van Leeuwen, F.W. (1996). Immunocytochemical evidence for the presence of vasopressin in intermediate sized neurosecretory granules of solitary neurohypophyseal terminals in the homozygous Brattleboro rat. *Neuroscience* **72**, 225–231.

Sonnenberg, H. (1994). The role of atrial natriuretic factor in salt balance. In *Perspectives in Comparative Endocrinology* (edited by K.G. Davey, R.E. Peter, and S.S. Tobe), pp. 85–88. Ottawa: National Research Council of Canada.

Soontjens, C.D., Rafter, J.J., and Gustafsson, J.-A. (1996). Ligands for orphan receptors. *J. Endocr.* **150**, S241–S257.

Sorenson, P.W., Scott, A.P., Stacey, N.E., and Bowdin, L. (1995). Sulfated 17,20β-dihydroxy-4-pregnen-3-one functions as a potent and specific olfactory stimulant with pheromonal actions in goldfish. *Gen. Comp. Endocr.* **100**, 128–142.

Souza, S.C., Frick, G.P., Wang, X., Kopchick, J.J., Lobo, R.B., and Goodman, H.M. (1995). A single arginine residue determines species specificity of the human growth hormone receptor. *Proc. Natl. Acad. Sci. USA* **92**, 959–963,

Sower, S.A., Chiang, Y.-C., Lovas, S., and Conlon, J.M. (1993). Primary structure and biological activity of a third gonadotropin-releasing hormone from lamprey brain. *Endocrinology* **132**, 1125–1131.

Sower, S.A., Takahashi, A., Nozaki, M., Gorbman, A., Youson, J.H., Joss, J., and Kawauchi, H. (1995). A novel glycoprotein in the olfactory and pituitary systems of larval and adult lampreys. *Endocrinology* **136**, 349–356.

Spallanzani (1784). *Dissertations Relative to the Natural History of Animals and Vegetables 2.* Trans. from the Italian, London. Quoted by F.H.A. Marshall, 1965.

Spanos, E., Colston, K.W., Evans, I.M.S., Galante, L.S., MacAuley, S.J. and MacIntyre, I. (1976a). Effect of prolactin on vitamin D metabolism. *Mol. Cell Endocr.* **5**, 163–167.

Spanos, E., Pike, J.W., Haussler, M.R., Colston, K.W., Evans, I.M.A., Goldner, A.M., McCain, T.A., and MacIntyre, I. (1976b). Cloning 1α, 25-dihydroxyvitamin D in the chicken: enhancement by injection of prolactin and during egg laying. *Life Sci.* **19**, 1751–1756.

Specker, J.L. (1988). Preadaptive role of thyroid hormones in larval and juvenile salmon: growth, the gut and evolutionary considerations. *Am. Zool.* **28**, 337–349.

Specker, J.L., Kishida, M., Huang, L., King, D.S., Nagahama, Y., Ueda, H., and Anderson, T.R. (1993). Immunocytochemical and immunogold localization of two prolactin isoforms in the same pituitary cells in the same granules in tilapia (*Oreochromis mossambicus*). *Gen. Comp. Endocr.* **89**, 29–38.

Specker, J.L. and Moore, F.L., (1980). Annual cycle of plasma androgens and testicular composition in the rough-skinned newt. *Taricha granulosa. Gen. Comp. Endocr.* **42**, 297–303.

Specker, J.L. and Sullivan, C.V. (1994). Vitellogenesis in fishes: status and perspectives. In *Perspectives in Comparative Endocrinology* (edited by K.G. Davey, R.E. Peter, and S.S. Tobe), pp. 304–315. Ottawa: National Research Council of Canada.

Spedding, M., Ouvry, C., Millan, M., Duhault, J., Dacquet, C., and Wurtman, R. (1996). Neural control of dieting. *Nature, Lond.* **380**, 488.

Speers, G.M., Percy, D.Y.E., and Brown, D.M. (1970). Effect of ultimobranchialectomy in the laying hen. *Endocrinology* **87**, 1292–1297.

Speight, A., Popkin, R., Watts, A.G., and Fink, G. (1980). Oestradiol-17β increases pituitary responsiveness by a mechanism that involves the release and the priming effect of luteinizing hormone releasing factor. *J. Endocr.* **88**, 301–308.

Spencer, E.M. and Tobiassen, O. (1981). The mechanism of the action of growth hormone on vitamin D metabolism in the rat. *Endocrinology* **108**, 1064–1070.

Spies, H.G. and Chappel, S.C. (1984). Mammals: non-human primates. In *Marshall's Physiology of Reproduction*, 4th edn (edited by G.E. Lamming), vol. 1, *Reproductive Cycles of Vertebrates*, pp. 659–712. Edinburgh: Churchill Livingstone.

Srivastava, A.K. and Meier, A.H. (1972). Daily variation in concentration of cortisol in plasma in intact and hypophysectomized gulf killifish. *Science* **177**, 185–187.

Srivastav, A.K. and Rani, L. (1989). Influence of calcitonin administration on serum calcium and inorganic phosphate level in the frog, *Rana tigrina. Gen. Comp. Endocr.* **74**, 14–17.

Srivastav, A.K., Srivastav, S.K., Sasayama, Y., and Suzuki, N. (1996). Corpuscles of Stannius-extract induced rapid but transient hypocalcemia and hyperphosphatemia in stingray. *Dasyatis akajei. Gen. Comp. Endocr.* **104**, 37–40.

Stacey, N.E., Cardwell, J.R., Liley, N.R., Scott, A.P., and Sorenson, P.W. (1994). Hormones as sex pheromones in fish. In *Perspectives in Comparative Endocrinology* (edited by K.G. Davey, R.E. Peter, and S.S. Tobe), pp. 438–448. Ottawa: National Research Council of Canada.

Stacey, N.E., Cook, A.F., and Peter, R.E. (1979). Ovulatory surge of gonadotropin in the goldfish. *Carassius auratus. Gen. Comp. Endocr.* **37**, 246–249.

Stallone, J.N. and Braun, E.J. (1985). Contribution of glomerular and tubular mechanisms to antidiuresis in conscious domestic fowl. *Am. J. Physiol.* **249**, F842–F850.

Stannius, H. (1839). Die Nebennieren bei Knochenfischen. *Arch. Anat. Physiol.* **97**, 97–101.

Steele, M.K. (1992). The role of brain angiotensin II in the regulation of luteinizing hormone and prolactin secretion. *Trend. Endocr. Metab.* **3**, 295–301.

Steele, N.C. and Evock-Clover, C.M. (1993). Role of growth hormone in

growth of homeotherms. In *The Endocrinology of Growth, Development, and Metabolism in Vertebrates* (edited by M.P. Schreibman, C.G. Scanes, and P.K.T. Pang), pp. 73–91. San Diego, CA: Academic Press.

Stehle, J.H., Foulkes, N.A., Molina, C.A., Simonneaux, V., Pevet, P., and Sassone-Corsi, P. (1993). Adrenergic signals direct rhythmic expression of transcriptional repressor CREM in the pineal gland. *Nature, Lond.* **365**, 314–320.

Steiner, D.F., Kemmler, W., Tager, H.S., and Peterson, J.D. (1974). Proteolytic processing in the biosynthesis of insulin and other proteins. *Fedn. Proc.* **33**, 2105–2115.

Steiner, D.F., Quin, P.S., Chan, S.J., Marsh, J., and Tager, H.S. (1980). Processing mechanisms in the biosynthesis of proteins. *Ann. N.Y. Acad. Sci.* **343**, 1–16.

Steiner, D.F. Smeekens, S.P., Ohagi, S., and Chan, S.J. (1992). The new enzymology of precursor processing endoproteases. *J. Biol. Chem.* **267**, 23435–23438.

Stenzel-Poor, M.P., Heldwein, K.A., Stenzel, P., Lee, S., and Vale, W. (1992). Characterization of the genomic corticotropin-releasing factor (CRF) gene from *Xenopus laevis*: two members of the CRF family exist in amphibians. *Mol. Endocr.* **6**, 1716–1724.

Stephens, T.W., Basinski, M., Bristow, P.K., Bue-Valleskey, J.M., Burgett, S.G., Craft, L., Hale, J., Hoffmann, J., Hsiung, H.M., Kriauciunas, A., MacKellar, W., Rosteck, P.R., Schoner, B., Smith, D., Tinsley, F.C., Zhang, X.-Y., and Heiman, M. (1995). The role of neuropeptide Y in the antiobesity action of the *obese* gene product. *Nature, Lond.* **377**, 530–532.

Sterling, K., Brenner, M.A., and Saldanha, V.F. (1973). Conversion of thyroxine to triiodothyronine by cultured human cells. *Science* **179**, 1000–1001.

Stern, J.M. and Lehrman, D.S. (1969). Role of testosterone in progesterone-induced incubation behaviour in male ring doves (*Streptopelia risoria*). *J. Endocr.* **44**, 13–22.

Stetson, M.H. and Erickson, J.E. (1972). Hormonal control of photoperiodically

induced fat deposition in white-crowned sparrows. *Gen. Comp. Endocr.* **19**, 355–362.

Stevens, C.E. and Hume, I.D. (1995). *Comparative Physiology of the Vertebrate Digestive System*, 2nd edn. Cambridge: Cambridge University Press.

Stewart, A.D. (1973). Sensitivity of mice to (8-arginine)- and (8-lysine)-vasopressins as antidiuretic hormones. *J. Endocr.* **59**, 195–196.

Stewart, D.R., Nevins, B., Hadas, E., and Vandlin, R. (1991). Affinity purification and sequence determination of equine relaxin. *Endocrinology* **129**, 375–383.

Stiffler, D.F. (1995). Active calcium transport in the skin of the frog *Rana pipiensis*: kinetics and seasonal rhythms. *J. Exp. Biol.* **198**, 967–974.

Stoff, J.S., Rosa, R., Hallac, R., Silva, P., and Epstein, F.H. (1979). Hormonal regulation of active chloride transport in the dogfish rectal gland. *Am. J. Physiol.* **237**, F138–F144.

Strauss, J.S. and Ebling, F.J. (1970). Control and function of skin glands in mammals. *Mem. Soc. Endocr.* **18**, 341–368.

Stuart, S. (1988). Characterization of a novel insulin receptor from the stingray liver. *J. Biol. Chem.* **263**, 7881–7886.

Sudoh, T., Kangawa, K., Minamino, N., and Matsuo, H. (1988). A new natriuretic peptide in porcine brain. *Nature, Lond.* **332**, 78–81.

Sudoh, T., Minamino, N., Kangawa, K., and Matsu, H. (1990). C-type natriuretic peptide (CNP): a new member of the natriuretic peptide family identified in porcine brain. *Biochem. Biophys. Res. Commun.* **168**, 863–870.

Suga, S.-I., Nakao, K., Hosoda, K., Mukoyama, M., Ogawa, Y., Shirakami, G., Arai, H., Saito, Y., Kambayashi, Y., Inouye, K., and Imura, H. (1992). Receptor selectivity of natriuretic peptide family, atrial natriuretic peptide, brain natriuretic peptide, and C-type natriuretic peptide. *Endocrinology* **130**, 229–239.

Sumpter, J.P., Denning-Kendall, P.A., and Lowry, P.J. (1984). The involvement of melanotrophin in physiological colour change in the dogfish, *Scyliorhinus canicula. Gen. Comp. Endocr.* **56**,

360–365.

Sumpter, J.P., Follett, B.K., Jenkins, N., and Dodd, J.M. (1978). Studies on the purification and properties of gonadotrophin from ventral lobes of the pituitary gland of the dogfish (*Scyliorhinus canicula* L.). *Gen. Comp. Endocr.* **36**, 264–274.

Sumpter, J.P., LeBail, P.Y., Pickering, A.D., Pottinger, T.G., and Carragher, J.F. (1991). The effects of starvation on growth and plasma growth hormone concentrations of the rainbow trout, *Oncorhynchus mykiss. Gen. Comp. Endocr.* **83**, 94–102.

Sundby, F., Frandsen, E.K., Thomsen, J., Kristiansen, K., and Brunfeldt, K. (1972). Crystallization and amino acid sequence of duck glucagon. *FEBS Lett.* **26**, 289–293.

Sundell, K. and Björnsson, B.Th. (1990). Effects of vitamin D_3, 25(OH)vitamin D_3, and 1,25$(OH)_2$vitamin D_3 on the *in vitro* intestinal calcium absorption in the marine teleost, Atlantic cod (*Gadus morhua*). *Gen. Comp. Endocr.* **78**, 74–79.

Sundell, K., Björnsson, B.Th., Itoh, H., and Kawauchi, H. (1992). Chum salmon (*Oncorhynchus keta*) stanniocalcin inhibits *in vitro* intestinal calcium uptake in Atlantic cod (*Gadus morhua*). *J. Comp. Physiol. B* **162**, 489–495.

Sundell, K., Norman, A.W., and Björnsson, B.Th. (1993). 1,25$(OH)_2$ vitamin D_3 increases ionized plasma calcium concentrations in the immature Atlantic cod *Gadus morhua. Gen. Comp. Endocr.* **91**, 344–351.

Suzuki, M., Kubokawa, K., Nagasawa, H., and Urano, A. (1995). Sequence analysis of vasotocin cDNAs of the lamprey, *Lampetra japonica*, and the hagfish, *Eptatretus burgeri*: evolution of cyclostome vasotocin precursors. *J. Mol. Endocr.* **14**, 67–77.

Suzuki, R., Takahashi, A., and Takei, Y. (1992). Different molecular forms of C-type natriuretic peptide isolated from the brain and heart of an elasmobranch, *Triakis scyllia. J. Endocr.* **135**, 317–323.

Suzuki, R., Togashi, K., Ando, K., Takei, Y. (1994). Distribution of molecular forms of C-type natriuretic peptide in plasma and tissue of a dogfish, *Triakis scyllia. Gen. Comp. Endocr.* **96**, 378–384.

Suzuki, S., Gorbman, A., Rolland, M., Montfort, M.-F., and Lissitzky, S. (1975). Thyroglobulins of cyclostomes and elasmobranchs. *Gen. Comp. Endocr.* **26**, 59–69.

Suzuki, S. and Kondo, Y. (1973). Thyroidal morphogenesis and biosynthesis of thyroglobulin before and after metamorphosis in the lamprey, *Lampetra reissneri. Gen. Comp. Endocr.* **21**, 451–460.

Swaminathan, R., Bates, R.F.L., and Care, A.R. (1972). Fresh evidence for a physiological role for calcitonin in calcium homeostasis. *J. Endocr.* **54**, 525–526.

Swanson, P., Dickoff, W.W., and Gorbman, A. (1987). Pituitary thyrotropin and gonadotropin of coho salmon (*Oncorhyncus kisutch*): separation by chromatofocusing. *Gen. Comp. Endocr.* **65**, 269–287.

Swanson, P., Suzuki, K., Kawauchi, H., and Dickoff, W.W. (1991). Isolation and characterization of two coho salmon gonadotropins, GTH I and GTH II. *J. Reprod.* **44**, 29–38.

Swift, D.R. and Pickford, G.E. (1965). Seasonal variations in the hormone content of the pituitary gland of the perch *Perca fluviatilis* L. *Gen. Comp. Endocr.* **5**, 354–365.

Sweeting, R.M. and McKeown, B.A. (1987). Growth hormone and seawater adaptation in coho salmon, *Oncorhynchus kisutch. Comp. Biochem. Physiol. A* **88**, 147–151.

Taché, Y., Holzer, P., and Rosenfeld, M.G. (eds.) (1992). Calcitonin gene-related peptide. *The first decade of a novel pleiotropic neuropeptide.* (Series of papers.) *Ann. N.Y. Acad. Sci.* vol. 657.

Tait, J.F. and Tait, S.A.S. (1979). Recent perspectives on the history of the adrenal cortex. *J. Endocr.* **83**, 3p–24p.

Takada, M. and Shomazaki, S. (1988). Effect of prolactin on transcutaneous Na transport in the Japanese newt, *Cynops pyrrhogaster. Gen. Comp. Endocr.* **69**, 141–145.

Takahashi, A., Amemiya, Y., Sarashi, M., Sower, S.A., and Kawauchi, H. (1995). Melanotropin and corticotropin are encoded by two distinct genes in the lamprey, the earliest evolved extant vertebrate. *Biochem. Biophys. Res. Commun.* **213**, 490–498.

Takahashi, N., Yoshihama, K., Kikuyama, S., Yamamoto, K., Wakabayashi, K., and Kato, Y. (1990). Molecular cloning and nucleotide sequence analysis of complementary DNA for bullfrog prolactin. *J. Mol. Endocr.* **5**, 281–287.

Takahashi, T. (1993). Circadian clocks a la CREM. *Nature, Lond.* **365**, 299–300.

Takasugi, N. and Bern, H.A. (1962). Experimental studies on the caudal neurosecretory system in *Tilapia mossambica. Comp. Biochem. Physiol.* **6**, 289–303.

Takayama, Y., Ono, M., Rand-Weaver, M., and Kawauchi, H. (1991). Greater conservation of somatolactin, a presumed pituitary hormone of the growth hormone/prolactin family, than of growth hormone in teleost fish. *Gen. Comp. Endocr.* **83**, 366–374.

Takei, Y. (1977). Angiotensin and water intake in the Japanese quail (*Coturnix coturnix japonica*). *Gen. Comp. Endocr.* **31**, 364–372.

Takei, Y. (1994). Structure and function of natriuretic peptides in vertebrates. In *Perspectives in Comparative Endocrinology* (edited by K.G. Davey, R.E. Peter, and S.S. Tobe), pp. 155–165. Ottawa: National Research Council of Canada.

Takei, Y. and Balment, R.J. (1993). Natriuretic factors in non-mammalian vertebrates. In *Natriuretic Factors in Non-mammalian Vertebrates: New Insights into Vertebrate Kidney Function* (edited by A. Brown, C. Rankin, and R.J. Balment), pp. 351–385. Cambridge: Cambridge University Press.

Takei, Y., Hasegawa, Y., Watanabe, T.X., Nakajima, K., and Hazon, N. (1993). A novel angiotensin I isolated from an elasmobranch fish. *J. Endocr.* **139**, 281–285.

Takei, Y., Takahashi, A., Watanabe, T.X., Nakajima, K., and Ando, K. (1994a). Eel ventricular natriuretic peptide: isolation of low molecular size form and characterization of plasma form by homologous radioimmunoassay. *J. Endocr.* **141**, 81–89.

Takei, Y., Takahashi, A., Watanabe, T.X., Nakajima, K., and Sakakibara, S. (1989). Amino acid sequence and relative biological activity of eel atrial natriuretic peptide. *Biochem. Biophys, Res. Commun.* **164**, 537–543.

Takei, Y., Takahashi, A., Watanabe, T.X., Nakajima, K., Sakakibara, S., Sasayama, Y., Suzuki, M., and Oguru, C. (1991). New calcitonin isolated from the ray, *Dasyatis akajei. Biol. Bull.* **180**, 485–488.

Takei, Y., Takano, M., Itahara, Y., Watanabe, T.W., Nakajima K., Conklin, D.J., Duff, D.W., and Olson, K.R. (1994b). Rainbow trout ventricular natriuretic peptide: isolation, sequencing, and determination of biological activity. *Gen. Comp. Endocr.* **96**, 420–426.

Takei, Y., Ueki, M., and Nishizawa, T. (1994c). Eel ventricular natriuretic peptide: cDNA cloning and mRNA expression. *J. Mol. Endocr.* **13**, 339–345.

Talamantes, F. (1975). Comparative study of the occurrence of placental prolactin in mammals. *Gen. Comp. Endocr.* **27**, 115–121.

Talamantes, F., Ogren, I., Markoff, E., Woodard, S., and Madrid, J. (1980). Phylogenetic distribution, regulation of secretion, and prolactin-like effects of placental lactogens. *Fedn. Proc.* **39**, 2582–2587.

Tamarkin, L., Westrom, W.K., Hamill, A.I., and Goldman, B.D. (1976). Effect of melatonin on the reproductive systems of male and female syrian hamsters: a diurnal rhythm in sensitivity to melatonin. *Endocrinology* **99**, 1534–1541.

Tanabe, Y., Ishii, T., and Tamaki, Y. (1969). Comparison of thyroxine-binding plasma proteins of various vertebrates and their evolutionary aspects. *Gen. Comp. Endocr.* **13**, 14–21.

Tanaka, Y., Frank,. H., and DeLuca, H.F. (1973). Intestinal calcium transport: stimulation by low phosphorous diets. *Science* **181**, 564–566.

Tartaglia, L.A., Dembski, M., Weng, X., Deng, N.H., Culpepper, J., Devos, R., Richards, G.J., Campfield, L.A., Clark, F.T., Deeds, J., Muir, C. Sanker, S., Moriarty, A., Moore, K.G., Smutko, J.S., Mays, G.G., Woolf, E.A., Monroe, C.A., and Tepper, R.I. (1995). Identification and expression cloning of a leptin receptor, OB-R. *Cell* **83**, 1263–1271.

Tashjian, A.H., Levine, L., and Wilhelmi, A.E. (1965). Immunochemical relatedness of porcine, bovine, ovine and primate growth hormones. *Endocrinology* **77**, 563–573.

Tatemoto, K., Carlquist, M., and Mutt, V. (1982). Neuropeptide Y – a novel brain peptide with structural similarities to peptide YY and pancreatic polypeptide. *Nature, Lond.* **296**, 659–660.

Tatemoto, K., Rökenhaues, A., Jornvall, H., McDonald, T.J., and Mutt, V. (1983). Galanin – a novel biologically active peptide from porcine intestine. *FEBS. Lett.* **164**, 124–128.

Taylor, J.D. and Bagnara, J.T. (1972). Dermal chromatophores. *Am. Zool.* **12**, 43–62.

Tecott, L.H., Sun, L.M., Akana, S.F., Strack, A.M., Lowenstein, D.H., Dallman, M.F., and Julius, D. (1995). Eating disorder and epilepsy in mice lacking 5-HT$_{2C}$ serotonin receptors. *Nature, Lond.* **374**, 542–546.

Temple, S.A. (1974). Plasma testosterone titers during the annual reproductive cycle of starlings (*Sturnus vulgaris*). *Gen. Comp. Endocr.* **22**, 470–479.

Tepperman, J. and Tepperman. H.M. (1970). Gluconeogenesis, lipogenesis and the Sherringtonian metaphor. *Fedn. Proc.* **29**, 1284–1293.

Tewary, P.D. and Farner, D.S. (1973). Effect of castration and estrogen administration on the plumage pigment of the male house finch (*Carpodacus mexicanus*). *Am. Zool.* **13**, 1278.

Thody, A.J., Cooper, M.F., Bowden, P.E., Meddis, D., and Shuster, S. (1976). Effect of α-melanocyte-stimulating hormone and testosterone on cutaneous and modified sebaceous glands in the rat. *J. Endocr.* **71**, 279–288.

Thomas, D.H. and Phillips, J.G. (1975). Studies in avian adrenal steroid function. I. Survival and mineral balance, following adrenalectomy in domestic ducks (*Anas platyrhynchos* L.). *Gen. Comp. Endocr.* **26**, 394–403.

Thomas, P. (1994). Hormonal control of final oocyte maturation in sciaenid fishes. In *Perspectives in Comparative Endocrinology* (edited by K.G. Davey, R.E. Peter, and S.S. Tobe), pp. 619–625. Ottawa: National Research Council of Canada.

Thompson, A.J. and Sargent, J.R. (1977). Changes in the levels of chloride cells and Na–K-dependent ATPase in the gills of yellow and silver eels adapting to sea water. *J. Exp. Zool.* **200**, 33–40.

Thorburn, G.D. and Liggins, G.C. (1994). Role of the fetal pituitary–adrenal axis and placenta in the initiation of parturition. In *Marhsall's Physiology of Reproduction*, 4th edn (edited by G.E. Lamming), vol. 3, *Pregnancy and Lactation*, Pt. 2, *Fetal Physiology, Parturition and Lactation*, pp. l003–1036. London: Chapman & Hall.

Thorens, B., Roth, J., Norman, A.W., Perrelet, A., and Orci, L. (1982). Immunocytochemical localization of the vitamin D-dependent calcium-binding protein in chick duodenum. *J. Cell. Biol.* **93**, 201–206.

Thorndyke, M.C. (1978). Evidence for a 'mammalian' thyroglobulin in endostyle of the ascidian *Stylea clava. Nature, Lond.* **271**, 61–62.

Thorndyke, M.C., and Georges, D. (1988). Functional aspects of peptide neurohormones in protochordates. In *Neurohormones in Invertebrates* (edited by M.C. Thorndyke and G.J. Goldsworthy), pp. 235–258. Cambridge: Cambridge University Press.

Thunhorst, R.L. and Fitts, D.A. (1994). Peripheral angiotensin causes salt appetite in rats. *Am. J. Physiol.* **267**, R171–R177.

Tierney, M., Takei, Y., and Hazon, N. (1997). The presence of angiotensin II receptors in elasmobranchs. *Gen. Comp. Endocr.* **105**, 9–17.

Tonon, M.C., Leroux, P., Stoeckel, M.E., Jegou, S., Pelletier, G., and Vaudry, H. (1983). Catecholaminergic control of α-melanocyte-stimulating hormone (αMSH) release by frog neurointermediate lobe *in vitro*: evidence for direct stimulation of αMSH release by thyrotropin-releasing hormone. *Endocrinology* **112**, 133–141.

Toop, T., Donald, J.A., and Evans, D.H. (1995a). Localization and characteristics of the natriuretic peptide receptors in the gills of the Atlantic hagfish, *Myxine glutinosa* (Agnatha). *J. Exp. Biol.* **198**, 117–126.

Toop, T., Donald, J.A., and Evans, D.H. (1995b). Natriuretic peptide receptors in the kidney and the ventral and dorsal aortae of the Atlantic hagfish *Myxine glutinosa* (Agnatha). *J. Exp. Biol.* **198**, 1875–1882.

Torresani, J., Gorbman, A., Lachiver, F., and Lissitzky, S. (1973). Immunological cross-reactivity between thyroglobulins of mammals and reptiles. *Gen. Comp. Endocr.* **21**, 530–535.

Torrey, T.W., (1971). *Morphogenesis of the Vertebrates*, 3rd edn, pp. 44–45. New York: Wiley.

Toyoda, F., Matsuda, K., Yamamoto, K., and Kikuyama, S. (1996). Involvement of endogenous prolactin in the expression of courtship behavior in the newt, *Cynops pyrrhogaster. Gen. Comp. Endocr.* **102**, 191–196.

Trant, J.M. and Thomas, P. (1989). Isolation of a novel maturation-inducing steroid produced *in vitro* by ovaries of Atlantic croaker. *Gen. Comp. Endocr.* **75**, 397–404.

Treacy, G.B., Shaw, D.C., Griffiths, M.E., and Jeffrey, P.D. (1989). Purification of a marsupial insulin: amino-acid sequence of insulin from the Eastern grey kangaroo *Macropus giganteus. Biochim. Biophys. Acta.* **990**, 263–268.

Tregear, G.W., Rietschoten, J.V., Greene, E., Keutmann, H.T., Niall, H.D., Reit, B., Parsons, J.A., and Potts, J.T. (1973). Bovine parathyroid hormone: minimum chain length of synthetic peptide required for biological activity. *Endocrinology* **93**, 1349–1353.

Trinh, K.-Y., Wang, N.C., Hew, C.L., and Crim, L.W. (1986). Molecular cloning and sequencing of salmon gonadotropin β subunit. *Eur. J. Biochem.* **159**, 619–624.

Tsafriri, A., Reich, R., and Abisogun, A.O. (1994). The ovarian egg and ovulation. In *Marshall's Physiology of Reproduction*, 4th edn (edited by G.E. Lamming), vol. 3, *Pregnancy and Lactation*, Pt 1, *Ovulation and Early Pregnancy*, pp. 1–92. London: Chapman & Hall.

Tsai, M.-J. and O'Malley, B.W. (1994). Molecular mechanisms of action of steroid/thyroid receptor superfamily members. *Annu. Rev. Biochem.* **63**, 451–486.

Turek, F.W. and van Cauter, E. (1994). Rhythms in reproduction. In *The Physiology of Reproduction*, 2nd edn

(edited by E. Knobil and J.D. Neill) vol. 2, pp. 487–540. New York: Raven Press.

Turton, M.D., O'Shea, D., Gunn, I., Beak, S.A., Edwards, C.M.D., Meeran, K., Choi, S.J., Taylor, G.M., Heath, M.M., Lambert, P.D., Wilding, J.P.H., Smith, D.M., Ghatei, M.A., Herbert, J., and Bloom, S.R. (1996). A role for glucagon-like peptide-1 in the central regulation of feeding. *Nature, Lond.* **379**, 69–72.

Tyler, C. R., Sumpter, J.P., Kawauchi, H., and Swanson, P. (1991). Involvement of gonadotropin in the uptake of vitellogenin into vitellogenic oocytes of the rainbow trout. *Oncorhynchus mykiss. Gen. Comp. Endocr.* **84**, 219–299.

Tyndale-Biscoe, C.H. and Evans, S.M. (1980). Pituitary–ovarian interactions in marsupials. In *Comparative Physiology: Primitive Mammals* (edited by K. Schmidt-Nielsen, L. Bolis, and C.R. Taylor), pp. 259–268. Cambridge: Cambridge University Press.

Tyndale-Biscoe, C.H. and Renfree, M.B. (1987). *Reproductive Physiology of Marsupials*. Cambridge: Cambridge University Press.

Uchiyama, M. and Murakami, T. (1994). Effects of AVT and vascular antagonists on kidney function and smooth muscle contraction in the river lamprey, *Lampetra japonica. Comp. Biochem. Physiol. A* **107**, 493–499.

Uchiyama, M., Saito, N., Shimada, K., and Murakami, T. (1994). Pituitary and plasma vasotocin levels in the lamprey, *Lampetra japonica. Comp. Biochem. Physiol. A* **107**, 23–26.

Underwood, H. and Hyde, L.L. (1989). The effect of daylength on the pineal melatonin rhythm of the lizard *Anolis carolinensis. Comp. Biochem. Physiol. A* **94**, 53–56.

Urano, A., Hyodo, S., and Suzuki, M. (1992). Molecular evolution of neurohypophysial hormone precursors. *Prog. Brain Res.* **92**, 39–46.

Urano, A., Kubokawa, K., and Hiraoka, S. (1994). Expression of the vasotocin and isotocin gene family in fish. In *Fish Physiology*, vol. XIII, *Molecular Endocrinology of Fish* (edited by N.M. Sherwood and C.L. Hew), pp. 101–132. San Diego, CA: Academic Press.

Urist, M.R. (1962). The bone-body fluid continuum: calcium and phosphorous in the skeleton and blood of extinct and living vertebrates. *Perspectus Biol. Med.* **6**, 75–115.

Urist, M.R. (1963). The regulation of calcium and other ions in the serums of hagfish and lampreys. *Proc. N.Y. Acad. Sci.* **109**, 294–311.

Urist, M.R. (1973). Testosterone-induced development of limb gills of the lungfish, *Lepidosiren paradoxa. Comp. Biochem, Physiol. A* **44**, 131–135.

Urist, M.R. (1976). Biogenesis of bone: calcium and phosphorous in the skeleton and blood in vertebrate evolution. *In Handbook of Physiology*, Sect. 7, *Endocrinology*, vol. 7, *The Parathyroid Gland*, pp. 183–213.

Urist, M.R., Uyeno, S., King, E., Okada, M., and Applegate, S. (1972). Calcium and phosphorous in the skeleton and blood of the lungfish, *Lepidosiren paradoxa*, with comment on humoral factors in calcium homeostasis in the Osteichthyes. *Comp. Biochem. Physiol. A* **42**, 393–408.

Utida, S., Hirano, T., Ando, M., Johnson, D.W., and Bern, H.A. (1972). Hormonal control of the intestine and urinary bladder in teleost osmoregulation. *Gen. Comp. Endocr.* **Suppl. 3**, 317–327.

Uva, B., Masini, M.A., Hazon, N., O'Toole, L.B., Henderson, I.W., and Ghiani, P. (1992). Renin and angiotensin converting enzyme in elasmobranchs. *Gen. Comp. Endocr.* **86**, 407–412.

Vale, W., Bilezikjian, L.M., and Rivier, C. (1994). Reproductive and other roles of inhibins and activins. In *The Physiology of Reproduction*, 2nd edn (edited by E. Knobil and J.D. Neill) vol. 2, pp. 1861–1878. New York: Raven Press.

Vale, W., Rivier, C., and Brown, M. (1977). Regulatory peptides of the hypothalamus. *Annu. Rev. Physiol.* **39**, 473–527.

Vale, W., Rivier, C., Hsueh, A., Campen, C., Meunier, H., Bicsak, T., Vaughan, J., Corrigan, A., Bardin, W., Sawchenko, P., Petraglia, F., Yu, J., Plotsky, P., Spiess, J., and Rivier, J. (1988). Chemical and biological characterization of the inhibin family of protein hormone. *Rec. Prog. Horm. Res.* **44**, 1–34.

Vale, W., Speiss, J., Rivier, C., and Rivier, J. (1981). Characterization of a 41-residue ovine hypothalamic peptide that stimulates secretion of corticotropin and β-endorphin. *Science* **213**, 1394–1397.

Valentijn, J.A., Vaudry, H., Kloas, W., and Cazin, L. (1994). Melanostatin (NPY) inhibited electrical activity of melanotrophs through modulation of K^+, Na^+ and Ca^{2+} current. *J. Physiol. Lond.* **475**, 185–195.

Valverde-R, C., Croteau, W., Lafleur, G.J., Orozco, A., and St Germain, D.L. (1997). Cloning and expression of a 5'-iodothyronine deiodinase from the liver of *Fundulus heteroclitus. Endocrinology*, **138**, 642–648.

van Cauter, E. (1989). Physiology and pathology of circadian rhythms. In *Recent Advances in Endocrinology and Metabolism* No. 3 (edited by C.R.W. Edwards and D.W. Lincoln), pp. 109–134. Edinburgh: Churchill Livingstone.

van der Kraak, G., Suzuki, K., Peter, R.E. Itoh, H., and Kawauchi, H. (1992). Properties of common carp gonadotropin I and gonadotropin II. *Gen. Comp. Endocr.* **85**, 217–229.

van Haasteren, G.A.C., Linkels, E., van Toor, H., Klootwijk, W., Kaptein, E., de Jong, F.H., Reymond, M.J., Visser, T.J., and de Greef, W.J. (1996). Effects of long-term food reduction on the hypothalamus–pituitary–thyroid axis in male and female rats. *J. Endocr.* **150**, 169–178.

van Kesteren, R.E., Smith, A.B., de With, N.D., van Minnen, J., Dirks, R.W., van der Schors, R.C., and Joosse, J. (1992a). A vasopressin-related peptide in the mollusc *Lymnaea stagnalis*: peptide structure, prohormone organization, evolutionary and functional aspects of *Lymnaea* conopressin. *Prog. Brain Res.* **92**, 47–57.

van Kesteren, R.E., Smit, A.B., Dirks, R.W., de With, N.D., Geraerts, W.P.M., and Joose, J. (1992b). Evolution of the vasopressin/oxytocin superfamily: characterization of a cDNA encoding a vasopressin-related precursor, preproconopressin, from the mollusc *Lymnaea stagnalis. Proc. Natl. Acad. Sci. USA* **89**, 4593–4597.

van Tienhoven, A., and Planck, R.J. (1973). The effect of light on avian reproductive activity. In *Handbook of Physiology*, Sect. 7 *Endocrinology*, vol. II. *Female reproductive tract* (Pt. 1), pp. 79–107. Washington, DC: American Physiological Society.

van Zoest, I.D., Heijman, P.S., Cruijsen, P.M.J.M., and Jenks, B.G. (1989). Dynamics of background adaptation in *Xenopus laevis*: role of catecholamines and melanocyte-stimulating hormone. *Gen. Comp. Endocr.* **76**, 19–28.

Vari, R.C., Adkins, S.D., and Samson, W.K. (1996). Renal effects of adrenomedullin in the rat. *Proc. Soc. Exp. Biol. Med.* **211**, 178–183.

Varma, M.M. (1977). Ultrastructural evidence for aldosterone and corticosterone-secreting cells in the adrenocortical tissue of the American bullfrog (*Rana catesbeiana*). *Gen. Comp. Endocr.* **33**, 61–75.

Vasilatos-Younken, R., Anderson, B.J., Rosebrough, R.W., McMurtry, J.P., and Bacon, W.L. (1991). Identification of circulating growth hormone-binding proteins in domestic poultry: an initial characterization. *J. Endocr.* **130**, 115–122.

Vaudry, H., Lamacz, M., Desrues, L., Louiset, E., Valentijn, J., Mei, Y.A., Chartrel, N., Conlon, J.M., Cazin, L., and Tonon, M.C. (1994). The melanotrope cell of the frog pituitary as a model of neuroendocrine integration. In *Perspectives in Comparative Endocrinology* (edited by K.G. Davey, R.E. Peter, and S.S. Tobe), pp. 5–11. Ottawa: National Research Council of Canada.

Vaughan, J., Donaldson, C., Bittencourt, J., Perrin, M.H., Lewis, K., Sutton, S., Chan, R., Turnbull, A.V., Lovejoy, D., Rivier, C., Rivier, J., Sawchenko, P.E., and Vale, W. (1995). Urocortin, a mammalian neuropeptide related to fish urotensin I and to corticotropin-releasing hormone. *Nature, Lond.* **378**, 287–292.

Vaughan, J.M. Fischer, W.H., Hoeger, C., Rivier, J., and Vale, W. (1989). Characterization of melanin-concentrating hormone from rat hypothalamus. *Endocrinology*, **125**, 1660–1665.

Verbalis, J.G., McCann, M.J., McHale, C.M., and Stricker, E.M. (1986). Oxytocin secretion in response to cholecystokinin and food: differentiation of nausea from satiety. *Science* **232**, 1417–1419.

Verbost, P.M., Butkus, A., Willems, P., and Wendelaar Bonga, S.E. (1993). Indications for two bioactive principles in the corpuscles of Stannius. *J. Exp. Biol.* **177**, 243–252.

Verbost, P.M. and Fenwick, J.C. (1995). N-terminal and C-terminal fragments of the hormone stanniocalcin show differential effects in eels. *Gen. Comp. Endocr.* **98**, 185–192.

Verkman, A.S., van Hoek, A.N., Ma, T., Frigeri, A., Skach, W.R., Mitra, A., Tamarappoo, B.K., and Farinas, J. (1996). Water transport across mammalian cell membranes. *Am. J. Physiol.* **270**, C12–C30.

Vermeirssen, E.L.M., and Scott, A.P. (1996). Excretion of free and conjugated steroids in rainbow trout (*Oncorhychus mykiss*): evidence for branchial excretion of the maturation-inducing steroid, 17,20β-dihydroxy-4-pregnen-3-one. *Gen. Comp. Endocr.* **101**, 180–194.

Vesely, D.L., Douglass, M.A., Dietz, J.R., Gower, W.R., McCormick, M.T., Rodriguez-Paz, G., and Schocken, D.D. (1994). Three peptides from the atrial natriuretic factor prohormone amino terminus lower blood pressure and produce diuresis, natriuresis, and/or kaliuresis in humans. *Circulation* **90**, 1129–1140.

Vesely, D.L., Gower, W.R., Giordano, A.T., and Friedl, F.E. (1993). Atrial natriuretic peptides in the heart and homolymph of the oyster, *Crassostrea virginica*: a comparison with vertebrates. *Comp. Biochem. Physiol. B* **106**, 535–546.

Vigier, B., Forest, M.G., Eychenne, B., Bézard, J., Garrigour, O., Robel, P., and Josso, N. (1989). Anti-Müllerian hormone produces endocrine sex reversal of fetal ovaries. *Proc. Natl. Acad. Sci. USA* **86**, 3684–3688.

Vigna, S.R. (1979). Distinction between cholecystokinin-like and gastrin-like biological activities extracted from gastrointestinal tissues of some lower vertebrates. *Gen. Comp. Endocr.* **39**, 512–520.

Vigna, S.R. (1986). Gastrointestinal tract.

In *Vertebrate Endocrinology: Fundamentals and Biomedical Implications*, vol. 1, *Morphological Considerations* (edited by P.K.T. Pang and M.P. Schreibman), pp. 261–278. Orlando, FL: Academic Press.

Vigna, S.R., Thorndyke, M.C., and Williams, J.A., (1986). Evidence for a common evolutionary origin of brain and pancreas cholecystokinin receptors. *Proc. Natl. Acad. Sci. USA* **83**, 4355–4359.

Vijayan, M.M., Mommsen, T.P., Glémet, H.C., and Moon, T.W. (1996). Metabolic effects of cortisol treatment in a marine teleost, the sea raven. *J. Exp. Biol.* **199**, 1509–1514.

Vinson, G.P., Whitehouse, B., and Hinson, J. (1992). *The Adrenal Cortex.* pp. 140–195. Englewood Cliffs NJ: Prentice Hall.

Wachtel, S.S., Wachtel, G., and Nakamura, D. (1991). Sexual differentiation. In *Vertebrate Endocrinology: Fundamentals and Biomedical Implications*, vol. 4, Pt B, *Reproduction* (edited by P.K.T. Pang and M.P. Schreibman), pp. 149–180. San Diego, CA: Academic Press.

Wada, M., Kobayashi, H., and Farner, D.S. (1975). Induction of drinking in the white-crowned sparrow, *Zonotrichia leucophrys gambelli*, by intracranial injection of angiotensin II. *Gen. Comp. Endocr.* **26**, 192–197.

Wade-Smith, J., Richmond, M.E., Mead, R.A., and Taylor, H. (1980). Hormonal and gestational evidence for delayed implantation in the striped skink, *Mephitis mephitis. Gen. Comp. Endocr.* **42**, 509–515.

Wagner, G.F. (1994). The molecular biology of the corpuscles of Stannius and regulation of stanniocalcin gene expression. In *Fish Physiology*, vol. XIII, *Molecular Endocrinology of Fish* (edited by N.M. Sherwood and C.L. Hew), pp. 273–306. San Diego, CA: Academic Press.

Wagner, G.F., Fenwick, J.C., Park, C.M., Milliken, C., Copp, D.H., and Friesen, H.G., (1988). Comparative biochemistry and physiology of teleocalcin from sockeye and coho salmon. *Gen. Comp. Endocr.* **72**, 237–246.

Wagner, G.F., Guiraudon, C.C., Milliken, C., and Copp, D.H. (1995). Immunological and biological evidence for a stanniocalcin-like hormone in human kidney. *Proc. Natl. Acad. Sci. USA* **92**, 1871–1875.

Wahli, W., Dawid, I.B., Ryffel, G.U., and Weber, R. (1981). Vitellogenesis and the vitellogenin gene family. *Science* **212**, 298–304.

Wako, H., Ishii, S. (1995). Secondary structure prediction of β-subunits of the gonadotropin–thyrotropin family from its aligned sequences using environment-dependent amino-acid substitution tables and conformational propensities. *Biochim. Biophys. Acta* **1247**, 104–112.

Waldo, C.M. and Wislocki, G.B. (1951). Observations on the shedding of the antlers of the virginia deer (*Odocoileus virginianus borealis*). *Am. J. Anat.* **88**, 351–395.

Walker, J.M., Akil, H., and Watson, S.J. (1980). Evidence for homologous actions of pro-opiocortin products. *Science* **210**, 1247–1249.

Wallis, M. (1992). The expanding growth hormone/prolactin family. *J. Mol. Endocr.* **9**, 185–188.

Wang, Y., and Conlon, J.M. (1995). Purification and structural characterization of vasoactive intestinal polypeptide from the trout and bowfin. *Gen. Comp. Endocr.* **98**, 94–101.

Wank, S.A. (1995). Cholecystokinin receptors. *Am. J. Physiol.* **269**, G628–G646.

Waring, C.P., Moore, A., and Scott, A.P. (1996). Milt and endocrine responses of mature male Atlantic salmon (*Salmo salar* L.) parr to water-borne testosterone, 17,20β-dihydroxy-4-pregnen-3-one-20-sulfate, and the urines from adult female and male salmon. *Gen. Comp. Endocr.* **103**, 142–149.

Waring, H. (1936). Colour changes in the dogfish (*Scyllium canicula*). *Proc. Liverpool Biol. Soc.* **49**, 17–64.

Waring, H. (1938). Chromatic behaviour of elasmobranchs. *Proc. Roy. Soc. Lond. Ser. B* **125**, 264–282.

Waring, H. (1942). The co-ordination of vertebrate melanophore responses. *Biol. Rev.* **17**, 120–150.

Waring, H. (1963). *Color Change Mechanisms in Cold-blooded Vertebrates.* London: Academic Press.

Warne, J.M., Hazon, N., Rankin, J.C., and Balment, R.J. (1994). A radioimmunoassay for the determination of arginine vasotocin (AVT): plasma and pituitary concentrations in fresh- and seawater fish. *Gen. Comp. Endocr.* **96**, 438–444.

Warren, W.C., Laing, R., Krivi, G.G., Siegel, N.R., and Anthony, R.V. (1990). Purification and structural characterization of ovine placental lactogen. *J. Endocr.* **126**, 141–149.

Wasserman, R.H., Henion, J.D., Haussler, M.R., and McCain, T.A. (1976). Calcinogenic factor in *Solanum malacoxylon*: evidence that it is 1,25-diyhdroxyvitamin D_3-glycoside. *Science* **194**, 853–855.

Wasserman, R.H. and Taylor, A.N. (1966). Vitamin D_3-induced calcium-binding protein in chick intestinal mucosa. *Science* **152**, 791–793.

Wathes, D.C. and Swann, R.W. (1982). Is oxytocin an ovarian hormone? *Nature, Lond.* **297**, 225–227.

Waugh, D., Anderson, G., Armour, K.J., Balment, R.J., Hazon, N., and Conlon, J.M. (1995a). A peptide from the caudal neurosecretory system in the dogfish *Scyliorhinus canicula* that is structurally related to urotensin I. *Gen. Comp. Endocr.* **99**, 333–339.

Waugh, D. and Conlon, J.M. (1993). Purification and characterization of urotensin II from the brain of a teleost (trout, *Oncorhynchus mykiss*) and an elasmobranch (skate, *Raja rhina*). *Gen. Comp. Endocr.* **92**, 419–427.

Waugh, D., Youson, J., Mims, S.D., Sower, S., and Conlon, J.M. (1995b). Urotensin I from the river lamprey (*Lampetra fluviatilis*), and the paddlefish (*Polyodon spathula*). *Gen. Comp. Endocr.* **99**, 323–332.

Weatherhead, B. and Logan, A. (1981). Interaction of α-melanocyte-stimulating hormone, melatonin, cyclic AMP and cyclic GMP in the control of melanogenesis in hair follicle melanocytes *In vitro. J. Endocr.* **90**, 89–96.

Weinstein, B. (1972). A generalized homology correlation for various hormones and proteins. *Experientia* **28**, 1517–1522.

Weisbart, M. and Idler, D.R. (1970).

Re-examination of the presence of corticosteroids in two cyclostomes, the Atlantic hagfish (*Myxine glutinosa* L.) and the sea lamprey (*Petromyzon marinus* L.) *J. Endocr.* **46**, 29–43.

Weisinger, R.S., Coghlan, J.P., Denton, D.A., Fan, J.S.K., Hatzikostas, S., McKinley, M.J., Nelson, J.F., and Scoggins, B.A. (1980). ACTH-elicited sodium appetite in sheep. *Am. J. Physiol.* **239**, E45–£50.

Weisinger, R.S., Denton, D.A., Di Nicolantonio, R., McKinely, M.J., Muller, A.F., and Tarjan, E. (1987). Role of angiotensin in sodium appetite of sodium-deplete sheep. *Am. J. Physiol.* **253**, R482–R488.

Weisinger, R.S., Denton, D.A., McKinely, M.J., Miselis, R.R., Park, R.G., and Simpson, J.B. (1993). Forebrain lesions that disrupt water homeostasis do not eliminate the sodium appetite of sodium deficiency in sheep. *Brain Res.* **628**, 166–178.

Weiss, M. (1980). Adrenocorticosteroids in prototherian, metatherian and eutherian mammals. In *Comparative physiology: primitive mammals* (edited by K. Schmidt-Nielsen, L. Bolis, and C.R. Taylor), pp. 285–96. Cambridge: Cambridge University Press.

Weiss, M. and Carson, R.S. (1987). Induction of adrenocortical special zone in the male possum (*Trichosurus vulpecula*). *Comp. Biochem. Physiol. A* **86**, 361–365.

Weiss, M., Edgar, J.A., Than, K.A., and Young, I.R. (1989). Identification of 5α-androstone-3α-17α-diol in adrenal venous plasma of female possum (*Trichosurus vulpecula*). *J. Steroid Biochem.* **32**, 591–597.

Weiss, M. and McDonald, I.R. (1965). Corticosteroid secretion in the monotreme *Tachyglossus aculeatus*. *J. Endocr.* **33**, 203–210.

Welshons, W.V., Lieberman, M.E., and Gorski, J. (1984). Nuclear localization of unoccupied oestrogen receptors. *Nature, Lond.* **307**, 747–749.

Wenberg, G.M. and Holland, J.C. (1973). The circannual variations of some of the hormones of the woodchuck (*Marmota monax*). *Comp. Biochem, Physiol. A* **46**, 523–535.

Wendelaar Bonga, S.E. (1978). The effects of changes in external sodium, calcium, and magnesium concentrations on prolactin cells, skin, and plasma electrolytes of *Gasterosteus aculeatus*. *Gen. Comp. Endocr.* **34**, 265–275.

Wendelaar Bonga, S. and Pang, P.K.T. (1986). Stannius corpuscles. In *Vertebrate Endocrinology: Fundamentals and Biomedical Implications*, vol. 1, *Morphological Considerations* (edited by P.K.T. Pang and M.P. Schreibman), pp. 439–464. Orlando, FL: Academic Press.

Wendelaar Bonga, S.E. and Pang, P.K.T. (1989). Pituitary hormones. In *Vertebrate Endocrinology: Fundamentals and Biomedical Implications*, vol. 3, *Regulation of Calcium and Phosphate* (edited by P.K.T. Pang and M.P. Schreibman), pp. 105–137. San Diego, FL: Academic Press.

Wennstrom, K.L. and Crews, D. (1995). Making males from females: the effects of aromatase inhibitors on a parthenogenetic species of whiptail lizard. *Gen. Comp. Endocr.* **99**, 316–322.

West, G.B. (1995). The comparative pharmacology of the suprarenal medulla. *Quart. Rev. Biol.* **30**, 116–137.

Westenfelder, C., Birch, F.M., Baranowski, R.L., Rosenfeld, M.J., Shiozawa, D.K., and Kablitz, G. (1988). Atrial natriuretic factor and salt adaptation in the teleost fish *Gila atraria*. *Am. J. Physiol.* **255**, F1281–F1286.

Wetsel, W.C., Valença, M.M., Merchenthaler, I., Liposits, Z., López, F.J., Weiner, R.I., Mellon, P.L. and Negro-Vilar, A. (1992). Intrinsic pulsatile secretory activity of immortalized luteinizing hormone-releasing hormone-secreting neurons. *Proc. Natl. Acad. Sci. USA* **89**, 4149–4153.

White, A.W. and Harrop, C.J.F. (1975). The islets of Langerhans of the pancreas of macropodid marsupials: a comparison with eutherian species. *Aust. J. Zool.* **23**, 309–319.

White, S.A., Bond, C.T., Francis, R.C., Kasten, T.L., Fernald, R.D. and Adelman, J.P. (1994). A second gene for gonadotropin-releasing hormone: cDNA and expression pattern in the brain. *Proc. Natl. Acad. Sci. USA* **91**, 1423–1427.

Whittier, J.M. (1994). Seasonal reproduction in squamate reptiles: endocrine patterns and variations. In *Perspectives in Comparative Endocrinology* (edited by K.G. Davey, R.E. Peter, and S.S. Tobe), pp. 586–593. Ottawa: National Research Council of Canada.

Wibbels, T. and Crews, D. (1994). Putative aromatase inhibitor induces male sex determination in a female unisexual lizard and in a turtle with temperature-dependent sex determination. *J. Endocr.* **141**, 295–299.

Williams, G.R. (1994). Solving the specificity puzzle. *Nature, Lond.* **370**, 330–331.

Williams, L.T., Lefkowitz, R.J., Watanabe, A.M., Hathaway, D.R., and Besch, H.R. (1977). Thyroid hormone regulation of β-adrenergic receptor number. *J. Biol. Chem.* **252**, 2787–2789.

Williams, T.D. and Sharp, P.J. (1993). Plasma prolactin during the breeding season in adult and immature macaroni (*Eudyptes chrysolophus*) and gentoo (*Pygoscelis papua*) penguins. *Gen. Comp. Endocr.*, **92**, 339–346.

Wilson, A.J., Carlson, S.S., and White, T.J. (1977). Biochemical evolution. *Annu. Rev. Biochem.* **46**, 573–639.

Wilson, F.E. and Reinert, B.D. (1993). The thyroid and photoperiodic control of seasonal reproduction in American tree sparrows. (*Spizella arborea*). *J. Comp. Physiol. B* **163**, 563–573.

Wilson, J.F. and Dodd, J.M. (1973a). The role of the pineal complex and lateral eyes in the colour change response of the dogfish, *Scyliorhinus canicula* L. *J. Endocr.* **58**, 591–598.

Wilson, J.F. and Dodd, J.M. (1973b). The role of melanophore-stimulating hormone in melanogenesis in the dogfish, *Scyliorhinus canicula* L. *J. Endocr.* **58**, 685–686.

Wilson, J.X. (1984). The renin–angiotensin system in nonmammalian vertebrates. *Endocr. Rev.* **5**, 45–61.

Wilson, J.X. (1989). The renin–angiotensin system in birds. In *Progress in Avian Osmoregulation* (edited by M.R. Hughes and A. Chadwick), pp. 61–79. Leeds: Leeds Philosophical and Literary Society.

Wilson, J.X., Van Pham, D., and Tan Wilson, H.I. (1985). Angiotensin and converting enzyme regulate extrarenal salt excretion in ducks. *Endocrinology* **117**, 135–140.

Wilson, S.C. and Cunningham, F.J. (1980). Effects of increasing day length and intermittent lighting schedules in the domestic hen on plasma concentrations of luteinizing hormone (LH) and the LH response to exogenous progesterone. *Gen. Comp. Endocr.* **41**, 546–553.

Wimalawansa, S.J. (1996). Calcitonin gene-related peptide and its receptors: molecular genetics, physiology, pathophysiology, and therapeutic potentials. *Endocrine Rev.* **17**, 533–585.

Wingfield, J.C. (1994). Hormone–behaviour interactions and mating systems in male and female birds. In *The Differences Between the Sexes* (edited by R.V. Short and A. Ballaban), pp. 303–330. Cambridge: Cambridge University Press.

Wingfield, J.C. and Kenagy, G.J. (1991). Natural regulation of reproductive cycles. In *Vertebrate Endocrinology: Fundamentals and Biomedical Implications*, vol. 4, Pt B, *Reproduction* (edited by P.K.T. Pang and M.P Schreibman), pp. 181–241. San Diego, CA: Academic Press.

Wingfield, J.C., Schwabl, H. and Mattocks, P.W. (1990). Endocrine mechanisms of migration. In *Bird Migration* (edited by E. Gwinner), pp. 231–256. Berlin: Springer-Verlag.

Wingstrand, K.G. (1951). *The Structure and Development of the Avian Pituitary*. Lund: C.W.K. Gleerup.

Wingstrand, K.G. (1966). Comparative anatomy and evolution of the hypophysis. In *The Pituitary Gland* (edited by G.W. Harris and B.T. Donovan), vol. 1, pp. 58–146. Berkeley: University of California Press.

Wittle, L.W., Augostini, R.S. and Chizmar, W.S. (1990). The occurrence of chronic hypocalcemia following parathyroidectomy in the green frog, *Rana clamitans. Gen. Comp. Endocr.* **80**, 419–426.

Wittle, L.W. and Dent, J.N. (1979). Effects of parathyroidectomy and of parathyroid extract on levels of calcium and phosphate in the blood and urine of the red-spotted newt. *Gen. Comp. Endocr.* **37**, 428–439.

Wolf, B.A., Colca, J.R., Turk, J.,

Florholmen, J., and McDaniel, M.L. (1988). Regulation of Ca^{2+} homeostasis by islet endoplasmic reticulum and its role in insulin secretion. *Am. J. Physiol.* **254**, E121–E136.

Wolff, E. and Wolff, E. (1951). The effects of castration on bird embryos. *J. Exp. Zool.* **116**, 55–97.

Wolin S.L. and Walter, P. (1993). Discrete nascent chain lengths are required for the insertion of presecretory proteins into microsomal membranes. *J. Cell Biol.* **121**, 1211–1219.

Wong, A. O.-L., Chang, J.P., and Peter, R.E. (1992). Dopamine stimulates growth hormone release from the pituitary of goldfish, *Carassius auratus*, through the dopamine D_1 receptors. *Endocrinology* **130**, 1201–1210.

Wong, C.C., Lam, K.Y., and Chiu, K.W. (1993). The extrathyroidal conversion of T_4 to T_3 in the striped racer snake, *Elaphe taeniura. J. Comp. Physiol. B* **163**, 212–218.

Wong, G.L., Luben, R.A., and Cohn, D.V. (1977). 1,25-Dihydroxycholecalciferol and parathormone: effects on isolated osteoclast-like and osteoblast-like cells. *Science* **197**, 663–665.

Woods, A.J. and Stock, M.J. (1996). Leptin activation of the hypothalamus. *Nature, Lond.* **381**, 745.

Woolley, P. (1957). Colour change in a chelonian. *Nature, Lond.* **179**, 1255–1256.

Wourms, J.P. and Lombardi, J. (1992). Reflections on the evolution of piscine viviparity. *Am. Zool.* **32**, 276–293.

Wright, A., Chester Jones, I., and Phillips, J.G. (1957). The histology of the adrenal gland of prototheria. *J. Endocr.* **15**, 100–107.

Wright, G. (1984). Immunocytochemical study of growth hormone, prolactin, and thyroid-stimulating hormone in the adenohypophysis of the sea lamprey. *Petromyzon marinus* L., during its upstream migration. *Gen. Comp. Endocr.* **55**, 269–274.

Wright, J.W., Krebs, L.T., Stobb, J.W., and Harding, J.W. (1995). The angiotensin IV system: functional implications. *Frontiers Neuroendocrinol.* **16**, 23–52.

Wunder, B.A. (1979). Hormonal mechanisms. In *Comparative Mechanisms*

of Cold Adaptation (edited by L.S. Underwood, L.L. Tieszen, A.B. Callahan, and G.E. Folk), pp. 143–158. New York: Academic Press.

Wurtman, R.J., Axelrod, J., and Kelly, D.E. (1968). *The Pineal*. New York: Academic Press.

Xavier, F. (1974). La pseudogestation chez *Nectophrynoïdes occidentalis* ANGEL. *Gen. Comp. Endocr.* **22**, 98–115.

Xavier, F. and Ozon, R. (1971). Recherches sur l'activité endocrine de l'ovaire de *Nectophrynoïdes occidentalis* ANGEL (amphibien anoure vivipare). ii. Synthèse *in vitro* de stéroids. *Gen. Comp. Endocr.* **16**, 30–40.

Xiong, F., Suzuki, K., and Hew, C.L. (1994). Control of teleost gonadotropin gene expression. In *Fish Physiology*, vol. XIII, *Molecular Endocrinology of Fish* (edited by N.M. Sherwood and C.L. Hew), pp. 135–158. San Diego, CA: Academic Press.

Yada, T., Hirano, T., and Grau, E.G. (1994). Changes in plasma levels of two prolactins and growth hormone during adaption to different salinities in the euryhaline tilapia, *Oreochromis mossambicus. Gen. Comp. Endocr.* **93**, 214–223.

Yagil, R., Etzion, Z., and Berlyne, G.M. (1973). The effect of *d*-aldosterone and spironolactone on the concentration of sodium and potassium in the milk of rats. *J. Endocr.* **59**, 633–636.

Yamamoto, T. (1953). Artificially induced sex-reversal in genotypic males of the medaka (*Oryzias latipes*). *J. Exp. Zool.* **123**, 571–594.

Yamamoto, T. (1958). Artificial induction of functional sex-reversal in genotypic females of the medaka (*Oryzias latipes*). *J. Exp. Zool.* **137**, 227–263.

Yamano, K., Tagawa M., de Jesus, E.G., Hirano, T., Miwa, S., and Inui, Y. (1991). Changes in whole body concentrations of thyroid hormones and cortisol in metamorphosing conger eel. *J. Comp. Physiol. B* **161**, 371–375.

Yamashita, Y., Matsuda, K., Hayashi, H., Hanaoka, Y., Tanaka, S., Yamamoto, K., and Kikuyama, S. (1993). Isolation and characterization of two forms of *Xenopus* prolactin. *Gen. Comp. Endocr.* **91**, 307–317.

Yamauchi, K., Kasahara, T., Hayashi, H., and Horiuchi, R. (1993). Purification and characterization of a 3,5,3'-L-triiodothyronine-specific binding protein from bullfrog tadpole plasma: a homolog of mammalian transthyretin. *Endocrinology* **132**, 2254–2261.

Yaron, Z. (1972). Endocrine aspects of gestation in viviparous snakes. *Gen. Comp. Endocr.* **Suppl. 3**, 663–673.

Yen, P.M. and Chin, W.W. (1994). New advances in understanding the molecular mechanisms of thyroid hormone action. *Trend. Endocr. Metab.* **5**, 65–72.

Ymer, S.I. and Herington, A.C. (1985). Evidence for the specific binding of growth hormone to a receptor-like protein in rabbit serum. *Mol. Cell. Endocr.* **41**, 153–161.

Yokota, S.D. (1990). Glomerular filtration dynamics in reptiles: endocrine effects. In *Progress in Comparative Endocrinology* (edited by A. Epple, C.G. Scanes, and M.H. Stetson), pp. 565–571. New York. Wiley-Liss.

Yorio, T. and Bentley, P.J. (1977). Asymmetrical permeability of the integument of tree frogs (HYLIDAE). *J. Exp. Biol.* **76**, 197–204.

Young, G., Björnsson, B.Th., Prunet, P., Lin, R.J., and Bern, H.A. (1989). Smoltification and seawater adaptation in coho salmon (*Oncorhynchus kisutch*): plasma prolactin, growth hormone, thyroid hormones, and cortisol. *Gen. Comp. Endocr.*, **74**, 335–345.

Young, J.Z. (1935). The photoreceptors of lampreys. ii. The function of the pineal complex. *J. Exp. Biol.* **12**, 254–270.

Youson, J.H. (1994). Environmental and hormonal cues and endocrine glands during lamprey metamorphosis. In *Perspectives in Comparative Endocrinology*

(edited by K.G. Davey, R.E. Peter, and S.S. Tobe), pp. 400–407. Ottawa: National Research Council of Canada.

Youson, J.H., Elliot, W.M., Beamish, R.J., and Wang, D.W. (1988). A comparison of endocrine pancreatic tissue in adults of four species of lampreys in British Columbia: a morphological and immunohistochemical study. *Comp. Biochem. Physiol.* **70**, 247–261.

Youson, J.H., Plisetskaya, E.M., and Leatherland, J.F. (1994). Concentrations of insulin and thyroid hormones in the serum of landlocked lampreys (*Petromyzon marinus*) of three larval year classes, in larvae exposed to two temperature regimes, and in individuals during and after metamorphosis. *Gen. Comp. Endocr.* **94**, 294–304.

Yu, J.Y.L., Dickoff, W.W., Swanson, P., and Gorbman, A. (1981). Vitellogenesis and its hormonal regulation in the Pacific hagfish. *Eptatretus stouti* L. *Gen. Comp. Endocr.* **43**, 492–502.

Yulis, C.R. and Lederis, K. (1988). Occurrence of anterior spinal, cerebrospinal fluid contacting, uretensin II neuronal system in various fish species. *Gen. Comp. Endocr.* **70**, 301–311.

Zadunaisky, J.A. (1984). The chloride cell: the active transport of chloride and the paracellular pathways. In *Fish Physiology* (edited by W.S. Hoar and D.J. Randall), vol. X, *Gills* Pt B, *Ion and Water Transfer*, pp. 129–176. Orlando, FL: Academic Press.

Zadunaisky, J.A. (1996). Chloride cells and osmoregulation. *Kidney Int.* **49**, 1563–1567.

Zadunaisky, J.A. and Degnan, K.J. (1980). Chloride active transport and osmoregulation. In *Epithelial Transport in Lower Vertebrates* (edited by B. Lahlou),

pp. 185–196. Cambridge: Cambridge University Press.

Zawalich, W.S. and Zawalich, K.G. (1996). Regulation of insulin secretion by phospholipase C. *Am. J. Physiol.* **271**, E407–E416.

Zhang, Y., Proenca, R., Maffel, M., Barone, M., Leopold, L., and Friedman, J.M. (1994). Positional cloning of the mouse *obese* gene and its human homologue. *Nature, Lond.* **372**, 425–432.

Zheng, T., Villalobos, C., Nusser, K.D., Gettys, T.W., Faught, W.J., Castano, J.P., and Frawley, L.S. (1997). Phenotypic characterization and functional correlation of α-MSH binding to pituitary cells. *Am. J. Physiol.* **272**, E282–E287.

Zhu, Y. and Thomas, P. (1996). Elevations of somatolactin in plasma and pituitaries and increased α-MSH cell activity in red drum exposed to black background and decreased illumination. *Gen. Comp. Endocr.* **101**, 21–31.

Zhu, Y. and Thomas, P. (1997). Effects of somatolactin to melanosome aggregation in the melanophores of red drum (*Sciaenops ocellatus*) scales. *Gen. Comp. Endocr.* **105**, 127–133.

Zimmermann, U., Fischer, J.A., and Muff, R. (1995). Adrenomedullin and calcitonin gene-related peptide interact with the same receptor in cultured human neuroblastoma SK-N-MC cells. *Peptides* **16**, 421–424.

Zuber-Vogeli, M. and Xavier, F. (1973). Les modifications cytologique de l'hypophyse distale des femelles de *Nectrophrynoides occidentalis* Angel après ovariectomie. *Gen. Comp. Endocr.* **20**, 199–213.

Index